International Conference on Differential Equations

ACADEMIC PRESS RAPID MANUSCRIPT REPRODUCTION

Proceedings of an International
Conference on Differential Equations
Held at the University of Southern
California, September 3-7, 1974

International Conference on Differential Equations

edited by

H. A. Antosiewicz

Department of Mathematics
University of Southern California,
Los Angeles, California

Academic Press, Inc.
New York San Francisco London 1975
A Subsidiary of Harcourt Brace Jovanovich, Publishers

ACADEMIC PRESS, INC.
111 Fifth Avenue, New York, New York 10003

United Kingdom Edition published by
ACADEMIC PRESS, INC. (LONDON) LTD.
24/28 Oval Road, London NW1

Library of Congress Cataloging in Publication Data

International Conference on Differential Equations, University of Southern California, 1974.
International Conference on Differential Equations.

Bibliography: p. 00
Includes index.
1. Differential equations–Congresses.
I. Antosiewicz, H. A. II. Title.
QA371.I54 1974 515'.35 75-6645
ISBN 0-12-059650-4

PRINTED IN THE UNITED STATES OF AMERICA

CONTENTS

*For papers with more than one author, an asterisk follows the name of the author who presented the paper at the conference.

CONTENTS

CONTENTS

CONTENTS

CONTENTS

INVITED ADDRESSES AND RESEARCH REPORTS

Balakrishnan, A. V., Department of Mathematics, University of California, Los Angeles, Los Angeles, California 90024

Banks, H. T., Division of Applied Mathematics, Brown University, Providence, Rhode Island 02912

Berkovitz, L. D., Department of Mathematics, Purdue University, West Lafayette, Indiana 47907

Braaksma, B. L. J., Mathematics Institute, University of Groningen, Postbus 800, Groningen, Netherlands

Brauer, F., Department of Mathematics, University of Wisconsin, Madison, Wisconsin 53706

Bucy, R. S., Department of Aerospace Engineering, University of Southern California, Los Angeles, California 90007

Cellina, A., Istituto Matematico "U. Dini", Viale Morgagni 67/A, Florence, Italy

Cesari, L., Department of Mathematics, University of Michigan, Ann Arbor, Michigan 48104

Chandra, J., U. S. Army Research Office, Mathematics Division, Box CM, Duke Station, Durham, North Carolina 27706

Chang, K. W., Department of Mathematics, University of Calgary, Calgary, Alberta T2N IN4, Canada

Coddington, E. A., Department of Mathematics, University of California, Los Angeles, Los Angeles, California 90024

Coles, W. J., Department of Mathematics, University of Utah, Salt Lake City, Utah 84112

Conti, R., Istituto Matematico "U. Dini", Viale Morgagni 67/A, Florence, Italy

Coppel, W. A., Department of Mathematics, I.A.S., Australian National University, P.O. Box 4, Canberra, A.C.T. 2600, Australia

Cronin, J., Department of Mathematics, Rutgers University, New Brunswick, New Jersey 08903

Dolecki, S., Department of Mathematics, University of Wisconsin, Madison, Wisconsin 53706

Erdélyi, A., Mathematical Institute, University of Edinburgh, Edinburgh EH1 1HZ, Scotland

Everitt, W. N., Department of Mathematics, University of Dundee, Dundee DD1 4HN, Scotland

Fleming, W. H., Department of Mathematics, Brown University, Providence, Rhode Island 02912

Hale, J. K., Department of Mathematics, Brown University, Providence, Rhode Island 02912

Halkin, H., Department of Mathematics, University of California, San Diego, La Jolla, California 92038

Harris, W. A., Jr., Department of Mathematics, University of Southern California, Los Angeles, California 90007

Hartman, P., Department of Mathematics, The Johns Hopkins University, Baltimore, Maryland 21218

Hoppensteadt, F., Courant Institute of Mathematical Sciences, 251 Mercer St., New York, New York 10012

Hsieh, P. F., Department of Mathematics, Western Michigan University, Kalamazoo, Michigan 49001

Imaz, C., Centro de Investigacion del Instituto Politecnico Nacional, Apartada Postal 14-740, Mexico 14, D.F.

Jackson, L. K., Department of Mathematics, University of Nebraska, Lincoln, Nebraska 68508

Jean, M., Laboratoire de Mecanique et d'Acoustique, 31 Chemin Joseph-Aiguier, 13009 Marseille, France

Kaplan, J. L., Department of Mathematics, Boston University, Boston, Massachusetts 02215

Kappel, F., Mathematisches Institut, University of Würzburg, Am Hubland, D 87 Würzburg, Germany

Kotin, L., U.S. Army Electronics Command, Attn: AMSEL-NL-H-2, Fort Monmouth, New Jersey 07703

Lakshmikantham, V., Department of Mathematics, University of Texas, Arlington, Texas 76019

Leung, A., Department of Mathematics, University of Cincinnati, Cincinnati, Ohio 45221

Levin, J. J., Department of Mathematics, University of Wisconsin, Madison, Wisconsin 53706

Lutz, D. A., Department of Mathematics, University of Wisconsin, Milwaukee, Wisconsin 53201

Martin, R. H., Department of Mathematics, North Carolina State University, Raleigh, North Carolina 27607

Mawhin, J., Institut Mathematique, Universite Catholique de Louvain, Louvain-la-Neuve, Belgium

Miller, R. K., Department of Mathematics, Iowa State University, Ames, Iowa 50010

Nohel, J. A., Department of Mathematics, Van Vleck Hall, University of Wisconsin, Madison, Wisconsin 53706

O'Malley, R. E., Department of Mathematics, University of Arizona, Tucson, Arizona 85721

Olech, C., Instytut Matematyczny PAN, ul. Sniadeckich 8, Skrytka Pocztowa 137, 00-950 Warszawa, Poland

Olver, F. W. J., Institute for Fluid Dynamics & Applied Mathematics, University of Maryland, College Park, Maryland 20742

Plis, A., Departmento de Matematicas, Centro de Investigacion I.P.N., Ap.P. 14140, Mexico 14, D.F.

Reid, W. T., Department of Mathematics, University of Oklahoma, Norman, Oklahoma 73069

Russell, D. L., Department of Mathematics, University of Wisconsin, Madison, Wisconsin 53706

Sacker, R. J., Department of Mathematics, University of Southern California, Los Angeles, California 90007

Schumitzky, A., Department of Mathematics, University of Southern California, Los Angeles, California 90007

Sell, G. R., School of Mathematics, University of Minnesota, Minneapolis, Minnesota 55455

Sibuya, Y., School of Mathematics, University of Minnesota, Minneapolis, Minnesota 55455

Yoshizawa, T., Mathematical Institute, Tohoku University, Sendai, Japan

Vorel, Z., Department of Mathematics, University of Southern California, Los Angeles, California 90007

Waltman, P., Department of Mathematics, University of Iowa, Iowa City, Iowa 52240

Warga, J., Department of Mathematics, Northeastern University, Boston, Massachusetts 02115

Wasow, W., Department of Mathematics, University of Wisconsin, Madison, Wisconsin 53706

Wong, J. S. W., Department of Mathematics, University of Iowa, Iowa City, Iowa 52240

CONTRIBUTED PAPERS

Bajaj, P. N., Department of Mathematics, Wichita State University, Wichita, Kansas 67208

Boehmer, K., Department of Mathematics, University of Wisconsin, Madison, Wisconsin 53606

Butler, G. J., Department of Mathematics, University of Alberta, Edmonton T69 291, Alberta, Canada

Cesari, L., Department of Mathematics, University of Michigan, Ann Arbor, Michigan 48104

Choudhury, A. K., Department of Electrical Engineering, School of Engineering, Howard University, Washington, D.C. 20001

Coleman, C. S., Department of Mathematics, Harvey Mudd College, Claremont, California 91711

Cooke, K. J., Department of Mathematics, Pomona College, Claremont, California 91711

Danchick, R., 11177 Montana Avenue, Los Angeles, California 90049

Datko, R., Department of Mathematics, Georgetown University, Washington, D.C. 20007

Friedland, S., The Institute for Advanced Study, Princeton, New Jersey 08540

Galperin, C., Department of Applied Mathematics, Israel Institute of Technology, Technion City, Haifa, Israel

Goodman, G. S., Istituto Matematico "U. Dini", Viale Morgagni 67/A, Florence, Italy

Grossberg, S., 2-832 MIT, Cambridge, Massachusetts 02139

Habets, P., Institut Mathematique, Université Catholique de Louvain, Chemin du Cyclotron, 2, 1348 Louvain-La-Neuve, Belgium

Hall, L. M., Department of Mathematics, University of Nebraska, Lincoln, Nebraska 68508

Howes, Fred, Courant Institute, 251 Mercer Street, New York, New York 10012

London, S., Helsinki Technical University, Helsinki, Finland

Lovelady, D. L., Department of Mathematics, Florida State University, Tallahassee, Florida 32306

Morris, G. R., Mathematics Department, University of New England, Armidale, Armidale, N.S.W. 2351, Australia

Newton, A., Department of Mathematics, University of Southern California, Los Angeles, California 90007

Obi, C., Department of Mathematics, University of Lagos, Lagos, Nigeria

Popov, V. M., Department of Mathematics, University of Florida, Gainesville, Florida 32601

Puttaswamy, T. K., Department of Math Sciences, Ball State University, Muncie, Indiana 47306

Rao, D. R. K., Faculty of Science, Jundi Shapur University, Ahwaz, Iran

Schaar, R. J., Department of Mathematics, University of Southern California, Los Angeles, California 90007

Schaffer, J. J., Department of Mathematics, Carnegie-Mellon University, Pittsburgh, Pennsylvania 15213

Suarez Cortes, R., Escuela Superior de Fisica y Matematicas del Instituto Politecnico Nacional, Mexico, D.F.

Weiss, L., University of Maryland, College Park, Maryland 20742

PREFACE

An International Conference on Differential Equations was held at the University of Southern California, September 3-7, 1974. The aim of the conference was to feature recent advances in the qualitative-analytic theory of differential equations and to stimulate discussion of new directions for future research. Three broad areas were highlighted in particular: analytic theory – singular perturbations, qualitative theory – boundary value problems, and mathematical control theory – variational methods.

The present volume constitutes the proceedings of this conference. It consists of the papers that were delivered as invited addresses and research reports, and of the abstracts of all contributed papers.

It is a pleasure to acknowledge the financial support received from the National Science Foundation, the Office of Naval Research, and the U. S. Army Research Office, Durham, which made the conference possible.

I also wish to record my special thanks to my colleagues W. A. Harris, Jr. and R. J. Sacker for their assistance in the planning and organization of the conference, and to my administrative assistant, Doris Tyler, for her untiring help throughout. Thanks are also due Elsie Walker and Nancy Hori for the excellent typing of the manuscripts.

International Conference on Differential Equations

CONTINUOUS EXTENSIONS: THEIR CONSTRUCTION AND THEIR APPLICATION IN THE THEORY OF DIFFERENTIAL EQUATIONS

H. A. Antosiewicz*
and
A. Cellina

A famous theorem of Tietze states that every real valued function which is continuous on a closed subset of a metric space can be extended continuously over the whole space (see, e.g., [3]). Dugundji proved the following more general result [3]:

Let X be a metric space, Z a locally convex topological linear space, and suppose $A \subset X$ is closed and $f: A \longrightarrow Z$ is continuous. Then there exists a mapping $g: X \longrightarrow Z$ which is a continuous extension of f such that $g(X) \subset \overline{\text{co}}\, f(A)$.

Extension theorems such as these have had many applications in various areas of differential equations. We would like to mention three examples.

1) Let I_1, I_2 be compact intervals in $\mathbb{R}$, and suppose $f: I_1 \times I_2 \to \mathbb{R}$ is such that (i) $t \longrightarrow f(t,x)$ is measurable for every $x \in I_2$ and (ii) $x \longrightarrow f(t,x)$ is continuous for every $t \in I_1$. A theorem of Scorza Dragoni [6] asserts that, for every $\varepsilon > 0$, there exists a closed set $I_\varepsilon \subset I_1$ with

* This work was done with partial support from the U.S. Army Research Office (Durham).

$$G(x) = \bigcup_{\alpha \in L_0(x)} V(F \circ g(\alpha); r_\alpha(x)) \cap H.$$

Here g is a mapping of the complement of A into A, (r_α) a suitable family of continuous non-negative real valued functions, and H the closed convex hull of the range of F; V is the closed neighborhood of $F \circ g(\alpha)$ consisting of all points with distance from $F \circ g(\alpha)$ not exceeding $r_\alpha(x)$.

Clearly, the sets $F \circ g(\alpha)$ play the role of the points y_α. Thus, in order that G have the desired properties, the following conditions must hold:

(i) If $p_\alpha(x)$ is small, $r_\alpha(x)$ is small so that new sets entering into the union are as small as possible.

(ii) For every $\bar{x}$, there exists an $\bar{\alpha} \in L_0(\bar{x})$ such that $V(F \circ g(\bar{\alpha}); r_{\bar{\alpha}}(\bar{x}))$ contains $F \circ g(\alpha)$ for every $\alpha \in L_0(x)$ whenever x is sufficiently close to $\bar{x}$.

(iii) If $\bar{x}$ is close to A, $r_\alpha(\bar{x})$ is small whatever $\alpha \in L_0(\bar{x})$. This will imply that, since $F \circ g(\alpha)$ is within some ε-ball of a fixed set, so is G(x), and hence G will be continuous at the boundary of A.

These conditions are stated precisely in [1] where the construction of the radii $r_\alpha(x)$ is given and all further details of the proof may be found.

It should be noted that the images of the extended mapping are closed subsets of the closed convex hull H although convex combinations are nowhere taken. It does not seem possible to preserve continuity using a a smaller set than H as range.

$\mu(I_\varepsilon) \geq \mu(I_1)-\varepsilon$ such that the restriction of f to $I_\varepsilon \times I_2$ is continuous.

If we wish to apply this theorem in proving the existence of solutions to the Cauchy problem for the differential equation

$$\dot{x} = f(t, x)$$

under Carathéodory conditions, we may use Tietze's theorem to provide a continuous extension f_ε of the restriction of f to the closed set $A = I_\varepsilon \times I_1$ and then consider the Cauchy problem for the differential equation

$$\dot{x} = f_\varepsilon(t, x).$$

Applying Peano's theorem to this equation and letting $\varepsilon > 0$ tend to zero, we can easily deduce the desired result. We remark that, in order to obtain solutions of the approximate problems that are defined in a common interval, we must be able not only to extend a function continuously but do so and preserve the bounds for its norm as well.

This approach has been used by Goodman [4].

2) Let $I = [t_0, T[$ be a non-empty interval in $\mathbb{R}$, let $q > 0$ be given, and suppose $u: [t_0-q, T[\rightarrow \mathbb{R}^n$ is a continuously differentiable mapping for which $\lim_{t \to T} u(t)$ does not exist. A theorem of Yorke [7] asserts that there exists a continuous mapping $G: \mathcal{C}([-q,0], \mathbb{R}^n) \rightarrow \mathbb{R}^n$ such that u is a non-continuable solution in I of the functional differential equation

$$\dot{x} = G(x_t)$$

where, as usual, $x_t(s) = x(t+s)$ for every $s \in [-q, 0]$.

The proof proceeds by showing that

$$A = \{u_t : t \in I\}$$

is a closed subset of $\mathcal{C}([-g, 0], \mathbb{R}^n)$ and that the mapping $g: A \to \mathbb{R}^n$ defined by setting $g(w) = w'(0)$ for every $w \in A$ is continuous. Thus g can be continuously extended to a mapping G of the whole space $\mathcal{C}([-g, 0], \mathbb{R}^n)$ into $\mathbb{R}^n$, which has the desired properties.

3) Let X be a non-reflexive Banach space. A theorem of Cellina [2] asserts that there exists a continuous mapping $f: \mathbb{R} \times X \to X$ for which the Cauchy problem

$$\dot{x} = f(t, x), \quad x(0) = 0$$

has no solution on any open interval in $\mathbb{R}$ that contains 0.

The proof proceeds by constructing a fixed point free continuous mapping g of the unit ball $B \subset X$ into itself and extending it to a continuous mapping $\bar{g}$ of X into B. A suitable f can then be defined in $\mathbb{R} \times X$ by setting $f(t, x) = 2t\,\bar{g}(\frac{x}{t^2})$ when $t \neq 0$ and $f(0, x) = 0$, whatever $x \in X$. It is crucial that the range of the extension $\bar{g}$ of g be contained in the closed convex hull of the range of g.

Let us now state a theorem on the continuous extension of multifunctions.

We recall that a mapping of a metric space X into the family $\mathcal{F}(Z)$ of non-empty closed bounded subsets of a normed space Z is called continuous if it is continuous as a mapping of X into the metric space $\mathcal{J}$ with Hausdorff distance h.

Theorem [1]. Let X be a metric space, Z a normed space, and suppose $A \subset X$ is closed and $F: A \to \mathcal{J}(Z)$ is continuous. Then there exists a continuous mapping $G: X \to \mathcal{J}(Z)$ such that $G(a) = F(a)$ for every $a \in A$ and $G(x) \subset \operatorname{co} \bigcup \{F(a): a \in A\}$ for every $x \in X$.

Our proof in [1], as Dujundji's in [3], is divided into three steps: first, we construct a suitable locally finite open covering of the complement of A; then we assign a "value" (i.e. a non-empty closed bounded subset of Z) to each element of the covering; and finally we "interpolate" continuously among the various assigned values. Of course, the crucial step is to make precise what we mean by interpolating among various sets.

In [3] the extended mapping is defined by the formula

$$f(x) = \sum_{p_\alpha(x) > 0} p_\alpha(x)\, y_\alpha$$

where (p_α) is the continuous partition of unity subordinated to the given covering (O_α) and each y_α is the image of a point in A "not too far" from O_α.

Two properties of this formula are noteworthy:

(i) The interpolation among the values y_α is obtained by taking convex combinations with continuous coefficients. This accounts for the continuity of f at points in the complement of A.

(ii) If x is "very close" to A and the points y_α entering in the sum are within an ε-ball of some point y, then $f(x)$ as convex combination of these y_α's also will be in the same ε-ball. This accounts for the continuity of f at the boundary of A.

Let us introduce the index set $L_0(x) = \{\alpha : p_\alpha(x) > 0\}$ so that $L_0(x_0) \subset L_0(x)$ for x sufficiently close to x_0 and

$$f(x) = \sum_{\alpha \in L_0(x)} p_\alpha(x)\, y_\alpha .$$

The mapping $x \to L_0(x)$ is not continuous in general. However, f is continuous since new vectors appear under the summation in the form $p_\alpha(x) y_\alpha$ with coefficients $p_\alpha(x)$ that are very close to zero, therefore modifying by little the value of the summation. Linearity of the space is clearly essential here.

In the case of a multifunction the most immediate analogue of a summation is the formation of a union. However, if we think of taking a union over the index set $L_0(x)$, as above, we find that at certain points new sets have to be added into the union but that there is no way of shrinking the size of these sets as there was before, by multiplying the point y_α by a small scalar $p_\alpha(x)$.

Our proof of the extension theorem rests upon the simple observation that, while it is impossible to reduce the size of the sets entering into the union, it is possible to increase the size of the sets already in the union. More precisely, we define the extension G of F by the formula

For an extension theorem for upper semi-continuous multifunctions we refer to [5].

REFERENCES

[1] H. A. Antosiewicz and A. Cellina. "Continuous extensions of multifunctions." Ist. Mat. "U. Dini". Universita Firenze, 1973/74/19. 8 pp.

[2] A. Cellina, "On the nonexistence of solutions of differential equations in nonreflexive of solutions of differential equations in nonreflexive spaces." Bull. Amer. Math. Soc. 78(1972), 1069-1072.

[3] J. Dujundji, Topology, Allyn and Bacon, Inc. Boston, 1966.

[4] G. S. Goodman, "On a theorem of Scorza-Dragoni and its applications to optimal control." pp. 222-233 in Math. Theory of Control. (A. V. Balakrishnan and L. W. Neustadt, ed.) Academic Press, N. Y. 1967.

[5] Tsoy-Wo Ma. "Topological degrees of set-valued compact fields in locally convex spaces." Dissertationes Math. Rozprawy Mat. 92(1972), 43 pp.

[6] G. Scorza Dragoni. Un teorema sulle funzioni continue rispetto ad una e misurabili rispetto ad un' altra variable. Red. Sem. Mat. Univ. Padova. 17 (1948) 102-108.

[7] J. A. Yorke "Noncontinuable solutions of differential-delay equations." Proc. Amer. Math. Soc. 21(1969), 648-652.

Department of Mathematics, University of Southern California
Los Angeles, California
and

Istituto Matematico "U. Dini"
Universita di Firenze, Florence

BOUNDARY CONTROL OF THE DIFFUSION EQUATION
A SEMIGROUP THEORETICAL APPROACH

By

A. V. Balakrishnan

An approach to boundary control of partial differential equations is outlined which is based on the theory of semigroups of operators, in contrast to the Lions-Magenes variational theory, both in techniques and results.

University of California, Los Angeles, California

AN ABSTRACT FRAMEWORK FOR APPROXIMATE SOLUTIONS TO OPTIMAL CONTROL PROBLEMS GOVERNED BY HEREDITARY SYSTEMS

H. T. Banks* and J. A. Burns**

1. Introduction.

In this note we develop a general theoretical framework for obtaining approximate solutions of linear functional differential equation (FDE) systems. By an appropriate choice of spaces we show in Sec. 2 that the nonhomogeneous linear FDE can be investigated via an equivalent abstract equation in Banach space. Then the theory of semigroups may be exploited and in particular approximation results

*This research was supported in part by the Office of Army Research under DA-ARO-D-31-124-73-G130, in part by the National Science Foundation under GP 28931X3, and in part by the Air Force Office of Scientific Research under AFOSR 71-2078C.

**This research was supported in part by the National Science Foundation under GP 28931X3 and in part by the Air Force Office of Scientific Research under AFOSR 71-2078C.

(such as Trotter's theorem [11]) for semigroups lead to approximate solutions of the original equation. Convergence results are established in Sec. 3. Also presented is one example of the type of approximations often used and which can be included as a special case of our general development here. Finally, in Sec. 4 we indicate how one may apply these ideas to obtain approximate solutions for optimal control problems with FDE systems.

The presentation given here, while similar in spirit to that developed in [1], is in fact quite distinct and results in a substantially different (from a computational viewpoint) sequence of finite-dimensional approximating control problems. Approximation ideas similar to those discussed below are employed in [9] to treat certain control problems involving partial differential equations.

A number of normed linear spaces will arise in the discourse below. We shall not distinguish between the various norms in statements which, when taken in context, make it clear what is meant. Thus, if x is an element of the space X with norm $|\cdot|$, we shall write simply $|x|$ for the norm of x.

2. Abstract representation for solutions of FDEs

For $1 \leq p < \infty$ we denote the usual Lebesgue spaces of R^ν-valued "functions" on $[a, b]$ by $L_p([a, b], R^\nu)$.

Throughout we shall use $L_p(a, b)$ to represent this space when $\nu = n$ is the dimension of our underlying functional differential equation system. For fixed $r > 0$ and $x: [-r, \infty) \to R^n$, $x_t: [-r, 0] \to R^n$ is defined for $t \geq 0$ by $x_t(s) = x(t+s)$. Let L be a linear transformation with domain in the linear space of R^n-valued Lebesgue-measurable functions on $[-r, 0]$ such that L restricted to the Banach space of continuous functions $\mathcal{C} = C([-r, 0], R^n)$ is a bounded linear operator. We also require that L satisfy the following assumptions (see [2]):

(A) If $t_1 > 0$ and $x \in L_p(-r, t_1)$, then the function $t \to L(x_t)$ is defined almost everywhere on $[0, t_1]$ and depends only on the equivalence class of the function x. Furthermore, $t \to L(x_t)$ is in $L_1(0, t_1)$ and there exists a continuous function Γ such that

$$\int_0^t | L(x_s) | \, ds \leq \Gamma(t) \left(\int_{-r}^t | x(s)|^p ds \right)^{1/p}$$

for all $t \in [0, t_1]$.

We consider the system described by the retarded functional differential equation

(2.1) $$\dot{x}(t) = L(x_t) + f(t), \quad t \geq 0,$$

and the initial data

(2.2) $$x(0) = \eta, \; x_0 = \phi .$$

where $\eta \in R^n$, $\phi \in L_p(-r, 0)$, $f \in L_p(0, t_1)$ for each $t_1 > 0$.

A solution to system (2.1) - (2.2) is a function x, in $L_p(-r, t_1)$ for each $t_1 > 0$, such that x is absolutely continuous (A.C.) for $t \geq 0$, x satisfies (2.1) a.e. on $[0, t_1]$, $x(0) = \eta$, and $x(s) = \phi(s)$ for almost every s in $[-r, 0]$.

It can be shown (again, see [2]) that system (2.1)-(2.2) has a unique solution defined on $[-r, \infty)$ which depends continuously on the initial data $(\eta, \phi) \in R^n \times L_p(-r, 0)$. Throughout the remainder of this paper, we denote by Z tha space $Z = R^n \times L_p([-r, 0], R^n)$ taken with the obvious norm $|(\eta, \phi)|^2 = |\eta|^2 + |\phi|^2$. For $t \geq 0$, define $S(t): Z \to Z$ by $S(t)(\eta, \phi) = (x(t), x_t)$, where x is the solution to the homogeneous $(f \equiv 0)$ equation (2.1) with initial data (2.2). It follows that $\{S(t)\}_{t > 0}$ is a strongly continuous semigroup and its infinitesimal generator, $\mathcal{A}$, has domain given by $\mathcal{D}(\mathcal{A}) = \{(\eta, \phi): \phi \text{ is A. C.}, \dot{\phi} \in L_p(-r, 0), \text{ and } \phi(0) = \eta\}$. For $(\eta, \phi) \in \mathcal{D}(\mathcal{A})$, one has $\mathcal{A}(\eta, \phi) = (L(\phi), \dot{\phi})$.

Furthermore, there are constants γ and M such that the spectrum of $\mathcal{A}$ lies on the left half-plane $\mathrm{Re}(\lambda) < \gamma$ and S(t) satisfies $|S(t)(\eta, \phi)| \leq Me^{(\gamma+\varepsilon)t}|(\eta, \phi)|$ for each $\varepsilon > 0$.

For (η, ϕ) fixed in Z and $f \in L_p(0, t_1)$, we define $z(\cdot\,; f): [0, t_1) \to Z$ by

$$(2.3) \qquad z(t; f) = S(t)(\eta, \phi) + \int_0^t S(t-\sigma)(f(\sigma), \theta)d\sigma,$$

where θ denotes the zero function in $L_p(-r, 0)$ and where we shall suppress the notation showing dependence of $z(\cdot\,; f)$ on the initial data unless it is explicitly needed.

The following improves and extends previous results of Borisovič and Turbabin [2].

Theorem 2.1. If $(\eta, \phi) = (\phi(0), \phi) \in \mathcal{D}(\mathcal{A})$, $t_1 > 0$, and $f \in L_p(0, t_1)$, then $z(t; f) = (x(t; f), x_t(f))$, where $x(\cdot\,; f)$ is the solution to system (2.1)-(2.2).

Proof. First suppose that ϕ, f are C^1. Then $\sigma \to F(\sigma) = (f(\sigma), \theta)$ is C^1 Z-valued and it follows, since $(\phi(0), \phi) \in \mathcal{D}(\mathcal{A})$ ([6, p. 135; p. 47]), that $t \to z(t; \phi, f)$ defined by (2.3) is the unique classical solution to

$$(2.4) \qquad \begin{aligned} \dot{z}(t) &= \mathcal{A}z(t) + F(t) \\ z(0) &= (\phi(0), \phi). \end{aligned}$$

However, it is not difficult to verify directly that $t \to w(t; \phi, f) \equiv (x(t; \phi, f), x_t(\phi, f))$, x the solution of (2.1)-(2.2), also is a classical solution of (2.4) and hence by uniqueness must agree with that given by (2.3).

Observe next that the solutions (with $\eta = \phi(0)$) to (2.1)-(2.2) and (2.3) are defined for $\phi \in$ A.C. ($\equiv$ absolutely continuous functions on $[-r, 0]$) and $f \in L_p(0, t_1)$. Furthermore, for fixed t the mapping $(\phi(0), \phi, f) \to z(t; \phi, f)$ from $\mathcal{D}(\mathcal{A})$ (with Z-norm) $\times L_p$ to Z and $(\phi(0), \phi, f) \to (x(t; \phi, f), x_t(\phi, f))$ from $R^n \times$ A.C. (with supremum $\mathcal{C}$ norm) $\times L_p$ to Z are continuous. Then for any ϕ, f with $(\phi(0), \phi) \in \mathcal{D}(\mathcal{A})$ and $f \in L_p(0, t_1)$, we choose $\{\phi^k\}, \{f^k\}$ in C^1 such that $|\phi^k - \phi|_{\mathcal{C}} \to 0$ (which implies $(\phi^k(0), \phi^k) \to (\phi(0), \phi)$ in Z-norm) and $|f^k - f|_{L_p} \to 0$. From the continuity properties pointed out above plus $z(t; \phi^k, f^k) = (x(t; \phi^k, f^k), x_t(\phi^k, f^k))$ for each k we are thus able to reach the desired conclusion.

In the sequel, we shall present an approximating system for (2.3) and thus obtain one for (2.1)-(2.2). We shall for ease of exposition restrict our consideration hereafter to the case $p = 2$ (hence the Hilbert space case). Analogous results can be obtained by appropriate modifications for the cases $1 < p < \infty$.

3. The approximating system and convergence results

Let Z^N be a sequence of finite-dimensional subspaces of Z and $P^N: Z \to Z^N$ a sequence of projections onto Z^N, $N = 1, 2, \ldots$. We also assume there exists a sequence of linear operators $\mathcal{A}^N: Z^N \to Z^N$ with the following hypothesis obtaining:

H: (i) there exist $\alpha > 0$, $K > 0$ such that $|e^{\mathcal{A}^N t}| \leq Ke^{\alpha t}$ for $N = 1, 2, \ldots$;

(ii) $|P^N(\xi,\psi) - (\xi,\psi)| \to 0$ for all $(\xi,\psi) \in Z = R^n \times L_2(-r,0)$;

(iii) $|\mathcal{A}^N P^N(\xi,\psi) - \mathcal{A}(\xi,\psi)| \to 0$ for all $(\xi,\psi) \in \mathcal{D}(\mathcal{A})$.

The approximating system for (2.3) is defined by the ordinary vector differential equation

(3.1) $\quad \dot{z}(t) = \mathcal{A}^N z(t) + P^N(f(t),\theta), \qquad t \geq 0,$

and the initial data

(3.2) $\quad z(0) = P^N(\eta,\phi).$

Since Z^N is finite-dimensional it follows that the solution to (3.1)-(3.2) is given by $z^N(t;f) = (x^N(t;f), y^N(t;f))$ where

(3.3) $\quad z^N(t;f) = S^N(t)P^N(\eta,\phi) + \int_0^t S^N(t-\sigma)P^N(f(\sigma),\theta)d\sigma,$

and $S^N(t) = e^{\mathcal{A}^N t}$.

Applying the well-known convergence results of Trotter (see Trotter [11] and Kato [5]) in the above setting yields

<u>Lemma 3.1.</u> If hypothesis H is satisfied and $t \geq 0$, then $|S^N(t)P^N(\xi,\psi) - S(t)(\xi,\psi)| \to 0$ for all $(\xi,\psi) \in Z$. Moreover, for each $t \geq 0$ and $f \in L_2(0,t)$, it follows that $|z^N(t;f) - z(t;f)| \to 0$.

Define the operators $T^N(t): R^n \to Z^N$ and $T(t): R^n \to Z$ by $T^N(t)\xi = S^N(t)P^N(\xi,\theta)$ and $T(t)\xi = S(t)(\xi,\theta)$ respectively, and note that $|T^N(t)| \leq |S^N(t)P^N| \leq Ke^{\alpha t}$.

<u>Lemma 3.2.</u> If the hypothesis H is satisfied, then the operators $T^N(t)$ converge in the uniform operator norm to $T(t)$ for each $t \geq 0$. In particular, if $t \geq 0$, then

$$\int_0^t |T^N(t-\sigma) - T(t-\sigma)|^2 \, d\sigma \to 0 \text{ as } N \to \infty.$$

Proof. To establish the desired convergence for $T^N(t)$ it is enough to observe that $|T^N(t)\xi - T(t)\xi| = |(S^N(t)P^N - S(t))(\xi, \theta)|$ for $\xi \in R^n$. Pointwise convergence of $S^N(t)P^N$ to $S(t)$ plus the fact that $T^N(t)$, $T(t)$ are defined on R^n allows one to conclude $|T^N(t) - T(t)| \to 0$. The second assertion of the lemma then follows, of course, from the dominated convergence theorem.

It is also easy to argue that the functions $t \to T^N(t)$ and $t \to T(t)$ are continuous with respect to the uniform operator topology. As a direct consequence of the previous developments we have the following theorem.

Theorem 3.1. Suppose $t_1 > 0$, $t \in [0, t_1]$, and hypothesis H obtains. Then

(i) The operator $F(t)$: $L_2(0, t_1) \to Z$ defined by

$F(t)f = \int_0^t S(t-\sigma)(f(\sigma), \theta)d\sigma$ is a compact operator.

(ii) The solutions z^N of (3.3) satisfy $z^N(t;f) \to z(t;f)$ for each $t \in [0, t_1]$, and the convergence is uniform in f for f in bounded subsets of $L_2(0, t_1)$.

Proof. Defining $F^N(t)$: $L_2 \to Z^N$ by $F^N(t)f = \int_0^t T^N(t-\sigma)f(\sigma)d\sigma$, one easily sees that $F^N(t)$ is a compact linear transformation. For fixed $t \in [0,t_1]$,

$$
\begin{aligned}
|F^N(t)f - F(t)f| &= \left| \int_0^t (T^N(t-\sigma) - T(t-\sigma))f(\sigma)d\sigma \right| \\
&\leq \int_0^t |T^N(t-\sigma) - T(t-\sigma)| \, |f(\sigma)| d\sigma \\
&\leq \left(\int_0^t |T^N(t-\sigma) - T(t-\sigma)|^2 d\sigma \right)^{1/2} |f|_{L_2} ,
\end{aligned}
$$

and it follows that $F^N(t) \to F(t)$ in the uniform operator topology. The conclusion (i) then follows from the compactness of $F^N(t)$, $N = 1, 2, \ldots$. Statement (ii) follows from these arguments taken with the fact that $S^N(t)P^N(\eta, \phi) \to S(t)(\eta, \phi)$ for $(\eta, \phi) \in Z$.

Using the inequality

$$
|z^N(t;f^k) - z(t;f)| \leq |z^N(t;f^k) - z(t;f^k)| + |z(t;f^k) - z(t;f)|
$$

with the above theorem, one obtains immediately

Corollary 3.1 . If hypothesis H holds and f^k converges weakly to f in $L_2(0, t_1)$, then for each fixed $t \in [0, t_1]$ we have $z^N(t;f^k) \to z(t;f)$ strongly in Z as $N, k \to \infty$. In particular, $z^N(t;f^N) \to z(t;f)$.

Actually, one is able to prove a much stronger result concerning the solution operator (2.3) than that indicated in Theorem 3.1. For fixed $(\eta, \phi) \in Z$, let $\mathscr{S}: L_2(0, t_1) \to \mathcal{C} = C([0, t_1], Z)$ be defined by $[\mathscr{S}(f)](t) = z(t;f)$, where $\mathcal{C}$ is the space of Z-valued continuous functions on $[0, t_1]$ with supremum norm. Then we have

Theorem 3.2 . If hypothesis H is satisfied, then $\mathscr{S}$ is a compact affine operator.

Proof. Letting $\mathcal{F}$ be a bounded subset of $L_2(0, t_1)$, we need only show that $\mathcal{S}(\mathcal{F})$ is relatively compact in $\mathcal{C}$, i.e. that $\{z(\cdot;f) | f \varepsilon \mathcal{F}\}$ is relatively compact. Since for each fixed t, $\{z(t;f) | f \varepsilon \mathcal{F}\}$ is a relatively compact subset of Z, our result follows immediately from the Ascoli theorem [8] if we establish equicontinuity of $\{z(\cdot;f) | f \varepsilon \mathcal{F}\}$. To do this it suffices to consider solutions with zero initial data $(\eta, \phi) = (0, \theta)$ and note then that

$$|z(t;f)-z(\tau;f)| = \left| \int_0^t T(t-\sigma)f(\sigma)d\sigma - \int_0^\tau T(\tau-\sigma)f(\sigma)d\sigma \right| .$$

Standard arguments taken with the uniform continuity of $t \to T(t)$ on $[0, t_1]$ then can be made to obtain the desired conclusions.

We next consider an example of the type of approximations which are included as special cases of the above discussions. We specialize (2.1) to the vector retarded equation

$$\dot{x}(t) = Ax(t) + Bx(t-r) + f(t), \quad t \geq 0, \tag{3.4}$$

where A and B are $n \times n$ constant matrices. Partition the interval $[-r, 0]$ by $-r = t_N^N < t_{N-1}^N < \ldots < t_0^N = 0$, where $t_j^N = \frac{-jr}{N}$, N a fixed positive integer. Let χ_j^N denote the characteristic function of $[t_j^N, t_{j-1}^N]$, $j = 1, 2, \ldots, N$. The space $Z^N \subseteq Z$ is defined by

$$Z^N = \{(\eta, \phi) \varepsilon Z | \phi = \sum_{j=1}^N v_j^N \chi_j^N, v_j^N \varepsilon R^n\},$$

and the projections $P^N: Z \to Z^N$ are given by

$$P^N(\eta, \phi) = (\eta, \sum_{j=1}^{N} \phi_j^N \chi_j^N)$$

where $\phi_j^N \equiv \frac{N}{r} \int_{t_j^N}^{t_{j-1}^N} \phi(s)ds.$ For $(\eta, \psi) \in Z^N$ with

$\psi = \sum_{j=1}^{N} v_j^N \chi_j^N$, the operators $\mathcal{A}^N$ are defined by

$$\mathcal{A}^N(\eta, \psi) = (A\eta + Bv_N^N, \frac{N}{r}(\eta - v_1^N)\chi_1^N + \sum_{j=2}^{N} \frac{N}{r}(v_{j-1}^N - v_j^N)\chi_j^N).$$

It can then be shown that Z^N, P^N and $\mathcal{A}^N$ satisfy the conditions of hypothesis H. That the sequence of averaging projections $\{P^N\}$ has the desired convergence properties is a rather standard result which is easily verified. That property (iii) of H holds for these $\mathcal{A}^N$ can be argued without difficulty using arguments similar to those found in [10], where those authors used these particular approximations to study controllability of (3.4)

Choosing a "basis" $\{e^0, e^1, \ldots, e^N\}$ for Z^N, with $e^0 = (1, \theta)$, $e^j = (0, \chi_j^N)$, $j = 1, 2, \ldots, N$, and writing $z^N(t;f) = \sum_{j=0}^{N} w_j(t)e^j$, we find that the approximating system (3.3) then becomes the N+1 vector equations

$$\dot{w}(t) = A^N w(t) + P^N(f(t), \theta)$$

with data

$$w(0) = \text{col}[\phi(0), \phi_1^N, \ldots, \phi_N^N].$$

Here we have $w(t) = \mathrm{col}[w_0(t), w_1(t), \ldots, w_N(t)]$, $P^N(f(t), \theta) = \mathrm{col}[f(t), 0, \ldots, 0]$ and

$$A^N = \begin{bmatrix} A & 0 & \cdots & & 0 & B \\ \frac{N}{r}I & -\frac{N}{r}I & 0 & & & 0 \\ 0 & \frac{N}{r}I & -\frac{N}{r}I & 0 & & 0 \\ \vdots & & & \ddots & & \vdots \\ & & & & \ddots & 0 \\ 0 & \cdots & & 0 & \frac{N}{r}I & -\frac{N}{r}I \end{bmatrix}$$

with I the $n \times n$ identity matrix.

We remark that this approximating system has been widely used in a number of investigations concerning the differential-difference equation (3.4) (for example, see [4], [7] among others). The principal convergence results usually alluded to in these efforts can, of course, be seen to be a special case of the results presented above.

Other types of projections which lead to an approximating system in the context of the framework put forth in this note have been investigated by the authors and will be discussed in detail in a future paper.

4. Application to optimal control problems.

We turn now to the application of the preceding results to establish a theoretical framework for approximate solutions of control problems. We discuss only the simplest of problems here to illustrate our ideas.

Define $J(x, u, \xi, \psi) \equiv g_0(\xi) + g_1(\psi) + J_1(x, u)$, where $g_0: R^n \to R^1$, $g_1 : L_2(-r, 0) \to R^1$, and $J_1 : L_2([0, t_1], R^n) \times L_2([0, t_1], R^m) \to R^1$ are continuous. Given a closed convex set $\mathcal{U}$ in $L_2([0, t_1], R^m)$, we wish to minimize the functional $\Phi(u) \equiv J(x(\cdot\,; u), u, x(t_1; u), x_{t_1}(u))$ over the set $\mathcal{U}$, where we now denote by $x(\cdot\,; u)$ the solution to system (2.1)-(2.2) with (η, ϕ) fixed in $\mathcal{D}(\mathcal{A})$ and $f(t) = C(t)u(t)$, C an $n \times m$ matrix-valued bounded measurable function and $u \in \mathcal{U}$.

The approximating control problem is that of minimizing $\Phi^N(u) \equiv J(x^N(\cdot\,; u), u, x^N(t_1; u), y^N(t_1\,; u))$ over the set $\mathcal{U}$. For simplicity, we shall assume that J is convex and that $\Phi^N(u^N) \to +\infty$ if $|u^N| \to \infty$. (In fact, much weaker continuity and convexity assumptions are sufficient to carry out the familiar arguments below. See, for example, [1] and [3].) An example of such a cost functional is

$$\begin{aligned}\Phi(u) &= J(x(\cdot\,; u), u, x(t_1; u), x_{t_1}(u)) \\ &= |x(t_1; u) - \zeta(0)|^2 + \int_{-r}^{0} |x_{t_1}(u)(s) - \zeta(s)|^2 ds \\ &\quad + \int_0^{t_1} [x^*(s;u)\, \mathcal{W} x(s;u) + u^*(s)\, \mathcal{R} u(s)] ds\end{aligned}$$

where $\mathcal{W} \geq 0$, $\mathcal{R} > 0$, and ζ is a given function.

It is known ([1], [3]) that optimal controls $\bar{u}, \bar{u}^N$ exist for the original problem and the approximating problems respectively under the above assumptions. In addition, if Φ is strictly convex, then the optimal controls are unique.

Theorem 4.1. Suppose hypothesis H is satisfied and $\bar{u}^N$ are optimal controls for the approximating problems. Then there is a subsequence $\{\bar{u}^{N_k}\}$ converging weakly in $L_2([0, t_1], R^m)$ to an admissible control u^0 which is an optimal control for the original problem with $\Phi^{N_k}(\bar{u}^{N_k}) \to \Phi(u^0)$. If Φ is strictly convex, then $\{\bar{u}^N\}$ converges weakly to the unique optimal control u^0 for the original problem and $\Phi^N(\bar{u}^N) \to \Phi(u^0)$.

Proof. The sequence $\{\bar{u}^N\}$ must be bounded. Otherwise $\Phi^N(\bar{u}^N) \to +\infty$ which is contradicted by the result $\Phi^N(\bar{u}^N) \leq \Phi^N(\bar{u}) \to \Phi(\bar{u}) < \infty$ following from Lemma 3.1. Let $\{\bar{u}^{N_k}\}$ be a subsequence converging weakly to some u^0 which is in $\mathcal{U}$, since $\mathcal{U}$ is weakly closed. By Theorem 3.1 and its corollary, the fact that the convergence is dominated, the convexity of J, and Lemma 3.1, we find for any $u \varepsilon \mathcal{U}$,

$$
\begin{aligned}
\Phi(u^0) &\leq \liminf J(x^{N_k}(\cdot;\bar{u}^{N_k}), \bar{u}^{N_k}, x^{N_k}(t_1;\bar{u}^{N_k}), y^{N_k}(t_1;\bar{u}^{N_k})) \\
&= \liminf \Phi^{N_k}(\bar{u}^{N_k}) \\
&\leq \limsup \Phi^{N_k}(\bar{u}^{N_k}) \\
&\leq \limsup \Phi^{N_k}(u) \\
&= \Phi(u) .
\end{aligned}
$$

Hence u^0 is optimal and, in particular, with $u = u^0$ above we see that $\Phi^{N_k}(\bar{u}^{N_k}) \to \Phi(u^0)$.

The above simple illustrative discussion indicates theoretically how one may use the results of this note to develop numerical methods to solve optimal control problems. Our early investigations of the efficiency of such procedures on examples indicate the likelihood of obtaining rapidly convergent, easily implementable methods. A detailed discussion of convergence properties along with numerical results will appear in a forthcoming paper.

REFERENCES

[1] H. T. Banks and A. Manitius, Projection series for retarded functional differential equations with applications to optimal control problems, J. Differential Equations, to appear.

[2] J. G. Borisovič and A. S. Turbabin, On the Cauchy problem for linear nonhomogeneous differential equations with retarded argument, Soviet Math. Doklady 10 (1969), 401-405.

[3] J. A. Burns, Existence theorems and necessary conditions for a general formulation of the minimum effort problem, J. Optimization Theory Appl., to appear.

[4] K. Ichikawa, Pontryagin's maximum principle in optimizing time-delay systems, Electrical Engineering in Japan 87 (1967), 75-83.

[5] T. Kato, Perturbation Theory for Linear Operators, Springer-Verlag, New York, 1966.

[6] S. G. Krein, Linear Differential Equations in Banach Space, Trans. Math. Monographs Vol. 29, Amer. Math. Soc., Providence, 1971.

[7] Iu. M. Repin, On the approximate replacement of

systems with lag by ordinary dynamical systems, Prikl. Mat. Mech. 29 (1965), no. 2, 226-235; English translation in Applied Math. Mech. 29 (1965), 254-264.

[8] H. L. Royden, Real Analysis, 2d edition, MacMillan, New York, 1968.

[9] H. Sasai and E. Shimemura, On the convergence of approximating solutions for linear distributed parameter optimal control problems, SIAM J. Control 9 (1971), 263-273.

[10] H. Sasai, T. Fukuda, and H. Ishigaki, On an approximate approach to controllability of delay-differential systems, to appear.

[11] H. F. Trotter, Approximation of semi-groups of operators, Pacific J. Math. 8 (1958), 887-919.

Brown University, Providence, Rhode Island
Brown University, Providence, Rhode Island and
Virginia Polytechnic Institute and State University, Blacksburg

LOWER CLOSURE AND EXISTENCE THEOREMS IN OPTIMAL CONTROL

Leonard D. Berkovitz

1. Introduction.

The problem of showing that an optimal control exists is that of showing that a certain functional attains a minimum on a given subset $\mathcal{Y}$ of its domain of definition $\mathcal{D}$. One would therefore expect to use the following general procedure to establish existence. One first shows that for an appropriate topology on $\mathcal{D}$ the set $\mathcal{Y}$ is compact. One then shows that the functional is lower semicontinuous on $\mathcal{Y}$ with respect to the topology in question. It turns out that in control problems with non-linear state equations the lower semicontinuity must be replaced by a concept known as "lower closure". In this note we summarize some lower closure theorems that we have recently obtained for a very general class of control problems. This class includes most problems of interest. The proofs are not difficult and are applications of Mazur's theorem, Fatou's lemma and Filippov's implicit functions theorem. Results similar to ours have also been announced by Cesari [5] and Bidaut [3] .

This research was supported by National Science Foundation Grant No. GP-33551X.

2. Notation and Formulation of Problem.

We shall use single letters to denote vectors, we shall use subscripts to distinguish vectors, and we shall use superscripts to denote components of vectors. The letter t will denote a vector $(t^1, \ldots, t^\nu)$, in real euclidean space R^ν, $\nu \geq 1$, the letter x will denote a vector $(x^1, \ldots, x^n)$ in real euclidean space R^n, $n \geq 1$, and the letter w a vector in R^m, $m \geq 1$. The euclidean norm of a vector x will be denoted by $|x|$. The inner product of two vectors x_1 and x_2 will be written as $\langle x_1, x_2 \rangle$. Thus $|x| = \langle x, x \rangle^{1/2}$.

Let $f^o : (t, x, w) \to f^o(t, x, w)$ be a real valued function defined on $R^\nu \times R^n \times R^m$ and let $f: (t, x, w) \to f(t, x, w)$ be a vector valued function on $R^\nu \times R^n \times R^m$ with range in R^r. Let G be a bounded region of the t space R^ν, let X be a region of the x-space R^n. Let $\mathcal{R}$ denote the Cartesian product $\bar{G} \otimes \bar{X}$, where $\bar{G}$ denotes the closure of G and $\bar{X}$ denotes the closure of X. Let Ω be a mapping that assigns to each point (t, x) in $\mathcal{R}$ a subset $\Omega(t, x)$ of the w-space R^m.

As usual, let $L_{p_i}(G)$, $p_i \geq 1$, denote the Banach space of real valued measurable functions z^i defined on G such that $\int_G |z^i|^{p_i} dt < \infty$, and let

$$\|z^i\|_{p_i} = \{\int_G |z^i|^{p_i} dt\}^{1/p_i} \qquad p_i \geq 1.$$

Let $\mathcal{Y}$ and $\mathcal{Z}$ denote the following Banach spaces:

$$\mathcal{Y} = \prod_{i=1}^{r} L_{q_i}(G) \qquad \mathcal{Z} = \prod_{i=1}^{n} L_{p_i}(G) \qquad q_i, p_i \geq 1.$$

The norm of an element $z = (z^1, \ldots, z^n)$ in $\mathcal{Z}$ is given by

$$\|z\| = \{ \sum_{i=1}^{n} \|z^i\|_{p_i}^2 \}^{1/2} .$$

A similar formula gives the norm of an element $y = (y^1, \ldots, y^r)$ in $\mathcal{Y}$.

Let $\mathcal{T}$ be a Banach space and let the norm of an element ϕ in $\mathcal{T}$ be denoted by $\|\phi\|$. Let M be a mapping from $\mathcal{T}$ to $\mathcal{Z}$ and let N be a mapping from $\mathcal{T}$ to $\mathcal{Y}$. Thus the image under N of an element ϕ in $\mathcal{T}$ is an element $N\phi = y = (y^1, \ldots, y^r)$ in $\mathcal{Y}$, where each y^i is in $L_{q_i}(G)$. Similarly the image under M of an element ϕ in $\mathcal{T}$ is an element $M\phi = z = (z^1, \ldots, z^n)$ in $\mathcal{Z}$, where each z^i is in $L_{p_i}(G)$. Note that N and M need not be linear.

Let $\mathcal{M}$ denote the set of all measurable functions u defined on G with range in R^m. Thus if $u \in \mathcal{M}$ then $u = (u^1, \ldots, u^m)$, u^i real valued and measurable in G.

DEFINITION 2.1. An element ϕ in $\mathcal{T}$ is said to be an admissible trajectory if there is function u in $\mathcal{M}$ such that the following hold:

$$\text{(2.1)} \quad \begin{array}{ll} \text{(i)} & (t, (M\phi)(t)) \in \mathcal{R} \text{ a.e. in } \overline{G} \\ \text{(ii)} & (N\phi)(t) = f(t, (M\phi)(t), u(t)) \text{ a.e. in } G \\ \text{(iii)} & u(t) \in \Omega(t, (M\phi)(t)) \text{ a.e. in } G \end{array}$$

(iv) the mapping $t \to f^o(t, (M\phi)(t), u(t))$ is in $L_1(G)$.

The function u is said to be an admissible control and the pair (ϕ, u) is said to be an admissible pair. The set of all admissible pairs will be denoted by $\mathcal{A}$.

The optimal control problem is to minimize the functional

(2.2) $$J(\phi, u) = \int_G f^o(t, (M\phi)(t), u(t))\, dt$$

in a given class $\mathcal{A}_o \subseteq \mathcal{A}$ of admissible pairs. That is, we are to find a (ϕ^*, u^*) in $\mathcal{A}_o$ such that $J(\phi^*, u^*) \leq J(\phi, u)$ for all (ϕ, u) in $\mathcal{A}_o$. Such a pair (ϕ^*, u^*) is called an optimal pair. The element ϕ^* is called an optimal trajectory; the element u^* is called an optimal control.

A motivation for the abstract control problem formulated above is given in Berkovitz [1]. Additional examples of problems included in this formulation can be found in Cesari [4].

3. Outline of Proof of Existence.

The proof of the existence of an optimal pair in a class $\mathcal{A}_o \subseteq \mathcal{A}$ proceeds in general as follows. Let

$$\mu = \inf \{ J(\phi, u) : (\phi, u) \in \mathcal{A}_o \} . \tag{3.1}$$

If $\mu = +\infty$, there is nothing to prove. If $\mu < \infty$, then there exists a minimizing sequence such that $J(\phi_k, u_k) \to \mu$. Conditions are then placed on the problem to ensure that $\mu > -\infty$. Additional conditions are placed on the problem to ensure that the sequence of trajectories $\{\phi_k\}$ is conditionally compact in some sense. From experience in less abstract problems we know that conditional weak compactness of the minimizing sequence will suffice.

For different special classes of problems different Banach spaces $\mathcal{T}$ are appropriate. Therefore, in treating a problem as general as ours it is not too fruitful to impose a condition that will be usable and that will ensure conditional weak compactness of minimizing sequences. We leave this to the specific applications.

Assuming that we guarantee the weak compactness of $\{\phi_k\}$, we select a subsequence, again labeled $\{\phi_k\}$, that converges weakly to an element ϕ^* in $\mathcal{T}$. Any boundary conditions or other conditions on the trajectories ϕ that are used to define the subclass $\mathcal{A}_o$ must be such that if $\phi_k \to \phi^*$ weakly, then ϕ^* must satisfy these conditions.

At this point a "**lower-closure**" theorem is used. This theorem states that, under appropriate hypotheses, if $\phi_k \to \phi^*$ weakly in $\mathcal{J}$, then there is a u^* in $\mathcal{M}$ such that (ϕ^*, u^*) is admissible and

$$(3.2) \qquad J(\phi^*, u^*) \leq \lim\ \inf\ J(\phi_k, u_k).$$

If we can also show that $(\phi^*, u^*) \in \mathcal{O}_0$, then $J(\phi^*, u^*) \geq \mu$. Hence (ϕ^*, u^*) is an optimal pair.

The difficult step in proving existence is the proof of the lower closure property. In the next section we shall give several lower closure theorems that cover most cases of interest. Theorem 4.2 below is the simplest, is the easiest to apply, and covers many special cases of interest.

DEFINITION 3.1 . A sequence of admissible pairs $\{(\phi_k, u_k)\}$ is said to be weakly lower closed whenever the following holds. If $\phi_k \to \phi$ weakly in $\mathcal{J}$ then there exists a measurable function $u: G \to R^m$ such that (ϕ, u) is admissible and

$$(3.3) \qquad \liminf_{k \to \infty} J(\phi_k, u_k) \geq J(\phi, u).$$

4. Lower Closure Theorems.

We first list a set of assumptions that will be in force for all of our theorems. For ease of reference we label this set Assumption I.

ASSUMPTION I. (1) For each t in G the function $\hat{f} = (f^o, f)$ is a continuous function of (x, w) in R^{n+m} and for each (x, w) in R^{n+m} the function $\hat{f}$ is measurable with respect to t in G. (2) The set

$$\mathcal{D} = \{ (t, x, w) : (t, x) \in \mathcal{R}, \quad w \in \Omega(t, x) \}$$

is closed. (3) For each (t, x) in $\mathcal{R}$ the set

$$Q^+(t, x) = \{ (\eta, \xi) : \eta \geq f^o(t, x, w), \ \xi = f(t, x, w), \ w \in \Omega(t, x) \}$$

is closed and convex. (4) There exists a real valued function ψ in $L_1(G)$ such that $\psi(t) \geq 0$ on G and a constant r-vector b such that

$$f^o(t, x, w) - \langle b, f(t, x, w) \rangle \geq -\psi(t)$$

for all (t, x, w) in $\mathcal{D}$. (5) If $\{\phi_k\}$ is a sequence of elements in $\mathcal{J}$ that converges weakly to an element ϕ in $\mathcal{J}$, then $M\phi_k \to M\phi$ strongly in $\mathcal{Q}$.

In our first two lower closure theorems we shall suppose that the sets $\Omega(t, x)$ are independent of x; i. e. $\Omega(t, x) = \Omega(t, x')$ for all x, x' in $\overline{X}$. In the first lower closure theorem we impose a generalized Holder type condition in the state variables on the function $\hat{f}$.

In Theorem 4.1, which follows, $p = \min\{p_i, \ i=1, \ldots, n\}$

and q is defined by the relation $1/p + 1/q = 1$ if $p > 1$ and $q = \infty$ if $p = 1$.

THEOREM 4.1 . Let Assumption I hold. Let the sets $\Omega(t, x)$ be independent of x. Let there exists a non-decreasing function μ defined on $[0, \infty)$ and a non-negative function H defined on $\overline{G} \times R^m$ such that the following hold. (i) $\lim_{\delta \to 0} \mu(\delta) = 0$. (ii) There exists a $\delta_0 > 0$ such that for $\delta > \delta_0$, $\mu(\delta) \leq \delta$. (iii) For all (t, x, w) and (t, x', w) in $\mathcal{D}$.

$$(4.1) \qquad |\hat{f}(t, x, w) - \hat{f}(t, x', w)| \leq \mu(|x - x'|)\{H(t, w) + K|x - x'|^{p-1}\}$$

where K is a non-negative constant. Let $\{(\phi_k, u_k)\}$ be a sequence of admissible pairs such that $\lim\inf J(\phi_k, u_k) < \infty$ and for all k

$$(4.2) \qquad \|H_k\|_q \leq A,$$

where A is a constant, $H_k(t) = H(t, u_k(t))$, and $\|\ldots\|_q$ denotes the L_q norm, $1 \leq q \leq \infty$. Then the sequence $\{(\phi_k, u_k)\}$ is weakly lower closed.

We remark that if $\hat{f}$ is uniformly continuous on $\mathcal{D}$, which occurs if $\hat{f}$ is continuous and $\mathcal{D}$ is compact, then (4.1) holds with $H \equiv 1$ and μ the modulus of continuity, suitably defined for large values of δ. If $\hat{f}$ is Lipschitz in x, then (4.1) holds with $\mu(\delta) = \delta$, $p = 1$, and H equal to the Lipschitz constant. Condition (4.2) also holds in these cases.

In the linear plant quadratic criterion problem we have

$$f(t, x, w) = A(t)x + B(t)w$$

and

$$f^o(t, x, w) = \langle x, Q(t)w \rangle + \langle x, P(t)w \rangle + \langle w, R(t)w \rangle ,$$

where A and B have entries in $L_2[G]$ and P, Q, R have entries in $L_\infty[G]$. It is easily verified that (4.1) holds in this problem and that if all the controls $\{u_k\}$ lie in a ball in $[L_2[G]]^m$ then (4.2) holds.

In the next theorem we dispense with the Holder-type condition and require instead that the controls $\{u_k\}$ of the sequence in question all lie in a ball in some L_p-space.

THEOREM 4.2. Let Assumption I hold. Let the sets $\Omega(t, x)$ be independent of x. Let $\{(\phi_k, u_k)\}$ be a sequence of admissible pairs such that $\liminf J(\phi_k, u_k) < \infty$ and such that the controls $u_k = (u_k^1, \ldots, u_k^m)$ are in the Banach space

$$\mathcal{U} = \prod_{i=1}^{m} L_{s_i}(G) \qquad 1 \leq s_i \leq \infty$$

and satisfy $\|u_k\| \leq R$ for some $R > 0$. Then the sequence

$\{(\phi_k, u_k)\}$ is weakly lower closed.

Among all of our theorems, Theorem 4.2 is the easiest to apply. In many problems of interest in applications, the controls $\{u_k\}$ belonging to a minimizing sequence all lie in a ball in some L_p space, $1 \leq p \leq \infty$.

Out next theorem, Theorem 4.3, is a slight extension of a theorem first given by Cesari [4] and was established by us in [1] by considerably simpler arguments than those first given by Cesari [4]. Independent of our work, Cesari [5] has also obtained the extension given here and has also given a simpler proof based on Mazur's theorem. In this connection also see Bidaut [3].

In Theorem 4.3 the sets $\Omega(t,x)$ need not be independent of x. On the other hand, a certain regularity in the behavior of the mapping $(t,x) \to Q^+(t,x)$ must be imposed. Let $\delta > 0$, let (t_0, x_0) belong to $\mathcal{R}$ and let $N_x(t_o, x_o, \delta)$ denote the set of points (t_0, x) in $\mathcal{R}$ such that $|x - x_0| < \delta$. Let

$$Q^+(N_x(t_0, x_0, \delta)) = \bigcup \{Q^+(t_0, x) : (t_0, x) \in N_x(t_0, x_0, \delta)\}.$$

For a set A let $c\ell$ co A denote the closure of the convex hull of A. The mapping Q^+ is said to satisfy the <u>weak</u>

Cesari property (Q^*) at (t_0, x_0) if

$$Q^+(t_0, x_0) = \bigcap_{\delta>0} \, c\ell \; co \; Q^+(N_x(t_0, x_0, \delta)) .$$

THEOREM 4.3 . Let Assumption I hold. Let the mapping Q^+ satisfy property (Q^*) at all points of $\mathcal{R}$ with the possible exception of a set of points whose t-coordinates lie in a set of measure zero in G. Let $\{(\phi_k, u_k)\}$ be a sequence of admissible pairs such that $\liminf J(\phi_k, u_k) < +\infty$. Then the sequence $\{(\phi_k, u_k)\}$ is weakly lower closed.

In all three lower closure theorems Assumption I holds. Thus, the sets $Q^+(t,x)$ are always assumed to be closed and convex. If the sets $Q^+(t,x)$ are not convex then we can replace the original problem with a "relaxed problem" in which the vectors $(N\phi)(t)$ lie in the convex hull of $Q^+(t,x)$. If we assume that the sets $\Omega(t,x)$ are compact, then as a consequence of the continuity of $\hat{f}$ in w it follows that the sets co $Q^+(t,x)$ are closed. The relaxed problem is cast as a new problem in which the set that plays the role of $Q^+(t,x)$ is the set co $Q^+(t,x)$, which is closed and convex. Thus we can use Theorems 4.1 to 4.3 to obtain lower closure theorems for relaxed control problems.

6. Outline of Proof.

Complete proofs of Theorems 4.1 and 4.3 are given in [1] and the proof of Theorem 4.2 is given in [2]. The basic idea underlying all of the proofs is the same, and we shall sketch it here and refer the reader to [1] and [2] for details.

Let

$$\gamma = \lim\ \inf\ J(\phi_k, u_k).$$

By hypotheses $\gamma < +\infty$. Using (4) of Assumption I it can be shown that $\gamma > -\infty$. We now let $\phi_k \rightarrow \phi$ weakly in $\mathcal{J}$. Then there is a subsequence of $\{(\phi_k, u_k)\}$ which we again label as $\{(\phi_k, u_k)\}$ such that $J(\phi_k, u_k) \rightarrow \gamma$. Since $y_k = N\phi_k$ converges to $y = N\phi$ weakly in $\mathcal{Y}$, it follows from Mazur's theorem that there is a sequence $\{\psi_j\}$ of convex xombinations of the y_k that converges strongly to y in $\mathcal{Y}$. Hence there is a subsequence of the ψ_j that converges a.e. to y. We then use the same convex combinations of the functions $t \rightarrow f^o(t, y_k(t), u_k(t))$ as were used to define the ψ_j to define a sequence $\{\lambda_j\}$. We then use Fatou's lemma and the convergence of $J(\phi_k, u_k)$ to γ to show that the function $\gamma = \lim\ \inf\ \lambda_j$ is integrable and that $\int_G \lambda\ dt \leq \gamma$. In Theorem 4.3 we use property (Q^*) to show that for a.e. t in G,

$(\lambda(t), y(t)) \in Q^+(t, z(t))$. In Theorems 4.1 and 4.2 we can define a sequence ω_j and a sequence σ_j such that for a.e. t,

$$\omega_j(t) - \lambda_j(t) \to 0$$

$$\sigma_j(t) - \psi_j(t) \to 0$$

and $(\omega_j(t), \sigma_j(t)) \in Q^+(t, z(t))$. It then follows that in this case also, $(\lambda(t), y(t)) \in Q^+(t, z(t))$. Thus there is a function $v : G \to R^m$ such that $y(t) = f(t, z(t), v(t))$ and $\lambda(t) \geq f^o(t, z(t), v(t))$. We then use an extension of Filippov's lemma to show that we can replace v by a measurable function u. Thus (ϕ, u) is admissible and $\lambda(t) \geq f^o(t, z(t), u(t))$. Integration of this inequality and the previously established relation $\int_G \lambda \; dt \leq \gamma$ then give the theorem.

REFERENCES

[1] L. D. Berkovitz, Existence and lower closure theorems for abstract control problems, SIAM J. Control 12(1974), 27-42.

[2] ________________, A lower closure theorem for abstract control problems with L_p-bounded controls, J. Optimization Theory Appl. To appear.

[3] M-F. Bidaut, Quelques résultats d'existence pour des problèmes de contrôle optimal. C. R. Acad. Sc.

Paris Ser A. 274(1972), 62-65.

[4] L. Cesari, Existence theorems for abstract multidimensional control problems, J. Optimization Theory Appl. 6(1970), 210-236.

[5] _______________, Closure theorems for orientor fields, Bull. Amer. Math. Soc. 79(1973), 684-689.

Purdue University, West Lafayette, Indiana

A SINGULAR CAUCHY PROBLEM AND GENERALIZED TRANSLATIONS

B. L. J. Braaksma

0. Introduction.

Generalized translations have been introduced by Delsarte [2] by means of a generalized Taylor formula. Formally his procedure is as follows. Let L be a suitable linear operator on a space S of functions defined on $(0,\infty)$. Let $\varphi(x,\lambda)$ be defined on $(0,\infty)\times\mathbb{C}$ such that $\varphi(x,\lambda)$ is holomorphic in λ, $L_x\varphi(x,\lambda) = \lambda\varphi(x,\lambda)$, $\varphi(0,\lambda) = 1$, $\varphi(x,\lambda) = \Sigma_{n=0}^{\infty}\varphi_n(x)\lambda^n$. Then the generalized translation operator T^y is defined as

$T^y = \Sigma_{n=0}^{\infty}\varphi_n(y)L^n = \varphi(y,L)$ ($L = \frac{d}{dx}$ gives ordinary translation.) An easy consequence of the definition is $L_x\varphi_n = \varphi_{n-1}$, $\varphi_n(0) = 0$ if $n \geq 1$, $L_x\varphi_o = 0$, $\varphi_o(0) = 1$. So if $u(x,y) = T^y f(x)$ then $u(x,0) = f(x)$ and

$L_y u(x,y) = \Sigma_{n=0}^{\infty} L_y\varphi_n(y)\, L_x^n f(x) = \Sigma_{n=1}^{\infty}\varphi_{n-1}(y) L_x^n f(x) = L_x u(x,y)$. Hence the translation of f satisfies

$$L_x u = L_y u, \quad u(x,0) = f(x). \tag{0.1}$$

In this paper we consider translations associated with the

differential operator

$$(0.2) \qquad L_x = \frac{\partial^2}{\partial x^2} + \frac{2p+1}{x} \frac{\partial}{\partial x} - q(x) .$$

Delsarte considered the case $q \equiv 0$. Levitan [6] and others investigated the case $p = -\frac{1}{2}$. One of the applications which Levitan made is the solution of the converse Sturm-Liouville problem.

In section 2 we consider uniqueness and existence of the solutions of a singular Cauchy-problem. In section 3 we give norm estimates for the solutions of this Cauchy problem. Next in section 4 generalized translations associated with (0.2) and a corresponding convolution algebra are treated. Part of these results have been obtained in cooperation with H. S. V. de Snoo and published in [1]. I express my thanks to H. S. V. de Snoo for his help.

1. The singular Cauchy problem.

We consider a generalization of the differential equation (0.1) with (0.2). Let $Q(x, y)$ be defined and continuous for $0 < y \leq x$. Define

$$(1.1) \qquad L_Q = \frac{\partial^2}{\partial x^2} - \frac{\partial^2}{\partial y^2} + (2p+1)\left(\frac{1}{x}\frac{\partial}{\partial x} - \frac{1}{y}\frac{\partial}{\partial y}\right) - Q(x, y).$$

We consider the Cauchy problem

$$L_Q u(x, y) = 0 \text{ if } 0 < y \leq x, \; u(x, 0) = f(x), \; u_y(x, 0) = 0 \text{ if } x > 0 ,$$

where $u(x, 0)$ and $u_y(x, 0)$ are interpreted as limits of $u(x, y)$ and $u_y(x, y)$ as $y \downarrow 0$. In the investigation of this problem it is useful to introduce the coordinates $X = x + y$, $Y = x - y$. Let $\tilde{u}(X, Y) = u(x, y)$ and similarly for other functions. Then $L_Q u(x,y) = 4\tilde{L}_Q \tilde{u}(X,Y)$ where

$$\tilde{L}_Q = \frac{\partial^2}{\partial X \partial Y} - \frac{2p+1}{X^2-Y^2}\left(Y\frac{\partial}{\partial X} - X\frac{\partial}{\partial Y}\right) - \frac{1}{4}\tilde{Q}(X,Y). \tag{1.2}$$

In case $Q = 0$ we have the following solution according to Delsarte [2]:

$$\tilde{u}_o(X,Y) = \int_Y^X \tilde{w}_o(X,Y,\xi) f(\xi)\, d\xi, \quad \text{where} \tag{1.3}$$

$$\tilde{w}_o(X,Y,\xi) = \tfrac{1}{2} B(p+\tfrac{1}{2}, p+\tfrac{1}{2})(X^2-Y^2)^{-2p}(X^2-\xi^2)^{p-\frac{1}{2}}(\xi^2-Y^2)^{p-\frac{1}{2}}\xi. \tag{1.4}$$

More precisely:

1) Suppose f is continuous. If $p > \frac{1}{2}$, then $\tilde{u}_o$ is solution of

$$\tilde{L}_o \tilde{u} = o, \ \tilde{u}(X,X) = f(X), \ \left(\frac{\partial \tilde{u}}{\partial X} - \frac{\partial \tilde{u}}{\partial Y}\right)(X,Y) = \frac{o(1)}{X-Y}, \quad Y \uparrow X. \tag{1.5}$$

If $p > \frac{3}{2}$, then $L_o u_o = 0$.

2) Suppose f is continuously differentiable. If $p > -\frac{1}{2}$, then $\tilde{u}_o$ is a solution of (1.5) where the expression $\frac{o(1)}{X-Y}$ is now replaced by $o(1)$. If $-\frac{1}{2} < p \leq \frac{3}{2}$, then $L_o u_o = 0$.

3) Suppose f is twice continuously differentiable. If $p > -\frac{1}{2}$, then $L_o u_o = 0$.

Now we consider general case. In the following we assume that the following condition is satisfied.

There exists a continuous function $Q(y)$ on $y > 0$ such that $|Q(x,y)| \leq Q(y)$ if $0 < y \leq x$ and

$$y^{1+p-|p|} Q(y) \in L(0,1), \ p > -\frac{1}{2}, \ p \neq 0. \tag{1.6}$$

Then we have the following result.

<u>Theorem 1</u>. Assume $p > -\frac{1}{2}$, $p \neq 0$, $0 \leq Y_0 < X_0$. Let

$f(X)$ be continuous on $[Y_o, X_o]$ if $p > \frac{1}{2}$ or let $f(X)$ be continuously differentiable on $[Y_o, X_o]$ if $p > -\frac{1}{2}$, $p \neq 0$. Suppose there exists a solution of

(1.7) $\tilde{L}_Q \tilde{u} = 0$ on $Y_o \leq Y < X \leq X_o$, $\tilde{u}(X, X) = f(X)$ on $Y_o \leq X \leq X_o$ and

(1.8) $$\left(\frac{\partial}{\partial X} - \frac{\partial}{\partial Y}\right) \tilde{u}\ (X, Y) = o(1)(X - Y)^{-1-p+|p|}$$

as $Y \to X$ uniformly on $Y_o \leq Y < X \leq X_o$. Then this solution is unique.

We only sketch the proof. Let $\tilde{u}_o$ be defined by (1.3). Then $\tilde{u}_1 = \tilde{u} - \tilde{u}_o$ satisfies (1.7) and (1.8) with $f \equiv 0$. Now we use the Riemann function A_Q associated with L_Q. This function has been constructed from the Riemann function A_o associated with L_o and estimated in terms of A_o in [1], theorem 1. Now it is easily verified that

(1.9) $$\begin{aligned} u_1(x, y) &= \tfrac{1}{2} A_Q(x - y + \varepsilon, \varepsilon; x, y)\, u_1(x - y + \varepsilon, \varepsilon) + \\ &+ \tfrac{1}{2} A_Q(x + y - \varepsilon, \varepsilon; x, y)\, u_1(x + y - \varepsilon, \varepsilon) \\ &- \tfrac{1}{2} \iint_{\Delta(\varepsilon)} Q(\xi, \eta)\, u_o(\xi, \eta)\, A_Q(\xi, \eta; x, y)\, d\xi\, d\eta + \\ &+ \tfrac{1}{2} \int_{x-y+\varepsilon}^{x+y-\varepsilon} \Big[\frac{\partial u_1}{\partial \varepsilon}(\xi, \varepsilon)\, A_Q(\xi, \varepsilon; x, y) + \\ &+ u_1(\xi, \varepsilon) \Big\{\frac{2p+1}{\varepsilon} A_Q(\xi, \varepsilon; x, y) - \frac{\partial}{\partial \varepsilon} A_Q(\xi, \varepsilon; x, y)\Big\}\Big]\, d\xi . \end{aligned}$$

Here $\Delta(\varepsilon)$ is the triangle with vertices $(x - y + \varepsilon, \varepsilon)$, $(x + y - \varepsilon, \varepsilon)$, (x, y), where $Y_o \leq Y < X \leq X_o$, $\varepsilon > 0$. Using the estimates for A_Q mentioned above we may show that in the righthand side of (1.9) all terms tend to zero as $\varepsilon \downarrow 0$ except for the double integral:

$$u_1(x, y) = \frac{1}{2} \iint_{\Delta(o)} Q(\xi, \eta)\, u_o(\xi, \eta)\, A_Q(\xi, \eta; x, y)\, d\xi d\eta, \tag{1.10}$$

and uniqueness follows.

Here and in the following we omit the case $p = 0$. This case is more complicated and in the estimates now logarithmic terms occur (cf. [1]). Furthermore, we remark a difference in the conditions for the cases $p > 0$ and $p < 0$. The solutions in these cases are connected as follows. If u is a solution of $L_Q u = 0$ for some value of p then $v(x, y) = (xy)^{2p}\, u(x, y)$ is a solution of the same equation with p replaced by $-p$. The condition $u(x,0) = f(x)$ is now transformed into a condition of an other type viz. $v(x, y) \sim (xy)^{2p} f(x)$ as $y \downarrow 0$. So besides (1.7) and (1.8) one may also study another type of Cauchy problem.

The existence of a solution of (1.7) and (1.8) may be proved under somewhat stronger assumptions on Q. We now start from (1.10). By substituting the integral (1.3) for $\tilde{u}_o$ we obtain

$$\tilde{u}(X, Y) = \int_Y^X \tilde{w}_Q(X, Y, \xi)\, f(\xi)\, d\xi\,, \tag{1.11}$$

where

$$\tilde{w}_Q(X,Y,\xi) = \tilde{w}_o(X,Y,\xi) - \frac{1}{4} \int_\xi^X d\alpha \int_Y^\xi d\beta\, \tilde{w}_o(\alpha, \beta, \xi) \tilde{Q}(\alpha, \beta) \tilde{A}_Q(\alpha, \beta; X Y), \tag{1.12}$$

and we have to investigate when this function $\tilde{u}$ satisfies (1.7) and (1.8). The result is

<u>Theorem 2</u>. Suppose $0 \le Y_o < X_o$, $y^c Q(y) \in L(o,1)$ where

$c = \min(1, p+\frac{1}{2})$. Assume one of the following two conditions is satisfied

i) f is continuous on $Y_o \le X \le X_o$, $p > \frac{1}{2}$.

ii) f is continuously differentiable on $Y_o \le X \le X_o$, $p > -\frac{1}{2}$, $p \ne 0$.

Then the Cauchy problem (1.7) and (1.8) possesses exactly one solution. Moreover, in case i) we have

$$(1.13)\quad \left(\frac{\partial}{\partial X} - \frac{\partial}{\partial Y}\right)\tilde{u}(X,Y) = o(1)(X-Y)^{-1}, \text{ as } Y \to X \text{ on } Y_o \le Y < X \le X_o$$

whereas in case ii) we have

$$(1.14)\quad \left(\frac{\partial}{\partial X} - \frac{\partial}{\partial Y}\right)\tilde{u}(X,Y) = o(1) + 0(1)\int_o^{\frac{X-Y}{2}} Q(t)\left(\frac{t}{X-Y}\right)^c dt, \text{ as } Y \to X \text{ on}$$

$Y_o \le Y < X \le X_o$.

So, if for instance $Q(y) = 0(y^{-\gamma}), y \downarrow 0, \gamma < c+1$, then in case ii)

$$\left(\frac{\partial}{\partial X} - \frac{\partial}{\partial Y}\right)\tilde{u}(X,Y) = 0(1)(X-Y)^{1-\gamma} + o(1), \text{ as } Y \to X \text{ on } Y_o \le Y < X \le X_o.$$

The original differential equation is satisfied if somewhat stronger conditions hold.

<u>Theorem 3</u>. Suppose $0 \le Y_o < X_o$ and $y^d Q(y) \in L(o,1)$ where $d = \min(1, p-\frac{1}{2})$. Assume $\tilde{Q}(X,Y)$ possesses continuous derivatives of the first order on $Y_o \le Y < X \le X_o$, and

$$(1.15)\qquad (X-\sigma)^{p+\frac{1}{2}} \frac{\partial \tilde{Q}}{\partial X}(X,\sigma) \in L(Y_o, X),$$

$$(1.16)\qquad (\sigma - Y)^{p+\frac{1}{2}} \frac{\partial \tilde{Q}}{\partial Y}(\sigma, Y) \in L(Y, X_o).$$

Then the Cauchy problem

(1.17) $L_Q u(x,y) = 0,\ Y_o \leq x - y < x + y \leq X_o,\ u(x,0) = f(x),\ u_y(x,y) =$

$= o(1)y^{-1-p+|p|}$ as $y \downarrow 0,\ Y_o < x < X_o$

possesses exactly one solution which is given by (1.11) in the following cases:

1) $f \in C_o[Y_o, X_o],\ p > \frac{3}{2}$; 2) $f \in C_1[Y_o, X_o], \frac{1}{2} < p \leq \frac{3}{2}$; 3) $f \in C_2[Y_o, X_o]$, $-\frac{1}{2} < p \leq \frac{1}{2},\ p \neq 0$.

Moreover, (1.13) and (1.14) hold in the cases i) and ii) of theorem 2.

This result is related to a theorem of Diaz and Ludford [3]. The proofs of these theorems will be published elsewhere.

2. Estimates for the solution of the Cauchy problem.

In this section we give estimates for the function u defined by (1.11) where we assume that f is continuous and Q satisfies (1.6). To that end we first estimate the kernel in (1.11). It may be shown that there exists a positive constant M such that

(2.1) $|\tilde{w}_Q(X,Y,\xi) - \tilde{w}_o(X,Y,\xi)| \leq M\tilde{w}_o(X,Y,\xi)\{-1+\exp M\rho_1(y)\}$ if $p>0$ and

(2.2) $|\tilde{w}_Q(X,Y,\xi) - \tilde{w}_o(X,Y,\xi)| \leq M\xi^{2p+1}(X^2-\xi^2)^{-p-\frac{1}{2}}(\xi^2-Y^2)^{-p-\frac{1}{2}}$

$\rho_{2p+1}(y)\{-1+\exp M\rho_1(y)\}$, if $-\frac{1}{2} < p < 0$,

where

(2.3) $\rho_\alpha(y) = \int_o^y t^\alpha Q(t)\,dt$.

Except for the values $p > \frac{1}{2}$ this has been proved in [1], theorem 2.

Using these estimates we may show that $\tilde{u}$ defined by (1.11) satisfies (cf. [1], theorem 3)

$$(2.4) \quad |u(x,y)| \le \sup |f(\xi)| \{M_1 + M_2 \exp M\rho_1(y)\} \text{ if } p>0,$$

$$(2.5) \quad (1+x^{-2p})^{-1}(1+y^{-2p})^{-1}|u(x,y)| \le \sup(1+\xi^{-2p})^{-1}|f(\xi)|$$

$$\{M_1 + M_2\, \rho_{2p+1}(y) \exp M\rho_1(y)\} \text{ if } -\frac{1}{2} < p < 0.$$

Here the sup is taken over $x - y \le \xi \le x + y$ and M, M_1 and M_2 are positive constants.

Moreover we may deduce from (2.1) and (2.2) the following norm estimates.

<u>Theorem 4.</u> Suppose $y_o \ge 0$ and $x^\alpha f(x) \in L_n(y_o < x < \infty)$. Let $u(x,y)$ be defined by (1.11). Then for every positive y we have

$$(2.6) \qquad x^\alpha u(x,y) \in L_n(y_o + y < x < \infty) \text{ if } n > 1,$$

$0 < \alpha < 2p+1$, $p > 0$ and also if $n = 1$, $0 < \alpha \le 2p+1$, $p > 0$. Moreover, there exists a positive constant M such that

$$(2.7) \qquad \|x^\alpha u(x,y)\|_n \le M\{\exp M\,\rho_1(y)\}\, \|x^\alpha f(x)\|_n$$

in these cases. Here the norm is taken in $L_n(y_o + y < x < \infty)$.

If $-\frac{1}{2} < p < 0$ and $n > 1$, $2p < \alpha < 1$ or $n = 1$, $2p < \alpha \le 1$ then (2.6) also holds and

$$(2.8) \quad \|x^\alpha u(x,y)\|_n \le M[1 + y^{-2p}\rho_{2p+1}(y)\{-1 + \exp M\rho_1(y)\}]\, \|x^\alpha f(x)\|_n.$$

3. Generalized translations.

We first extend the considerations of sections 1 and 2 by assuming that $Q(x, y)$ is defined and antisymmetric in the first quadrant $x > 0$, $y > 0$. Now we extend the function $u(x, y)$ defined by (1.11) if $0 < y \leq x$ by symmetry: $u(x, y) = u(y, x)$. Then $\tilde{u}(X,Y)$ and $u(x, y)$ are solutions of (1.7), (1.8) and (1.17) if $0 \leq |Y| < X$ and if $x > 0$, $y > 0$, if the assumptions of theorem 2 and 3 respectively are satisfied. Moreover, theorem 4 appears to hold with $y_o = 0$ and $L_n(y_o + y < x < \infty)$ replaced by $L_n(0 < x < \infty)$.

Next we consider the special case $Q(x, y) = q(x) - q(y)$ where $|q(x) - q(y)| \leq Q(y)$ if $0 < y \leq x$. If $q(x)$ and $Q(x)$ are continuous on the positive x-axis R_+ and $xQ(x) \in L(0, 1)$, we define the translation $T^y f$ of a locally integrable function f on R_+ associated with the operator (0.2) by

$$(3.1) \qquad (T^y f)(x) = \int_{x-y}^{x+y} f(\xi)\, w(x, y, \xi)\, d\xi \quad \text{if } 0 < y \leq x$$

and by

$$(3.2) \qquad (T^y f)(x) = (T^x f)(y) \quad \text{if } 0 < x \leq y .$$

Here $p > -\frac{1}{2}$ and $w(x, y, \xi)$ is the translation kernel defined by (1.12) with $Q(x, y) = q(x) - q(y)$.

First we remark that theorem 1 now implies the so-called product formula for the solution $\omega_\lambda(x)$ of

$$(3.3) \quad L_x \omega(x) = -\lambda^2 \omega(x), \text{ if } x>0,\ \omega(0)=1,\ \omega'(x)=0(1)\int_0^x (t/x)^{2p+1} q(t)dt$$

as $x \downarrow 0$, where L_x is given by (0.2). Then $\omega_\lambda(x)\omega_\lambda(y)$ is a solution of (1.7) and (1.8) with $f(x) = \omega_\lambda(x)$. Hence

$$(3.4) \quad \omega_\lambda(x)\omega_\lambda(y) = \int_{x-y}^{x+y} w(x,y,\xi)\omega_\lambda(\xi)d\xi, \quad 0 \leq y \leq x .$$

From the estimates in section 2 we may deduce that translation is a bounded operator on certain spaces. If $\psi(x)$ is a positive measurable function on R_+ we denote by $\Psi_\infty(R_+)$ the space of measurable functions on R_+ such that

$$\|[f]\|_\infty = \operatorname*{ess\ sup}_{x \in R_+} \left|\frac{f(x)}{\psi(x)}\right| < \infty .$$

Similarly $\Psi_\infty(R^2_+)$ is defined using $\psi \otimes \psi$ as comparison function. Then T^y is a bounded operator from $\Psi_\infty(R_+)$ to $\Psi_\infty(R^2_+)$ if (1.6) holds and $\psi(x) = 1$ if $p > 0$, $\psi(x) = 1 + x^{-2p}$ if $-\frac{1}{2} < p < 0$. (cf. [1], theorem 4).

From the extension of theorem 4 mentioned above we define

<u>Theorem 5.</u> Assume (1.6). If $p > 0$, then T^y is a bounded operator on $L_n(x^\beta dx;\ 0 < x < \infty)$ if $n > 1$, $0 < \beta < n(2p+1)$ or $n = 1$, $0 < \beta \leq 2p+1$. In these cases $\|T^y\| \leq M \exp M\rho_1(y)$ with a constant M independent of y. If $-\frac{1}{2} < p < 0$, then T^y is a bounded operator on $L_n(x^\beta dx;\ 0 < x < \infty)$ if $n > 1$, $2pn < \beta < n$ or $n = 1$, $2p < \beta \leq 1$, and

$$(3.5) \qquad \|T^y\| \leq M[1 + y^{-2p}\rho_{2p+1}(y)\{-1 + \exp M\rho_1(y)\}]$$

with some constant M independent of y.

Another case where the translation operator is bounded is given in [1], theorem 8. There the positivity of the translation which holds under general conditions has been used.

In [1] we have shown that if $q \in L(0,1)$ and

$$(3.6) \qquad \begin{aligned} K(x,y,z) &= z^{-2p-1} w(x,y,z) \text{ if } 0 \le x-y \le z \le x+y, \\ K(x,y,z) &= K(y,x,z) \text{ if } 0 \le y-x \le z \le x+y, \\ K(x,y,z) &= 0 \text{ otherwise,} \end{aligned}$$

then $K(x,y,z)$ is symmetric in its three variables. With $d\mu(z) = z^{2p+1} dz$ we now have

$$(3.7) \qquad (T^y f)(x) = \int_0^\infty f(z) K(x,y,z)\, d\mu(z).$$

Convolution associated with the generalized translation may be defined as follows

$$(3.8) \quad (f * g)(x) = \int_0^\infty f(y)(T^x g)(y) d\mu(y) = \int_0^\infty \int_0^\infty f(y) g(z) K(x,y,z) d\mu(y) d\mu(z),$$

where f and g are suitable functions. If $q \in L(0,1)$ it follows that this convolution product is commutative. Convolution algebras associated with generalized translation in case $p = -\frac{1}{2}$ have been studied a.o. by Hutson and Pym[5]. Their methods also apply to the case $p > -\frac{1}{2}$. Flensted — Jensen and Koornwinder [4] considered a convolution algebra associated with the Jacobi differential operator, a special case of L_x in (0.2). Their methods may be partially extended to the general operator L_x in (0.2). We mention some of the results which may be obtained in this way. Let the Fourier transform $f \to \hat{f}$ be defined as follows

$$(3.9) \qquad \hat{f}(\lambda) = \int_0^\infty f(z) \omega_\lambda(z)\, d\mu(z),$$

where ω_λ is defined by (3.3) and f is a suitable function. Using the product formula (3.4) we see that under general conditions for f and g and $q \in L(0,1)$

(3.10) $\widehat{f * g} = \hat{f}\,\hat{g}$.

From theorem 5 we may deduce

<u>Theorem 6</u>. Suppose $p > 0$ and $tQ(t) \in L(0,\infty)$. Let m, n and r be such that $1 \leq m, n, r \leq \infty$ and $\frac{1}{m} + \frac{1}{n} - 1 = \frac{1}{r}$. If $f \in L_m(d\mu)$, $g \in L_n(d\mu)$, then $f * g$ exists and belongs to $L_r(d\mu)$. Furthermore, there exists a constant M independent of f and g such that

(3.11) $\|f * g\|_r \leq M\|f\|_m \|g\|_n$.

In order to obtain a Banach algebra we consider the case $m = n = r = 1$ in theorem 6 and assume $Q(t) \in L(0,1)$, $tQ(t) \in L(1,\infty)$. Define $d\mu_1(z) = M d\mu(z)$. Then the Fourier transform appears to be injective and from (3.10) we deduce the associativity of the convolution. Hence $(L_1(d\mu_1)^*_{,})$ is now a semi-simple commutative Banach algebra. In addition to (3.10) we may show that the only complex homomorphisms of the algebra have the form $\chi(f) = \hat{f}(\lambda)$, where λ is such that $\omega_\lambda(x)$ is bounded.

REFERENCES

[1] B. L. J. Braaksma and H. S. V. de Snoo, Generalized translation operators associated with a singular differential operator, Proc. Conf. Theory of ordin. and part. diff. equat. in Dundee 1974, to appear in Springer Lecture Notes.

[2] J. Delsarte, Sur une extension de la formule de

Taylor, Journn. Math. Pure Appl. 17, 213-231 (1936).

[3] J. B. Diaz and G. S. S. Ludford, On the singular Cauchy problem for a generalization of the Euler-Poisson-Darboux equation in two space variables, Annali Math. Pura Appl., Serie Quarta, 38, 33-50 (1955).

[4] M. Flensted-Jensen and T. Koornwinder, The convolution structure for Jacobi function expansions, Arkiv Math. 11, 245-262 (1973).

[5] V. Hutson and J. S. Pym, Measure algebras associated with a second order differential operator, J. Funct. An. 12, 68-96 (1973).

[6] B. M. Levitan, Generalized translation operators and some of their applications, Jerusalem, 1964.

University of Groningen, Groningen, Netherlands

SOME MODELS FOR POPULATION GROWTH WITH HARVESTING

Fred Brauer* and David A. Sánchez**

1. Many models for the study of population growth are formulated in terms of first order ordinary differential equations, or systems of first order ordinary differential equations for populations of interacting species. The removal of members from a population at a constant time rate, as may occur when the species is hunted, can be built into such models by introducing a parameter into the differential equation. We are interested in the dependence of limits of solutions on this parameter, particularly in situations where there may be a discontinuity in the dependence. Such an occurence is called a catastrophe, as it may correspond to a biological catastrophe such as extinction of a species. We give general results for one-species and two-species models on the critical harvest rate which leads to catastrophe, and we indicate their application to some explicit models which have been used for population studies.

Another type of model which has been used to study some populations leads to a differential-difference equation.

---

*This research was partially supported by the National Science Foundation, Grant No. GP-28267.

**Sponsored by the U. S. Army, Contact No. DA-31-124-ARO-D-462.

We study the question of what range of values of the delay will preserve the stability of the system for a general delay model, and we also indicate how the maximum delay which preserves stability may depend on the harvesting rate.

This paper contains an outline of our results. Details of proofs and predictions from experimental data may be found in [4].

2. Let x represent the size of a population whose growth rate depends only on the population size. More precisely, we assume that $x(t)$ is a continuously differentiable non-negative function for $t \geq 0$, and that $x'(t)/x(t)$, the rate of growth of population per member in unit time, is a continuous function $f(x(t))$. Then the population size is governed by the first order differential equation

$$x' = xf(x) . \tag{1}$$

If members are removed from the population at a constant rate of E members per unit time, then the population size is governed by the differential equation

$$x' = xf(x) - E . \tag{2}$$

An equilibrium point would then be a solution $x_\infty(E) \geq 0$ of the equation

$$xf(x) - E = 0 , \tag{3}$$

and such an equilibrium point is asymptotically stable if $[xf(x)]'_{x=x_\infty(E)} < 0$. By simple arguments using the inverse function theorem, we can establish the following result.

<u>Theorem 1:</u> Suppose $f(x) > 0$ for $0 < x < \xi$ and $f(\xi) = 0$, so that ξ is a stable equilibrium point for (1). Then there

is an interval $\alpha < x \leq \xi$ on which $[xf(x)]' < 0$. If $[xf(x)]'_{x=\alpha} = 0$ but $[xf(x)]''_{x=\alpha} \neq 0$, then there is a stable equilibrium point $x_\infty(E) > 0$ of (2) which depends continuously on E for $0 \leq E \leq \alpha f(\alpha)$. For $E > \alpha f(\alpha)$, every solution of (2) tends to zero in finite time.

Corollary: The critical harvest rate $E_c = \alpha f(\alpha)$ is equal to the maximum rate of growth of the unharvested population governed by the equation (1).

For a specific function f(x), Theorem 1 gives a means of calculating the critical harvest, namely by finding the value of the function xf(x) at its maximum. The corollary indicated that the critical harvest can also be found by observing the growth of the population when it is not being harvested.

Next, we examine models for populations of two co-existing species, and study the effect on their equilibrium of harvesting one species at a constant rate. Thus we assume that the two population sizes are represented by continuously differentiable non-negative functions x(t) and y(t) for $t \geq 0$. We also assume that the growth rate of each population per member in unit time, given by $x'(t)/x(t)$ and $y'(t)/y(t)$ respectively, is a continuous function of the two population sizes, so that the population sizes are governed by the system of first order differential equations

(4) $$x' = xf(x, y), \quad y' = yg(x, y).$$

If members of the first population are harvested at a constant rate of E members per unit time while the second population is undisturbed, then the governing equations become

$$x' = xf(x, y) - E, \quad y' = yg(x, y) . \tag{5}$$

An equilibrium point of (5) is therefore a solution $(x_\infty(E), y_\infty(E))$, with $x_\infty(E) \geq 0$, $y_\infty(E) \geq 0$, of the pair of equations

$$xf(x, y) - E = 0, \quad yg(x, y) = 0 . \tag{6}$$

The two-species analogue of Theorem 1 is the following one.

Theorem 2: Let (ξ, η) be an equilibrium point of (4) and suppose $f_x(\xi,\eta)g_y(\xi,\eta)-f_y(\xi,\eta)g_x(\xi,\eta)>0$, $g_y(\xi,\eta) < 0$. Let (α, β) be the solution of the pair of equations

$$yg(x, y) = 0$$

$$\det \begin{bmatrix} f(x, y) + xf_x(x, y) & xf_y(x, y) \\ yg_x(x, y) & yg_y(x, y) \end{bmatrix} = 0 . \tag{7}$$

Then the critical harvest is given by $E_c = \alpha f(\alpha, \beta)$, and there is an asymptotically stable equilibrium point of (5) which depends continuously on E for $0 \leq E < E_c$, but not for $E \geq E_c$.

It is reasonable to assume $f_x(x, y) < 0$ and $g_y(x, y) < 0$. These assumptions merely say that increase of each population has an inhibiting effect on its own growth. There are two essentially different situations for the interaction of the two species. If $f_y(x, y) < 0$ and $g_x(x, y) < 0$, the growth of each species has an inhibiting effect on the growth of the other. In this case, we say that the two species are in competition. It is easy to show that under competition the effect of harvesting species x at a constant rate is to reduce the limiting value of x and to increase the limiting value of y (at least up to the critical harvest).

The other possibility is that $f_y(x,y) < 0$, but $g_x(x,y) > 0$, which indicates a predator-prey relationship with species x as prey and species y as predator. Then it is easily seen that the effect of harvesting species x at a constant rate is to decrease the equilibrium population of both species.

3. A choice of $f(x)$ which has been used in many single-species population models is $f(x) = \lambda - ax$, where λ and a are positive constants. This is the so-called logistic model of A. J. Lotka [8]. Application of Theorem 1 to this model gives a critical harvest of $E_c = \lambda^2/4a$. This model is simple enough that one may easily calculate directly that $x_\infty(E) = \frac{1}{2}x_\infty(0)[1+(1-E/E_c)^{1/2}]$ for $0 \leq E \leq E_c$, $x_\infty(0) = \lambda/a$, and $x_\infty(E_c) = \frac{1}{2}x_\infty(0)$. For $E > E_c$, the solutions of $x' = x(\lambda - ax) - E$ tend to zero in finite time, and the technique of separation of variables shows that the extinction time for a population with initial value λ/a is given by

$$T = \frac{4}{(4aE - \lambda^2)^{1/2}} \arctan \frac{\lambda}{(4aE - \lambda^2)^{1/2}} .$$

These results have been applied to experimental data such as the work of R. S. Miller and D. B. Botkin [10] on sandhill cranes (Grus Canadensis), and an excellent fit is obtained. Details may be found in [4].

The logistic model for two species corresponds to the choice $f(x,y) = \lambda - ax - by$, $g(x,y) = \mu - cx - dy$ in (4) or (5). In the competing species situation, the constants are all positive, while in the predator-prey situation, c is negative. The equilibrium points for the logistic model with harvesting are the intersections of the pair of lines $y(\mu - cx - dy) = 0$ with the hyperbola $x(\lambda - ax - by) = E$. For any $E > 0$ this hyperbola has the lines $x = 0$ and $ax + by = \lambda$ as asymptotes, opens downward, and its vertex moves toward the x-axis as E increases. Various possibilities arise, depending on the coefficients λ, μ, a, b, c, d, and they can be examined graphically by sketching the relevant lines and hyperbolas. Because of the relative simplicity of the equations determining the equilibrium points, they can also be treated analytically.

The unharvested logistic competing-species problem has been studied exhaustively; see for example [1], or [8]. Two particularly interesting cases and the effect of harvesting are described by the following results.

Theorem 3: Suppose $ad - bc > 0$, $c\lambda - a\mu > 0$, and $d\lambda - b\mu > 0$, (which implies that $y \to 0$ in the unharvested case). If, in addition, $2ad\mu - bc\mu - cd\lambda > 0$, then for harvests E in the interval

$$\frac{\mu(c\lambda - a\mu)}{c^2} < E < \frac{(d\lambda - b\mu)^2}{4d(ad-bc)}$$ the system has an

asymptotically stable critical point $(x_\infty(E), y_\infty(E))$ with $x_\infty(E) > 0$, $y_\infty(E) > 0$. If $E \leq \frac{\mu(c\lambda - a\mu)}{c^2}$, $y \to 0$ and x tends to a positive limit, while if $E > \frac{(d\lambda - b\mu)^2}{4d(ad-bc)}$, x tends to zero infinite time and y tends to a positive limit.

Theorem 4: Suppose $ad - bc > 0$, $c\lambda - a\mu < 0$, and $d\lambda - b\mu > 0$, (which implies that the unharvested system has an asymptotically stable equilibrium point). Then this equilibrium is maintained for

$0 \le E < \frac{(d\lambda - b\mu)^2}{4d(ad-bc)}$. If $E > \frac{(d\lambda - b\mu)^2}{4d(ad-bc)}$, x tends to zero in finite time and y tends to a positive limit.

The overharvesting of sardines off the California coast and the concomitant increase of the anchovy population is an observed phenomenon which is consistent with the above consequences of the logistic model.

Given that a species x is being harvested, it is entirely conceivable that it may be in competition with another species y without the harvester's knowledge. Simple calculation shows that the one-species critical harvest $\lambda^2/4a$ is less than the two-species critical harvest $\frac{(d\lambda - b\mu)^2}{4d(ad-bc)}$ for all values of the parameters for which $ad - bc > 0$, $d\lambda - b\mu > 0$. In practice, one would expect that if there is undiscernible competition, then the coefficients b and c in (12) and (13), which measure the degree of interaction, would be small compared to a and d. In this case, the difference between the two calculated critical harvests would be small, and predictions based on a one species model might not lead to catastrophe. However, in view of the ubiquitous character of competition, the inequality between one-species and two-species harvests suggest that any "real world" harvesting experiments based on one-species population models might have disastrous outcomes.

4. In their studies of competition between species, M.E. Gilpin and F.J. Ayala [5], and F.J. Ayala, M.E. Gilpin,

and J. G. Ehrehfeld [2] suggest that the logistic model fails to explain certain observed phenomena. They propose a model which for one species depends on three parameters λ, K, θ, with

$$f(x) = \lambda [1 - (\frac{x}{K})^{\theta}].$$

Application of Theorem 1 then gives $E_c = \frac{K\lambda\theta}{(1+\theta)^{1+1/\theta}}$, and $x_{\infty}(E_c) = \frac{K}{(1+\theta)^{1+1/\theta}}$, but we remark that it is considerably more difficult to give explicit expressions for limiting populations or extinction times for various harvesting rates with this model than with the logistic model. Numerical calculations are feasible, and it would be of interest to compare the results of such calculations with experimental data. This would be particularly valuable for the analogous two-species competition model

$$x' = x[\lambda - \lambda (\frac{x}{K_1})^{\theta_1} - by] - E$$
$$y' = y[\mu - cx - \mu (\frac{y}{K_2})^{\theta_2}] .$$

5. Another type of population model, suggested first by G. E. Hutchinson [7], takes the form

$$x'(t) = x(t) [\lambda - ax(t - T)], \quad T > 0 .$$

Such an equation might describe, for example, the growth of a herbivore feeding on vegetation which requires a time T to recover from grazing. It is known (see, for example, the article by R. M. May [9]) that $x(t) \to \lambda/a$ as $t \to \infty$ and that the equilibrium point λ/a is asymptotically stable provided $T < \pi/2\lambda$. If $T > \pi/2\lambda$, oscillations may be set up by small changes in the initial conditions.

We examine the stability question for the more general model

$$(8) \qquad x'(t) = x(t) f(x(t - T) - E$$

with harvesting, and with a growth rate describes by f as in Section 2. Corresponding to any harvest E, $0 \leq E \leq E_c$, there is a limiting value $x_\infty(E)$ of the solution of (8), which satisfies (3). By linearizing about this equilibrium point, taking Laplace transforms, using a theorem of N. D. Hayes [6] on the location of roots of the resulting transcendental characteristic, equation, and applying the standard stability theory for almost-linear differential equations [3;p. 118 & p. 336], we are led to the following result.

Theorem 5: For $0 \leq E < E_c$, the equilibrium point $x_\infty(E)$ of (8) is asymptotically stable if T satisfies

$$(9) \qquad f(x_\infty(E))T < 1$$

and

$$(10) \qquad [f(x_\infty(E)) - x_\infty(E) f'(x_\infty(E))]T < \pi/2 .$$

This result can be applied to the various models to give explicit bounds for T under which stability is preserved; and these bounds may increase with E. For example, for the logistic models, the theorem gives a uniform bound for $0 \leq E < E_c$ of $T < \pi/2\lambda$, which is the bound for the unharvested case. However, a more refined analysis of the proof of Theorem 5 applied to the logistic case shows that λT is actually required only to be bounded by a function which may be larger than $\pi/2$ for some values of E Numerical calculations indicate that there may indeed be values T_0 of the delay satisfying $\pi/2\lambda < T_0 < 2/\lambda$ for which the equilibrium point $x_\infty(E)$ is asymptotically stable if the harvest rate E is close enough to E_c. A possible qualitative explanation of this phenomenon is that with heavy harvesting more time may be available for the food supply to recover from foraging.

In [9], a model is studied which involves a predator-prey situation with a time delay. There should be no problem in extending the results to general predator-prey models with harvesting of one species. Competing species problems, on the other hand, lead to much more complicated questions on the location of roots of the appropriate transcendental equation, and no useful results are evident.

6. The methods describes above can be applied to various other questions, which we do not analyze in detail. The method of sliding hyperbolas described in Section 3 can easily be used to study predator-prey models under harvesting, as by pesticides, or under negative harvesting, as by the introduction of more predators to control a pest. The analysis is straight-forward, and no new problems or surprises are encountered in the study of such models. It would, of course, be of interest to make predictions for such models and to compare them with experimental data.

In view of the fact that all the models studied are only approximations, it would be more appropriate to study perturbation problems of the form

$$\begin{aligned} x' &= xf(x, y) - E + p(x, y) \\ y' &= yg(x, y) + q(x, y) , \end{aligned} \tag{11}$$

where $f(x, y)$ and $g(x, y)$ are assumed to have a specific form and where $p(x, y)$, $q(x, y)$ are unknown errors associated with the given form. If the unperturbed problem (5) has an asymptotically stable equilibrium, and if $|p(x, y)|$, $|q(x, y)|$ are small, then standard results such as those dealing with total stability show that for large t solutions of (11) are close to the solutiond predicted by (5). We suggest that bounds for $|p(x, y)|$, $|q(x, y)|$ in (11) and refined perturbation theorems may be more useful than more exact but more complicated unperturbed models.

REFERENCES

[1] F. J. Ayala, Competition, coexistence and evaluation, in "Essays in Evolution and Genetics in Honor of Theodosius Dobzhansky", M. K. Hecht and W. C. Steere, eds., Appleton-Century-Crofts, New York, 1970, 121-158.

[2] F. J. Ayala, M. E. Gilpin and J. G. Ehrenfeld, Competition between species: theoretical models and experimental tests, Theor. Pop. Biol. 4(1973), 331-356.

[3] R. Bellman and K. L. Cooke, "Differential Difference Equations", Academic Press, New York, 1963.

[4] F. Brauer and D. A. Sánchez, Constant rate population harvesting: Equilibrium and stability, Mathematics Research Center, University of Wisconsin, Technical Summary Report # 1461.

[5] M. E. Gilpin and F. J. Ayala, Global models of growth and competition, Proc. Nat. Acad. Sci. USA, 70 (1963),

3590-3593.

[6] N. D. Hayes, Roots of transcendental equation associated with a certain differential-difference equation, J. London Math. Soc. 25 (1950), 226-232.

[7] G. E. Hutchinson, Circular causal systems in ecology, Ann. N. Y. Acad. Sci. 50 (1948), 221-246.

[8] A. J. Lotka, "Elements of Physical Biology", Williams and Wilkins, Baltimore 1924. (Reissued as "Elements of Mathematical Biology", Dover, 1956).

[9] R. M. May, Time-delay versus stability in population models with two and three trophic levels, Ecology 54 (1973), 315-325.

[10] R. S. Miller and D. B. Botkin, Endangered species; models and predictions, Amer, Sci. 62 (1974), 172-181.

University of Wisconsin, Madison, Wisconsin
University of California, Los Angeles, California and
Mathematics Research Center, University of Wisconsin, Madison, Wisconsin

THE PROBLEM OF HILL FOR SYSTEMS

R. S. Bucy

For a second order linear scalar Hamiltonian system with periodic coefficients, Hill in [11], proposed a method for determining the characteristic exponents. His results identifies the characteristic exponents with the zeroes of an infinite determinant. Poincaré, in [13], mathematically justified Hill's results. In the special case of the Mathieu equation, the infinite determinant is the determinant of a continuant matrix, and continued fraction methods allow the systematic computation of the characteristic exponents, see [1]. The appropriate generalization of the infinite determinant for higher order scalar systems with periodic coefficients is reported in [14]. For periodic systems a perturbation method is described in [9], which is not completely effective.

For general linear periodic Hamiltonian systems we develop in this paper a new method which not only gives the characteristic exponents but also allows for the numerical determination of the associated constant Hamiltonian system given by the Floquet theory. Our results are based

This research was supported in part by the United States Air Force, Air Force Systems Command, under Grant AFOSR-AF-2141.

on finding <u>stable</u> periodic solutions of an associated matrix Riccati equation with periodic coefficient matrices. The determination of these periodic equilibrium solutions actually allows the determination of the constant coefficient matrices for the Riccati equation which is associated with the Floquet equivalent system.

To fully grasp our method in the simplest setting, consider the classical case of the scalar equation $\ddot{x} - q(t)x = 0$ with $q(t)$ positive bounded and periodic of period 2π. It is easily seen that

$$\frac{1}{2\pi} \int_0^{2\pi} p_+(s)ds \quad \text{and} \quad \frac{1}{2\pi} \int_0^{2\pi} p_-(s)ds$$

are the characteristic exponents, with $p_+(s)$ and $p_-(s)$ the periodic solutions, respectively positive and negative, of $\dot{p} = q(t) - p^2$. The solutions $p_+(s)$ and $p_-(s)$ exist in view of the theory of filtering, see [6], and $p_+(t)$ can be found by solving the above Riccati equation forward in time from any non-negative initial condition, the resulting solution will limit to $p_+(t)$. Further, the Floquet transformation is

$$\begin{pmatrix} y_1(t) \\ y_2(t) \end{pmatrix} = \begin{pmatrix} 1 & -\dfrac{1}{(p_+(t)-p_-(t))} \\ p_+(t) & \dfrac{-p_-(t)}{(p_+(t)-p_-(t))} \end{pmatrix} \begin{pmatrix} x(t) \\ \dot{x}(t) \end{pmatrix}$$

and

$$\frac{d}{dt} \begin{pmatrix} y_1 \\ y_2 \end{pmatrix} = \begin{pmatrix} -f & m \\ c & f \end{pmatrix} \begin{pmatrix} y_1 \\ y_2 \end{pmatrix}$$

where

where

$$f = -\frac{(p_+(0) + p_-(0))}{(p_+(0) - p_-(0))} \Delta^{\frac{1}{2}}$$

$$m = \frac{2}{(p_+(0) - p_-(0))} \Delta^{\frac{1}{2}}$$

$$c = -\frac{2p_+(0)p_-(0)}{(p_+(0) - p_-(0))} \Delta^{\frac{1}{2}}$$

and

$$\Delta^{\frac{1}{2}} = \frac{1}{2\pi} \int_0^{2\pi} p_+(s)ds = -\frac{1}{2\pi} \int_0^{2\pi} p_-(s)ds \ .$$

An easy way to verify the above is to note that $x_1(t) = \exp(\int_0^t p_+(s)ds)$ and $x_2(t) = \exp(\int_0^t p_-(s)ds)$ are a linearly independent set of solutions of $\ddot{x}-q(t)x = 0$ and $\overline{\lim}\, \frac{1}{t}\log|x_1(t)| = \Delta^{\frac{1}{2}}$. Now, from Floquet's theorem

$$\begin{pmatrix} x(t) \\ \dot{x}(t) \end{pmatrix} = S(t)\, e^{Ht}\, S^{-1}(0) \begin{pmatrix} x(0) \\ \dot{x}(0) \end{pmatrix} = R(t,0) \begin{pmatrix} x(0) \\ \dot{x}(0) \end{pmatrix}$$

with $H = \begin{pmatrix} -f & m \\ c & f \end{pmatrix}$, and $R(t, 0)$ can be expressed in terms of x_1 and x_2, since these are independent solutions. Now, for $Sp(2, R)$, the Lie algebra of $2x2$ real matrices of trace zero, the expodential map can be explicitly evaluated, see [3], Chapter 3, Problem 15, and hence f, m, and c determined as above, as well as the Floquet transformation $S(t)$. An indication of the validity of the above results is the method of Bruns for the restricted 3-body problem, described by Poincaré in [12], Volume 2, which concerns a method of obtaining a perturbation series for the characteristic exponent by solving a sequence of linear equations, which can be seen to be nothing more than a perturbation series for the equation $\dot{p} = q(t) - p^2$, actually the characteristic exponent is complex in Brun's case. Notice our method is dependent on numerically finding $p_+(s)$ and $p_-(s)$. The casual observes might ask why not simply numerically integrate $\ddot{x} - q(t)x = 0$? The answer is of course that finding p_+ is a numerical stable and quickly convergent process whereas integration of the original equation is a basically unstable procedure.

This paper will be concerned with the following problem; consider the 2n dimensional system

$$\frac{d}{dt}\begin{pmatrix} \underline{x} \\ \underline{y} \end{pmatrix} = H(t)\begin{pmatrix} \underline{x} \\ \underline{y} \end{pmatrix} \qquad \underline{x}, \underline{y} \in R^n$$

with $H(t)$ Hamiltonian and periodic, find the periodic Floquet transformation of coordinates reducing the above equation to a constant Hamiltonian system;

$$\frac{d}{dt}\begin{pmatrix} \underline{\eta} \\ \underline{\xi} \end{pmatrix} = H\begin{pmatrix} \underline{\eta} \\ \underline{\xi} \end{pmatrix} \qquad \underline{\eta}, \underline{\xi} \in R^n$$

and determine H and the characteristic exponents. This is accomplished by the study of the periodic solutions of

$$\dot{P} = [-P, I]\; H(t)\begin{bmatrix} I \\ P \end{bmatrix}$$

a matrix Riccati equation. Because of limitations of space, this paper will often only indicate and will present results in detail only for $n = 2$, the full theory will be presented in a future paper. In particular, a digital computer program is being developed to exploit this method, see [2].

1. Preliminary Results

Consider the generalized Hill equation

$$(1.0) \quad \begin{pmatrix} \dot{\underline{x}} \\ \dot{\underline{y}} \end{pmatrix} = H(t) \begin{pmatrix} \underline{x} \\ \underline{y} \end{pmatrix}$$

with $\underline{x}$ and $\underline{y}$ n-vectors and $H(t)$ a continuous periodic real $2n \times 2n$ matrix of period π. Further $H(t)$ is Hamiltonian, (i.e., $H(t)J' + JH(t) = 0$, where $J = \begin{pmatrix} 0 & I \\ -I & 0 \end{pmatrix}$ a $2n \times 2n$ matrix with $n \times n$ block entries). In view of Floquet's theorem, the fundamental matrix of (1.0), $S(t,\tau)$ factors as

$$(1.1) \quad S(t,\tau) = L(t)\, e^{K(t-\tau)}\, L^{-1}(\tau)\ ^{*}$$

with $L(t)$ a real $2n \times 2n$ matrix of period 2π and K a real $2n \times 2n$ matrix and $L(o) = I$.

Lemma 1. K is Hamiltonian and $L(t)$ is symplectic, $\{$i.e., $L'(t)JL(t) = J$ for all $t\}$

Proof. Since (1.0) is a Hamiltonian system, $S(t,o)$ is sympletic or $e^{K'2\pi}\, J\, e^{K2\pi} = J$. Now

$$e^{-K2\pi} = J^{-1}\, e^{K'2\pi} J = e^{J^{-1}KJ2\pi}$$

---

* Notice as K is the log $S(2\pi, o)$, K is not unique in (1.1); we will choose K later.

or

$K'J + KJ = 0$. Since K is Hamiltonian, it follows that e^{Kt} is symplectic, hence $L(t) = S(t,o)\, e^{-Kt}$ is symplectic, hence $L(t) = S(t,o)\, e^{-Kt}$ is symplectic as the symplectic matrices are closed under multiplication.

The system

$$(1.2) \qquad \begin{pmatrix} \dot{\underline{\eta}} \\ \dot{\underline{\xi}} \end{pmatrix} = K \begin{pmatrix} \underline{\eta} \\ \underline{\xi} \end{pmatrix}$$

will be called the Floquet equivalent system of (1.0). Notice that

$$\begin{pmatrix} \underline{x}(t) \\ \underline{y}(t) \end{pmatrix} = L(t) \begin{pmatrix} \underline{\eta}(t) \\ \underline{\xi}(t) \end{pmatrix}$$

defines a periodic transformation of the solutions (1.2) bijectively to solutions of (1.0). Both K and $H(t)$ can be represented as 2×2 matrices with $n\times n$ block matrix elements as

$$K = \begin{pmatrix} -F' & M \\ C & F \end{pmatrix}$$

$$H(t) = \begin{pmatrix} -F'(t) & M(t) \\ C(t) & F(t) \end{pmatrix}$$

<u>Lemma 1.</u> K <u>is Hamiltonian and</u> $L(t)$ <u>is symplectic</u>, $\{$i.e., $L'(t) J L(t) = J$ for all $t\}$

with the off diagonal blocks symmetric. We associate with (1.0) and (1.2) matrix Riccati equations

$$(1.3) \qquad \dot{S} = FS + SF' - SMS + C$$

$$\text{and} \qquad S(0) = \Gamma$$

$$(1.4) \qquad \dot{P} = F(t)P + PF'(t) - PM(t)P + C(t)$$

$$P(t_o) = \Lambda$$

Consider a real symplectic 2nx2n matrix $A = \begin{pmatrix} A_{11} & A_{12} \\ A_{21} & A_{22} \end{pmatrix}$

with A_{ij} ij = 1, 2 nxn matrices and consider the mapping T_A with domain $D_A = \{Z,$ nxn symmetric matrices $|$ with $\det |A_{11} + A_{12}Z| \neq 0\}$ defined as

$$T_A(Z) = (A_{21} + A_{22}Z)(A_{11} + A_{12}Z)^{-1} .$$

It is classical that

$$T_{A \cdot B}(Z) = T_Z(T_B(Z))$$

for suitable Z, see [15].

It is well known that solutions of (1.3) and (1.4) are given by $T_{e^{Kt}}(\Gamma)$ and $T_{S(t,t_o)}(\Lambda)$, respectively, on their intervals of existence.

<u>Lemma 2.</u> <u>There is a choice of K in the Floquet representation such that if $P_1(t)$ is a real symmetric solution of (1.4) of period 2π, then $P_1(0)$ is an equilibrium</u>

solution of (1.3) and $P_1(t) = T_{L(t)}(P_1(o))$.

Proof. Since $P_1(t)$ is periodic $P_1(o) = P_1(2\pi) = T_{S(2\pi,o)}(P_1(o)) = T_{e^{K2\pi}}(P_1(o))$. Notice that $e^{2\pi K} = C^2$ since H(t) has period π, see the discussion following Floquet theorem in [10]. Let $e^{2\pi K} = \begin{pmatrix} A & B \\ C & D \end{pmatrix}$ so that

$$P_1(o) = \{C + D\,P_1(o)\}\,\{A + B\,P_1(o)\}^{-1}$$

since $e^{2\pi K}$ is symplectic

$$e^{N^{-1}KN2\pi} = N^{-1}e^{K2\pi}N = \begin{pmatrix} D-P_1(o)B & 0 \\ B & A+B\,P_1(o) \end{pmatrix} =$$

$$\begin{pmatrix} \phi & 0 \\ \phi'^{-1}L & \phi'^{-1} \end{pmatrix}$$

with $N = \begin{pmatrix} 0 & I \\ I & P_1(o) \end{pmatrix}$, $L = L'$, with the last equality holding since $e^{N^{-1}KN2\pi}$ is symplectic.

Since

$$N^{-1}e^{K2\pi}N = N^{-1}C^2N = (N^{-1}CN)^2 = \begin{pmatrix} \phi & 0 \\ \phi'^{-1}L & \phi'^{-1} \end{pmatrix}$$

it follows that $\begin{pmatrix} \phi & 0 \\ \phi'^{-1}L & \phi'^{-1} \end{pmatrix}$ has a real logarithm and ϕ has a real logarithm. Let F* be any version of the logarithm, then

$$\begin{pmatrix} \phi & 0 \\ \phi'^{-1}L & \phi'^{-1} \end{pmatrix} = \exp\left[2\pi \begin{pmatrix} F^* & 0 \\ x & -F^{*\prime} \end{pmatrix}\right]$$

if $e^{2\pi F'^*} x \, e^{2\pi F^*} - x = F^{*\prime} L + LF^*$ can be solved for x. The last equation possesses a solution for arbitrary L iff $\lambda_i(F^*) \neq \sqrt{-1}\ \nu$ for $\nu = \pm 1, \pm 2 \cdots$. Take F_1 to be that version of the logarithm of ϕ satisfying the above condition.

Now

$$\exp\{2\pi K\} = N \exp\left\{2\pi \begin{pmatrix} F_1 & 0 \\ x & -F_1' \end{pmatrix}\right\} N^{-1}$$

with x satisfying $e^{2\pi F_1'} x \, e^{2\pi F_1} - x = F_1'L + LF_1$ and we take $K = N \begin{pmatrix} F_1 & 0 \\ x & -F_1' \end{pmatrix} N^{-1}$! This choice of K gives $F_1 = F - P_1(o)M$, $x = M$ and

$$F\,P_1(o) + P_1(o)\,F' - P_1(o)\,MP_1(o) + C = 0$$

where

$$K = \begin{pmatrix} -F' & M \\ C & F \end{pmatrix} .$$

Now

$$P_1(t) = T_{S(t,o)}(P_1(o)) = T_{L(t)}(T_{e^{Kt}}(P_1(o))) = T_{L(t)}(P_1(o))$$

as $P_1(o)$ is a fixed point of (1.3).

<u>Remarks.</u> To see the non-uniqueness problem, consider $e^{2\pi K} = I$, then $K = 0$, $K = J$, $K = 2J$, etc. are all logarithms of I. Out choice of K in the above lemma is the natural one for our purposes, and from now on K and L(t) are fixed as above. The characteristic exponents of (1.0) are just the characteristic roots of K which we denote by

$$\mu_1, -\mu_1, \mu_2, -\mu_2 \cdots \mu_n, -\mu_n, \text{ with } \mathrm{Re}(\mu_i) \leq 0.$$

It is clear from the above proof that each real symmetric fixed point of (1.3), $\bar{\Sigma}$, induces a periodic solution of (1.3), namely, $T_{L(t)}(\bar{\Sigma})$. If Σ_1 is a complex symmetric fixed point of (1.3), then Σ_2 the complex conjuate is also a fixed point of (1.3) and they induce a real almost periodic solution $T_{L(t)}(R(t))$, where R(t) is periodic of period $2\pi/\mathrm{Im}(\mu_k)$ for some k. This can be seen by examining the cross ratio of a solution of (1.4) and Σ_1 and requiring it to be periodic.

2. <u>Periodic Solutions and Characteristic Exponents</u>

We now suppose that (1.4) possesses two invertible periodic matrix solutions $P_+(s)$ and $P_-(s)$ of period 2π with $\Sigma(s) = P_+(s) - P_-(s)$ uniformly positive definite and $\Sigma^{-1}(s)$ uniformly positive definite, {i.e., there exist two real numbers α and β such that $0 < \alpha I \leq \Sigma(s) \leq \beta I$ in the ordering of the cone of positive semi-definite matrices}. Denote this assumption by A_1. Sufficient conditions for A_1 are (1) $M(s) \geq 0$; (2) $C(s) \geq 0$; and (3) controllability and observability hold for $H(t)$, actually (2) and (3) can be considerably weakened; see [8]. By A_2 we denote the condition that K be generic, {i.e., its characteristic roots are simple with real parts unequal to zero}. This second condition is necessary and sufficient for a modified structural stability of the Riccati equation (1.3) and is clearly required in order for the characteristic exponents to be effectively computable. Denote by $\psi(t,s)$ the fundamental matrix $F(t) - P_+(t)\,M(t) = \bar{F}_+(t)$. The following theorem relates the periodic solutions of (1.4) and the characteristic exponents as;

<u>Theorem 1.</u> <u>Suppose</u> A_1 <u>holds, then the symplectic transformation</u>

$$\underline{z}(t) = L(t)\begin{pmatrix}\underline{x}(t)\\ \underline{y}(t)\end{pmatrix}$$

<u>where</u>

$$L(t) = \begin{pmatrix} T'(t)^{-1} & T'^{-1}(t)\,S(t) \\ -P_+(t) & I \end{pmatrix} \quad \underline{\text{with}}\ \ T(t) = (I + P_+(t)S(t))$$

<u>and</u> $-P_-(t) = S^{-1}(t)$ gives

$$\underline{\dot{z}} = \begin{pmatrix} -\bar{F}_+' & 0 \\ 0 & \bar{F}_+ \end{pmatrix}\underline{z}. \tag{1.5}$$

Further, the characteristic exponents of (1.0) are

$\pm \ell n | \lambda_j(\psi(2\pi,o)) |$ $j = 1 \cdots n$, $(\lambda_i(A))$ denotes the eigenvalues of A.

Proof. A direct computation from the definition shows that L(t) is symplectic. Now

$$\dot{\underline{z}} = (\dot{L} + LH)\begin{pmatrix}\underline{x}\\ \underline{y}\end{pmatrix}$$

but
$$\dot{L} + LH = \begin{pmatrix} -\bar{F}_+' T^{-1'} & -\bar{F}_+' T^{-1'} S \\ -\bar{F}_+ P_+ & \bar{F}_+ \end{pmatrix} = \begin{pmatrix} -\bar{F}_+' & 0 \\ 0 & \bar{F}_+ \end{pmatrix}$$

hence
$$\dot{\underline{z}} = \begin{pmatrix} -\bar{F}_+' & 0 \\ 0 & \bar{F}_+ \end{pmatrix} \underline{z}.$$ Now denoting $S^*(t, o)$

as the fundamental matrix of the system (1.5), then

$$\underline{z}(t) = S^*(t, o)\, \underline{z}(0)$$

or

$$\underline{x}(t) = L^{-1}(t)\, S^*(t, o)\, L(o)\, \underline{x}(0)$$

consequently

$$L^{-1}(2\pi)\begin{pmatrix} \psi'(0, 2\pi) & 0 \\ 0 & \psi(2\pi, 0) \end{pmatrix} L(0) = S(2\pi, 0)\,.$$

Now the final relation gives the asserted result as L is periodic.

Remark. a) Theorem 1 is the time-varying version of Theorem 3.2 of [4].

b) $$\frac{1}{2\pi}\int_0^{2\pi} \text{trace}\ \bar{F}_+(s)ds = \sum_{i=1}^{n} \mu_i = \text{trace}\ (F - P_+(o)M)$$

in view of the Jacobi identity with μ_i the characteristic exponents, with real part negative. Real part of μ_i is non-positive since Σ^{-1} is a Liaponnof function for $F_+(s)$!

c) The first part of the theorem holds for the general time variable case.

Theorem 2. Let $P_\theta(t)$ be any periodic solution of (1.4) of period 2π, then the characteristic exponents of (1.0) coincide with the set

$$\pm \ell n\ [\lambda_i(\psi_\theta(2\pi, o))] \quad i = 1, 2 \cdots n$$

{where ψ_θ is the fundamental matrix of $F(t) - P_\theta(t)M(t)$}.

Proof. It can be easily established that

$$F_t^{-1}\ S(t, t_o)\ F_{t_o} = S^{**}(t, t_o)$$

where S^{**} is the fundamental matrix of

$$\begin{pmatrix} \bar{F}_\theta(t) & 0 \\ M(t) & -\bar{F}'_\theta(t) \end{pmatrix}$$

with $\bar{F}_\theta(t) = F(t) - P_\theta(t)\ M(t)$ and $F_t = \begin{pmatrix} 0 & I \\ I & P_\theta(t) \end{pmatrix}$.

Now, as $P_\theta(t)$ is periodic so is F_t and the characteristic roots of $S(2\pi,o)$ and $S^{**}(2\pi,o)$ coincide. But the characteristic roots of $S^{**}(2\pi, o)$ are $\lambda_i(\psi(2\pi, o))$ and $\dfrac{1}{\lambda_i(\psi(2\pi, o))}$ $i = 1 \ldots n$, so the result follows.

Remark.

$$\frac{1}{2\pi} \int_0^{2\pi} \text{trace } (\bar{F}_\theta(s) ds) = \sum_{i=1}^{n} \mu_i^*$$

where $\sum_{i=1}^{n} \mu_i^* = \text{trace } \{F - P_\theta(o)M\}$. In fact, since

$$\det(K - \lambda I) = \prod_{i=1}^{n} (\lambda^2 - \mu_i^2) = \prod_{i=1}^{n} (\lambda^2 - \mu_i^{*2}) \text{ each}$$

$\mu_i^* = \pm \mu_i$; see [7].

In view of these theorems, the general procedure is clear. Each periodic solution of (1.4) corresponds to an equilibrium solution of (1.3), and each equilibrium solution of (1.3) corresponds to a subset of n of the 2^n numbers $\pm \mu_i$ $i = 1 \ldots n$. The latter correspondence is explicitly $P_\theta(o) \leftrightarrow (\mu_i^*, \ldots, \mu_n^*)^*$ where μ_i^* are the characteristic roots of $F - P_\theta(o)M$. Our procedure, when μ_i are real, is then to find a distinguished set of n periodic solutions of (1.4) each of period 2π, $P_{\theta_j}(t)$ $j = 1 \ldots n$, computing the integrals

$$\frac{1}{2\pi} \int_0^{2\pi} \text{trace } \bar{F}_{\theta_j}(t)\, dt = \sum_{i=1}^{n} \mu_i(\theta_j)$$

*$(\mu_i^*, \ldots, \mu_n^*)$ must be the roots of a polynomial with real coefficients as $P_\theta(o)$ is the value at zero of a real periodic solution. When this is not the case, almost periodic solutions of (1.4) arise.

and solving for the μ_i. By distinguished set we mean as such that det $F \neq 0$, where $F_{ij} = \text{sgn}\ f(i, \theta_j)$ and $\mu_i(\theta_j) = \text{sgn}\ f(i, \theta_j)\ \mu_i$.

The last remarks follow from the following results;

<u>Theorem 3. Let $\bar{P}$ be a symmetric equilibrium of (1.3), then</u>

$$(-\bar{P}, I)\,\Delta(K) = 0$$

<u>where</u> $\det(K - \lambda I) = (-1)^n \Delta(\lambda)\,\Delta(-\lambda)$ <u>and</u> $\Delta(\lambda) = \det(F - \bar{P}M - \lambda I)$.

<u>In particular</u> Δ <u>has real coefficients if</u> $\bar{P}$ <u>is real</u>; (see [8], for a proof).

III. The Case of $n = 2$

We now consider in detail the special case when $n = 2$, under the assumptions A_1 and A_2. Consider first the case of real μ_i, so the characteristic exponents are $\mu_2 < \mu_1 < 0 < -\mu_1 < -\mu_2$. Because we are in the generic case, we have four equilibria of (1.3);

$$\begin{cases} P_+(o) & \leftrightarrow (\mu_1, \mu_2) \\ P_-(o) & \leftrightarrow (-\mu_1, -\mu_2) \\ P_{\theta_1}(o) & \leftrightarrow (-\mu_1, \mu_2) \\ P_{\theta_2}(o) & \leftrightarrow (\mu_1, -\mu_2) \end{cases}$$

which correspond to periodic solutions of (1.4)

$$P_+(t),\ P_-(t),\ P_{\theta_1}(t),\ P_{\theta_2}(t).$$

Now $P_+(t)$ and $P_-(t)$ can be determined numerically by finding Γ so that $\Pi(2\pi, \Gamma, o) \geq \Gamma$ with $\Pi(t, \Gamma, o)$ the solution of (1.4) then

$$\Pi(2\pi n, \Gamma, o) \underset{n\to\infty}{\longrightarrow} P_+(o)$$

and $\Pi(t, P_+(o), o) = P_+(t)$ and $\Pi(-2\pi n, \Gamma, o) \to P_-(o)$ and $\Pi(-t, P_-(o), o) = P_-(-t)$. Now $F(t)$ and $M(t)$ are given so one can compute:

$$(1.6)\quad \frac{1}{2\pi}\int_0^{2\pi} \text{trace}\,(F(t) - P_-(t)\,M(t))dt = -\mu_1 - \mu_2,$$

in view of theorem 2. Now, since the set $\det\,[P - P_-(t)] = 0$ with P a solution of (1.4) is an invariant set for (1.4), we shall solve (1.4) subject to the constraint $\det\,(P - P_-(t)) = 0$. Let $\Delta(t) = P - P_p(t)$ with $\Delta(o) = \underline{c}\,\underline{c}^*$ with $\underline{c}$ an n-vector

$$\mathring{\Delta}(t) = \bar{F}_-(t)\,\Delta(t) + \Delta(t)\,\bar{F}'_-(t) - \Delta(t)M(t)\,\Delta(t$$

then

$$\Delta(t) = \psi_-(t, o)\,\underline{c}\,\underline{c}'\,[I + \{\psi'_-(t, o)\,\Sigma^{-1}(t)\,\psi_-(t, o) - \Sigma^{-1}(o)\}\underline{c}\,\underline{c}']^{-1}\psi'(t, o)$$

or with $W(t) = \psi'_-(t, o)\,\Sigma^{-1}(t)\psi_-(t, o) - \Sigma^{-1}(o)'$

$$\Delta(t) = \psi_-(t, o)\,\underline{c}\,\underline{c}'\left[I - \left[\frac{W(t)\,\underline{c}\,\underline{c}'}{1 + \|\underline{c}\|^2_{W(t)}}\right]\right]\psi'(t, o)$$

using the Shur lemma, [6], and noting that

$$W(t) = \int_0^t \psi'_-(s, o)\,M(s)\,\psi_-(s, o)ds \geq 0.$$ Finally, $\Delta(t)$

can be expressed as

$$(1.7)\quad \Delta(t) = \frac{\psi_-(t,o)\,\underline{c}\,\underline{c}'\,\psi_-'(t,o)}{1 - \|\underline{c}\|^2_{\Sigma^{-1}(o)} + \|\psi(t,o)\underline{c}\|^2_{\Sigma^{-1}(t)}}$$

It is easy to see that

$$\Delta(2n\pi) \to \frac{\chi\,\underline{c}\,\underline{c}'\chi'}{\|\chi\underline{c}\|^2_{\Sigma^{-1}(o)}} \qquad \text{as } n \to \infty$$

where $\chi = \lim_{n\to\infty} e^{\mu_2(2n\pi)}\,\psi_-(2n\pi, o)$ and χ is the projection determined by the eigenvector of $\psi_-(2\pi,o)$ corresponding to the eigenvalue $-\mu_2$. Clearly, the initial condition

$$\Lambda^* = P_-(o) + \frac{\chi\underline{c}\,\underline{c}'\chi'}{\|\chi\underline{c}\|^2_{\Sigma^{-1}(o)}}$$

generates a periodic solution, which must be either $P_{\theta_1}(t)$ or P_{θ_2} but Λ^* must be $P_{\theta_1}(o)$ as this is stable, and hence $\Pi(t,\Lambda,o) = P_{\theta_1}(t)$. But again by theorem 2

$$(1.8)\quad \frac{1}{2\pi}\int_o^{2\pi} \text{trace}\,\{F(t) - P_{\theta_1}(t)M(t)\}\,dt = -\mu_1 + \mu_2$$

But equations (1.6) and (1.7) combined yield

$$(1.9)\quad \frac{1}{2\pi}\int_o^{2\pi} \text{trace}\,\{\Delta^*(t)\,M(t)\}\,dt = -2\mu_2$$

with $\Delta^*(t) = P_{\theta_1}(t) - P_-(t)$

and hence both μ_1 and μ_2 can be determined.

However,

$$\Delta(t) = \frac{\psi_-(t,o)\,\underline{d}\,\underline{d}'\,\psi'_-(t,o)}{1 - \|\underline{d}\|^2_{\Sigma^{-1}(o)} + \|\psi_-(t,o)\underline{d}\|^2_{\Sigma^{-1}(t)}}$$

with $d = \dfrac{\underline{c}\,\chi}{\|\chi\,\underline{c}\|_{\Sigma^{-1}(o)}}$ so that

$$\Delta^*(t) = \frac{\psi_-(t,o)\chi\, c\, c'\, \chi'\, \psi'_-(t,o)}{\|\psi_-(t,o)\,\chi\,\underline{c}\|^2_{\Sigma^{-1}(t)}}$$

and trace $\Delta^*(t)\,M(t) = \dfrac{\|\psi_-(t,o)\chi\,\underline{c}\|^2_{M(t)}}{\|\psi_-(t,o)\chi\,\underline{c}\|^2_{\Sigma^{-1}(t)}}$

but $\dfrac{d}{dt}\|\psi_-(t,o)\chi\,\underline{c}\|^2_{\Sigma^{-1}} = \|\psi_-(t,o)\,\chi\,\underline{c}\|^2_{M(t)}$

or

$$-2\mu_2 = \frac{1}{2\pi}\,\ln\,\frac{\|\psi_-(2\pi,o)\chi\,\underline{c}\|^2_{\Sigma^{-1}(o)}}{\|\chi\,\underline{c}\|^2_{\Sigma^{-1}(o)}}\;.$$

Now $\Delta^*(o) = \underline{\ell}\,\underline{\ell}' = P_{\theta_1}(o) - P_-(o)$ and since both $P_{\theta_1}(o)$ and $P_-(o)$ are fixed points of (1.3).

(2.0) $\quad \bar{F}_-\Delta(o) + \Delta(o)\bar{F}_-' - \Delta(o) M\Delta(o) = 0$

with $\bar{F}_- = F - P_-(o)\ M$ or

$$\bar{F}_-\underline{\ell} = \left\{ \|\ell\|_M^2 - \frac{\ell' \bar{F}_- \ell}{\|\ell\|^2} \right\} \underline{\ell}$$

Hence $\underline{\ell}$ is an eigenvector of $\bar{F}_-$ with eigenvalue $\frac{1}{2}\|\underline{\ell}\|_M^2$, which has the value $-\mu_2$. To see this, note that $\bar{F}_-\ell = \tau\ell$ and let $\ell_\perp$ be orthogonal to $\ell(\bar{F}_-'\ell_\perp, \ell) =$ $(\ell_\perp, F_-\ell) = \tau(\ell, \ell_\perp) = 0$ or

$$\bar{F}_-'\ell_\perp = \sigma\ell_\perp .$$

Now $\bar{F}_{\theta_1} = F - P_\theta(o)\ M = F - P_-(o)\ M - \ell\ell'\ M$ and therefore

$$\bar{F}_{\theta_1}\ell_\perp = \sigma\ell_\perp .$$

In view that σ is a common eigenvalue of $\bar{F}_-'$ and $\bar{F}_{\theta_1}'$ $\sigma = \mu_1$. Now, $\tau = \mu_2$

$$\bar{F}_{-}\underline{\ell}\,\underline{\ell}' + \underline{\ell}\,\underline{\ell}'\,\bar{F}_{\theta_1} = 0$$

$$\tau\,\ell\,\ell' + \ell\,\ell'\,\bar{F}_{\theta_1} = 0$$

$$\ell'\,\bar{F}'_{\theta_1} = -\tau\,\ell'$$

and the eigenstructure again gives the result.

Hence

$$\bar{F}_{-}\underline{\ell} = -\mu_2\underline{\ell} \tag{2.1}$$

and $-2\mu_2 = \dfrac{\|\chi\,\underline{c}\|^2_M}{\|\chi\,\underline{c}\|^2_{\Sigma^{-1}(o)}}$.

By the same argument, with time reversed and considering $\bar{F}_+$ instead of $\bar{F}_-$, there exists $\underline{k}$ determined analogously to $\underline{\ell}$ so that $\bar{F}_+\underline{k} = \mu_2\underline{k}$ and

$$\bar{F}'_+\,\underline{k}_\perp = \mu_1\,\underline{k}_\perp \quad \text{with } (k_\perp, k) = 0.$$

However, from the relation

$$\bar{F}_-\Sigma = -\Sigma\,\bar{F}'_+$$

it follows that

$$\bar{F}_-\Sigma\,\underline{k}_\perp = -\mu_1\Sigma\,\underline{k}_\perp \tag{2.2}$$

Now (2.1) and (2.2), determine $\bar{F}_-$ as the solution of

(2.3)
$$\bar{F}_- \Sigma \underline{k}_\perp = -\mu_1 \Sigma \underline{k}_\perp$$
$$\bar{F}_- \underline{\ell} = -\mu_2 \underline{\ell}$$

where, of course $\Delta^{**}(o) = -\underline{k}\,\underline{k}' = P_{\theta_2}(o) - P_+(o)$, is the limiting periodic solution evaluated at 0 of

$$\dot{\Delta}(t) = \bar{F}_+(t)\,\Delta(t) + \Delta(t)\,\bar{F}_+(t) - \Delta(t)\,M(t)\,\Delta(t)$$

with $\det \Delta(t) = 0$, solved backwards in time and $k_\perp$ is such that $\underline{k}'_\perp \underline{k} = 0$.

Once $\bar{F}_-$ is known, the following identities yield the other matrix elements of K

(2.4)
$$\bar{F}'_- \Sigma^{-1}(o) + \Sigma^{-1}(o)\, \bar{F}_- = M$$
$$\bar{F}_- + P_-(o)\,M = F$$

and
$$C = -\bar{F}_-\,P_-(o) - P_-(o)\,F'$$

periodic.

So far then, we have shown that the unique limit periodic solution

(2.5)
$$\Delta^*(t) = \frac{\psi_-(t,o)\chi\,\underline{c}\,\underline{c}'\,\chi\underline{\psi}'(t,o)}{\|\psi_-(t,o)\,\chi\underline{c}\|^2_{\Sigma^{-1}(t)}}$$

of

(2.6)
$$\dot{\Delta} = \bar{F}_-(t)\,\Delta + \Delta\,\bar{F}'_-(t) - \Delta M(t)\,\Delta$$

<u>on the subspace where</u> det $\Delta = 0$ <u>and any</u> $\Delta(o) = \underline{c}\,\underline{c}'$ <u>and</u> its dual $\Delta^{**}(o)$ together with $P_-(t)$ and $P_+(t)$, both of which we have previously indicated how to obtain numerically, suffices to determine the matrix K. The problem of solving (2.6) numerically is more delicate; we must eliminate one of the three equations of (2.6), by using the constraint, in order to converge to Δ^*, because even though det $\Delta = 0$ is an invariant set, roundoff error tends to throw the solution off of the set. Since $\Delta(o) \geq 0$ implies $\Delta(t) \geq 0$, the effective way to solve (2.6) is to replace

$$\dot{\Delta}_{11} = f^1(\Delta_{11}, \Delta_{12}, \Delta_{22})$$

$$\dot{\Delta}_{12} = f^2(\Delta_{11}, \Delta_{12}, \Delta_{22})$$

$$\dot{\Delta}_{22} = f^3(\Delta_{11}, \Delta_{12}, \Delta_{22})$$

by the system

$$\dot{\Delta}_{11} = f^1(\Delta_{11}, \Delta_{12}, \Delta_{12}^2/\Delta_{11})$$

$$\dot{\Delta}_{12} = f^2(\Delta_{11}, \Delta_{12}, \Delta_{12}^2/\Delta_{11}) ,$$

that is we use the constraint to eliminate a coordinate of Δ and reduce the order of the system. The method we have indicated works well, provided $\Delta_{11}(t)$ does not become too small, if it does, one can eliminate one of the other coordinates or change variables and eliminate. Of course, the solution of (1.4) subject to the constraint det $\{P(t) - P_+(t)\} = 0$ is analogous to (2.6).

EXAMPLE 1 $F(t) = \begin{pmatrix} 0 & 0 \\ 0 & 0 \end{pmatrix}$, $M(t) = \begin{pmatrix} 1 & 0 \\ 0 & 1 \end{pmatrix}$, and

$$C(t) = \begin{pmatrix} 2 & 1 \\ 1 & 2+\cos t \end{pmatrix}$$

then $\mu_1 = -1.7321$, $\mu_2 = -.9712$

$$\Sigma(o) = \begin{pmatrix} 2.7401 & .6741 \\ .6741 & 3.3394 \end{pmatrix}$$

$$\Delta^*(o) = \begin{pmatrix} 1.7238 & 1.8366 \\ 1.8366 & 3.1659 \end{pmatrix}$$

and $\Delta^{**}(o) = -\Delta^*(o)$.

In fact, since in this case if $P(t)$ is a solution of (1.4), then $-P(-t)$ is also. It is clear that $\underline{k} = \underline{\ell}$ and $F = 0$, so that (2.3) becomes

$$(\overline{P}_+(o))^{\frac{1}{2}} M(\overline{P}_+(o))^{\frac{1}{2}} (\overline{P}_+(o))^{-\frac{1}{2}}\underline{k} = -\mu_2(\overline{P}_+(o)^{-\frac{1}{2}}\underline{k}$$

$$(\overline{P}_+(o))^{\frac{1}{2}} M(\overline{P}_+(o))^{\frac{1}{2}} (\overline{P}_+(o))^{\frac{1}{2}}\underline{k}_\perp = -\mu_2(\overline{P}_+(o))^{\frac{1}{2}}\underline{k}_\perp$$

$$[\overline{P}_+(o)]^{\frac{1}{2}} M[\overline{P}_+(o)]^{\frac{1}{2}} = -\mu_2/2\,[\overline{P}_+(o)]^{-\frac{1}{2}}\,\underline{k}\,\underline{k}'[\overline{P}_+(o)]^{-\frac{1}{2}} - -\frac{\mu 1}{\|\underline{k}\|^2_{P_+(o)}} \cdot$$

$$[\overline{P}_+(o)]^{\frac{1}{2}}\,\underline{k}_\perp\underline{k}'_\perp\,[\overline{P}_+(o)]^{\frac{1}{2}}$$

then

$$C = \begin{pmatrix} 1.9813 & 1.0205 \\ 1.0205 & 2.36011 \end{pmatrix}$$

and

$$M = \begin{pmatrix} .9824 & .0520 \\ .0520 & .7856 \end{pmatrix}$$

These results were computed on a timesharing system and are accurate to 2 places approximately; see [2] for a more accurate determination of μ_1, and μ_2.

We now consider the remaining case where $\mu_1 = -\lambda + i\,\omega$ $\mu_2 = -\lambda - i\,\omega$ as in the case where both μ_1 and μ_2 are real,

$$(2.7) \quad \frac{1}{2\pi}\int_0^{2\pi} \operatorname{trace}\{F(t) - P_-(t)\,M(t)\}\,dt = -\mu_1 - \mu_2 = 2\lambda$$

Of course, $P_-(t)$ and $P_+(t)$ are determined exactly in the same way as they were previously. The following example shows the necessity for the assumption A_2;

EXAMPLE 2 Suppose $K = \begin{pmatrix} -F' & M \\ C & F \end{pmatrix}$

where $F = \begin{pmatrix} 0 & 1 \\ -1 & 0 \end{pmatrix}$ $C = \begin{pmatrix} 1 & 0 \\ 0 & 1 \end{pmatrix}$ and $M = \begin{pmatrix} a & 0 \\ 0 & 0 \end{pmatrix}$

It is easy to check that if $0 < a < 8$, no constant real equilibria exist besides P_+ and P_- which are given by

$$(2.8) \quad P_+ = \frac{1}{a}\begin{pmatrix} \alpha & \beta-1 \\ \beta-1 & \alpha\beta \end{pmatrix} \qquad P_- = \frac{1}{a}\begin{pmatrix} -\alpha & \beta-1 \\ \beta-1 & -\alpha\beta \end{pmatrix}$$

with $\beta^2 = 1 + a$, $\alpha^2 = 2\beta - 2 + a$, α and β positive.
If $a = 8$, one and only one other real solution besides those given by (2.8) exists; it is

$$P_\theta = \begin{pmatrix} 0 & -\frac{1}{2} \\ -\frac{1}{2} & 0 \end{pmatrix}$$

while for $a > 8$ two other equilibria exist besides those given by (2.8); they are

$$(2.9) \qquad P_{\theta_\pm} = \frac{1}{a}\begin{pmatrix} \pm\sqrt{\beta^2-2\beta-3} & -\beta-1 \\ -\beta-1 & \mp\beta\sqrt{\beta^2-3\beta-3} \end{pmatrix}$$

Of course, (2.9) represents complex equilibria for $0 < a < 8$, and it is easily seen that they induce a real periodic solution for this range of a, which is in the domain of attraction of the set of initial conditions Γ, such that $\det(\Gamma - P_-) = 0$. The period of this solution is $2\sqrt{\beta - \alpha^2/4}$ or $\underline{2\,\omega!}$, where $\pm\omega = \mathrm{Im}[\lambda_i(\overline{F}_-)]$. From this example, it is clear that case of K for $a = 8$ is not effectively computable in the sense that if for $\varepsilon > 0$ $a = 8 - \varepsilon$, the limit motion with $\det(\Gamma - P_-) = 0$ of the Riccati equation corresponding to K is periodic, while, if $a = 8 + \varepsilon$, the limit motion is constant. It is clear that A_2 is necessary to avoid the case $a = 8$ in this example since numerical roundoff errors make the situation of μ_i real numerically indistinguishable from that of μ_i complex ! For more details concerning this example, see [4] and [5].

Numerically, the case where K has complex exponents implies that the solution of (2.5), $\Delta^*(t, \underline{c}\ \underline{c}', o)$ does not possess a limit as $n \to \infty$, for $t - 2n\pi$, in contradiction to the case of real roots of K where a limit exists.

Consider the case where K has complex roots $\mu_1, \mu_2, -\mu_1, -\mu_2$ with $\mu_1 = -\lambda + i\omega$, $\mu_2 = \overline{\mu}_1$. We will study the complex symmetric solutions of

$$(3.0) \qquad \dot{\Delta} = \overline{F}_-(t)\,\Delta + \Delta\,\overline{F}_-'(t) - \Delta(t)\,M(t)\,\Delta(t)$$

on the surface $\det \Delta(t) = 0$, which is invariant for (3.0).

Consider a solution satisfying the constraint $\Delta(t) = X(t) + i\,Y(t)$, with $X(t)$ and $Y(t)$ real symmetric 2×2 matrices. We recall the notation previously introduced; $\psi_-(t,o)$ is the fundamental matrix of $\overline{F}_-(t) = F(t) - P_-(t)M(t)$ and

$$(3.1)\quad W(t) = \psi_-'(t,o)\Sigma^{-1}(t)\psi_-(t,o) - \Sigma^{-1}(o) = \int_o^t \psi_-'(s,o)M(s)\psi_-(s,o)ds$$

It will be convenient to consider the real representation map

$$\phi : \Delta(t) \rightarrow \begin{pmatrix} X(t) & Y(t) \\ -Y(t) & X(t) \end{pmatrix} = K(t)\ .$$

Now, as $\Delta(t)$ satisfies (3.0), it is easily verified that[2]

$$(3.2)\quad \frac{dK}{dt} = (\overline{F}_-(t) \oplus I)K + K(\overline{F}_-(t) \oplus I)' - K(M(t) \oplus I)K$$

and as $\det \Delta(t) = 0$, $\Delta(o) - (\underline{\ell} + i\,\underline{m})(\underline{\ell}' + i\,\underline{m}')$, so that

$$(3.3)\quad K(o) = \begin{pmatrix} \underline{\ell}\,\underline{\ell}' - \underline{m}\,\underline{m}' & \underline{\ell}\,\underline{m}' + \underline{m}\,\underline{\ell}' \\ -(\underline{\ell}\,\underline{m}' + \underline{m}\,\underline{\ell}') & \underline{\ell}\,\underline{\ell}' - \underline{m}\,\underline{m}' \end{pmatrix}$$

for $\underline{\ell}$ and $\underline{m}$ column vectors in R^2. Further, as ϕ is a homomorphism, $K(o) = V\,V^*$ with

$$V = \begin{pmatrix} \underline{\ell} & \underline{m} \\ -\underline{m} & \underline{\ell} \end{pmatrix} \quad \text{and} \quad V^* = \begin{pmatrix} \underline{\ell}' & \underline{m}' \\ -\underline{m}' & \underline{\ell}' \end{pmatrix}\ ,$$

4x2 and 2x4 matrices respectively. It is easily verified that a solution of (3.2) satisfying (3.3) is

---

[2] $\oplus$ denotes Kronecker product.

$$K(t) = [\psi_-(t,o) \oplus I] \; V \begin{pmatrix} 1 + \|\underline{\ell}\|^2_{W(t)} - \|\underline{m}\|^2_{W(t)} & 2\underline{\ell}' W(t) \underline{m} \\ -2\underline{\ell}' W(t) \underline{m} & 1 + \|\underline{\ell}\|^2_{W(t)} - \|\underline{m}\|^2_{W(t)} \end{pmatrix}^{-1}$$

$$V^* [\psi_-(t,o) \; I]' \tag{3.4}$$

whenever the indicated inverse exists; in analogy to (1.7). In particular, the 4 parameter family of solutions (3.4) contains a 2-parameter family of almost periodic solutions $K(t)$ when the additional constraints

$$(3.5) \qquad \begin{aligned} 1 &= \|\underline{\ell}\|^2_{\Sigma^{-1}(o)} - \|\underline{m}\|^2_{\Sigma^{-1}(o)} \\ \underline{m}' \, \Sigma^{-1}(o) \, \underline{\ell} &= 0 \end{aligned}$$

are imposed on the $\underline{\ell}$ and $\underline{m}$ specifying $K(o)$ in (3.3). The form of these solutions is

$$K(t) = (L_1(t) \; B(t) \; I) \; V \; A_t^{-1} \; V^* (L_1(t) \; B(t) \oplus I)' \tag{3.6}$$

where

$$A_t = \begin{pmatrix} \alpha_t & \beta_t \\ -\beta_t & \alpha_t \end{pmatrix}, \quad e^{-\lambda t} \psi_-(t,o) = L_1(t) \, B(t), \quad \lambda > 0$$

$$\alpha_t = \|L_1(t) B(t) \underline{\ell}\|^2_{\Sigma^{-1}(t)} - \|L_1(t) B(t) \underline{m}\|^2_{\Sigma^{-1}(t)}$$

$$\beta_t = 2 \, \underline{m}' \; B'(t) \, L_1'(t) \, \Sigma^{-1}(t) \; L_1(t) B(t) \underline{\ell} \quad .$$

In view of Floquet's theorem $L_1(t)$ is periodic with period 2π, while $B(t)$ is periodic of period $2\pi/\omega$. Every solution (3.4) is easily seen to limit to one of the family (3.6) as $t \to \infty$. In particular, when $\underline{\ell}_o$ and $\underline{m}_o$ are respectively the real and imaginary parts of an eigenvector of $\psi_-(2\pi,o)$ the members of the family (3.6) correspond to a periodic solution of (3.2) $\bar{\bar{K}}(t)$ of period 2π and conversely. Of course, each of the two 2π periodic solutions of (3.2), $\bar{\bar{K}}_1(t)$, $\bar{\bar{K}}_2(t)$ corresponds to periodic solutions $\Delta_1(t)$ and $\Delta_2(t)$ of period 2π of equation (3.0), with $\Delta_1(t)$ induced by $\underline{\ell}_o$, $\underline{m}_o$, while $\Delta_2(t)$ is induced by $\underline{\ell}_o$, $-\underline{m}_o$.

Using this last observation, $\bar{\bar{K}}_1(t)$ can be numerically determined by solving (3.2) on the subspace determined by (3.3) and (3.5), using the constraints exactly as in the real case to eliminate state variable in (3.2) and iterating to find the periodic solution, and hence determining $\underline{\ell}_o$ and $\underline{m}_o$ in the process. Now, once $\bar{\bar{K}}_1(t)$ is found; ω is determined from

$$(3.7)\quad \frac{1}{2\pi}\int_o^{2\pi} \text{trace } J_4\, \bar{\bar{K}}(t)\,\{M(t)\oplus I\}\, dt = 4\,\omega$$

and then $\bar{F}_-$ can be determined by the equations

$$(3.8)\quad \begin{aligned} \bar{F}_-(\underline{\ell}_o + i\,\underline{m}_o) &= (\lambda + i\omega)\,(\underline{\ell}_o + i\,\underline{m}_o) \\ \bar{F}_-(\underline{\ell}_o - i\,\underline{m}_o) &= (\lambda - i\omega)\,(\underline{\ell}_o - i\,\underline{m}_o) \end{aligned}$$

Once $\bar{F}_-$ is determined, F, M, and C, are determined as in the real case.

CONCLUSIONS

By developing a complete knowledge of the domains of attraction of symmetric real and complex equilibria of a regular autonomous Riccati equation, we have shown how to numerically determine both the characteristic exponents and the Floquet equivalent system of a generalized Hill equation. The main tool here is the theorem which relates to each periodic Riccati equation an autonomous Riccati equation in such a way that the equilibria of the latter equation stand in 1-1 correspondence with periodic solutions of the former equation. While, because of limitations of time and space, we have confined ourselves here to 4-dimensional systems, it is clear that our methods and results are general and we plan to present the complete theory elsewhere in the near future.

REFERENCES

[1] M. Abramowitz and I. A. Stegun, Handbook of Mathematical Functions, Dover, New York, 1965.

[2] E. Bekir, Thesis, Aerospace Engineering Department, U.S.C., Los Angeles, 1975.

[3] N. Bourbaki, Groups et Algebras de Lie, Chapters 1-3, Actualites Sci. et. Indus., #1285, Paris, 1971.

[4] R. S. Bucy, "The Riccati Equation and Its Bounds", Journal Computer and Syste Science, 6, # 4, 1972, 343-353.

[5] R. S. Bucy, "Structural Stability of the Raccati Equation", to appear SIAM Journal of Control, 1975.

[6] R. S. Bucy and P. D. Joseph, Filtering for Stochastic Processes with Applications to Guidance, Interscience, New York, 1968.

[7] J. R. Canabal, "Geometry of the Riccati Equation",

Stochastics, 1, #2, 1974, 129-149.

[8] J. R. Canabal, "Periodic Geometry of the Riccati Equation", to appear in Stochastics, 1, #4, 1974.

[9] L. Cesari, Sulla stabileta delle soluzioni dei sistemi di equazioni differentiali lineari a coefficienti periodici, Mem. Accad. Italia, 11, 1941, 633-695.

[10] E. Coddington and N. Levinson, Theory of Ordinary Differential Equation, McGraw Hill, New York, 1955.

[11] G. W. Hill, "Researches in Lunar Theory", American Journal of Math., 1, #5, 1878, 129-245.

[12] H. Poincaré, Les Methodes Nouvelles de la Mecanique Celeste, Vol. 2, Gauthier-Villars, Paris, 1892-99; Dover, New York, 1957.

[13] H. Poincaré, "Sur les determinants d'ordre infini", Bull. Soc. Math., France, 14, 1886, 77-90.

[14] F. Riesz, Les systems d'equations lineares a une infini d'inconues, Gauthier-Villars, Paris, 2nd ed., 1952.

[15] C. L. Siegel, "Symplectic Geometry", American Journal of Math., 65, 1943, 1-86.

University of Southern California, Los Angeles, California

ALTERNATIVE METHODS IN NONLINEAR ANALYSIS

Lamberto Cesari

Alternative methods, based on functional analysis, have recently undergone an intensive development through the work of many researchers. The purpose of this lecture is to describe the present state of the subject and to outline recent applications. In particular, we shall show the impact of different lines of thought: Minty's and Brézis's theory of maximal monotone operators in Hilbert spaces, Schauder's principle of invariance of domain, and invariant properties of topological degree. Applications will be outlined to high order nonlinear elliptic problems with nontrivial null space, nonlinear oscillations at resonance, and nonselfadjoint problems in ordinary and partial differential equations.

1. A canonic decomposition.

We are interested in solving an operator equation

(1) $$Fx = 0, \quad x \in S,$$

in a given space S, and when $F = E - N$ is the difference between a linear operator E (not necessarily bounded) and a possibly nonlinear operator N, and thus (1) reduces to $Ex = Nx$. If E^{-1} exists, or equivalently the null space of E is trivial, then this equation can be written as

$$x - E^{-1}Nx = 0,$$

or $(I + KN)x = 0$. This is a Hammerstein equation, and very extensive work has been done in this direction. For references to the wide literature one can see [20].

Whenever KN is compact, concepts and methods of topological degree theory in function spaces are relevant, and we should refer to the work of Leray-Schauder [45], Rothe [65] , and Nagumo [56, 57].

We are particularly interested in the case where E has a nontrivial null space, the case which is often mentioned as the "problem at resonance". Thus we shall not exclude that E may have a nonzero null space. This will be actually the most interesting case.

In the line of the bifurcation process of Poincaré [64], Lyapunov [49] and Schmidt [70] we should decompose (1) into a system of two equations, possibly in different spaces. Much work has been done in this direction, for which we refer to the very substantial papers [73] and [74].

The injection of ideas of functional analysis in the last years has made the process a remarkably fine tool of analysis, particularly in the difficult "problem at resonance" in the usual terminology.

The general theory which has ensued, with all its variants and ramifications, is often referred to as the bifurcation theory, or alternative methods. It is being used in theoretical existence analysis of the solutions, in methods of successive approximations of the solutions, and in estimating the error of approximate solutions.

In terms of functional analysis alone, equation (1) is being decomposed into a system of two equations, one of which, usually called the auxiliary equation, is in an ∞-dimensional space and is often uniquely solvable. By substitution, we are then reduced to the solution of the other equation, usually called the bifurcation, or determining, equation, which possibly lies in a finite dimensional space. We have thus reduced problem (1) to a theoretically much simpler <u>alternative</u> problem, in a finite dimensional space. (However the two equations may also be studied as a whole).

All this is apparent in the decomposition process proposed by Cesari in [7] for Banach spaces and in [8] for Hilbert spaces for selfadjoint problems (1963), as well as in the analogous or modified processes for nonselfadjoint problems which were proposed in short succession. We mention here the one of Locker [47], and those proposed by Hale [30, pp. 262-266], by Hale, Bancroft, and Sweet [31 and 30, pp. 278-287] and by Hale [26, pp. 6-14]. We should also mention here the modification proposed by Gustafson and Sather in [24, 25], and the most recent ones proposed by Osborn and Sather in [59, 60], and by Kannan and Locker [38].

The general lines of this uniform process seems to be the underlying principle in the papers by Antosiewicz [1, 2], Bartle [4], Cesari [14], Cronin [17, 18, 19], Friedrichs [22], Graves [23], Hale [27], Lewis [46], Nirenberg [58],

Perello [61], Rodionov [67], and Vainberg and Trenogin [73], Landesman and Lazer [41].

Recently great improvements in the theory have been made for selfadjoint problems by the use of the process in [7] and [8], maximal monotone operators, Schauder's principle of invariance of domain, properties of topological degree. In order to illustrate these new ideas with the necessary precision, we shall describe here briefly the process in [7] and [8], though referring to work in more general situations.

We shall assume that the operator F in (1) is of the form $F = E - N$, where $E: D(E) \to S$ is a linear operator not necessarily bounded with domain $D(E) \subset S$, and $N: D(N) \to S$ is an operator not necessarily linear with domain $D(N) \subset S$, $D(E) \cap D(N) \neq \emptyset$. If $D \subset D(E) \cap D(N)$, then $F: D \to S$, and equation (1) restricted to D, takes the form

$$Ex = Nx, \quad x \in D. \tag{2}$$

We may think of E as a linear differential operator with given (homogeneous) boundary conditions, and of $D(E)$ as the set of those elements $x \in S$ sufficiently smooth so that E can be applied. But (2) could as well be understood as the equation satisfied by weak solutions in S in a suitable sense.

The process we are presenting here depends first of all on the choice of a projection operator P leading to a decomposition of (1) into a system of two equations of which one is always solvable. We shall illustrate in Section 2 how such a projection operator could be chosen. Here we only show how such a P can be used in suitably decomposing (1) into a system.

Let $P: S \to S$ be a projection operator, that is, a continuous linear operator in S with $PP = P$, or P idempotent. If S_o is the range of P, and S_1 the null space of P, then $S = S_o + S_1$ (direct sum), and every element $x \in S$ has the representation $x = (x_o, x_1)$, or $x = x_o + x_1$, with $x_o = Px \in S_o$, $x_1 = (I - P)x \in S_1$, where I is the identity operator in S. Then obviously equation (1) decomposes into the system of two equations $PFx = 0$, $(I-P)Fx = 0$. If D_o denotes the projection of D on S_o, $D_o = PD \subset S_o$, then we are seeking pairs $x = (x_o, x_1)$, with $x_o \in D_o$, $x \in P^{-1}x_o$, $x_o = (x_o, x_1) \in D$, which satisfy (1). Here $P^{-1}x_o$ can also be written in the form $x_o + S_1$, the "fiber" of P through $x_o \in S_o$.

We assume here that a linear operator $H: S_1 \to S_1$ is known to exist for which we require the following properties:

(h_1) $H(I-P)\,Ex = (I-P)x$ for all $x \in D(E)$;

(h_2) $PEx = EPx$ for all $x \in D(E)$;

(h_3) $EH(I-P)Nx = (I-P)Nx$ for all $x \in S$.

If x is a solution of equation (1), that is, $Fx = 0$, or $Ex = Nx$, then by applying $H(I-P)$ we have $H(I-P)Ex = H(I-P)Nx$, and by force of (h_1) also $(I-P)x = H(I-P)Nx$, or

$$x = Px + H(I-P)Nx.$$

Thus, $x = Tx$, with $T = P + H(I - P)N$. In other words,

any solution of (1) is a fixed point of the transformation T.

Conversely, if x is any fixed point of T, that is, $x = Tx$, then by applying E we have $Ex = EPx + EH(I-P)Nx$, and by force of (h_2) and (h_3) also $Ex = PEx + (I-P)Nx$, or

$$Ex - Nx = P(Ex - Nx),$$

that is, $Fx = PFx$. In other words, any fixed point of T is a solution of (1) if and only if $PFx = 0$, or

$$P(Ex - Nx) = 0.$$

We conclude this analysis by saying that under hypotheses (h_1), (h_2), (h_3), or briefly hypotheses (h), problem (1), or $Ex = Nx$, is equivalent to the system of equations

(3) $$x = x^* + H(I - P)Nx,$$

(4) $$P(Ex - Nx) = 0.$$

Here x^* denotes any element of S_o, and thus $x = (x^*, z)$, $Px = x^*$, $x^* \in S_o$, $z \in S_1$. In other words, by defining T as the map

(5) $$Tx = x^* + H(I-P)Nx, \quad x \in S,$$

then T transforms the fiber $x^* + S_1$ into itself. By writing the unknown x in the form $x = (x^*, z)$, $x^* \in S_o$, $z \in S_1$, we see that equations (3), (4) form a system in the unknowns x^*, z, and this system is equivalent to equation (1).

As we shall see below, for a large class of problems, equation (3) is uniquely solvable for every $x^* \in S_o$. If $x = x(x^*)$, $x^* \in S_o$, is the unique solution of (3), then (4) reduces to an equation $Mx^* = 0$ in the unknown x^* in S_o. Often, S_o is a finite dimensional space, and thus the original problem (1) is reduced to an alternate problem

$Mx^* = 0$ in a finite dimensional space S_o (alternative methods).

Equation (3) is said to be the auxiliary equation, and (4) the bifurcation, or determining equation.

Actually, the auxiliary equation (3) is of the form

$$x - H(I-P)Nx = x^*,$$

or $(I + KN)x = x^*$, that is, again, a Hammerstein equation. Actually, $K = - H(I-P)$, is suitable spaces, is a compact operator (see section 2 below). Great many possibilities are at hand for the study of the auxiliary equation, and, as we shall see, of the bifurcation equation also.

The following main ideas have been repeatedly used in the study of the auxiliary and bifurcation equations:

1. Schauder's fixed point theorem (Cesari [7], 1963, Landesman and Lazer [41],1970);
2. Banach's fixed point theorem (Cesari [8], 1964, and much subsequent work);
3. Invariance properties of topological degree (Cesari [7],1963, Knobloch [39],1964, Mawhin [50], 1969);
4. Brouwer's fixed point theorem, particularly C. Miranda's equivalent form (Cesari [11], 1960, Knobloch [39], 1964);
5. Theory of monotone and maximal monotone operators in Banach spaces (Gustafson and Sather [24], 1972, Cesari and Kannan [16], 1974);

6. Schauder's principle of invariance of domain (Kannan [35], 1974];
7. Implicit function theorem and Newton's "polygon" method in Banach spaces (Sather [69],1970);
8. Decomposition of the main operators in products of operators (square root of a positive operator, Gustafson and Sather [24, 25], 1972; $E = TT^*$ with suitable boundary conditions on T, T^*, Kannan and Locker [38], 1974).

2. The auxiliary equation and Banach's fixed point theorem.

For the sake of simplicity we limit ourselves here to the case where S is a real separable Hilbert space (Cesari [7], [8]). Let $\langle x, y\rangle$ denote the inner product in S and $\|x\| = \langle x, x\rangle^{1/2}$ the norm. (In any application E may be a linear differential operator on some domain $G \subset E^{\nu}$ with given homogeneous linear boundary conditions on the boundary ∂G of G).

We assume here that the associated linear problem $Ex + \lambda x = 0$ has countably many real eigenvalues λ_i and eigenfunctions ϕ_i, $i = 1, 2, \ldots$, $\lambda_i \leq \lambda_{i+1}$, $E\phi_i + \lambda_i\phi_i = 0$, with $\lambda_i \to +\infty$, and $\{\phi_i, i = 1, 2, \ldots\}$ is a complete orthonormal system in S(then, in the situation mentioned above, $S = L_2(G)$)

Thus every element $x \in S$ has a Fourier series $x = \Sigma_1^{\infty} c_i\ \phi_i$, $c_i = \langle x, \phi_i\rangle$, which is convergent in S, and if $x = \Sigma_1^{\infty} d_i\,\phi_i$ is any other element, then

$$\langle x, y\rangle = \Sigma_1^\infty c_i d_i, \quad \|x\| = \langle x, x\rangle^{1/2} = (\Sigma_1^\infty c_i^2)^{1/2}.$$

Since $\lambda_1 \leq \lambda_2 \leq \cdots \leq \lambda_m \leq \cdots$, $\lambda_m \to +\infty$ as $m \to \infty$, there is some integer m so that $\lambda_{m+1} > 0$, and we shall define P by taking

$$Px = \Sigma_1^m c_i \phi_i, \quad c_i = \langle x, \phi_i\rangle.$$

Then the range $S_o = PS$ of P is the m-dimensional space $S_o = \{\phi_1, \ldots, \phi_m\}$ spanned by $\phi_1, \ldots, \phi_m$. If S_1 is the complementary space, then any element $x \in S_1$ has Fourier series $x = \Sigma_{m+1}^\infty c_i \phi_i$, and we can define $H\colon S_1 \to S_1$ by taking

$$\text{(6)} \qquad Hx = -\Sigma_{m+1}^\infty c_i \lambda_i^{-1} \phi_i, \quad x \in S_1.$$

This is possible since we choose m so that $\lambda_{m+1} > 0$ (and thus $\lambda_i > 0$ for all $i \geq m+1$). We have now

$$\text{(7)} \qquad \|Hx\| = (\Sigma_{m+1}^\infty c_i^2 \lambda_i^{-2})^{1/2} \leq \lambda_{m+1}^{-1} (\Sigma_{m+1}^\infty c_i^2)^{1/2} = \lambda_{m+1}^{-1} \|x\|,$$

or $\|Hx\| \leq k_o \|x\|$, $k_o = \lambda_{m+1}^{-1}$ for all $x \in S_1$. We see that $H\colon S_1 \to S_1$ is a linear bounded operator, and that H can always be made to be a contraction $(k_o < 1)$ on S_1 by taking m so that $\lambda_{m+1} > 1$.

It is apparent that P is a linear projector operator, $PP = P$, $\|P\| = 1$. If I denotes the identity operator in S, then $I - P$ is also a projector operator. Finally, it is apparent that H has the properties (h) of Section 1.

The operator H has other interesting properties we shall use later. First

$$(8) \quad \langle -Hx, x\rangle = \Sigma_{m+1}^{\infty} c_i^2 \lambda_i^{-1} \geq \lambda_{m+1} (\Sigma_{m+1}^{\infty} c_i^2 \lambda_i^{-2})$$

$$= \lambda_{m+1} \|Hx\|^2 \geq 0.$$

Moreover H, as defined by (6), is compact in the Hilbert space S (see, e.g., [66], p. 234).

(2.i) <u>Theorem</u> (Cesari [8]). Under the hypotheses above, if $N: S \to S$ is uniformly Lipschitzian of given constant L, that is, $\|Nx_1 - Nx_2\| \leq L\|x_1 - x_2\|$ for all $x_1, x_2 \in S$, then by taking m so that $\lambda_{m+1} > L$ the mapping T defined by (5) is a contraction map of $x^* + S_1$ into itself, and the auxiliary equation (3) is uniquely solvable for every $x^* \in S_o$.

Indeed, $Tx = x^* + H(I-P)Nx$, and thus T maps S into $x^* + S_1$. Then certainly T maps $x^* + S_1$ into itself as mentioned in Section 1. For any two elements $x_1, x_2 \in x^* + S_1$ we also have

$$\|Tx_1 - Tx_2\| = \|H(I-P)(Nx_1 - Nx_2)\| \leq \lambda_{m+1}^{-1} \|Nx_1 - Nx_2\|$$

$$\leq (L\lambda_{m+1}^{-1}) \|x_1 - x_2\|,$$

where $k_o = L\lambda_{m+1}^{-1} < 1$ because of the choice of m.

By Banach's fixed point theorem, there is one and only one element $x = x(x^*) \in x^* + S_1$, such that $x = Tx$, and $x(x^*)$ satisfies the auxiliary equation (3).

Of course, in most applications, N does not map all of S into S, nor N is uniformly Lipschitzian. Indeed, in most cases, N is the Nemitsky operator corresponding to some real function f which may be a polynomial. But then Theorem (2.i) can be easily turned into a theorem of local character by means of a number of devices, for example, the use of an associated second norm, say, a Sup norm, and by restricting the search of solutions in suitable domains of the function space S, say a ball in the Sup norm (Cesari [7, 8]).

This process yields naturally bounds for the solutions. It was then shown that, by displacing the origin at a given approximate solution x_o of the problem, the existence of an exact solution in a rather small ball around x_o could be proved, already for extremely low values of m, yielding therefore error bounds of the approximate solution.

The above theory and many examples have been presented in detail by Cesari in [7] and [8]. Another very simple example of estimation of an error bound of a given approximate solution with m = 1 is shown by Cesari, Borges, and Sanchez in [15].

Another device to overcome the difficulty that N may not map all of S into S and may not be uniformly Lipschitzian, has been used by Shaw [71] for boundary value problems with elliptic partial differential equations of order m, which are not necessarily selfadjoint. Let S be a suitable Sobolev space $W_p^{\ell}(G)$ of functions x(t),

$t \in G$, and let us search for solutions in a given ball $\|x\|_p^{\ell} \leq R$ in the Sobolev norm. Then the indices p, ℓ can be so chosen that by Sobolev's imbedding theorem the functions x and those derivatives of x which actually enter in the operator N are continuous on $G \cup \partial G$ and their Sup norms are not larger than some number R_1 which depends only on G, p, R and ℓ. Then again the essence of Theorem (2.i) can be applied.

Whenever T is a contraction, rapidly convergent methods have been devised for the approximation of both $x^* \in S_o$ and of the solution $x(x^*)$ of the auxiliary equation $Tx = x$, and therefore of the solution x of the original problem $Fx = 0$ (Sanchez [68]), thus improving previous work by Banfi and Casadei.

Finally, the following theoretically important result has been proved. Whenever the structure above is available and T is a contraction, then the Brouwer topological degree of the mapping corresponding to the finite dimensional bifurcation equation and the Leray-Schauder degree for the given problem coincide (Williams [76]).

3. Applications to ordinary differential equations.

Let us consider the problem of existence, approximation and stability of periodic solutions of given period of ordinary differential systems of the form

$$(9) \quad y' + Ay = \varepsilon f(t, y), \; y = (y_1, \ldots, y_n), \; -\infty < t < +\infty,$$

where f is periodic in t of some period L, where ε

is a small parameter, and A is a constant $n \times n$ matrix. The case where the homogeneous system $y' + Ay = 0$ has nontrivial periodic solutions of the same period L is the most important one (problem at resonance). This problem was treated in the years 1955-60 by Hale, Gambill, and others in a series of papers by a process which later was shown to be a particular case of the process of Section 1 above. In particular Hale showed, by this method, that for A and f satisfying suitable symmetry conditions, system (9) may have families of periodic solutions of the given period.

The problem of existence, approximation, and stability of subharmonic solutions of system (9) was treated by the same method, and so were the analogous problems for cycles of autonomous systems

$$y' + Ay = \varepsilon f(y).$$

We refer to the expositions in [10], [11], [30], and [28] for the extensive work on the subject.

Later, by using precisely the process of Section 1, Mawhin and Hale obtained sharper and more satisfactory results,. We refer to the expositions in [50] and [30] also for systems of the form

$$y' + Ay = f(t, y) \tag{10}$$

not necessarily of the perturbation type. We just note that the Fourier series mentioned in section 1 are here exactly the sine and cosine Fourier series

$$y(t) = 2^{-1} a_o + \Sigma_1^\infty (a_n \cos n\omega t + b_n \sin n\omega t),$$
$$\omega = 2\pi/L,$$

and Py is then the mth Fourier polynomial approximation of y:

The problem of periodic solutions is only a particular case of the general boundary value problem

$$(11) \quad \begin{array}{l} y' + A(t)y = f(t, y), \quad y = (y_1, \ldots, y_n), \quad a \leq t \leq b, \\ B_1 y(a) + B_2 y(b) = 0, \end{array}$$

where the second equation represents system of homogeneous linear boundary conditions at a and b. These problems are not necessarily selfadjoint, but the modified schemes of Hale and Locker apply, as well as other analogous schemes. Work has been done by Locker [47], Fabry [21], Nagle [55] and others in this direction, extending most of the results which had been obtained for periodic solutions only.

Also, work was done by Perello [61] and by Rodionov [67] for periodic solutions of differential equations with delay.

We mention here that Knobloch [39] by the process of Section 1, combined with the determination of a priori bounds, the concepts of upper and lower solutions, and the use of C. Miranda's form of Brouwer's fixed point theorem, proved simple qualitative criteria for the existence of periodic solutions of the differential equations

$$y'' = g(t, y, y'), \quad -\infty < t < +\infty .$$

Knobloch [40] used then the same technique also for comparison and oscillation theorems for nonlinear ordinary differential equations.

4. Mawhin's work on periodic solutions and extension to general equations with Fredholm operators.

Again we consider for a moment the problem of periodic solutions of system (10). Under hypotheses, it is possible to give suitable a priori bounds for the periodic solutions, say $|y| \leq R$. On the other hand, by the process of section 1, and by taking m fixed and sufficiently large so as to guarantee that $-H(I-P)N$ has the needed properties throughout the ball $|y| \leq R$, then the entire Leray-Schauder process based on topological degree theory is available to prove the existence of solutions. Mawhin developed this idea rather extensively, and we refer for details to his exposition [52].

More recently, Mawhin has shown that under some additional hypotheses besides those of Sections 1 and 2, and once the separated auxiliary and bifurcation equations (3) and (4) have been obtained, it is possible to combine them in a unique equation that he can discusses by the knowledge of an a priori bound and Leray-Schauder argument. He has applied this remark to the study of periodic solutions of functional differential equations [51].

Mawhin has also extended the above approach to the study of equations Ex = Nx where E is a linear Fredholm operator of any positive index [53]. See also Mawhin's expositions at the 1974 Conferences in Providence and in Los Angeles).

5. Application to ordinary differential equations in the complex field.

Harris, Sibuya, and Weinberg [33] have recently shown that, by the use of a decomposition scheme similar to the one of Section 1, the classical fundamental theorems of Cauchy, Frobenius, Perron, Lettenmeyer foe linear ordinary homogeneous differential equations and systems in the complex field become an immediate consequence of Banach's fixed point theorem on contraction mappings, and of very simple algebraic manipulations. In particular the rather tedious proof by majorants of the convergence of the formal series solutions is eliminated. Harris [32], by the same technique also proved nonlinear versions of the same theorems. Underlying this application is a Banach space of holomorphic functions u(z) in a disk, namely those u(z) which have power series expansion $u(z) = \Sigma_o^\infty c_n z^n$, with $\Sigma_o^\infty |c_n| \delta^n < \infty$ for a given $\delta > 0$, with norm

$$\|u\| = \Sigma_o^\infty |c_n| \delta^n.$$

The projection operator is then defined by $Pu = \Sigma_o^{m-1} c_n z^n$ for some fixed $m \geq 1$. The partial inverse operator $H: S_1 \Rightarrow S_1$ is the formal integration operation

$$H(\Sigma_m^\infty c_n z^n) = \Sigma_m^\infty (n+1)^{-1} c_n z^{n+1} ,$$

and then $\|H\| = (m+1)^{-1}$. By taking m sufficiently large the auxiliary equation is uniquely solvable by Banach's fixed point theorem on contraction maps. Here, the bifurcation equations are simply the usual first m recursive relations for the coefficients of the series solutions (plus the indicial equation for the regular singular case). Mawhin has just presented a very simple argument for the same classical theorems based on Fredholm index.

6. Application to periodic solutions of nonlinear hyperbolic partial differential equations.

Cesari [12] and Hale [29] have discussed the question of periodic solutions of the nonlinear wave equation

$$u_{xx} - u_{yy} = \varepsilon \ f(x, y, u, u_x, u_y)$$

$$u(0, y) = u(2\pi, y), \ u(x, 0) = u(x, 2\pi),$$

where f is 2π-periodic in x and y, and ε is a small parameter. In other words, the question of the existence of solutions of the equation above in a torus. Simple criteria for the existence of such solutions were obtained by essentially the method of Section 1. The difficulty here is that the bifurcation equation is not finite dimensional.

By the same method Petrovanu [62, 63] studied the periodic solutions of other hyperbolic differential equations and systems, and in particular, for Tricomi's differential systems. Hecquet extended this work to higher order differential equations and systems. We mention here also the relevant work of Haimovici along the lines mentioned above, and parallel work of Torelli, DeSimon, and Vejvoda.

7. Monotone and maximal monotone operators.

Let H be a given real Hilbert space,. An operator $T : D(T) \to H$ with domain $D(T) \subset H$ and range $R(T) \subset H$, is said to be monotone if

$$\langle x_1 - x_2, Tx_1 - Tx_2 \rangle \geq 0 \text{ for all } x_1, x_2 \in D(T),$$

where $\langle , \rangle$ denotes the inner product in H. If we allow T to be set valued (or multivalued), $T: D(T) \to 2^H$, each Tx, $x \in D(T)$, being some subset of H, then such a map $T: D(T) \to 2^H$ is said to be monotone provided

$$\langle x_1 - x_2, u_1 - u_2 \rangle \geq 0 \text{ for all } x_1, x_2 \in D(T),\ u_1 \in Tx_1,\ u_2 \in Tx_2 .$$

It is clear that this is actually a property of the graph $\Gamma = [(x, u),\ x \in D(T),\ u \in Tx] \subset H \times H$, as a subset of $H \times H$.

We shall say that any subset Γ of $H \times H$ is a monotone set provided $\langle x_1 - x_2, u_1 - u_2 \rangle \geq 0$ for all $(x_1, u_1), (x_2, u_2) \in \Gamma$. At this level of generality we see that any (set valued) map $T: D(T) \to 2^H$ has an inverse $T^{-1} : R(T) \to 2^H$ (also a set valued map), and that T is monotone if and only if T^{-1} is monotone, and if and only if their graph Γ is a monotone set.

A monotone set Γ is said to be maximal monotone if Γ is not a proper subset of another monotone set

$\Gamma' \supset \Gamma$ in $H \times H$. A (set valued) map $T: D(T) \to 2^H$ is said to be maximal monotone if its graph is a maximal monotone set.

In other words, if we partial order the subsets of $H \times H$ by set inclusion, then a set Γ is said to be maximal monotone if Γ is monotone, and in any orderd collection of subsets of $H \times H$, containing Γ and all monotone, then Γ is the maximal element. We see, by Zorn's lemma, that any monotone subset Γ of $H \times H$ is certainly contained in some maximal monotone set Γ' (and then $\Gamma' = \Gamma$ if Γ is maximal monotone).

The last remarks can be reworded by saying that for any map $T: D(T) \to 2^H$ which is monotone, but not maximal monotone, there is another map $T': D(T') \to 2^H$ which is maximal monotone, with $D(T) \subset D(T')$, and $Tx \subset T'x$ for all $x \in D(T)$.

Thus, a map $T: D(T) \to 2^H$ is maximal monotone if and only if its inverse T^{-1} is maximal monotone.

A map $T: D(T) \to 2^H$ is said to be strictly monotone if

$$\langle x_1 - x_2, u_1 - u_2 \rangle > 0 \text{ for all } x_1, x_2 \in D(T),\ u_1 \in Tx_1, u_2 \in Tx_2,\ x_1 \neq x_2.$$

A map $T: D(T) \to 2^H$ is said to be strongly monotone if there is a constant $c > 0$ such that

$$\langle x_1 - x_2, u_1 - u_2 \rangle \geq c \, \|x_1 - x_2\|^2 \text{ for all } x_1, x_2 \in D(T),\ u_1 \in Tx_1,\ u_2 \in Tx_2.$$

A map $T: D(T) \to 2^H$ is said to be coercive if $\|x\|^{-1}$ $\langle x, u \rangle \to +\infty$ as $\|x\| \to \infty$, $x \in D(T)$, $u \in Tx$; briefly, if $\|x\|^{-1} \langle Tx, x \rangle \to +\infty$ as $\|x\| \to \infty$. The definition of course is of interest only if $D(T)$ is unbounded.

It is easy to prove that a strongly monotone map is coercive. We also mention here that any (nonlinear) operator T, or N, is said to be bounded if it maps bounded sets into bounded sets.

For the theory of maximal monotone operators in Hilbert spaces we refer to Brézis [5]. We only mention here that a monotone map $T: D(T) \to 2^H$ is maximal monotone if and only if $R(I + T) = H$ (Minty [54]). Also, if $T: D(T) \to 2^H$ is maximal monotone and coercive, then $R(T) = H$.

The sum $A + B$ of monotone maps with $D(A) \cap D(B) \neq \emptyset$ is monotone as a map defined on $D(A) \cap D(B)$. If both A and B are maximal monotone, then $A + B$ is also maximal monotone under a set of alternate mild hypotheses, for instance, if $\text{Int}\, D(A) \cap D(B) \neq \emptyset$ (see Crandall, Pazy, and Brézis [6] for a recent proof).

A map $T: D(T) \to H$ is said to be hemicontinuous if T is continuous on every line segment s of $D(T)$ (with respect to the strong topology on s and the weak topology on the range). A monotone hemicontinuous map $T: H \to H$ is maximal monotone.

8. Auxiliary equation and monotone maps.

We shall write the auxiliary equation (3) in either form

(12) $$x - H(I - P)Nx = x^*,$$

(13) $$[I - H(I - P)N]\, x = x^*,$$

that is, in the form $x + KNx = x^*$ of a Hammerstein equation. As mentioned a number of results from Hammerstein equation theory are pertinent here. However, we present results obtained by maximal monotone operator theory. We use the notations of Section 2 and m is chosen so that $\lambda_{m+1} > 0$.

(8. i) <u>Theorem</u> (Cesari and Kannan [16]) Under hypotheses (h), if N is monotone and hemicontinuous, then the auxiliary equation (3) has always one and one solution $x = x(x^*)$, $x \in x^* + S_1$, for every $x^* \in S_o$.

<u>Proof.</u> We consider here the operators $N: S \to S$ and $-H(I-P): S \to S_1$. Actually, $-H$ is a one-to-one map from S_1 onto S_1. If for any $z \in S_1$ we write $z = -Hy$, $y \in S_1$, then $-H(I-P)(y+S_o) = z$, and $[-H(I-P)]^{-1}z = y+S_o$. Thus, $[-H(I-P)]^{-1}$ is a set-valued map defined for each $z \in S_1$ with values $y + S_o$. Also, we see that (12) certainly implies

(14) $$[-H(I - P)]^{-1}(x - x^*) + Nx \ni 0, \; x \in x^* + S_1.$$

Conversely, if (14) holds, then $x-x^* -H(I-P)Nx$ is in S_o and since $x-x^* \in S_1$ and $S_o \cap S_1 = 0$, we derive $x- x^* -H(I-P)Nx = 0$. Thus equations (12) and (14) are equivalent.

Equation (14) is of the form $(A + B)x \ni 0$, where $A: x^* + S_1 \to S$, $B: S \to S$ are the operators defined by $Ax = [-H(I-P)]^{-1}(x-x^*)$ for $x \in x^* + S_1$, and by $Bx = Nx$ for $x \in S$ and A is thus a set-valued map. Since $-H(I-P)$: $S \to S_1 \subset S$ is monotone and continuous, then $-H(I-P)$ is maximal monotone, and thus so is its inverse $A = [-H(I-P)]^{-1}: S_1 \to S$. By hypothesis, N is monotone and hemicontinuous. Thus, $B=N$ also is maximal monotone. Finally, $D(A) = S_1 \subset S = D(B)$, where S is the whole space. Then, $A+B$ is maximal monotone. Let us prove that A is coercive. Indeed, for $z \in Ay = [-H(I-P)]^{-1}y$ we have $y = -H(I-P)z$, $\|y\| = \|-H(I-P)z\|$, and by (8) also

$$\|y\|^{-1}\langle z, y\rangle = \|y\|^{-1}\langle z, -H(I-P)z\rangle \geq \|y\|^{-1}\lambda_{m+1}\|-H(I-P)z\| = \lambda_{m+1}\|y\| \to \infty$$

as $\|y\| \to \infty$. Since A is coercive, so is $A+B$. We conclude that $R(A+B) = S$. Hence, equation $(A+B)x \ni 0$ is solvable, that is, (14) is solvable, and (12) is solvable.

Let us prove that, for every $x^* \in S_0$, (12) has a unique solution. Indeed, if x_1, x_2 were two solutions, then

$$x_1 - x^* = H(I-P)(Nx_1), \qquad x_2 - x^* = H(I-P)(Nx_2), \quad \text{and}$$

$$\begin{aligned} 0 &\leq \langle Nx_1 - Nx_2, x_1 - x_2 \rangle \\ &= -\langle Nx_1 - Nx_2, -H(I-P)Nx_1 + H(I-P)Nx_2\rangle \\ &\leq -\lambda_{m+1}\|-H(I-P)Nx_1 + H(I-P)Nx_2\|^2 \\ &= -\lambda_{m+1}\|x_1 - x_2\|^2, \end{aligned}$$

where $\lambda_{m+1} > 0$ and we have used property (8) of the operator H. Thus, $x_1 = x_2$. We have proved that the auxiliary equation (12), or (13), has a unique solution $x \in x^* + S_1$ for every $x^* \in S_o$. We shall indicate it by

$$(15) \qquad x = [I - H(I - P)N]^{-1} x^*, \qquad x^* \in S_o.$$

(8.ii) <u>Theorem</u> (Cesari and Kannan [16]) Under the hypotheses of Sect. 2, if $\lambda_i > 0$ for all i, and $N:S \to S$ is monotone and hemicontinuous, then the given equation $Ex = Nx$ has a unique solution.

If $\lambda_i \geq 0$ and $N: S \to S$ is monotone, hemicontinuous, and coercive, then the given equation $Ex = Nx$ has at least one solution.

If some λ_i are negative, say $\lambda_1 \leq \cdots \leq \lambda_m \leq 0 < \lambda_{m+1} \leq \ldots$, if $N:S \to S$ is hemicontinuous and bounded, and if $\langle Nx-Ny, x-y\rangle \geq c\|x-y\|^2$ for all $x, y \in S$ and some constant $c > 0$, $c > -\lambda_1$, then the given equation $Ex = Nx$ again has at least one solution.

<u>Proof.</u> (a) If $\lambda_i > 0$ for all i, that is, $0 < \lambda_1 \leq \lambda_2 \leq \ldots, \lambda_m \to +\infty$ as $m \to \infty$, then we can take in Section 1, $m = 0$, $S_o = \{0\}$, $S_1 = S$, and the given equation $Ex = Nx$ is reduced to the only auxiliary equation. Then our statement is a corollary of (8.i).

(b) If some λ_i are zero, $0 = \lambda_1 \leq \lambda_2 \leq \ldots$, let m be the smallest integer such that $\lambda_{m+1} > 0$. We reduced

$Ex = Nx$ to the system of the auxiliary and bifurcation equations (3) and (4). For any $x^* \in S_o$ the auxiliary equation has a unique solution given by (15). Then the bifurcation equation $P(E-N)x = 0$ becomes

$$PN[I - H(I-P)N]^{-1} x^* - PEx = 0.$$

Since $PEx = EPx = Ex^* = PEx^*$, we have the equation

$$PN[I-H(I-P)N]^{-1} x^* - PEx^* = 0,$$

or $Tx^* = PEx^*$, where $T: S_o \to S_o$ is the operator defined by $T = PN[I-H(I-P)N]^{-1}$.

Let us prove that T is monotone. Indeed, if $x^*, y^* \in S_o$ and we take $u = [I-H(I-P)N]^{-1} x^*$, $v = [I-H(I-P)N]^{-1} y^*$, then $u - H(I-P)Nu = x^*$, $v-H(I-P)Nv = y^*$, and

$$\begin{aligned} (16) \qquad & \langle Tx^* - Ty^*, x^* - y^* \rangle = \langle PNu-PNv, x^*-y^* \rangle \\ & = \langle Nu-Nv, x^*-y^* \rangle = \langle Nu-Nv, u-v \rangle \\ & + \langle Nu-Nv, -H(I-P)Nu + H(I-P)Nv \rangle \geq 0, \end{aligned}$$

since both N and $-H(I-P)$ are monotone operators. Thus, T is monotone.

Let us prove that T is maximal monotone. It is enough to prove that $R(I+T) = S_o$. In other words, we have to prove that, for every $a^* \in S_o$ the equation

$$(17) \qquad x + PN[I - H(I-P)N]^{-1} x = a^*$$

is solvable in S_o. Actually, we shall first show that the equation

$$x - H(I-P)Nx = a^* - PNx \tag{18}$$

is solvable. Indeed, this equation is equivalent to

$$[-H(I-P) + P]^{-1}(x-a^*) + Nx \ni 0. \tag{19}$$

The same argument as for equation (14) shows that (19) and hence (18) is solvable. From (18) now, for $b^* = a^* - PNx$, we obtain $x - H(I-P)Nx = b^*$, and also

$$x = [I - H(I-P)N]^{-1} b^*,$$

$$PN[I - H(I-P)N]^{-1} b^* = PNx = a^* - b^*,$$

$$b^* + PN\,[I - H(I-P)N]^{-1} b^* = a^*.$$

Thus, equation (17) is solvable for every $a^* \in S_o$; hence, $R(I + T) = S_o$ and T is maximal monotone.

We have written the bifurcation equation in the form $(T-E)x^* = 0$, $x^* \in S_o$, where $T: S_o \to S_o$ is maximal monotone and so is $-E: S_o \to S_o$. Thus, $T-E: S_o \to S_o$ is maximal monotone, and it remains to prove that $T-E$ is coercive. For every $x^* \in S_o$ we have now

$$\langle Tx^* - Ex^*, x^*\rangle = \langle Tx^*, x^*\rangle + \langle +Ex^*, x^*\rangle \geq \langle Tx^*, x^*\rangle$$

$$= \langle PN[I-H(I-P)N]^{-1} x^*, x^*\rangle = \langle PNa, x^*\rangle,$$

where $a = [I-H(I-P)N]^{-1}x^*$, or $x^* = a - H(I-P)$, so that, since $x^* \in S_o$, we have $x^* = Px^* = Pa$. Thus

$$\langle Tx^* - Ex^*, x^* \rangle \geq \langle PNa, x^* \rangle = \langle Na, x^* \rangle$$

$$\geq \langle Na, a \rangle + \langle Na, -H(I-P)Na \rangle \geq \langle Na, a \rangle .$$

Also, $\|x^*\| = \|Pa\| \leq \|a\|$ so that

$$\|x\|^{-1} \langle Tx^* - Ex^*, x^* \rangle \geq \|a\|^{-1} \langle Na, a \rangle .$$

Since $\|x^*\| \to \infty$ implies that $\|a\| \to \infty$, and since N is coercive, we conclude that $\|x\|^{-1} \langle Tx^* - Ex^*, x^* \rangle \to \infty$ as $\|x\| \to \infty$. We have proved that T - E is maximal monotone and coercive, thus $R(T-E) = S_o$, and the equation $(T-E)x^* = 0$ is solvable.

(c) Let us assume now that some λ_i are negative, and as usual let m be the smallest integer such that $\lambda_{m+1} > 0$. The proof in (b) above remains essentially the same for what concerns T. Still we have to prove that T-E is monotone, maximal monotone, and coercive. From (16) we have now

$$\langle Tx^* - Ty^*, x^* - y^* \rangle \geq \langle Nu - Nv, u - v \rangle \geq c \|u-v\|^2,$$

with $u - H(I-P)Nu = x^*$, $v - H(I-P)Nv = y^*$, hence $(u-v) - H(I-P)(Nu-Nv) = x^*-y^*$, where $x^* - y^* \in S_o$ and $-H(I-P)(Nu-Nv) \in S_1$. Thus, $\|u-v\| \geq \|x^*-y^*\|$, and

$$\langle Tx^* - Ty^*, x^* - y^* \rangle \geq c \|x^* - y^*\|^2 .$$

On the other hand, for $x^* = \Sigma_1^m c_i \phi_i$ we have $Ex^* = - \Sigma_1^m c_i \lambda_i \phi_i$, thus $\langle -Ex^*, x^* \rangle = \Sigma_1^m c_i^2 \lambda_i \geq \lambda_1 \|x^*\|^2$, and finally

$$\langle -Ex^* + Ey^*, x^*-y^*\rangle \geq \lambda_1 \|x^*-y^*\|^2,$$

$$\langle (T-E)x^* - (T-E)y^*, x^*-y^*\rangle \geq (c + \lambda_1) \|x^*-y^*\|^2,$$

where $c + \lambda_1 > 0$. We have proved that $T-E$ is monotone and coercive. Note that T is continuous if N is bounded and E is certainly a continuous operator restricted to S_o. Thus $T-E: S_o \Rightarrow S_o$ is a continuous map defined on the whole of S_o. Thus, $T-E$ is maximal monotone, and coercive as proved above. Hence, $R(T-E) = S_o$, and equation $(T-E)x^* = o$ is solvable.

Remark 1. Both the second and third case in the theorem above allow for $\lambda_i = 0$ for some i. This is the "resonance case". The interest of the theorem (8.ii) is that such a resonance situation is treated exactly by the same general argument. This case, where some λ_i are zero is the most interesting one and is characterized by E having a nontrivial null space. This case occurs for a problem of the form $Ex + \lambda_m x = Nx$, where λ_m is an eigenvalue of E, and thus $E' = E + \lambda_m I$ has a nontrivial null space.

Osborn and Sather [59, 60] have just announced an analogue of Theorem (8.i) for nonselfadjoint problems.

9. Schauder's principle of invariance of domain.

Schauder's principle is generally stated as follows:

(9.i) If $T:H \to H$ is a weakly compact map, if $I + T$ is one-to-one, and $(I+T)^{-1}$ is bounded, then $R(I+T) = H$ and $(I+T)^{-1}$ is single valued.

Here H denotes a real Hilbert space, and by T compact we mean as usual that every bounded sequence $[u_n]$ is mapped into a relatively compact sequence $[Tu_n]$. By $(I+T)^{-1}$ bounded we mean that $(I+T)^{-1}$ maps bounded sets into bounded sets. The conclusion that $R(I+T) = H$ and that $(I+T)^{-1}$ is single valued can be rewarded by saying that for every $v \in H$ there is one and only one element $u \in H$ such that $(I+T)u = v$. The important statement (9.i) is essentially proved in the form above by Nagumo [57], but can be traced back to Schauder. The proof of (9.i) is based on the concept and properties of topological degree (see Nagumo [56, 57]).

10. The auxiliary equation and Schauder's principle.

(10.i) Theorem (Kannan [35]). Under the hypotheses of Sect. 2, if $N:S \to S$ is continuous and bounded, and if there are numbers p, q, m, $p > 0$, m integer, such that $\lambda_m < q \leq p < \lambda_{m+1}$, and

$$\langle Nx - Ny, x-y \rangle \geq -p \, \|x-y\|^2 \quad \text{for all } x, y \in S; \tag{20}$$

then the auxiliary equation is uniquely solvable. If in addition

$$\langle Nx - Ny, x^* - y^* \rangle \leq -q \, \|x^* - y^*\|^2 \tag{21}$$

for all $x^*, y^* \in S_o$ and corresponding solutions x, y of the auxiliary equation, then the given equation $Ex = Nx$ also has a unique solution.

Proof. (a) Let $T = -H(I-P)N : S \to S_1$. First we note that any bounded sequence $[x_n]$ is mapped by N into a bounded sequence $[Nx_n]$, and then by $-H(I-P)$ into a convergent sequence since H is compact. Thus, T is compact.

Let us prove that I+T is one-to-one. Indeed, let us assume, if possible, that we have $(I+T)x = (I+T)y$, or

$$x - H(I-P)Nx = y - H(I-P)Ny, \qquad x, y \in S.$$

Then, by (20) and (8), we have

$$\begin{aligned} -p\|x - y\|^2 &\leq \langle Nx - Ny, x - y\rangle \\ &= -\langle Nx - Ny, -H(I-P)Nx + H(I-P)Ny\rangle \\ &\leq -\lambda_{m+1} \|-H(I-P)Nx + H(I-P)Ny\|^2 \\ &= -\lambda_{m+1} \|x - y\|^2, \end{aligned}$$

a contradiction, since $-p > -\lambda_{m+1}$. Thus, I+T is one-to-one.

Let us prove that $(I+T)^{-1}$ is bounded. For $(I+T)^{-1}y = x$, we have $y = (I+T)x$, or $y = x - H(I-P)Nx$. Then, by (8) we have

$$\begin{aligned} \lambda_{m+1} \|y - x\|^2 = \lambda_{m+1} \|-H(I-P)Nx\|^2 &\leq \langle Nx, -H(I-P)Nx\rangle \\ &= \langle Nx, y-x\rangle = -\langle Ny-Nx, y-x\rangle + \langle Ny, y-x\rangle . \end{aligned}$$

By hypothesis we have now

$$\begin{aligned} \lambda_{m+1} \|y-x\|^2 &\leq p\|y-x\|^2 + \langle Ny, y-x\rangle, \\ (\lambda_{m+1} - p)\|y-x\|^2 &\leq \|Ny\| \; \|y-x\|, \\ \|y-x\| &\leq (\lambda_{m+1} - p)^{-1} \|Ny\|. \end{aligned}$$

Thus $\|x\| \leq \|y\| + (\lambda_{m+1} - p)^{-1} \|Ny\|$. Since N is bounded, then $\|y\| \leq R$ for some R implies $\|x\| \leq R'$ for some R', that is, $(I+T)^{-1}$, or $[I - H(I-P)N]^{-1}$, is a bounded map.

By force of Schauder's principle (9.i), the equation $x + Tx = x^*$, or $x - H(I-P)x = x^*$, has a unique solution x in S for every $x^* \in S_o$, and necessarily $x \in x^* + S_1$.

(b) From the previous part (a) we know that the auxiliary equation has a unique solution $x = x(x^*)$, $x \in x^* + S_1$, for every $x^* \in S_o$. It remains to show that the bifurcation equation has a unique solution.

Let $T': S_o \to S_o$ denote the operator defined by $T'x^* = Ex^* - PN[I-H(I-P)N]^{-1}x^*$, so that the bifurcation equation reduces to $T'x^* = 0$. Let x^*, y^* be any two elements of S_o and let x, y denote the corresponding solutions of the auxiliary equation. Thus

$$[I - H(I-P)N]^{-1}x^* = x, \qquad [I - H(I-P)N]^{-1}y^* = y,$$

$$\langle T'x^* - T'y^*, x^* - y^* \rangle = \langle Ex^* - Ey^*, x^* - y^* \rangle -$$

$$(22) \quad \langle PNx - PNy, x^* - y^* \rangle = \langle Ex^* - Ey^*, x^* - y^* \rangle$$

$$- \langle Nx - Ny, x^* - y^* \rangle$$

Note that, for $x^* \in S_o$, or $x^* = \Sigma_1^m c_i \phi_i$, $Ex^* = -\Sigma_1^m c_i \lambda_i \phi_i$, and $\langle Ex^*, x^* \rangle = -\Sigma_1^m c_i^2 \lambda_i \geq -\lambda_m \|x^*\|^2$. Hence, from (22) and (21) we derive $\langle T'x^* - T'y^*, x^* - y^* \rangle \geq -\lambda_m \|x^* - y^*\|^2 + q\|x^* - y^*\|^2 = (q - \lambda_m) \|x^* - y^*\|^2$, where $q - \lambda_m > 0$. We have proved that $T': S_o \to S_o$ is strongly monotone and therefore coercive.

Let us prove that T' is continuous on S_o. We have only to show that $[I - H(I-P)N]^{-1}$ is continuous. With the same notations as in (22) we have

$$[I - H(I-P)N]^{-1} x^* = x, \qquad [I - H(I-P)N]^{-1} y^* = y,$$

$$x - H(I-P)Nx = x^* \quad , \qquad y - H(I-P)Ny = y^* ,$$

$$x - y = (x^* - y^*) - H(I-P)(Nx - Ny),$$

where the two terms in the last member are in S_o and S_1 respectively. Hence,

$$\|x - y\|^2 = \|x^* - y^*\|^2 + \|-H(I-P)(Nx - Ny)\|^2. \tag{23}$$

By (20), (21), (8) and (23) we have now

$$\begin{aligned} -q\|x^* - y^*\|^2 &\geq \langle Nx - Ny, x^* - y^*\rangle \\ &= \langle Nx - Ny, x - y\rangle \\ &\quad + \langle Nx - Ny, -H(I-P)(Nx-Ny)\rangle \\ &\geq -p\|x - y\|^2 + \lambda_{m+1} \|-H(I-P)(Nx-Ny\|^2 \\ &= -p\|x^* - y^*\|^2 + (\lambda_{m+1} - p)\|-H(I-P)(Nx-Ny)\|^2. \end{aligned}$$

Thus,

$$\|-H(I-P)(Nx - Ny)\|^2 \leq (\lambda_{m+1} - p)^{-1}(p-q)\|x^* - y^*\|^2,$$

and again by (23) also

$$\|x-y\|^2 \leq [1 + (\lambda_{m+1} - p)^{-1}(p-q)]\|x^* - y^*\|^2.$$

We have proved that T' is continuous. Since $T' : S_o \to S_o$

is monotone and S_o is the whole space for T', we conclude that T' is maximal monotone, and, as we have proved above, T' is also coercive. Thus $R(T') = S_o$; hence, the equation $T'x^* = 0$ is solvable in S_o.

Let us prove that the bifurcation equation has only one solution. Indeed, if $x^*, y^* \in S_o$ were two solutions, that is, $T'x^* = 0$, $T'y^* = 0$, and we denote by x, y the corresponding solutions of the auxiliary equation, then

$$[I - H(I-P)N]^{-1}x^* = x, \qquad I - [H(I-P)N]^{-1}y^* = y,$$

$$Ex^* - PN[I - H(I-P)N]^{-1}x^* = 0, \quad Ey^* - PN[I - H(I-P)N]^{-1}y^*$$

$$= 0, \; Ex^* - Ey^* = PNx - PNy .$$

Hence,

$$-\lambda_m \|x^* - y^*\|^2 \leq \langle Ex^* - Ey^*, x^* - y^*\rangle$$

$$= \langle PNx - PNy, x^* - y^*\rangle$$

$$= \langle Nx - Ny, x^* - y^*\rangle \leq -q\|x^* - y^*\|^2,$$

a contradiction, since $-q < -\lambda_m$. This proves the uniqueness, and (10. i) is proved.

Remark 2. It may be of interest to consider the problem "at resonance"

(24) $$Ex + \lambda_m x = Nx,$$

where λ_m is any eigenvalue of the operator E. If $E' = E + \lambda_m I$, then by suitable indexing, we can assume that $\lambda_m < \lambda_{m+1}$, and then the eigenvalues of E' are

$\lambda_1 - \lambda_m \leq \cdots \leq \lambda_m - \lambda_m = 0 < \lambda_{m+1} - \lambda_m \leq \cdots$. Let $\varepsilon > 0$ be any number, $0 < \varepsilon < 2^{-1}(\lambda_{m+1} - \lambda_m)$, and take $q = \varepsilon$, $p = \lambda_{m+1} - \lambda_m - \varepsilon$. Then (24) has a unique solution if we require for N to satisfy

$$\langle Nx - Ny,\ x - y\rangle \geq -p \|x - y\|^2 \text{ for all } x, y \in S, \text{ and}$$

$$\langle Nx - Ny,\ x^* - y^*\rangle \leq -q \|x^* - y^*\|^2 \text{ for all } x^*, y^* \in S_o$$

and the same conventions as above.

In view of problems "at resonance", the following statement is of interest. We word it as a modification of Theorem (8.ii), and then we transfer the conclusion to the case (24) in the Remark which follows. The proof of this theorem makes straightforward use of properties of the topological degree.

(10.ii) Theorem (Kannan [35]). Under the hypotheses of Sect. 2, if $\lambda_1 \leq \cdots \leq \lambda_m \leq 0 < \lambda_{m+1} \leq \cdots$, if (a) $N: S \to S$ is bounded and continuous; (b) $\langle Nx - Ny,\ x-y\rangle \geq -c \|x-y\|^2$ for some $c < \lambda_{m+1}$ and all $x, y \in S$; (c) there is $R > 0$ such that $\langle Nx,\ x^*\rangle \leq 0$ for all $x^* \in S_o$ with $\|x^*\| = R$ and all $x \in S$ with $Px = x^*$, then the given equation $Ex = Nx$ has at least one solution.

Proof. Using the integer m above, we decompose $Ex = Nx$ in the usual auxiliary and bifurcation equations. Let us prove that the auxiliary equation is uniquely solvable for every $x^* \in S_o$. We write it in the usual form

$$-H(I-P)^{-1}(x - x^*) + Nx \ni 0,$$

we take $y = x - x^*$, and we denote by $T'' : S_1 \Rightarrow S$ the operator defined by

$$T''y = [-H(I-P)]^{-1} y + N_{x^*}(y),$$

where $N_{x^*}(y) = N(x^* + y)$. For $y_1 = -H(I-P)(x_1-x^*)$, $y_2 = -H(I-P)(x_2-x^*)$, we have as usual

$$\begin{aligned}\langle T''y_1 - T''y_2, y_1-y_2\rangle &= \langle x_1-x_2, -H(I-P)x_1 + H(I-P)x_2\rangle \\ &\quad + \langle N(x^*+y_1)-N(x^*+y_2), (x^*+y_1)-(x^*+y_2)\rangle \\ &\geq \lambda_{m+1} \|y_1-y_2\|^2 - c\|y_1-y_2\|^2 \\ &= (\lambda_{m+1}-c)\|y_1-y_2\|^2.\end{aligned}$$

Thus, T'' is strongly monotone and coercive. Moreover, T'' is the sum of $N_{x^*}: S \to S$, hemicontinuous, monotone, hence maximal monotone, and of $[-H(I-P)]^{-1} : S_1 \to S$, monotone, maximal monotone, with S_1 a subset of the whole space S. Hence, T'' is maximal monotone. Thus, the auxiliary equation is solvable. Since T'' is strongly monotone, the soluti on is unique.

Let us prove that the bifurcation equation also is solvable. The bifurcation equation (4), which we had written in the form

$$PN[I - H(I-P)N]^{-1} x^* - PEx^* = 0,$$

can also be written in either form

$$Ex^* - PN[I - H(I-P)N]^{-1}x^* = 0, \text{ or}$$

$$x^* + P[E - I - PN(I - H(I-P)N)^{-1}]x^* = 0,$$

that is, $(I+T_o)x^* = 0$, where $T_o : S_o \to S_o$ is a bounded map in the finite dimensional space S_o; hence, T_o is compact. Let V denote the closed ball $V = [x^* \in S_o \mid \|x^*\| \leq R]$. Let us prove that for $0 \leq t < 1$, the equations $x^* + tT_o x^* = 0$ cannot have solutions x^* with $\|x^*\| = R$. This is evident for $t = 0$. Suppose if possible that there is such a solution for $0 < t < 1$, say $x^* \in S_o$, $\|x^*\| = R$, $x^* + tT_o x^* = 0$. Then

$$(25) \qquad 0 = \langle 0, T_o x^* \rangle = \langle x^*, T_o x^* \rangle + t\langle T_o x^*, T_o x^* \rangle .$$

If $u = [I - H(I-P)N]^{-1} x^*$, then $Pu = x^*$, and

$$\langle T_o x^*, x^* \rangle = \langle Ex^*, x^* \rangle - \|x^*\|^2 - \langle PNu, x^* \rangle$$

$$\geq (-\lambda_m) \|x^*\|^2 - \|x^*\|^2 - \langle Nu, x^* \rangle .$$

Since $-\lambda_m \geq 0$, and $\langle Nu, x^* \rangle \leq 0$, we also have $\langle T_o x^*, x^* \rangle \geq -\|x^*\|^2$. By (25) we have then

$$0 \geq -\|x^*\|^2 + t\|T_o x^*\|^2, \quad \text{with } T_o x^* = -t^{-1} x^*.$$

Hence, $0 \geq -\|x^*\|^2 + t^{-1}\|x^*\|^2$, or $\|x^*\| = 0$, a contradiction.

We have proved that $x^* + tT_o x^* = 0$ has no solution x^* with $\|x^*\| = R$ for $0 \leq t < 1$. Thus, either $x^* + T_o x^* = 0$ for some $\|x^*\| = R$, or $x^* + TT_o x^* \neq 0$ for all $0 \leq t \leq 1$, $\|x^*\| = R$. In the latter case, by topological degree theory (see Nagumo [56, 57]), the equation $x^* + T_o x^* = 0$ must have a solution x^* with $\|x^*\| < R$. In any case, $(I + T_o)x^* = 0$ has a solution x^* with $\|x^*\| \leq R$.

Remark 3. For the problem "at resonance"

$$Ex + \lambda_m x = Nx, \tag{26}$$

where λ_m is any eigenvalue of E, we can state now that (26) has a solution provided (a) $N: S \to S$ is continuous and bounded, (b) $\langle Nx - Ny, x-y\rangle \geq -c\,\|x-y\|^2$ for some constant $c < \lambda_{m+1} - \lambda_m$ and all $x, y \in S$; (c) there is $R > 0$ such that $\langle Nx, x^*\rangle \leq 0$ for all $x^* \in S_0$ with $\|x^*\| = R$ and all $x \in S$ with $Px = x^*$.

Indeed, as in Remark 2, we can choose the labeling in such a way that the operator $E' = E + \lambda_m I$ has eigenvalues $\lambda_1 - \lambda_m \leq \cdots \leq \lambda_m - \lambda_m = 0 < \lambda_{m+1} - \lambda_m \leq \cdots$, and now we have only to apply Theorem (10. ii).

11. Application to oscillations of conservative systems.

We consider here the problem of existence of 2π-periodic solutions for the nonlinear differential system

$$x'' + \operatorname{grad} G(x) = p(t), \tag{27}$$

or

$$d^2x_i/dt^2 + \partial G/\partial x_i = p_i(t), \quad -\infty < t < +\infty,$$

$$i = 1, \ldots, n,$$

where $x = (x_1, \ldots, x_n)$, $p(t) = (p_1, \ldots, p_n)$ is continuous and 2π-periodic in t, and $G: R^n \to R$ is a real-valued function defined in R^n and of class C^2. These equations can be interpreted as the equations of motion of a mechanical system subject to conservative internal forces and a periodic external force of given period.

Let S be the space of n-vector functions whose components are L^2-integrable in $[0, 2\pi]$, with inner product defined by

$$\langle x, y\rangle = \int_0^{2\pi} x(t) \cdot y(t)dt,$$

where $x \cdot y$ denotes the usual inner product in R^n, $x \cdot y = x_1y_1 + \cdots + x_ny_n$. Let S_E denote the set of all $x \in S$ which are absolutely continuous in $[0, 2\pi]$ together with x', and with $x'' \in L_2[0, 2\pi]$. Let $E: S_E \to S$ denote the linear differential operator defined by $Ex = x''$, $x \in S_E$. Let $N: S \to S$ be defined by $Nx(t) = -\operatorname{grad} G(x(t)) + p(t)$ under hypotheses on G to be given later, which will guarantee first of all that $N: S \to S$. Then (27) can be written as

$$Ex = Nx,$$

where we associate to E the boundary conditions $x(0) = x(2\pi)$, $x'(0) = x'(2\pi)$. The corresponding homogeneous problem $Ex + \lambda x = 0$ with the same boundary conditions has a countable set of eigenvalues $0, 1^2, 2^2, \ldots$, each repeated a suitable number of times, say briefly, λ_i, $i = 1, 2, \ldots$, with eigenfunctions ϕ_i, $i = 1, 2, \ldots$, or $\phi_i = (\phi_{i1}, \ldots, \phi_{in})$, which form a complete orthonormal system in S. Thus, any element $x \in S$ has a Fourier series $x = \Sigma_1^\infty c_i\phi_i$ with $c_i = \langle x, \phi_i\rangle$. For any integer m, let P denote the projection operator defined by $Px = \Sigma_1^m c_i\phi_i$ so that $S_o = PS$, the range of P, is the m-dimensional space

S_o spanned by $\phi_1, \dots, \phi_m$. We shall take $m \geq 1$ so that all ϕ_i corresponding to certain eigenvalues $0, 1^2, \dots, h^2$ are all in S_o, $h \geq 0$. For what follows, we could well take $h = 0$.

If $S_1 = (I-P)S$, then any $x \in S_1$ has Fourier series $x = \Sigma_{m+1}^{\infty} c_i \phi_i$, and we define $H: S_1 \to S_1$ as the linear operator $Hx = -\Sigma_{m+1}^{\infty} c_i \lambda_i^{-1} \phi_i$, where $\lambda_{m+1} > 0$. It is easy to see that relations (h) hold.

From Theorem (10.ii) we could derive the following statement concerning 2π-periodic solutions for equations(27).

(11. i) Let us assume that for some integer M and real numbers p, q we have $M^2 < q \leq p < (M+1)^2$ and

$$(28) \qquad qI \leq (\partial^2 G(a)/\partial x_i \partial x_j) \leq pI$$

for all $a \in R^n$, and that $p(t) \in S$. Then (27) has a unique periodic solution of period 2π (Kannan [35, 36], Lazer and Sanchez [43], Leach [44], Loud [48], Lazer and Leach [42]).

Above we say that for two symmetric $n \times n$ matrices A, B we have $A < B$ [or $A \leq B$] if as usual $B - A$ is the matrix of the coefficients of a positive definite (positive semidefinite) quadratic form.

From the remark at the end of Section 10 we can derive an analogous statement for the problem at resonance

$$(29) \qquad x'' + M^2 x + \text{grad } G(x) = p(t) .$$

Again we take as before $Nx = -\text{grad}(x(t)) + p(t)$.

(11. ii) If (a) $N: S \to S$ is continuous and bounded; (b) $\langle Nx-Ny, x-y\rangle \geq -c\,\|x-y\|^2$ for some constant $c < (M+1)^2 - M^2 = 2M+1$ and all $x, y \in S$; (c) there is $R > 0$ such that $\langle Nx, x^*\rangle \leq 0$ for all $x^* \in S_o$ with $\|x^*\| = R$ and all $x \in S$ with $Px = x^*$, then (29) has at least one 2π-periodic solution. For $n = 1$ the following simple corollary may be of interest. It concerns the nonlinear differential equation

$$(30) \qquad x'' + m^2 x + h(x) = p(t) ,$$

where m is an integer.

(11. iii) If $p(t)$ is 2π-periodic and continuous, if h is of class C^1 with $|h(x)| \leq M$ for all real x, if $0 < h'(x) < b < (m+1)^2 - m^2 = 2m+1$, if there are constants $c < d$, $C < D$ such that $h(x) \leq C$ for $x \leq c$, $h(x) \geq D$ for $x \geq d$, and $(A^2 + B^2)^{1/2} < 2(D-C)$ where $A = \int_0^{2\pi} p(s) \cos ms\, ds$, $B = \int_0^{2\pi} p(s) \sin ms\, ds$, then (30) has a periodic solution of period 2π.

12. A nonlinear problem in potential theory and general elliptic boundary value problems.

We consider here as an example the problem of existence of solutions for the nonlinear boundary value problem

$$\Delta u + g(x, y, u) = 0, \quad (x, y) \in A = [x^2+y^2 < 1],$$

(31)
$$u = 0, \quad (x, y) \in \partial A = [x^2 + y^2 = 1],$$

under various hypotheses on g (a nonlinear Dirichlet problem).

Let S be the Hilbert space $L_2(A)$, and let $\langle , \rangle$ and $\| \ \|$ denote the usual inner product and norm in $S = L_2(A)$. The linear problem $\Delta u + \lambda u = 0$ in A, $u = 0$ on ∂A, has a well known fundamental system $\{\lambda_i\}$ and $\{\phi_i\}$ of eigenvalues and orthonormal eigenfunctions, with $0 < \lambda_1 < \lambda_2 < \dots$. Also $\{\phi_i\}$ is complete in S. Thus, every element $u \in S$ has Fourier series $u(x, y) = \Sigma_1^\infty c_i\phi_i$ with $c_i = \langle u, \phi_i \rangle$. For any $m \geq 0$ we define P by taking $Pu = \Sigma_1^m c_i\phi_i$, so that $S_o = PS$ is the m-dimensional linear space spanned by $\phi_1, \dots, \phi_m$. Let S' be the subset of S of all functions $u(x, y)$ that are essentially bounded in A. Also, let X be the set of all functions $u(x, y)$ such that u is continuous in $A \cup \partial A$ with $u = 0$ on ∂A, u has continuous first order partial derivatives in A, and Δu (computed in the sense of distributions) is a measurable, essentially bounded function defined a.e. in A. Then it was proved in [9] that $H: S' \to X$. Also $X \subset S'$, $\Delta : X \Rightarrow S' \subset S$, and relations (h) hold. We take for N the Nemitsky operator $Nu = -g(x, y, u(x, y))$. Many local results had been proved in [9] on the sole considerations of Sections 1 and 2. Now the results of Sections 8 and 10 apply and the following statements can be derived (see Kannan [37]).

(12.i) If $N: S \to S$ is continuous and monotone, then (31) has a unique solution.

(12.ii) If $N: S \to S$ is continuous and bounded, if

(a) $\langle Nu_1 - Nu_2, u_1 - u_2 \rangle \geq -p\|u_1 - u_2\|^2$ for all $u_1, u_2 \in S$ and some p, $0 < p < \lambda_{m+1}$; (b) $\langle Nu_1 - Nu_2, u_1^* - u_2^* \rangle \leq -q\|u_1^* - u_2^*\|^2$ for all $u_1^*, u_2^* \in S_o$, $u_1, u_2 \in S$ with $Pu_1 = u_1^*$, $Pu_2 = u_2^*$, and some $q > \lambda_m$, then (31) has a unique solution.

We consider now the problem "at resonance"

(32) $$\Delta u + \lambda_m u + g(x, y, u) = 0, \ (x, y) \in A; \ u = 0, \ (x, y) \in A.$$

(12.iii) If $N: S \to S$ is continuous and bounded, if

(a) $\langle Nu_1 - Nu_2, u_1 - u_2 \rangle \leq b\|u_1 - u_2\|^2$ for all $u_1, u_2 \in S$ and some b, $0 < b < \lambda_{m+1} - \lambda_m$; (b) there is some $R > 0$ such that $\langle Nu, u^* \rangle \leq 0$ for all $u^* \in S_o$ with $\|u^*\| = R$ and all $u \in S$ with $Pu = u^*$.

The results of Sections 8 and 10 are quite general, and they can be applied to any selfadjoint elliptic problem.

For instance, for the nonlinear Neumann problem

(33) $$\begin{aligned} &\Delta u + g(x, y, u) = 0, \quad (x, y) \in A = [x^2 + y^2 < 1], \\ &\partial u/\partial n = 0, \quad (x, y) \in \partial A = [x^2 + y^2 = 1], \end{aligned}$$

the eigenvalues are now $0 = \lambda_1 < \lambda_2 \leq \lambda_3 \leq \cdots$, so that problem (33) is a problem "at resonance".

Under the conditions of (12. i) we can use the second part of Theorem (8. ii) and conclude that problem (33) has at least one solution.

Statement (12. i) is still valid, and we can apply it with $m = 1$ and thus take $0 = \lambda_1 < q < p < \lambda_2$, or for any $m \geq 1$ and take $\lambda_m < q < p < \lambda_{m+1}$.

Statement (12. iii) holds for problem (33) with Neumann boundary conditions with $m = 1$. Alternatively it can be applied to problem $\Delta u + \lambda_m u + g(x, y, u) = 0$ in A, $\partial u/\partial n = 0$ pn ∂A, for $m \geq 1$.

The considerations above hold for general selfadjoint uniformly elliptic linear operators E under usual mild conditions (boundedness and measurability of the coefficients,

Gårding inequality) so as to guarantee the properties mentioned in Sections 1 and 2.

Finally, in connection with problems at resonance, let us consider the boundary value problem

$$
\text{(34)} \qquad \begin{aligned} Eu + \alpha u + g(u) &= h(x), && x \in G, \\ u &= 0, && x \in \partial G, \end{aligned}
$$

where G is a bounded domain in E^ν with smooth boundary ∂G, and where E is a second order uniformly elliptic operator on G defined by

$$E = \sum_{i,j=1}^{n} (\partial/\partial x_i) a_{ij}(x)(\partial/\partial x_j).$$

Here $a_{ij}(x)$ are real bounded measurable functions on G, $a_{ij} = a_{ji}$, and there is a constant $c > 0$ such that for all $x \in G$ and $(\xi_1, \ldots, \xi_n) \in E^n$ we have

$$\sum_{i,j=1}^{n} a_{ij}(x)\xi_i \xi_j \geq c \sum_{i=1}^{n} \xi_i^2 .$$

We assume that α is an eigenvalue for the linear operator E, with null space of dimension one, so that all solutions of $Eu + \alpha u = 0$ in G, $u = 0$ on ∂G are given by $aw(x)$, $x \in G$, a real, where $w(x)$ is a given function on G. In (34) we assume that h is an element of $L_2(G)$, and that g is a real-valued continuous function on the real line with finite limits

$$r = \lim_{s \to -\infty} g(s) \qquad R = \lim_{s \to -\infty} g(s),$$

and such that $r \leq g(s) \leq R$ for all s real. Let G^+, G^- denote the subsets of G where $w(x) \geq 0$ or $w(x) \leq 0$

respectively, and let

$$w^+ = \int_{G^+} |w| dx \quad , \quad w^- = \int_{G^-} |w| dx.$$

The following very sharp result holds:

(12. iv) Theorem (Landesman and Lazer [41]). The inequalities

$$rw^+ - Rw^- \leq \langle h, w \rangle \leq Rw^+ - rw^-$$

are necessary for (34) to have a (weak) solution. The slightly stronger inequalities

$$rw^+ - Rw^- < \langle h, w \rangle < Rw^+ - rw^- \tag{35}$$

are sufficient for (34) to have a weak solution. If we know that $r < g(s) < R$ for all s real, then the inequalities (35) are both necessary and sufficient for (34) to have a weak solution.

For the sufficiency part of the proof of (12. iv) Landesman and Lazer have used a decomposition of the functional relation representing the weak solutions of (34) into a system of two equations, corresponding to the auxiliary and bifurcation equations (3), (4). They have then shown that the system has a solution by a subtle application of Schauder fixed point theorem. Landesman and Lazer pointed out in the same paper [41] various extensions of their own result. Much work has been originated by Landesman's and Lazer's research, in particular Mawhin has shown some extensions of (12. iv) by topological degree arguments.

13. Strong nonlinearities.

All theorems in Sections 8 and 10, however general, they all contain the provision that $N:S \to S$ maps S into S. If say $S = L_2(G)$, it is clear that N need not have this property. For instance if $G = [0,1]$, $f(u) = u^3$, $(Nu)(t) = f(u(t))$, and $u(t) = t^{-1/3}$ for $0 < t \leq 1$, then $u \in L_2[0,1]$, but $(Nu)(t) = t^{-1}$, $0 < t \leq 1$, is not even L-integrable.

As mentioned in Section 2, there was an analogous difficulty in the application of Theorem (2.i) based on contraction mappings, and suitable devices were needed to overcome the difficulty.

Gustafson and Sather in [24, 25] used the device of decomposing the positive definite operator $K = -H(I-P)$ into the product $K = K^{1/2} K^{1/2}$, and this device allowed them to include a large class of strong nonlinearities. This idea had been used also by Vainberg and Lavrentiev [75] in Hammerstein equation theory. Actually Gustafson and Sather were very successful with this device and could apply the theory to the Hartree helium equation. The technical requirement for the success of the device is that $D(N) \supset D(K^{1/2})$.

This point of view has been extended by Kannan and Locker [38]. For instance, they consider the case in which the differential operator E itself may have a natural decomposition $E = TT^*$ with suitable boundary conditions on T and T^*. In this case a decomposition $K=-H(I-P)=J^*J$ is automatically induced, and the domain and ranges of these operators J, J^* are well known. This is relevant since it is still essential that $D(N) \supset D(J^*)$.

Another device is connected with the use of suitable Sobolev spaces.

Problems of periodic solutions of systems of periodic differential equations, and elliptic boundary value problems, have been treated by these devices, and corresponding sharp results will be reported elsewhere.

REFERENCES

[1] H. A. Antosiewicz, Boundary value problems, for nonlinear ordinary differential equations, Pac. Journ. Math. 17, 1966, 191-197.

[2] H. A. Antosiewicz, Un analogue du principe du point fixe de Banach. Annali di Mat. pura appl. (4)74, 1966, 61-64.

[3] H. Amann, Existence theorems for equations of Hammerstein type. Applicable Analysis 2, 1973, 385-397.

[4] R. G. Bartle, Singular points of functional equations, Trans. Amer. Math. Soc. 75, 1953, 366-384.

[5] H. Brézis, Operateurs maximaux monotones, Amer. Elsevier, New York 1973.

[6] H. Brézis, M. Crandall, A. Pazy, Perturbation of nonlinear maximal monotone sets. Comm. Pure Appl. Math. 23, 1970, 123-144.

[7] L. Cesari, Functional analysis and periodic solutions of nonlinear differential equations. Contributions to differential equations 1, 149-187, Wiley 1963.

[8] L. Cesari, Functional analysis and Galerkin's method. Mich. Math. Journ. 11, 1964, 385-414.

[9] L. Cesari, A nonlinear problem in potential theory, Mich. Math. Journ. 16, 1969, 3-20.

[10] L. Cesari, Nonlinear Mechanics, Lecture Notes C.I.M.E., Bressanone 1972. Cremonese, Roma 1973, 3-95.

[11] L. Cesari, Existence theorems for periodic solutions of nonlinear Lipschitzian differential systems and fixed point theorems. Contributions to the Theory of Nonlinear Oscillations 5, 1960, 115-172. Annals of Mathematics Study, Princeton No. 45.

[12] L. Cesari, Existence in the large of periodic solutions of hyperbolic partial differential equations. Arch. Rat. Mech. Anal. 20, 1965, 170-190.

[13] L. Cesari, Asymptotic Behavior and Stability Problems in Ordinary Differential Equations. Springer Verlag, 3d edit. 1971.

[14] L. Cesari, Sulla stabilità delle soluzioni dei sistemi di equazioni differenziali lineari a coefficienti periodici. Mem. Accad. Italia (6) 11, 1941, 633-695.

[15] L. Cesari, C. A. Borges, D.A. Sanchez, Functional analysis and the method of harmonic balance, Quaterly of Applied Mathematics. To appear.

[16] L. Cesari and R. Kannan, Functional analysis and nonlinear differential equations, Bull. Amer. Math. Soc. 79, 1973, 1216-1219.

[17] J. Cronin, Branch points of solutions of equations in Banach spaces, Trans. Amer. Math. Soc. 69, 1950, 208-231.

[18] J. Cronin, Fixed points and topological degree in nonlinear analysis, Amer. Math. Soc., Providence 1964.

[19] J. Cronin, Equations with bounded nonlinearities. Journ. Differential Equations 14, 1973, 581-596.

[20] C. L. Dolph and G. J. Minty, On nonlinear integral equations of the Hammerstein type. Integral Equations, Madison, University of Wisconsin Press 1964, 99-154.

[21] C. Fabry, Boundary value problems for weakly nonlinear ordinary differential equations. Annales Soc. Sci. Bruxelles 85, 1971, 221-238.

[22] K. Friedrichs, Special topics in analysis. Lecture Notes, New York Univ. 1953-54.

[23] L. M. Graves, Remarks on singular points of functional equations. Trans. Amer. Math. Soc. 79, 1955, 150-157.

[24] K. Gustafson and D. Sather, Large nonlinearities and monotonicity, Arch. Rat. Mech. Anal. 48, 1972, 109-122.

[25] K. Gustafson and D. Sather, Large nonlinearities and closed linear operators, Arch. Rat. Mech. Anal. 52, 1973, 10-19.

[26] J. K. Hale, Applications of alternative problems. Lecture Notes. Brown University 1971

[27] J. K. Hale, Periodic solutions of nonlinear systems of differential equations. Riv. Mat. Univ. Parma 5, 1954, 281-311.

[28] J. K. Hale, Oscillations in Nonlinear Systems. McGraw-Hill, New York 1963.

[29] J. K. Hale, Periodic solutions of a class of hyperbolic equations containing a small parameter. Arch. Rat. Mech. Anal. 23, 1967, 380-398.

[30] J. K. Hale, Ordinary Differential Equations, Interscience 1969.

[31] J. K. Hale, S. Bancroft, D. Sweet, Alternative problems for nonlinear equations. Journ. Differential Equations 4, 1968, 40-56.

[32] W. A. Harris, Holomorphic solutions of nonlinear differential equations at singular points, pp. 184-187. Advances in Differential and Integral Equations, SIAM Studies in Applied Math., No. 5, 1969.

[33] W. A. Harris, Y. Sibuya, L. Weinberg, Holomorphic solutions of linear differential systems at singular points. Arch. Rat. Mech. Anal. 35, 1969, 245-248.

[34] P. Hess, On nonlinear equations of Hammerstein type in Banach spaces. Proc. Amer. Math. Soc. 30, 1971, 308-312.

[35] R. Kannan, Periodically disturbed conservative systems, Journ. Differential Equations. To appear.

[36] R. Kannan, Existence of periodic solutions of differential equations, Trans. Amer. Math. Soc. To appear.

[37] R. Kannan, Existence of solutions of a nonlinear problem in potential theory, Mich. Math. Journ. To appear.

[38] R. Kannan and J. Locker, Operators J*J and nonlinear Hammerstein equations. To appear.

[39] H. W. Knobloch, Eine neue Methode zur Approximation von periodischen Lösungen nicht linear Differentialgleichungen zweiter Ordnung. Math. Zeit. 82, 1963, 177-197.

[40] H. W. Knobloch, Comparison theorem for nonlinear second order differential equations. Journ. Differential Equations 1, 1965, 1-25.

[41] E. M. Landesman and A. Lazer, Nonlinear perturbations of linear elliptic boundary value problems at resonance. J. Math. and Mech. 19, 1970, 609-623.

[42] A. C. Lazer and D. E. Leach, Bounded perturbations of forced harmonic oscillations at resonance. Annali di Mat. pura appl. 72, 1969, 49-68.

[43] A. C. Lazer and D. A. Sanchez, On periodically perturbed conservative systems, Mich. Math. Journ. 16, 1969, 193-200.

[44] D. E. Leach, On Poincaré's perturbation theorem and a theorem of W. S. Loud, Journ. Differential Equations 7, 1970, 34-53.

[45] J. Leray and J. Schauder, Topologie et équations fonctionnelles, Ann. Sci. Ecole Norm. Sup. 51, 1934, 45-78.

[46] D. C. Lewis, On the role of first integrals in the perturbation of periodic solutions. Annals of Math. 63, 1956, 535-548.

[47] J. Locker, An existence analysis for nonlinear equations in Hilbert spaces. Trans. Amer. Math. Soc. 128, 1967, 403-413.

[48] W. S. Loud, Periodic solutions of nonlinear differential equations of Duffing type. Differential and Functional Equations. Benjamin, New York 1967.

[49] A. M. Lyapunov, Sur les figures d'equilibre peu différentes des ellipsoids d'une masse liquide homogène douée d'un mouvement de rotation. Zap. Akad. Nauk St. Petersburg 1, 1906, 1-225.

[50] J. Mawhin, Le problème des solutions périodiques en mécanique non linéaire. Thése Univ. Liége 1969.

[51] J. Mawhin, Periodic solutions of nonlinear functional differential equations. Journ. Differential Equations 10, 1971, 240-261.

[52] J. Mawhin, Degré topologique et solutions périodiques des systèmes différentiels nonlinéaires. Bull. Roy. Soc. Liège, 38, 1969, 308-398.

[53] J. Mawhin, Nonlinear perturbations of Fredholm mappings in normed spaces and applications to differential equations. Trabalho de matematica no. 58 Univ. of Brasilia 1974.

[54] G. Minty, On a monotonicity method for the solution of nonlinear equations in Banach spaces. Proc. Nat. Acad. Sci. U.S.A., 50, 1963, 1038-1041.

[55] K. Nagle, Boundary value problems for nonlinear ordinary differential equations. Univ. of Michigan thesis 197 .

[56] M. Nagumo, A theory of degree of mappin infinitesimal analysis, Amer. J. Math. 73, 1951, 485-496.

[57] M. Nagumo, Degree of mappints in convex linear topological spaces, Amer. J. Math. 73, 1951, 497-511.

[58] L. Nirenberg, Functional Analysis, Lecture Notes, New York Univ. 1960-61.

[59] J. E. Osborn and D. Sather, Alternative problems for nonlinear equations. Journ. Differential Equations. To appear.

[60] J. E. Osborn and D. Sather, Alternative problems and monotonicity, Journ. Differential Equations. To appear.

[61] C. Perello, Periodic solutions of differential equations with time lag containing a small parameter. Journ. Differential Equations 4, 1968, 160-175.

[62] D. Petrovanu, Solutions périodiques pour certain équations hyperboliques. Analele Stintifice Iasi 14, 1968, 327-357.

[63] D. Petrovanu, Periodic solutions of the Tricomi problem. Mich. Math. Journ. 16, 1969, 331-348.

[64] H. Poincaré, Les méthodes nouvelles de la mécanique céleste, Gauthier-Villars, Paris 1892-1899. Dover, New York 1957.

[65] E. Rothe, The theory of topological order in some linear topological spaces, Iowa State College J. Sci., 13, 1939, 373-390.

[66] F. Riesz and B. Sz. Nagy, Functional Analysis, Ungar, New York 1955.

[67] A. M. Rodionov, Periodic solutions of nonlinear differential equations with time lag. Trudy Sem. Differential Equations Lumumba Univ. Moscow 2, 1963, 200-207.

[68] D. A. Sanchez, An iterative scheme for boundary value problems by alternative methods, Univ. of Wisconsin Math. Res. Ctr., Report 1412, Jan. 1974.

[69] D. Sather, Branching of solutions of an equation in Hilbert space. Arch. Rat. Mech. Anal. 36, 1970.

[70] E. Schmidt, Zur Theorie der Linearen und nichtlinearen Integralgleichungen und der Verzweigung ihrer Lösungen, Math. Ann. 65, 1908, 370-399.

[71] H. Shaw, Elliptic problems of order m with nonzero null space. Univ. of Mich. thesis 1975.

[72] M. M. Vainberg, Variational methods for the study of nonlinear operators, Holden-Day, San Francisco 1964.

[73] M. M. Vainberg and V. A. Trenogin, The methods of Lyapunov and Schmidt in the theory of nonlinear equations and their further development. Russian Math. Surveys 17, 1962, 1-60.

[74] M. M. Vainberg and P. G. Aizengendler. The theory and methods of investigation of branch points of solutions. Progress in Mathematics, II. Plenum Press, New York 1968, 1-72.

[75] M. M. Vainberg and I. M. Lavrentiev, Nonlinear equations of Hammerstein type with potential and monotone operators in Banach spaces. Math. Sbornik USSR 16, 1972, 333-347.

[76] S. Williams, A connection between the Cesari and Leray-Schauder methods. Michigan Math. Journ. 15, 1968, 441-448.

University of Michigan, Ann Arbor, Michigan

MINIMUM PRINCIPLES AND POSITIVE SOLUTIONS FOR A CLASS OF NONLINEAR DIFFUSION PROBLEMS

Jagdish Chandra, Paul Wm. Davis and B. A. Fleishman

1. Introduction.

The minimum (or maximum) principle is a powerful and widely used instrument in the study of differential equations and inequalities. To paraphrase Protter and Weinberger [7, p. v], a function which satisfies a differential inequality in a domain, and consequently takes its minimum on the boundary of the domain, is said to satisfy a minimum principle. In the theory of ordinary and partial differential equations, such principles have been used to discover a variety of qualitative features of solutions, including *a priori* bounds and positivity.

In this paper we consider a broad class of nonlinear ordinary differential equations and related inequalities of the form

$$(1) \qquad F(x, u, u', Du') \equiv R(x, u, u', Du') - S(x, u, u', Du') - f(x, u) = 0,$$

subject to a variety of nonlinear boundary conditions. (Equations such as (1) might also be described as steady-state diffusion equations.) Here Du denotes any one of the four Dini derivatives, e.g. [4, p. 7], while u' denotes the usual two-sided derivative. At the end of an interval, however, Du and u' are to be interpreted as appropriate one-sided derivatives.

It is shown here that if $R(x, u, u', Du')$ satisfies a minimum principle (MinP) on an interval, then under certain conditions on S and f, F obeys a similar principle. We are thus able to establish non-negativity of solutions of (1) subject to certain nonlinear boundary conditions. This modifies principle is also exploited to derive a priori bounds on solutions.

In the terminology of one of the problems that motivated this study, R is a diffusion term, S is a convection term, and f is a source term. The hypotheses imposed on S and f are physically reasonable in this context.

In section 4, we exhibit a specific diffusion term R which obeys a minimum principle. (Of course, [5], [8], [9], [11], etc. suggest many other forms that R might take.) Since this term arises in the examples considered in sections 6 and 7 (models of enzyme diffusion-kinetics and gas lubrication theory), we can apply the general results for operators of the form (1) to establish lower bounds for the solutions of these particular boundary value problems.

These examples also demonstrate some advantage in including a middle term like $S(x, u, u', Du')$ in the general differential expression F. In addition, the first example, which involves the derivative of a function of $|u'|$, indicates the value of having Du', rather than u'', as one of the arguments in F.

Here our purpose is to present simple arguments in support of useful conclusions, not the most general extension in one dimension of earlier studies of nonlinear maximum principles. However, Remark 5 below does exhibit lower bounds for the solutions of a problem to which the usual ellipticity criteria (existence of $\partial F/\partial u''$ or monotonicity or linearity in u'') do not apply.

Most of the results described here have multi-dimensional and parabolic analogues. Some are obvious, but others require more complicated arguments. Rather than obscure the simple derivation of the useful results that were our original motivation, we shall discuss the multi-variable version of these ideas elsewhere.

2. Notation and Assumptions.

We denote the intervals $(0,1)$ and $[0,1]$ by I and $\bar{I}$, respectively.

Let $R(x,u,v,w)$, $S(x,u,v,w)$ and $f(x,u)$ be real-valued functions defined for $x \in I$ and $-\infty < u, v, w < \infty$.

We make the following assumptions:

(A-1) For $x \in \bar{I}$ and $-\infty < u, w < \infty$, $S(x,u,0,w) = 0$.

(A-2) For $x \in \bar{I}$ and $u \neq 0$, $f(x,u)u > 0$.

The differential expression $R(x,u,u',Du')$ is said to satisfy a minimum principle (MinP) if, for any non-constant function $u(x)$ for which u' and Du' are defined,

(i) $R(x,u,u'Du') \leq 0$, $x \in I$, implies that u does not assume a local minimum in I,

(ii) $R(x,u,u'Du') \leq 0$, $x \in [0,1)$ $(x \in (0,1])$, implies that $u'(0) > 0$ $(u'(1) < 0)$ if u assumes a local minimum at

$x = 0$ $(x = 1)$.

The corresponding maximum principle (MaxP) is obtained by reversing all inequalities and replacing "minimum" with "maximum" in the statements above.

It is usually much easier to establish that R obeys a minimum principle if the hypotheses of (i) and (ii) above are strengthened to $R < 0$. Theorems 2 and 3 and the supporting lemmas are still valid in this case.

3. Basic Lemmas.

Lemma 1: Let $R(x, u, u', Du')$ satisfy MinP. Further, let $S(x, u, u', Du')$ and $f(x, u)$ satisfy, respectively, (A-1) and (A-2). Suppose $u(x)$ is a function for which $R(x, u, u', Du')$ is an upper semi-continuous function of x on $\bar{I}$ and for which $F(x, u, u', Du') \leq 0$, $x \in \bar{I}$. Then $u(x)$ can not assume a negative local minimum in I unless it is identically constant there.

Proof: Suppose that u is not identically constant and that it assumes a negative local minimum at $c \in I$. Since $u'(c) = 0$, $u(c) < 0$ and $F \leq 0$, hypotheses (A-1) and (A-2) yield $R(c, u(c), 0, Du'(c)) < 0$. Since R is an upper semi-continuous function of x, there is a neighborhood of c in which $R(x, u, u', Du')$ is non-positive. But the existence of a minimum within this neighborhood contradicts (i) of MinP.

Lemma 2: Let the hypotheses of Lemma 1 hold. Suppose $u(x)$ is not identically constant. It $u(x)$ assumes a negative local minimum at $x = 0$ $(x = 1)$, then $u'(0) > 0$ $(u'(1) < 0)$.

Proof: Suppose the non-constant function $u(x)$ assumes a

negative minimum at $x = 0$. Clearly, $u'(0) \geq 0$. If $u'(0) = 0$, then from (A-1), (A-2), and $F \leq 0$, we get $R(0, u(0), 0, Du'(0)) < 0$, and we arrive at a constrdiction of MinP(ii) as in the proof of Lemma 1. A similar argument applies to the minimum at $x = 1$.

Remark 1: If the inequality in (A-2) is reversed, then the statements of Lemmas 1 and 2 are valid for positive rather than negative minima. Furthermore, if we combine the two versions of (A-2), that is, if instead of (A-2) we assume that $f(x, u) < 0$ for $x \in \bar{I}$ and $u \neq 0$, then Lemmas 1 and 2 hold for both positive and negative minima (e.g., $f = -u^2$).

4. A Nonlinear Diffusion Operator.

Let $r(x, u, v)$ be a real-valued function, continuous and strictly positive for $x \in \bar{I}$, $-\infty < u, v < \infty$.

Theorem 1: The differential expression $D(r(x, u, u')u')$ obeys MinP for all functions $u(x)$ which are continuously differentiable on $\bar{I}$ and for which $D(r(x, u, u')u')$ exists on $\bar{I}$.

Proof: Suppose that

$$(2) \qquad D(r(x, u, u')u') \leq 0, \quad x \in I,$$

for some function $u(x)$ which is continuously differentiable and non-constant on $\bar{I}$. Since $r(x, u, u')u'$ is continuous, (2) implies that it is non-increasing on any closed subinterval of I (e.g., [4, Lemma 1.2.1.]).

Suppose that $u(x)$ assumes a minimum at $c \in I$; then $u'(c) = 0$ and the non-increasing character of $r(x, u, u')u'$ yield

$$r(x, u, u')u'(x) < 0 \qquad \text{for } c \leq x < 1.$$

Consequently, $u'(x) \leq 0$ for $c \leq x < 1$, because $r(x, u, u$ is strictly positive. Since c is a minimum point, $u'(x) = 0$ for $c \leq x < 1$. Applying the same argument on $0 < x \leq c$, we get $u'(x) = 0$ for $x \varepsilon I$. Since $u'(x)$ is continuous, $u'(x) \equiv 0$ on $\bar{I}$, in violation of the assumption that u is non-constant. Therefore, $u(x)$ has no local minimum in I unless it is constant on $\bar{I}$.

Now suppose that (2) holds on $0 \leq x < 1$ and that a non-constant function $u(x)$ assumes a local minimum at $x = 0$. Clearly, $u'(0) \geq 0$. If $u'(0) = 0$, a slight modification of the preceding argument will force the contradiction that $u'(x) \equiv 0$ on $\bar{I}$. A similar argument at $x = 1$ completes the proof.

Notice that Theorem 1 is not a special case of the results in [5], [8], [9] or [11] because we have not assumed that $r(x, u, u')$ is differentiable.

5. A Priori Bounds and Positivity.

In Theorem 2 below we derive bounds similar to those in [3].

Theorem 2 (*A priori* bounds): Let the real-valued function $g(x, u)$ satisfy

$$(3) \qquad g(x, u) \leq 0 \quad \text{when } x \varepsilon \bar{I} \text{ and } u \leq m,$$

for some constant $m \leq 0$. Let $R(x, u, u', Du')$ obey MinP, and let $S(x, u, u', Du')$ and $f(x, u)$ satisfy (A-1) and (A-2), respectively. Let $u(x)$ be a differentiable function for which R is an upper semi-continuous function of x on I for which

$$F(x, u, u', Du') \leq g(x, u) \quad \text{for } x \in I. \tag{4}$$

Then

$$\min[m, u(0), u(1)] \leq u(x) \quad \text{for } x \in \bar{I}. \tag{5}$$

Proof: Let $J = \{x \in \bar{I} \mid u(x) \leq m\}$, and suppose J is non-empty. (When J is empty, (5) is trivially true.) Since $u(x)$ is differentiable and, hence, continuous on I, J is a union of closed intervals $[a_i, b_i]$, some of which may consist of single points. (If $J \neq \bar{I}$, there are at most a countable number of such intervals because the complement of J in $\bar{I}$ is an open set, which may be written as a countable union of disjoint open intervals.)

Lemma 1 prevents $u(x)$ from attaining a negative minimum in $[a_i, b_i]$, so that $u(x)$ is bounded below on $[a_i, b_i]$ by $u(a_i)$ and/or $u(b_i)$. But all these values are just m, $u(0)$, or $u(1)$, depending on the location of $[a_i, b_i]$ in $\bar{I}$. The conclusion of the theorem follows immediately.

By employing a maximum principle, we can obtain a similar result for an a priori upper bound if $g(x, u) \geq 0$ when $x \in \bar{I}$ and $u \geq M$ for some constant $M \geq 0$. The inequality in (4) must also be reversed.

Theorem 3 (Positivity): Let the hypotheses of Lemma 1 hold. Further, let one of the following sets of boundary inequalities be satisfied:

$$u(0) \geq a(u(0)), \quad u(1) \geq b(u(1)), \tag{6a}$$

$$\begin{cases} \alpha u(0) - \beta u'(0) \geq a(u(0)) \\ \\ \gamma u(1) + \delta u'(1) \geq b(u(1)) \end{cases} \tag{6b}$$

with $\alpha, \gamma \geq 0$, $\alpha^2 + \gamma^2 > 0$, $\beta, \delta > 0$,

(6c) $$u(0) \geq a(u(0)), \quad u'(1) \geq b(u(1)),$$

(6d) $$-u'(0) \geq a(u(0)), \quad u(1) \geq b(u(1)),$$

where a and b are real-valued functions satisfying

(7) $$a(u) \geq 0, \quad b(u) \geq 0 \quad \text{for} \quad u < 0.$$

Then $u(x) \geq 0$ for $x \in \bar{I}$.

Proof: By taking $g \equiv 0$ and $m = 0$, we obtain from Theorem 2

$$\min[0, u(0), u(1)] \leq u(x) \quad \text{for } x \in \bar{I}.$$

In case (6a) the conclusion is an immediate consequence of (7).

Consider now (6b). If u is constant, that constant cannot be negative without violating (6b). If u is non-constant and is negative-valued somewhere, it must assume a negative minimum either at the end points of I or in its interior. Suppose this minimum occurs at $x = 0$. Then Lemma 2 implies that $u'(0) > 0$, but (6b) forces $u'(0) \leq 0$. A similar contradiction occurs at $x = 1$. Since Lemma 1 prohibits a negative minimum in I, the non-negativity of u on $\bar{I}$ is assured.

A suitable combination of the preceding arguments shows that both (6c) and (6d) lead to the same conclusion. This completes the proof of Theorem 3.

Remark 2: Without changing either the statement of Theorem 3 or its proof, we may regard α, β, γ and δ in (6b) as functions of u and u' at $x = 0$ and $x = 1$, provided they retain the sign properties required in (6b) for all u and u'. Such a situation, in fact, arises quite naturally in the model of enzyme diffusion-kinetics which will be discussed in the next section. Likewise, the functions a and b can depend on both u and u′ if (7) remains valid for $u < 0$ and $-\infty < u' < \infty$.

Remark 3: If $R(x, u, 0, 0) = 0$, then the condition $\alpha^2+\gamma^2 > 0$ may be dropped, for in this case, if u is a negative constant, $F \leq 0$ implies $f(x, u) \geq 0$, in violation of (A-2).

Remark 4: The function $f(x,u)$ in (1) can obviously be multiplied by any positive-valued function of u, u', Du' without affecting the preceding results.

Remark 5: The hypotheses on F used in Theorem 3 are weaker than the usual ellipticity criteria, such as in [5], [8] and [11]. Thus, for example, Theorem 3 ensures that every solution of

(8) $$F(x, u, u', u'') \equiv u'' + (1 + \sin u'')u' - (1+|u''|)u \leq 0,$$

subject to the boundary inequalities

(9) $$u(0) \geq 1, \quad u(1) \geq 1,$$

is non-negative on $\bar{I}$. (One such solution is $u \equiv 1$.) The differential expression F in (8), however, can not be tested for ellipticity in any of the usual ways, e.g. [5], [11], because neither does $\partial F/\partial u''$ exist at $u'' = 0$ nor is F monotone in u''.

6. Application: Enzyme Diffusion-Kinetics.

In this section and the next we describe some problems which may be considered special cases of (1). The first example involves models for steady-state, one-dimensional enzyme diffusion-kinetics, described by equations of the form [6]

$$\varepsilon(r(c, |c'|)c')' = f(c) \tag{10}$$

where ε is a parameter depending on the (small) diffusion coefficient and the (large) number of bacteria, or other absorption sites, between which diffusion of substrate occurs under the influence of the nonlinear diffusion coefficient $r(c, |c'|)$. The unknown function $c(x)$ represents substrate concentration, and the given function $f(c)$ is a source or expected reaction rate term. All quantities in (10) are dimensionless and real-valued.

While the functional form of the diffusion coefficient may vary from problem to problem, it is generally continuous and positive [6].

Consider a one-dimensional cell covering the interval $\overline{I}$. If the substrate region outside the cell is maintained at unit concentration, then a typical boundary condition is

$$\varepsilon r(c(1), |c'(1)|)c'(1) = h(1 - c(1)), \tag{11}$$

where $h \geq 0$ is the permeability coefficient of the membrane [6]. This boundary condition is of the type (6b). (See Remark 2.) The extremes of permeability lead to simpler conditions of the form (6c) and (6a, d) respectively. An impermeable membrane ($h = 0$) at $x = 1$ implies $c'(1) = 0$, while a totally porous membrane ($h \to \infty$) requires $c(1) = 1$. Identical considerations apply at $x = 0$, with the reminder that the outward gradient there is $-c'(0)$.

A typical form for $f(c)$ is the approximation of Michaelis-Menton kinetics,

$$f(c) = c/(c + K_d),$$

where K_d is the (positive) dissociation constant for the reaction forming the complex from enzyme and substrate. In any case, f(c) is a positive, "invaribly monotonically increasing function of c" [6]. Therefore, f(c) certainly satisfies (A-2) for $c \geq 0$.

For negative values of c, the source term f(c) lacks physical significance in the context of enzyme kinetics. To meet (A-2), we shall require that f(c) for $c < 0$ be the odd extension of f(c) for $c > 0$. This extension offers an appealing symmetry of source and sink, but it is only a formalization designed to accomodate the mathematical possibility of negative concentration. Any extension yielding $f(c) < 0$ for $c < 0$ would suffice.

Theorems 1 and 3 now assure us that the model (10-11) will not yield meaningless, negative values for concentration: c(x) is non-negative throughout $\bar{I}$. The source term f(c) remains on its physically reasonable branch.

In the case of a spherically symmetric cell of unit radius, the multi-variable conservation equation can be reduced to the form

$$(12) \qquad (r(c, |c'|)c')' + 2r(c, c')c'/\rho = f(c)/\varepsilon ,$$

where $c' = dc/d\rho$ and ρ is the radial coordinate. For the middle term of (12) to be well-defined at $\rho = 0$, we must impose the usual boundary condition $c'(0) = 0$. At $\rho = 1$, we can impose (11) or any of its variants. Since (A-1) is evidently satisfied by the middle term of (12), we can appeal to Lemma 1 and Theorems 1 and 3 to conclude that $c(\rho) \geq 0$, $0 \leq \rho \leq 1$. (See [6].)

If $r(c,|c'|)$ is differentiable, we can avoid the use of Theorem 1 by writing (12) as

$$r(c,|c'|)c'' + [(r(c,|c'|))' + 2r(c,|c'|)/\rho]c' = f(c)/\varepsilon .$$

The leading term obeys a MinP by the usual arguments, e.g., [5], [11], but an appeal to Lemma 1 with

$$S(\rho, c, c', Dc') \equiv [(r)' + 2r/\rho]c'$$

is now essential if we are to conclude that c is non-negative. Notice that the expression Dc' in S arises in computing $|c'|'$ at $c' = 0$.

7. Application: Gas-Lubricated Slider Bearing.

The one-dimensional slider bearing can be pictured as the plane $z = 0$ moving in the positive x direction while the infinite strip $z = h(x) > 0$, $0 \leq x \leq 1$, $-\infty < y < \infty$, remains fixed. The velocity of the plane $z = 0$ is proportional to a positive constant Λ. If the region around the bearing is filled with a gas, then the relative motion of the lower plane can build up sufficient pressure under the fixed strip to permit movement of light loads with little frictional resistance.

If the gas is compressible and isothermal, then the pressure $p(x)$ under the strip $z = h(x)$ is governed by the Reynold's equation

$$(h^3(x)uu')' - \Lambda h(x)u' - \Lambda h'(x)u = 0 \qquad (13)$$

subject to the boundary conditions

$$u(0) = u(1) = 1 . \qquad (14)$$

Here pressure has been averaged in the z direction and normalized against ambient pressure. See [2] for details of the bearing geometry and a derivation of (13-14).

We shall consider the converging wedge bearing; that is, we assume $h'(x) < 0$ for $x \in \bar{I}$. In [10] it was shown that the boundary value problem (19-14) has a unique, twice continuously differentiable solution $u(x)$ which satisfies

(15) $$0 < h(1)/h(0) \leq u(x) \quad \text{for } x \in \bar{I}.$$

Consequently, $h^3(x)u$ is strictly positive, and according to Theorem 1 the differential expression $(h^3(x)uu')'$ obeys MinP. We now apply Lemmas 1 and 2, modified as in Remark 1, to conclude that $u(x) \geq 1$ for $x \in \bar{I}$ and that $u'(0) > 0$, $-u'(1) > 0$. (Note that (15) does not yield $u \geq 1$ because $h(1) < h(0)$.) Indeed, since a positive local minimum can not occur in I, $u(x) > 1$ in I. In other words, the inward pressure gradient is strictly positive at both the leading and trailing edges of the converging wedge bearing, and the film pressure inside the bearing is always above ambient pressure.

Alternatively, we can avoid using the bound (15) by a change of dependent variable $v = u^2/2$, transforms (13-14) into

$$(h^3(x)v')' - \Lambda h(x)(2v)^{-1/2}v' - \Lambda h'(x)(2v)^{1/2} = 0$$

$$v(0) = v(1) = 1/2.$$

We now recover the result described in the preceding paragraph by a direct appeal to Lemmas 1 and 2, Remark 1, and the inverse transform $u = +(2v)^{1/2}$, for the differential expression $(h^3(x)v')'$ obeys MinP by familiar linear considerations [7].

The superambient character of the film pressure can also be established from a monotone iteration scheme converging to the solution of the boundary value problem (13-14) [1]. However, the positivity of the inward pressure gradient at x = 0 and x = 1 is not immediately evident from this approach.

REFERENCES

[1] J. Chandra and P. W. Davis, A monotone method for quasilinear boundary value problems, Arch, Rational Mech. Anal., 54(1974), 257-266.

[2] W. A. Gross, Gas Film Lubrication, John Wiley and Sons, New York, 1962.

[3] H. B. Keller, Elliptic boundary value problems suggested by nonlinear diffusion processes, Arch. Rational Mech. Anal. 35 (5) (1969), 363-381. MR 41 # 639.

[4] V. Lakshmikantham and S. Leela, Differential and Integral Inequalities, v.I, Academic Press, New York, 1969.

[5] A. McNabb, Strong comparison theorems for elliptic equations of second order, J. Math. Mech. 10 (3) (1961), 431-440. MR 26 # 448.

[6] J.D. Murray, A simple method for obtaining approximate solutions for a class of diffusion-kinetics enzyme problems: I. General class and illustrative examples, Math. Biosci, 2 (1968), 379-411.

[7] M.H. Protter and H. F. Weinberger, Maximum Principles in Differential Equations, Prentice-Hall, Inc., Englewood Cliffs, New Jersey, 1967.

[8] R.M. Redheffer, An extension of certain maximum principles, Montash. Math. 66 (1962), 32-42. MR 26 # 446.

[9] J. Serrin, On the strong maximum principle for quasilinear second order differential inequalities, J. Functional Analysis 5 (1970), 184-193. MR 41 #3966.

[10] W.J. Steinmetz, On a nonlinear boundary value problem in gas lubrication theory, SIAM J. Appl. Math., to appear.

[11] W. Walter, appendix to Differential and Integral Inequalities, Springer-Verlag, Berlin, 1970.

U.S. Army Research Office, Durham, North Carolina
Worchester Polytechnic Institute, Worchester, Massacusetts
Rensselaer Polytechnic Institute, Troy, New York

DIAGONALIZATION METHOD IN SINGULAR PERTURBATIONS

K. W. Chang

1. Introduction.

Consider the quasilinear boundary value problem

(1) $$\varepsilon y'' + A(t,\varepsilon,y)\, y' = f(t,\varepsilon,y)\ ,$$

(2) $$y(0,\varepsilon) = \alpha(\varepsilon)\,,\ y(1,\varepsilon) = \beta(\varepsilon)\ ,$$

where $\varepsilon > 0$ is a small parameter. For scalar functions y, f, α, β, A this boundary problem has been studied by many authors such as Coddington and Levinson [5], Wasow [2] and, most recently, by Howes [9] (see the references at the end). However, for real-valued n-dimensional vector functions y, f, α, β and $n \times n$ matrix function A, the boundary problem (1), (2) seems open. It is not apparent that the techniques employed in the papers quoted above for solving scalar boundary problem can be extended to deal with the vector boundary problem.

It is our intention here to obtain existence results for this vector boundary problem which are analogous to those for the scalar boundary problem. That is, under the assumption that the generate problem

(3) $$A(t, 0, u)u' = f(t, 0, u) ,$$

(4) $$u(1) = \beta(0) ,$$

has a solution $u(t)$ and under additional assumptions, it will be proved that for all sufficiently small ε, there exists $\delta_0 > 0$, independent of ε, such that if $|\alpha(\varepsilon) - u(0)| \le \delta_0$ the vector boundary problem (1), (2) has a solution $y=y(t,\varepsilon)$ on $[0,1]$ such that

(5) $$\begin{aligned} y(t,\varepsilon) &= u(t) + O(\varepsilon) + O(e^{-\mu t/\varepsilon}) \\ y'(t,\varepsilon) &= u'(t) + O(\varepsilon) + O(\varepsilon^{-1} e^{-\mu t/\varepsilon}) \end{aligned}$$

where the standard order symbol holds uniformly in t as $\varepsilon \to 0$; the last term (with $\mu > 0$) in each expression denotes the boundary layer estimate.

To prove the above results, we split up the difference $y - u(t)$ into the so-called outer and inner corrections, and transform each resulting vector differential equation into a canonical or diagonalized system of two first order equations in such a way that the proof can be based on two Volterra integral equations with simple explicitly given kernels. This is achieved essentially through the use of the solution of a matrix Riccati equation. We give this fundamental transformation first in the next section.

A converse or complementary result (Theorem 2) will be given in the last section. Here one assumes the existence of a solution $y(t,\varepsilon)$ of (1), (2) and aims to show that the limiting function $\lim_{\varepsilon \to 0} y(t,\varepsilon) = u(t)$ exists on $(0,1]$ and, in fact, solves the degenerate problem (3). For a recent result of this nature for the scalar problem, see [7], where the authors made use of the maximum principle and the concept of weak solutions. An earlier result was already stated (without proof) by Bris [1]. Our approach here is to use the solution of an auxiliary problem in such a way that we can express the solution of (1), (2) in a simple integral form from which the required results can be deduced.

2. Diagonalization.

We describe first the transformation of the linear problem consisting of the vector differential equation

$$(6) \qquad \varepsilon x'' + C(t,\varepsilon)x' + D(t,\varepsilon)x = g(t)$$

and one of the boundary conditions

$$(7) \qquad \begin{array}{lll} \text{(i)} & x(1,\varepsilon) = \beta_1, & x'(0,\varepsilon) = 0, \\ \text{(ii)} & x(0,\varepsilon) = \alpha_1, & x(1,\varepsilon) = 0 \end{array}$$

into an analogous problem for a block diagonalized system of two first order equations. We assume

(2.I) $C(t,\varepsilon) = C(t,0) + 0(\varepsilon)$, and $D(t,\varepsilon) = D(t,0) + 0(\varepsilon)$

are continuous $n \times n$ matrix functions for $0 \le t \le 1$ and $\varepsilon > 0$. Moreover, every eigenvalue of $C(t) = C(t,0)$ has a real part $\ge 8\mu > 0$ for $0 \le t \le 1$.

Then it is well-known (cf. [4]) that for sufficiently small ε the linear equation

$$(8) \qquad \varepsilon x' + C(t,\varepsilon)x = 0$$

has a fundamental matrix solution $X(t) = X(t,\varepsilon)$ such that

$$(9) \qquad |X(t)X^{-1}(s)| \leq \ell \exp(-2\mu(t-s)/\varepsilon) \text{ for } 1 \geq t \geq s \geq 0,$$

where $\ell > 0$ is independent of ε .

We now state the following results.

Lemma 1. Let (2.I) hold. Then there exists $\varepsilon_0 > 0$ such that for $0 < \varepsilon \leq \varepsilon_0$ and $0 \leq t \leq 1$ the following matrix equations

$$(10) \qquad \varepsilon P' = -C(t,\varepsilon)P - \varepsilon P^2 - D(t,\varepsilon),$$

$$(11) \qquad \varepsilon Q' = \varepsilon P(t,\varepsilon)Q + Q[C(t,\varepsilon) + \varepsilon P(t,\varepsilon)] - I,$$

have, respectively, solutions $P = P(t,\varepsilon)$ and $Q = Q(t,\varepsilon)$ which are uniformly bounded and $P(0,\varepsilon) = 0$*, $Q(1,\varepsilon) = 0$.

Moreover, the change of variables

$$(12) \qquad \begin{bmatrix} x \\ x' \end{bmatrix} = \begin{bmatrix} I & -\varepsilon Q \\ P & I-\varepsilon PQ \end{bmatrix} \begin{bmatrix} w \\ z \end{bmatrix}$$

with $P = P(t,\varepsilon)$, $Q = Q(t,\varepsilon)$ transforms (6), (7) into the diagonalized system

$$(13) \qquad \begin{aligned} w' &= P(t,\varepsilon)w + Q(t,\varepsilon)g(t) \\ \varepsilon z' &= -[C(t,\varepsilon) + \varepsilon P(t,\varepsilon)]z + g(t) \end{aligned}$$

and one of the boundary conditions

$$(14) \qquad \begin{aligned} &\text{(i)} \quad w(1,\varepsilon) = \beta_1, \qquad z(0,\varepsilon) = 0, \\ &\text{(ii)} \quad w(0,\varepsilon) - \varepsilon Q(0,\varepsilon)z(0,\varepsilon) = \alpha_1, \quad w(1,\varepsilon) = 0. \end{aligned}$$

*Zero, the zero vector or a zero matrix will all be denoted by 0.

Proof. The existence of a unique solution $P(t,\varepsilon)$ of the matrix Riccati equation (10) such that $P(0,\varepsilon) = 0$ and $\|P(t,\varepsilon)\| \leq \rho$ where $\rho = \mu^{-1}\|D\|/2$ can be proved in the same manner as the proof of the theorem in [3], by considering the integral equation $P(t,\varepsilon) = TP(t,\varepsilon)$ where

$$TP(t,\varepsilon) = -\int_0^t X(t)X^{-1}(s)[P^2(s,\varepsilon) + \varepsilon^{-1}D(s,\varepsilon)]ds.$$

To obtain a bounded solution of (11), let $W(t) = W(t,\varepsilon)$ be the fundamental matrix of the linear equation

$$w' = P(t,\varepsilon)w$$

such that $W(0) = I$. If $\|P(t,\varepsilon)\| \leq \rho$ we have

$$|W(t)W^{-1}(s)| \leq \exp(\rho|t-s|).$$

Also, by Theorem 2 [6] there exists $\varepsilon_1 > 0$ such that for $0 < \varepsilon \leq \varepsilon_1$ the equation

$$\varepsilon z' = -[C(t,\varepsilon) + \varepsilon P(t,\varepsilon)]z$$

has a fundamental matrix $Z(t) = Z(t,\varepsilon)$, with $Z(0) = I$, such that

$$|Z(t)Z^{-1}(s)| \leq \ell_1 \exp(-\mu(t-s)/\varepsilon) \text{ for } 1 \geq t \geq s \geq 0,$$

where $\ell_1 > 0$ is independent of ε.

It is readily verified by differentiation that

$$Q(t,\varepsilon) = \varepsilon^{-1}\int_t^1 W(t)W^{-1}(s)Z(s)Z^{-1}(t)ds \tag{15}$$

is a solution of (11) and for $0 < \varepsilon \leq \mu/2\rho$

$$|Q(t,\varepsilon)| \leq \varepsilon^{-1}\ell_1 \int_t^1 \exp[(\rho - \mu/\varepsilon)(s-t)]ds$$

$$\leq \ell_1(\mu - \varepsilon\rho)^{-1} \leq 2\ell_1/\mu.$$

Thus $Q(t,\varepsilon)$ is bounded.

A straightforward calculation shows that the transformation (12) takes (6), (7) into (13), (14).

Remarks. It will be required in the proof of Theorem 1 that the inverse matrix $Q^{-1}(0,\varepsilon)$ exists for sufficiently small ε. In fact, it follows from (15) that for $0 \leq t \leq 1$

$$\lim_{\varepsilon \to 0} Q(t,\varepsilon) = C^{-1}(t,0) .$$

To see this, we use the identity (which can be checked by integration by parts)

$$H(t) = \varepsilon^{-1} \int_t^1 W(t)W^{-1}(s)H(t)[C(s,\varepsilon)+\varepsilon P(s,\varepsilon)]Z(s)Z^{-1}(t)ds$$

$$+ \int_t^1 W(t)W^{-1}(s)P(s,\varepsilon)H(t)Z(s)Z^{-1}(t)ds$$

$$+ W(t)W^{-1}(1)H(t)Z(1)Z^{-1}(t)$$

and set $H = C^{-1}$. (More details are given in the proof of Lemma 2.)

Let us hasten now to point out that the diagonalized system (13) is more tractable than the original vector equation (6). In fact, the general solution $w = w(t,\varepsilon)$, $z = z(t,\varepsilon)$ of (13) is

$$w(t,\varepsilon) = W(t)p_1 + \int_0^t W(t)W^{-1}(s)Q(s,\varepsilon)g(s)ds ,$$

$$z(t,\varepsilon) = Z(t)p_2 + \varepsilon^{-1} \int_0^t Z(t)Z^{-1}(s)g(s)ds .$$

It only remains to choose the arbitrary constant vectors p_1, p_2 so as to satisfy the boundary conditions (14).

3. Statement and Proof of Theorem 1.

We suppose that the following three assumptions hold:

(3.I) The degenerate problem

$$A(t,0,u)u' = f(t,0,u), \qquad u(1) = \beta(0)$$

has a solution $u = u(t) \in C^2$ for $0 \le t \le 1$ and $\beta(\varepsilon) = \beta(0) + O(\varepsilon)$.

(3.II) For some $\delta > 0$ and $\varepsilon^* > 0$ the function

$$F(t,\varepsilon,y,y') \equiv f(t,\varepsilon,y) - A(t,\varepsilon,y)y'$$

and its partial derivatives F_ε, F_y are continuous in (t,ε,y,y') for

$$0 \le t \le 1,\ 0 < \varepsilon \le \varepsilon^*,\ |y-u(t)| \le \delta,\ |y'-u'(t)| \le \delta(1+\varepsilon^{-1}e^{-\mu t/\varepsilon})$$

such that there exists a constant $k > 0$ with

$$|A(t,\varepsilon,u(t)+y) - A(t,\varepsilon,u(t)+y^*)| \le k|y-y^*|,$$

$$|F_y(t,\varepsilon,u(t)+y,\ u'(t)) - F_y(t,\varepsilon\ u(t)+y^*,\ u'(t))| \le k|y-y^*|.$$

(3.III) Every eigenvalue of $A(t,0,u(t))$ has a real part $\ge 8\mu > 0$ for $0 \le t \le 1$.

We now state and prove the following theorem.

Theorem. Let (3.I), (3.II), (3.III) hold. Then there exists a positive constant δ_0 (independent of ε) such that for $|\alpha(\varepsilon) - u(0)| \le \delta_0$ and for ε sufficiently small, the vector boundary problem (1), (2) has a solution $y(t,\varepsilon)$ on $[0,1]$ satisfying

$$\begin{aligned} y(t,\varepsilon) &= u(t) + O(\varepsilon) + O(e^{-\mu t/\varepsilon}), \\ y'(t,\varepsilon) &= u'(t) + O(\varepsilon) + O(\varepsilon^{-1}e^{-\mu t/\varepsilon}). \end{aligned} \tag{5}$$

Proof. The existence of the solution $y(t,\varepsilon)$ with the required properties will be proved in two steps as in [8]. That is, we set

$$y - u(t) = \xi + \zeta$$

and require that ξ (the 'outer correction') satisfies

$$\varepsilon\xi'' + A(t,\varepsilon,u(t)+\xi)(u' + \xi') = f(t,\varepsilon,u(t)+\xi) - \varepsilon u''$$

or

(16)
$$\varepsilon\xi'' = F(t,\varepsilon,u(t)+\xi, u'(t)+\xi') - \varepsilon u'',$$
$$\xi(1,\varepsilon) = \beta(\varepsilon) - u(1) = \beta(\varepsilon) - \beta(0), \ \xi'(0,\varepsilon) = 0,$$

while ζ (the 'inner correction') satisfies

(17)
$$\varepsilon\zeta'' = F(t,\varepsilon,u(t)+\xi+\zeta, u'(t)+\xi'+\zeta') - F(t,\varepsilon,u(t)+\xi, u'(t)+\xi'),$$
$$\zeta(0,\varepsilon) = \alpha(\varepsilon) - u(0) - \xi(0,\varepsilon), \qquad \zeta(1,\varepsilon) = 0.$$

(i) We consider the terminal value problem (16) first. This problem has the form of the problem (6), (7)(i) if we express the equation of (16) as

$$\varepsilon\xi'' + C_1(t,\varepsilon)\xi' + D_1(t,\varepsilon)\xi = g(t,\varepsilon,\xi,\xi')$$

with

$$C_1(t,\varepsilon) = A(t,\varepsilon,u(t))$$

$$D_1(t,\varepsilon) = -F_y(t,\varepsilon,u(t),u'(t))$$

and

$$g(t,\varepsilon,\xi,\xi') = F(t,\varepsilon,u(t)+\xi, u'(t)+\xi') + C_1(t,\varepsilon)\xi' + D_1(t,\varepsilon)\xi - \varepsilon u''.$$

By assumptions (3.II) and (3.III) the coefficients C_1, D_1 satisfy hypothesis (2.I) of Lemma 1. Denote by $P_1(t,\varepsilon)$, $Q_1(t,\varepsilon)$ the bounded solutions of (10), (11) with C_1, D_1

replacing C,D respectively. Then by the change of variables (12) with $x = \xi$, $P = P_1(t,\varepsilon)$, $Q = Q_1(t,\varepsilon)$, we see that w, z satisfy the diagonalized system

$$w' = P_1(t,\varepsilon)w + Q_1(t,\varepsilon)h(t,\varepsilon,w,z)$$

$$\varepsilon z' = -[C_1(t,\varepsilon) + \varepsilon P_1(t,\varepsilon)]z + h(t,\varepsilon,w,z)$$

and the boundary conditions

$$w(1,\varepsilon) = \beta(\varepsilon) - \beta(0), \qquad z(0,\varepsilon) = 0.$$

Here

$$h(t,\varepsilon,w,z) = g(t,\varepsilon,\xi,\xi').$$

Equivalently, w, z satisfy the integral equations

$$\begin{aligned} w(t,\varepsilon) &= U(t)U^{-1}(1)[\beta(\varepsilon)-\beta(0)] \\ &\quad - \int_t^1 U(t)U^{-1}(s)Q_1(s,\varepsilon)h[s,\varepsilon,w(s,\varepsilon),z(s,\varepsilon)]ds \end{aligned} \tag{18}$$

$$z(t,\varepsilon) = \varepsilon^{-1}\int_0^t V(t)V^{-1}(s)h[s,\varepsilon,w(s,\varepsilon),z(s,\varepsilon)]ds,$$

where $U(t) = U(t,\varepsilon)$, $U(0) = I$, is the fundamental matrix of

$$w' = P_1(t,\varepsilon)w$$

and $V(t) = V(t,\varepsilon)$, $V(0) = I$, is the fundamental matrix of

$$\varepsilon z' = -[C_1(t,\varepsilon) + \varepsilon P_1(t,\varepsilon)]z.$$

Since $P_1(t,\varepsilon)$ is bounded, by choosing $L > 1$ large enough we may assume that

$$|U(t)U^{-1}(s)| \le L$$

(19)

$$|V(t)V^{-1}(s)| \le L\exp[-\mu(t-s)/\varepsilon] \text{ for } 1 \ge t \ge s \ge 0.$$

Now, by assumption (3.I) given $m > 0$ there exists $\varepsilon_2 < 1$ such that $|\beta(\varepsilon) - \beta(0)| \leq m\varepsilon$ for $0 < \varepsilon \leq \varepsilon_2$. Write

$$K = 3k(1 + \|P_1\| + \|Q_1\| + \|P_1\| \|Q_1\|)^2 .$$

Then on applying the mean value theorem twice we obtain, in view of (3.II) and (3.III)

$$|h(t, \varepsilon, w, z) - h(t, \varepsilon, w^*, z^*)| \leq K\pi(w, w^*, z, z^*), \tag{20}$$

where $\pi(w, w^*, z, z^*)$ is the largest of the three values

$$|w-w^*| \max(|w|, |w^*|, |z|, |z^*|),$$

$$\varepsilon|z-z^*| \max(|w|, |w^*|, |z|, |z^*|),$$

$$|z-z^*| \max(|w|, |w^*|, \varepsilon|z|, \varepsilon|z^*|).$$

Also,

$$h(t, \varepsilon, 0, 0) = -\varepsilon u''.$$

The existence of the solution of the integral equations (18) will now be shown by the contraction mapping principle. Let (τ, η) be any pair of bounded continuous functions with norm $\|(\tau, \eta)\| = \|\tau\| + \|\eta\|$ and define the mapping T by $T(\tau, \eta) = (w, z)$, where

$$\begin{aligned} w(t, \varepsilon) &= U(t)U^{-1}(1)[\beta(\varepsilon) - \beta(0)] \\ &\quad - \int_t^1 U(t)U^{-1}(s)Q_1(s, \varepsilon)h[s, \varepsilon, \tau(s, \varepsilon), \eta(s, \varepsilon)]ds, \end{aligned} \tag{21}$$

$$z(t, \varepsilon) = \varepsilon^{-1} \int_0^t V(t)V^{-1}(s)h[s, \varepsilon, \tau(s, \varepsilon), \eta(s, \varepsilon)]ds.$$

From (19), (20) and (21) we have

$$\|w(t, \varepsilon) - w^*(t, \varepsilon)\| \leq LK\|Q_1\| \|\pi(\tau(t, \varepsilon), \tau^*(t, \varepsilon), \eta(t, \varepsilon), \eta^*(t, \varepsilon))\|$$

$$\|z(t, \varepsilon) - z^*(t, \varepsilon)\| \leq LK\mu^{-1}\| \pi(\tau, \varepsilon), \tau^*(t, \varepsilon), \eta(t, \varepsilon), \eta^*(t, \varepsilon))\|$$

and hence

$$\|T(\tau(t,\varepsilon), \eta(t,\varepsilon)) - T(\tau^*(t,\varepsilon), \eta^*(t,\varepsilon))\|$$

$$\leq LK(\|Q_1\| + \mu^{-1})\|\pi(\tau(t,\varepsilon), \tau^*(t,\varepsilon), \eta(t,\varepsilon), \eta^*(t,\varepsilon))\|.$$

Choose $\varepsilon_2 < 1$ so small that

$$4\varepsilon KL^2(\|Q_1\| + \mu^{-1})[m + (\|Q_1\| + \mu^{-1})\|u''\|] \leq 1 \text{ for } 0 < \varepsilon \leq \varepsilon_2.$$

If $\|\tau\|, \|\tau^*\|, \|\eta\|, \|\eta^*\| \leq \gamma\varepsilon$, where

$$\gamma = 2L[m + (\|\tilde{Q}_1\| + \mu^{-1})\|u''\|],$$

then

$$\|\pi(\tau, 0, \eta, 0)\| \leq \gamma^2\varepsilon^2$$

$$\|\pi(\tau, \tau^*, \eta, \eta^*)\| \leq \gamma\varepsilon(\|\tau - \tau^*\| + \|\eta - \eta^*\|).$$

It follows that for $0 < \varepsilon \leq \varepsilon_2$

$$\|T(\tau(t,\varepsilon), \eta(t,\varepsilon)) - T(\tau^*(t,\varepsilon), \eta^*(t,\varepsilon))\|$$

$$\leq \frac{1}{2}[\|\tau(t,\varepsilon) - \tau^*(t,\varepsilon)\| + \|\eta(t,\varepsilon) - \eta^*(t,\varepsilon)\|].$$

Similarly,

$$\|w(t,\varepsilon)\| \leq Lm\varepsilon + L\|Q_1\| \|h(t,\varepsilon,\tau(t,\varepsilon), \eta(t,\varepsilon))\|$$

$$\leq Lm\varepsilon + L\|Q_1\| [K\|\pi(\tau(t,\varepsilon), 0, \eta(t,\varepsilon), 0)\| + \varepsilon\|u''\|]$$

$$\leq (m + \|Q_1\| \|u''\|) L\varepsilon + LK\|Q_1\| \gamma^2\varepsilon^2$$

$$\|z(t,\varepsilon)\| \leq L\mu^{-1}(K\gamma^2\varepsilon^2 + \varepsilon\|u''\|)$$

and

$$\|T(\tau(t,\varepsilon), \eta(t,\varepsilon))\| \leq L\varepsilon(m + \|Q_1\| \|u''\| + \mu^{-1}\|u''\|)$$

$$+ LK(\|Q_1\| + \mu^{-1})\gamma^2\varepsilon^2$$

$$\leq \frac{\gamma\varepsilon}{2} + \frac{\gamma\varepsilon}{2} = \gamma\varepsilon .$$

Then T is a contraction mapping on $S = \{(\tau, \eta): \|(\tau, \eta)\| \leq 2\gamma\varepsilon\}$, and so T has a unique fixed point in S. Therefore (18) has a unique solution $w(t,\varepsilon)$, $z(t,\varepsilon)$ such that

$$\|w(t,\varepsilon)\| + \|z(t,\varepsilon)\| \leq 2\gamma\varepsilon .$$

It follows that the problem (16) has a unique solution $\xi = \xi(t,\varepsilon)$ on $[0,1]$ such that

$$\|\xi(t,\varepsilon)\| = \|w(t,\varepsilon) - \varepsilon Q_1(t,\varepsilon) z(t,\varepsilon)\| = O(\varepsilon),$$

$$\|\xi'(t,\varepsilon)\| = \|P_1(t,\varepsilon)\xi(t,\varepsilon) + z(t,\varepsilon)\| = O(\varepsilon) .$$

(ii) Next we consider the problem (17), where $\xi = \xi(t,\varepsilon)$ is the solution just obtained. This problem can be put in the form of the problem (6), (7)(ii) by writing the equation (17) as

$$\varepsilon\zeta'' + C_2(t,\varepsilon)\zeta' + D_2(t,\varepsilon)\zeta = g^*(t,\varepsilon,\zeta,\zeta')$$

with

$$C_2(t,\varepsilon) = A[t,\varepsilon, u(t) + \xi(t,\varepsilon)]$$

$$D_2(t,\varepsilon) = -F_y[t,\varepsilon, u(t) + \xi(t,\varepsilon), u'(t) + \xi'(t,\varepsilon)]$$

and

$$g^*(t,\varepsilon,\zeta,\zeta') = F(t,\varepsilon, u(t) + \xi(t,\varepsilon) + \zeta, u'(t) + \xi'(t,\varepsilon) + \zeta')$$

$$- F(t,\varepsilon, u(t) + \xi(t,\varepsilon), u'(t) + \xi'(t,\varepsilon))$$

$$+ C_2(t,\varepsilon)\zeta' + D_2(t,\varepsilon)\zeta .$$

Clearly,

$$C_2(t,\varepsilon) = A(t,\varepsilon,u(t)) + O(\varepsilon)$$

$$D_2(t,\varepsilon) = -F_y[t,\varepsilon,u(t),u'(t)] + O(\varepsilon)$$

and therefore C_2, D_2 also satisfy hypothesis (2.I) of the Lemma. Denote by $P_2(t,\varepsilon)$, $Q_2(t,\varepsilon)$ the bounded solutions of (10), (11) with C_2, D_2 in place of C, D respectively. Applying the change of variables (12) with $x = \zeta$, $w = w^*$, $z = z^*$, $P = P_2(t,\varepsilon)$, $Q = Q_2(t,\varepsilon)$, we obtain the diagonalized system

$$w^{*\prime} = P_2(t,\varepsilon)w^* + Q_2(t,\varepsilon)G(t,\varepsilon,w^*,z^*)$$

$$\varepsilon z^{*\prime} = -[C_2(t,\varepsilon) + \varepsilon P_2(t,\varepsilon)]z^* + G(t,\varepsilon,w^*,z^*)$$

with the boundary conditions

$$w^*(0,\varepsilon) - \varepsilon Q_2(0,\varepsilon)z^*(0,\varepsilon) = \alpha(\varepsilon) - u(0) - \xi(0,\varepsilon),$$

$$w^*(1,\varepsilon) = 0 .$$

Here,

$$G(t,\varepsilon,w^*,z^*) = g^*(t,\varepsilon,\zeta,\zeta').$$

Equivalently, we get the integral equations

$$(22)\quad \begin{aligned} w^*(t,\varepsilon) &= -\int_t^1 U_1(t)U_1^{-1}(s)Q_2(s,\varepsilon)G[s,\varepsilon,w^*(s,\varepsilon),z^*(s,\varepsilon)]ds \\ \varepsilon z^*(t,\varepsilon) &= Q_2^{-1}(0,\varepsilon)Z_1(t)[w^*(0,\varepsilon) - \alpha(\varepsilon) + u(0) + \xi(0)] \\ &\quad + \int_0^t Z_1(t)Z_1^{-1}(s)G[s,\varepsilon,w^*(s,\varepsilon),z^*(s,\varepsilon)]ds, \end{aligned}$$

where $U_1(t)$, $U_1(0) = I$, is the fundamental matrix of $w' = P_2(t,\varepsilon)w$ and $Z_1(t)$, $Z_1(0) = I$, is the fundamental matrix of

$$\varepsilon z' = -[C_2(t,\varepsilon) + \varepsilon P_2(t,\varepsilon)]z .$$

Since $P_2(t,\varepsilon)$ is bounded, $U_1(t)$, $Z_1(t)$ satisfy the inequalities (19) if we choose $L > 1$ large enough. Note also that

$$G(t,\varepsilon,0,0) = 0$$

and moreover, G also satisfies the inequality (20) where, for convenience, we use the same constant K although it is now defined in terms of P_2, Q_2.

We now prove by the method of successive approximations that the integral equations (22) has a solution if $|\alpha(\varepsilon) - u(0)| \leq \delta_0$ where

$$\delta_0 = \frac{1}{2c_1} - \|\xi(0,\varepsilon)\|$$

and

$$c_1 = 2\mu^{-1}LK\|Q_2^{-1}(0,\varepsilon)\| [1 + \|Q_2\|(1+\|Q_2^{-1}(0,\varepsilon)\|)] .$$

Since $|\xi(0,\varepsilon)| = O(\varepsilon)$, it follows that $\delta_0 > 0$ for ε small enough. Then

$$c = |\alpha(\varepsilon) - u(0) - \xi(0,\varepsilon)| \leq 1/2c_1 . \tag{23}$$

Set $(w_0^*, z_0^*) = (0,0)$ and $(w_n^*, z_n^*) = (T_1 w_{n-1}^*, T_2 z_{n-1}^*)$, $n=1,2,\ldots$ where

$$\begin{aligned} T_1 w^*(t,\varepsilon) &= -\int_t^1 U_1(t)U_1^{-1}(s)Q_2(s,\varepsilon)G[s,\varepsilon,w^*(s,\varepsilon), z^*(s,\varepsilon)]\,ds , \\ T_2 z^*(t,\varepsilon) &= \varepsilon^{-1}Q_2^{-1}(0,\varepsilon)Z_1(t)[T_1 w^*(0,\varepsilon) - \alpha(\varepsilon) + u(0) + \xi(0,\varepsilon)] \end{aligned} \tag{24}$$

$$+ \varepsilon^{-1} \int_0^t Z_1(t) Z_1^{-1}(s) G[s, \varepsilon, w^*(s,\varepsilon), z^*(s,\varepsilon)]\, ds\,.$$

By (19), (20) and (24) we obtain

$$|w^*_{n+1}(t,\varepsilon) - w^*_n(t,\varepsilon)|$$

$$\leq L \|Q_2\| K \int_t^1 \pi[w^*_n(s,\varepsilon), w^*_{n-1}(s,\varepsilon), z^*_n(s,\varepsilon), z^*_{n-1}(s,\varepsilon)]\, ds$$

$$\varepsilon |z^*_{n+1}(t,\varepsilon) - z^*_n(t,\varepsilon)| \leq \|Q_2^{-1}(0,\varepsilon)\| e^{-\mu t/\varepsilon} |w^*_{n+1}(t,\varepsilon) - w^*_n(t,\varepsilon)|$$

$$+ K \int_0^t e^{-\mu(t-s)/\varepsilon} \pi[w^*_n(s,\varepsilon), w^*_{n-1}(s,\varepsilon), z^*_n(s,\varepsilon), z^*_{n-1}(s,\varepsilon)]\, ds$$

and

$$w^*_1(t,\varepsilon) = 0$$

$$\varepsilon |z^*_1(t,\varepsilon)| \leq \|Q_2^{-1}(0,\varepsilon)\| c e^{-\mu t/\varepsilon}$$

It can be proved by induction that

$$|w^*_n(t,\varepsilon) - w^*_{n-1}(t,\varepsilon)|,\ \varepsilon |z^*_n(t,\varepsilon) - z^*_{n-1}(t,\varepsilon)|$$

$$\leq \| Q_2^{-1}(0,\varepsilon)\| c (cc_1)^{n-1} e^{-\mu t/\varepsilon}$$

$$|w^*_n(t,\varepsilon)|,\ \varepsilon |z^*_n(t,\varepsilon)| \leq 2 \|Q_2^{-1}(0,\varepsilon)\| c e^{-\mu t/\varepsilon}$$

with c defined by (23). Therefore, since $cc_1 \leq 1/2$, the series

$$\Sigma_{n=1}^{\infty} [w^*_n(t,\varepsilon) - w^*_{n-1}(t,\varepsilon)],\ \Sigma_{n=1}^{\infty} [z^*_n(t,\varepsilon) - z^*_{n-1}(t,\varepsilon)]$$

converge uniformly in (t,ε) to a solution $w^*(t,\varepsilon)$, $z^*(t,\varepsilon)$ of (22) and

$$|w^*(t,\varepsilon)| \leq \Sigma_{n=1}^{\infty} |w^*_n(t,\varepsilon) - w^*_{n-1}(t,\varepsilon)| \leq 2 \|Q_2^{-1}(0,\varepsilon)\| c e^{-\mu t/\varepsilon}$$

$$|z^*(t,\varepsilon)| \leq \Sigma_{n=1}^{\infty} |z^*_n(t,\varepsilon) - z^*_{n-1}(t,\varepsilon)| \leq 2 \|Q_2^{-1}(0,\varepsilon)\| c \varepsilon^{-1} e^{-\mu t/\varepsilon}\,.$$

Thus the problem (17) has a solution $\zeta(t,\varepsilon) = w^*(t,\varepsilon) - \varepsilon Q_2(t,\varepsilon)z^*(t,\varepsilon)$. Returning to the original problem (1), (2) we obtain a solution $y(t,\varepsilon)$ on $[0,1]$ which satisfies

$$y(t,\varepsilon) - u(t) = \xi(t,\varepsilon) + \zeta(t,\varepsilon) = O(\varepsilon) + O(e^{-\mu t/\varepsilon})$$

$$y'(t,\varepsilon) - u'(t) = \xi'(t,\varepsilon) + \zeta'(t,\varepsilon) = O(\varepsilon) + O(\varepsilon^{-1}e^{-\mu t/\varepsilon}).$$

This completes the proof.

4. A Converse Result

We proceed now to prove the following converse or complementary result.

Theorem 2. Suppose $f(t,\varepsilon,y)$, $A(t,\varepsilon,y)$ and their partial derivatives with respect to ε, y are continuous in (t,ε,y) for $0 < \varepsilon \leq \varepsilon^+$, $0 \leq t \leq 1$, $|y| \leq \delta^+$ (ε^+, δ^+, > 0), and suppose $\alpha(\varepsilon)$, $\beta(\varepsilon)$ are continuous for $0 \leq \varepsilon \leq \varepsilon^+$. Moreover, suppose the problem (1), (2) has a solution $y(t,\varepsilon)$ for $0 \leq t \leq 1$, $0 < \varepsilon \leq \varepsilon^+$ such that

(4.I) Every eigenvalue of $A(t,\varepsilon,y(t,\varepsilon))$ has real part $\geq 2\mu > 0$, and (4.II) $f(t,\varepsilon,y(t,\varepsilon))$ is uniformly bounded.

Then there exists a sequence $\{\varepsilon_n\}$ such that $u(t) = \lim_{\varepsilon_n \to 0} y(t,\varepsilon_n)$ exists for $0 < t_0 \leq t \leq 1$. Moreover, $u(t)$ satisfies the degenerate problem

$$A(t,0,u(t))u'(t) = f(t,0,u(t)), \qquad 0 \leq t_0 < t \leq 1,$$

(25)

$$u(1) = \beta(0)$$

and

$$y(t,\varepsilon) \to u(t), \qquad y'(t,\varepsilon) \to u'(t)$$

uniformly on any subinterval $0 < t_0 \le t \le 1$.

In the remaining of this section some notations have a different meaning from similar ones in earlier sections but should cause no confusion. We first prove the following elementary results about an auxiliary problem.

Lemma 2. Let $y(t,\varepsilon)$ be the given solution of Theorem 2 such that (4.I) hold. Then the linear problem

$$(26) \qquad \varepsilon Q' = QA(t,\varepsilon,y(t,\varepsilon)) - I, \qquad Q(1,\varepsilon) = 0,$$

has a solution $Q = Q(t,\varepsilon)$ which is uniformly bounded for $0 \le t \le 1$, $0 < \varepsilon \le \varepsilon^+$ and, moreover,

$$\text{(i)} \qquad \lim_{\varepsilon\to 0} Q(t,\varepsilon) = A^{-1}(t,0,y(t,\varepsilon)) \quad \text{for} \quad 0 \le t < 1.$$

If $y(t,\varepsilon) \to u(t)$, then

$$\text{(ii)} \qquad \lim_{\varepsilon\to 0} Q(t,\varepsilon) = A^{-1}(t,0,u(t)) \quad \text{for} \quad 0 \le t < 1.$$

Proof. By assumption (4.I) it follows that $A(t,\varepsilon,y(t,\varepsilon))$ has an inverse A^{-1} and moreover, the linear equation

$$\varepsilon v' = -A(t,\varepsilon,y(t,\varepsilon))v$$

has a fundamental matrix $V(t) = V(t,\varepsilon)$ such that

$$|V(t)V^{-1}(s)| \le \ell_2 e^{-\mu(t-s)/\varepsilon} \quad \text{for} \quad 0 \le s \le t \le 1,$$

where ℓ_2 is a positive constant independent of ε. It can easily be seen by differentiation that

$$Q(t,\varepsilon) = \varepsilon^{-1}\int_t^1 V(s)V^{-1}(t)\,ds$$

is the solution of (26) and is uniformly bounded, since

$$|Q(t,\varepsilon)| \leq \varepsilon^{-1} \int_t^1 \ell_2 e^{-\mu(s-t)/\varepsilon}\, ds \leq \ell_2/\mu .$$

From the identity

$$I = \varepsilon^{-1} \int_t^1 A(s,\varepsilon,y(s,\varepsilon))V(s)V^{-1}(t)ds + V(1)V^{-1}(t)$$

we get

$$Q(t,\varepsilon) - A^{-1}(t,0,y(t,\varepsilon))$$

$$= \varepsilon^{-1}A^{-1}(t,0,y(t,\varepsilon)) \left\{ \int_t^1 [A(t,0,y(t,\varepsilon)) - A(s,\varepsilon,y(s,\varepsilon))]\, V(s)V^{-1}(t)ds - \varepsilon V(1)V^{-1}(t) \right\} .$$

The last term is $O(\varepsilon e^{-\mu(1-t)/\varepsilon})$. Write the integral as the sum $\int_{t+\delta}^1 + \int_t^{t+\delta}$. Then

$$\int_{t+\delta}^1 [A(t,0,y(t,\varepsilon)) - A(s,\varepsilon,y(s,\varepsilon))]V(s)V^{-1}(t)ds = O(\varepsilon e^{-\delta/\varepsilon})$$

and

$$\int_t^{t+\delta} [A(t,0,y(t,\varepsilon)) - A(s,\varepsilon,y(s,\varepsilon))]V(s)V^{-1}(t)ds$$

$$\leq \sup_{t\leq s\leq t+\delta} \|A(t,0,y(t,\varepsilon)) - A(s,\varepsilon,y(s,\varepsilon))\| \ell_2 \mu^{-1} \varepsilon = o(\varepsilon),$$

by choosing $\delta > 0$ small enough. Therefore we obtain (i). Similarly, we obtain result (ii) if $y(t,\varepsilon) \to u(t)$.

Proof of Theorem 2. We can now give a simple integral representation for the given solution $y(t,\varepsilon)$. Let $Q(t,\varepsilon)$ be the function satisfying (26). Then an easy calculation shows that the transformation

$$\begin{pmatrix} y \\ y' \end{pmatrix} = \begin{pmatrix} I & -\varepsilon Q(t,\varepsilon) \\ 0 & I \end{pmatrix} \begin{pmatrix} w \\ z \end{pmatrix}$$

brings (1), (2) into the following equations

$$w' = Q(t,\varepsilon) f(t,\varepsilon,y(t,\varepsilon)),$$

$$\varepsilon z' = -A(t,\varepsilon, y(t,\varepsilon))z + f(t,\varepsilon,y(t,\varepsilon)),$$

$$w(1,\varepsilon) = \beta(\varepsilon),$$

$$w(0,\varepsilon) - \varepsilon Q(t,\varepsilon) z(0,\varepsilon) = \alpha(\varepsilon).$$

That is, the components $w(t,\varepsilon)$, $z(t,\varepsilon)$ of the solution $y(t,\varepsilon)$ have the integral representation

$$w(t,\varepsilon) = \beta(\varepsilon) - \int_t^1 Q(s,\varepsilon) f(s,\varepsilon,y(s,\varepsilon))ds$$

$$\varepsilon z(t,\varepsilon) = Q^{-1}(0,\varepsilon)V(t)[w(0,\varepsilon) - \alpha(\varepsilon)] + \int_0^t V(t)V^{-1}(s) f(s,\varepsilon,y(s,\varepsilon))ds.$$

Clearly, $w(t,\varepsilon)$, $\varepsilon z(t,\varepsilon)$ and hence $y(t,\varepsilon)$ are uniformly bounded for ε small enough, in view of assumptions (4.I), (4.II) and the fact that Q is uniformly bounded. Also

$$y'(t,\varepsilon) = z(t,\varepsilon) = O(\varepsilon^{-1} e^{-\mu t/\varepsilon}) + O(1).$$

It follows that on the interval $[t_0, 1]$, where $t_0 > 0$, the sequence of functions $\{y(t,\varepsilon)\}$ are uniformly bounded and equicontinuous. Hence there exists a sequence $\{\varepsilon_n\}$ such that the functions $y(t,\varepsilon_n)$ converge uniformly as $\varepsilon_n \to 0$ to a continuous function on $[t_0, 1]$. Write

$u(t) = \lim_{\varepsilon_n \to 0} y(t,\varepsilon_n)$ for $0 < t_0 \le t \le 1$. Clearly, on $[t_0, 1]$

$$\lim_{\varepsilon \to 0} \varepsilon z(t,\varepsilon) = 0$$

and so

$$u(t) = \lim_{\varepsilon_n \to 0} y(t, \varepsilon_n) = \lim_{\varepsilon_n \to 0} w(t, \varepsilon_n) .$$

That is, for $0 < t_0 \le t \le 1$

$$u(t) = \lim_{\varepsilon_n \to 0} \beta(\varepsilon_n) - \lim_{\varepsilon_n \to 0} \int_t^1 Q(s, \varepsilon_n) f(s, \varepsilon_n, y(s, \varepsilon_n)) ds$$

$$= \beta(0) - \int_t^1 A^{-1}(s, 0, u(s)) f(s, 0, u(s)) ds ,$$

by virtue of result (ii) of Lemma 2 and the continuity of f. On differentiation, $u(t)$ satisfies (25). Consequently, by the results (5) of Theorem 1 we obtain $y(t, \varepsilon) \to u(t)$, $y'(t, \varepsilon) \to u'(t)$ uniformly on any subinterval $0 < t_0 \le t \le 1$.

REFERENCES

[1] Briš, N.I., On boundary problems for the equation $\varepsilon y'' = f(x, y, y')$ for small ε's., Dokl. Akad. Nauk S.S.S.R. 95 (1954), 429-432.

[2] Chang, K. W., Approximate solutions of nonlinear boundary value problems involving a small parameter, SIAM J. Appl. Math. 26 (1974).

[3] Chang, K. W., Remarks on a certain hypothesis in singular perturbations, Proc. Amer. Math. Soc. 23 (1969), 41-45.

[4] Chang, K. W. and Coppel, W. A., Singular perturbations of initial value problems over a finite interval, Arch. Rational Mech. Anal. 32 (1969), 268-280.

[5] Coddington, E.A. and Levinson, N., A boundary value problem for a nonlinear differential equation with a small parameter, Proc. Amer. Math. Soc. 3 (1952), 73-81.

[6] Coppel, W.A., Dichotomies and reducibility, J. Differential Equations, 3 (1967), 500-521.

[7] Dorr, F.W., Parter, S.V. and Shampine, L.F., Applications of the maximum principle to singular perturbation problems, SIAM Review 15 (1973), 43-88.

[8] Erdélyi, A., On a nonlinear boundary value problem involving a small parameter, J. Austral. Math. Soc. 2 (1962), 425-439.

[9] Howes, F.A., Singular perturbations and differential inequalities, Ph.D. Thesis, University of Southern California, 1974.

[10] O'Malley, R.E., Jr., Introduction to singular perturbations, Academic Press, New York, 1974.

[11] Sibuya, Y., On perturbations of discontinuous solutions of ordinary differential equations, Natural Science Report of Ochanomizu Univ. 11 (1960), 1-18.

[12] Wasow, W., Singular perturbations of boundary value problems for nonlinear differential equations of the second order, Comm. Pure Appl. Math. 9 (1956), 93-113.

University of Calgary, Calgary, Alberta - Canada

EIGENFUNCTION EXPANSIONS FOR SELFADJOINT SUBSPACES GENERATED BY SYMMETRIC ORDINARY DIFFERENTIAL OPERATORS[1]

Earl A. Coddington

1. Introduction. This report summarizes several results which will appear, among others, in a joint work with A. Dijksma, [1]. We consider a differential expression

$$L = \sum_{k=0}^{n} p_k D^k, \quad D = d/dx,$$

on an arbitrary open real interval $\iota = (a, b)$, where the p_k are complex-valued functions of class $C^k(\iota)$, and $p_n(x) \neq 0$, $x \in \iota$. We assume L is formally symmetric, that is, $L = L^+$, where

$$L^+ = \sum_{k=0}^{n} (-1)^k D^k \bar{p}_k .$$

Operators (single-valued functions) in the Hilbert space $\mathfrak{H} = \mathcal{L}^2(\iota)$ will be identified with their graphs in $\mathfrak{H}^2 = \mathfrak{H} \oplus \mathfrak{H}$. Associated with L in $\mathfrak{H}$ are two closed operators, the minimal operator S_0 and its adjoint S_0^*, which is the

1 This work was supported in part by the National Science Foundation under NSF Grant GP-33696X.

maximal operator for L in $\mathfrak{H}$. The operator S_0 may be defined as the closure in $\mathfrak{H}^2$ of the set of all $\{f, Lf\}$, $f \varepsilon C_0^n(\iota)$, the set of all functions in $C^n(\iota)$ with compact support. The maximal operator S_0^* is the set of all $\{f, Lf\} \varepsilon \mathfrak{H}^2$ such that $f \varepsilon C^{n-1}(\iota)$ and $f^{(n-1)}$ is locally absolutely continuous on ι. Since $L = L^+$ we have $S_0 \subset S_0^*$, that is, S_0 is symmetric.

For illustrative purposes, let us suppose that we are in the regular case where a, b are finite, and

$p_k \varepsilon C^k(\bar{\iota})$, $p_n(x) \neq 0$, $x \varepsilon \bar{\iota}$, where $\bar{\iota}$ is the closure of ι. Then we may write

$$S_0^* = \{\{f, Lf\} \varepsilon \mathfrak{H}^2 \mid f \varepsilon C^{n-1}(\bar{\iota}),\ f^{(n-1)} \varepsilon AC(\bar{\iota})\},$$

$$S_0 = \{\{f, Lf\} \varepsilon S_0^* \mid \tilde{f}(a) = \tilde{f}(b) = 0\},$$

where $\tilde{f}(x)$ is the matrix with rows $f(x), f'(x), \ldots, f^{(n-1)}(x)$. A typical eigenvalue problem for L in $\mathfrak{H}$ is given by

$$Lf = \lambda f,$$

(1)

$$b_j(f) \equiv \sum_{k=1}^{n} m_{jk} f^{(k-1)}(a) + n_{jk} f^{(k-1)}(b) = 0, \quad j = 1, \ldots, p,$$

where $m_{jk}, n_{jk} \varepsilon \mathbb{C}$, the complex numbers. Associated with (1) is the operator A in $\mathfrak{H}$ defined by

$$A = \{\{f, Lf\} \varepsilon S_0^* \mid b_j(f) = 0,\ j = 1, \ldots, p\},$$

and clearly $S_0 \subset A \subset S_0^*$. The problem (1) is just the problem of computing the eigenvalues and eigenfunctions of A. The boundary functionals b_j are linear on the domain of S_0^*, $\mathcal{D}(S_0^*)$, but they are not continuous there. However, they are continuous linear functionals on S_0^*, considered as a subspace (closed linear manifold) in $\mathfrak{H}^2$. This implies that there exist $\{\sigma_j, \tau_j\} \in \mathfrak{H}^2$, $j = 1, \ldots, p$, such that

$$b_j(f) = (\{f, S_0^* f\}, \{\sigma_j, \tau_j\}) = (f, \sigma_j) + (Lf, \tau_j), \quad f \in \mathcal{D}(S_0^*).$$

Then, if B is the span of $\{\sigma_1, \tau_1\}, \ldots, \{\sigma_p, \tau_p\}$, we have $A = S_0^* \cap B^\perp$, where $B^\perp = \mathfrak{H}^2 \ominus B$.

The functionals b_j are not the only type of continuous linear functionals on S_0^*. For example, if $\mu_1, \ldots, \mu_n$ are functions of bounded variation on $\bar{\imath}$, then the b given by

$$b(f) = \sum_{k=1}^{n} \int_a^b f^{(k-1)}(x) d\mu_k(x), \quad f \in \mathcal{D}(S_0^*),$$

is continuous on S_0^*. Of course, the most general finite set of continuous linear functionals on S_0^* is prescribed by an arbitrary subspace $B \subset \mathfrak{H}^2$, $\dim B = p < \infty$, and it is natural to consider, for such a B, the operator $A = S_0^* \cap B^\perp$. Since $S_0 \subset S_0^*$, we have $S_0 \cap B^\perp \subset A$. Now A need not be densely defined, and hence its adjoint A^*, defined by

$$A^* = \{\{h, k\} \in \mathfrak{H}^2 \mid (g, h) = (f, k), \text{ all } \{f, g\} \in A\},$$

need not be (the graph of) an operator, and, even if $\mathcal{D}(A)$ is dense in $\mathfrak{H}$, A^* need not be a differential operator. More generally, therefore, we should study subspaces $A \subset \mathfrak{H}^2$ satisfying $S_0 \cap B^\perp \subset A$, and this is what we do.

2. The Basic Operator and Its Selfadjoint Subspace Extensions.

We now consider an L in the general (not necessarily regular) case on an arbitrary open interval ι, and let S_0 be the minimal operator for L in $\mathfrak{H} = \mathcal{L}^2(\iota)$. Let B be a subspace of $\mathfrak{H}^2$, $\dim B = p < \infty$, and put $S = S_0 \cap B^\perp$. We shall assume a nontriviality condition, $B \cap S_0^\perp = \{\{0, 0\}\}$, which can always be achieved. Note that if $\{\sigma, \tau\} \in B \cap S_0^\perp$ then $(f, \sigma) + (S_0 f, \tau) = 0$ for all $f \in \mathcal{D}(S_0)$, and such an element gives no restriction on S_0. This S is our basic operator. It is symmetric, $S \subset S^*$, and one can show that S^* is an algebraic sum

$$S^* = S_0^* \dotplus (-B^{-1}) = \{\{f, S_0^* f\} + \{\tau, -\sigma\} \mid f \in \mathcal{D}(S_0^*),$$
$$\{\sigma, \tau\} \in B\}.$$

We are interested in the selfadjoint subspaces $H = H^*$ such that $S \subset H$. For such an H, let $H_\infty = \{\{0, g\} \in H\}$ and $H(0) = \{g \in \mathfrak{H} \mid \{0, g\} \in H\}$. Then $H_s = H \ominus H_\infty$ is (the graph of) a closed densely defined selfadjoint operator on $H(0)^\perp = \mathfrak{H}^2 \ominus H(0)$. This allows a spectral analysis of the subspace H. If S has a selfadjoint extension H in $\mathfrak{H}^2$, then $S \subset H \subset S^*$. It turns out that S has a selfadjoint extension H in $\mathfrak{H}^2$ if and only if S_0 does, that is, if and only if S_0 has equal deficiency indices. One can characterize all such H, when they exist, by generalized boundary conditions. These involve boundary values and integral terms, and integro-boundary terms also enter into the description of the operator part H_s of H.

The symmetric operator S always has selfadjoint extensions in $\mathcal{K}^2$, $\mathcal{K} \subset \mathcal{K}$, for an appropriate Hilbert space $\mathcal{K}$. Let H be any such. Then H_s is a selfadjoint operator on $H(0)^{\perp} = \mathcal{K}^2 \ominus H(0)$, and it has the spectral resolution

$$H_s = \int_{-\infty}^{\infty} \lambda \, dE_s(\lambda),$$

where $E_s = \{E_s(\lambda) | \lambda \in \mathbb{R}\}$ is the <u>spectral family</u> of projections in $H(0)^{\perp}$ for H_s. The <u>resolvent</u> R_H of H is the operator-valued function defined by $R_H(\ell) = (H - \ell I)^{-1}$, $\ell \in \mathbb{C}_0 = \mathbb{C} \setminus \mathbb{R}$. It has the representation

$$R_H(\ell) = \int_{-\infty}^{\infty} \frac{dE(\lambda)}{\lambda - \ell}, \quad \ell \in \mathbb{C}_0,$$

where $E(\lambda)f = E_s(\lambda)f$, $f \in H(0)^{\perp}$, and $E(\lambda)f = 0$, $f \in H(0)$. The set $E = \{E(\lambda) | \lambda \in \mathbb{R}\}$ is the <u>spectral family</u> of projections in $\mathcal{K}$ for H. Let P denote the orthogonal projection of $\mathcal{K}$ onto $\mathcal{K}$, and let

$$R(\ell) f = PR_H(\ell)f, \quad f \in \mathcal{K}, \ \ell \in \mathbb{C}_0.$$

The operator-valued function R is called the generalized resolvent of S corresponding to H. It is easy to see that

(2) $$(R(\ell)f, f) = \int_{-\infty}^{\infty} \frac{d(F(\lambda)f, f)}{\lambda - \ell}, \quad f \in \mathfrak{K}, \quad \ell \in \mathbb{C}_0,$$

where $F(\lambda)f = PE(\lambda)f$, $f \in \mathfrak{K}$. The set $F = \{F(\lambda) \mid \lambda \in \mathbb{R}\}$ is called the generalized spectral family of S corresponding to H. The main problem is to compute F, and this can be accomplished by an analysis of the nature of $R(\ell)$ and an inversion of the formula (2). This leads to an eigenfunction expansion result for each selfadjoint subspace extension H of S.

3. Eigenfunction Expansions. Let $\sigma = (\sigma_1, \ldots, \sigma_p)$, $\tau = (\tau_1, \ldots, \tau_p)$, where $\{\sigma_1, \tau_1\}, \ldots, \{\sigma_p, \tau_p\}$ is a basis for B, and let $c \in \iota$ be fixed. We define $s^1 = (s_1, \ldots, s_n)$, $u = (u_{n+1}, \ldots, u_{n+p})$, to be the unique vector-valued functions on $\iota \times \mathbb{C}$ satisfying

(3) $$\begin{cases} (L - \ell)\, s^1(\ell) = 0, & \hat{s}^1(c, \ell) = I_n, \\ (L - \ell)u(\ell) = \sigma + \ell\tau, & \hat{u}(c, \ell) = O_n^p, \end{cases}$$

where I_n is the $n \times n$ identity matrix, and O_n^p is the $n \times p$ zero matrix. Finally, we let

(4) $$s(\ell) = (s^1(\ell) : u(\ell) + \tau),$$

the $1 \times (n+p)$ matrix obtained by placing the elements of $u(\ell) + \tau$ next to those of $s^1(\ell)$ in the order indicated.

Theorem 1. Let H be any selfadjoint subspace extension of S in $\mathcal{K}^2$, $\mathcal{H} \subset \mathcal{K}$, with generalized spectral family F. If s is defined via (3), (4), then there exists an $(n+p) \times (n+p)$ matrix-valued function ρ on $\mathbb{R}$ which is hermitian, nondecreasing, of bounded variation on each finite interval, such that

$$(F(\lambda) - F(\mu))f = \int_{\mu}^{\lambda} s(\nu)d\rho(\nu)\hat{f}(\nu), \quad f \varepsilon C_0(\iota),$$

where λ, μ are continuity points of F, and

$$\hat{f}(\nu) = \int_a^b s^*(x,\nu)f(x)dx.$$

The matrix-valued function ρ is essentially uniquely determined by F, and is called the spectral matrix for H and F.

Let ζ, η be functions from $\mathbb{R}$ into $\mathbb{C}^{n+p}$, considered as $(n+p) \times 1$ matrices, and define

$$(\zeta, \eta) = \int_{-\infty}^{\infty} \eta^*(\nu)d\rho(\nu)\zeta(\nu), \quad \|\zeta\| = (\zeta,\zeta)^{1/2}.$$

Then we can form the Hilbert space $\hat{\mathcal{H}} = \mathcal{L}^2(\rho) = \{\zeta \mid \|\zeta\| < \infty\}$.

Theorem 2. Let H, s, ρ be as in Theorem 1. For $f \varepsilon \mathcal{H}$,

$$\hat{f}(\nu) = \int_a^b s^*(x,\nu) f(x) dx$$

converges in norm in $\hat{\mathfrak{H}}$, and

(5) $$F(\infty)f = \int_{-\infty}^{\infty} s(\nu) d\rho(\nu) \hat{f}(\nu),$$

where the integral converges in norm in $\mathfrak{H}$. The map V from $\mathfrak{H}$ into $\hat{\mathfrak{H}}$, defined by $Vf = \hat{f}$, is a contraction, $\|Vf\| \leqq \|f\|$. It is an isometry, $\|Vf\| = \|f\|$, for $f \in \mathfrak{H} \cap H(0)^{\perp} = \mathfrak{H} \ominus PH(0)$, and $F(\infty) f = f$ for such f.

It is easy to see that PH(0) has a finite dimension, and (5) represents an eigenfunction expansion for $f \in \mathfrak{H} \ominus PH(0)$, a space of finite codimension.

Theorem 3. We have $V(\mathfrak{H} \ominus PH(0)) = \hat{\mathfrak{H}}$ if and only if F is the spectral family for a selfadjoint subspace extension of S in $\mathfrak{H}^2$ itself.

REFERENCE

[1] E. A. Coddington and A. Dijksma, Selfadjoint subspaces and eigenfunction expansions for ordinary differential subspaces, to appear.

University of California, Los Angeles, California

A NONLINEAR OSCILLATION THEOREM

W. J. Coles

1. Introduction.

The equation

$$(1) \qquad y'' + q(t,y) = 0 \qquad (q \text{ continuous})$$

is said to be oscillatory if all solutions which exist on right hald-lines have arbitrarily large zeros. For an extensive bibliography of papers considering oscillatory properties of (1), see for instance Wong [13]; here we briefly sketch some developments in the direction of the theorems in §2.

In case (1) is linear $(q(t,y) = f(t)y)$, the well-known Wintner [10] theorem says that (1) is oscillatory if

$$(2) \qquad \int^{\infty} f = \infty.$$

A Natural extension is to the nonlinear case

$$(3) \qquad y'' + f(t)|y|^{\alpha} \operatorname{sgn} y = 0 .$$

For $f \geq 0$ and α an odd integer greater than 1, Atkinson [1] proved that (3) is oscillatory iff

$$(4) \qquad \int^{\infty} t\, f(t)\, dt = \infty;$$

the same theorem for arbitrary $\alpha > 1$ follows from results of Waltman [8] and also Wong [12].

In the sublinear case $(0 < \alpha < 1)$, Belohorec [3] proved that if $f > 0$ then (3) is oscillatory iff

$$(5) \qquad \int^{\infty} t^{\alpha} f(t)\, dt = \infty .$$

In case $f \ngeq 0$, Waltman [9] proved that (3) is oscillatory for $\alpha > 1$ if (2) holds; results of Bhatia[2] and Wong[11] imply Waltman's theorem for the sublinear case as well. Kamenev [5] showed that if $g \in C'$, $y\, g(y) > 0$ for $y \neq 0$, $g' \geq 0$, and

$$\int^{\infty} \frac{1}{g} < \infty,$$

then (4) implles oscillation for

$$(6) \qquad y'' + f(t)\, g(y) = 0 .$$

For $\alpha > 1$, this result applies to (3), but, since $f \ngeq 0$, neither (2) nor (4) implies the other. Belohorec [4] proved that (3) is oscillatory for $0 < \alpha < 1$ if there is a $\beta \in [0, \alpha]$ such that

$$(7) \qquad \int^{\infty} t^{\beta} f(t)\, dt = \infty .$$

Wong [13] showed that (2) implies oscillation for (1) provided

$$(8) \qquad \frac{q(t,y)}{g(y)} \geq f(t) ,$$

where $y\, g(y) > 0$ for $y \neq 0$, and g' exists on $(-\infty, \infty)$. Other sufficient conditions for oscillation of (1) are couched in terms of weighted means of weighted integrals

$$\int^{t} \rho(s)\, f(s)\, ds$$

(e.g., see Wong [13], [14] and Kamenev [6], [7]), but we do not pursue this direction here. Rather, we note that conditions mentioned above are both necessary and sufficient if $f \geq 0$, but are known only to be sufficient if $f \ngeq 0$. Thus if $f \ngeq 0$ it makes sense for (6) (and, more generally, for (1) with q satisfying (8)) to seek other weight functions ρ such that

(9) $$\int^{\infty} \rho f = \infty$$

implies oscillation. That such results are of interest follows from the fact that if ρ_1 and ρ_2 are two non-negative functions, neither of which is a constant multiple of the other, then there is a continuous function f such that

$$\int^{\infty} \rho_1 f = \infty \quad \text{and} \quad \underline{\lim}_{t\to\infty} \int^{t} \rho_2 f < \infty ;$$

i.e., essentially different weight functions in (9) test different classes of coefficient functions for (6).

2. Conditions sufficient for oscillation.

Our overriding assumptions on (1) are

(10) $$\begin{cases} \dfrac{q(t,y)}{g(y)} \geq f(t); \ g \in C'; \ g' > 0 \text{ and } y\, g(y) > 0 \text{ for } y \neq 0; \\ G(y) = \displaystyle\int_0^y \frac{1}{g} \text{ exists.} \end{cases}$$

The final assumption implies that our results will be extensions of the sublinear case for (3).

THEOREM 1. If there is a $\rho \in C''$ such that $\rho > 0$, $\rho' \leq 0$, $\rho'' \geq 0$, and

(11) $$\int^{\infty} \rho(t)\, f(t)\, dt = \int^{\infty} \frac{1}{\rho(t)} \int^{t} \rho(s)\, f(s)\, ds\, dt = \infty ,$$

then (1) is oscillatory.

THEOREM 2. Let $c > 0$ and g be such that

$$(12) \qquad \frac{G''(y)\,G(y)}{G'^2(y)} \leq -\frac{1}{c} .$$

If there is a $\rho \in C''$ such that $\rho > 0$, $\rho' > 0$, (11) holds, and

$$(13) \qquad \frac{\rho''(t)\,\rho(t)}{\rho'^2(t)} \leq -c ,$$

then (1) is oscillatory.

PROOF OF THEOREMS 1 AND 2. Suppose that (1) has a nonoscillatory solution y, which may be supposed positive on some right half-line; henceforth we stay on this half-line.

To isolate the term $\int^t \rho f$, let $u = y'/g(y)$ and use (1) and (8) to get

$$u' = \frac{-q(t,y)}{g(y)} - u^2 g'(y) \leq -f - u^2 g'(y).$$

Multiplication by ρ and integration by parts give

$$(14) \qquad \rho u \leq \rho_0 u_0 + \int_{t_0}^{t} \rho' u - \int_{t_0}^{t} \rho f - \int_{t_0}^{t} g'(y)\,\rho u^2 .$$

Since $G(y) = \int_0^y \frac{1}{g}$, integration of $\int_{t_0}^{t} \rho' u$ by parts gives

$$(15) \qquad \int_{t_0}^{t} \rho' u = \rho' G(y) - \rho_0' G(y_0) - \int_{t_0}^{t} \rho'' G(y).$$

Let $a + b = 1$; write

$$\int_{t_0}^{t} \rho' u = a \int_{t_0}^{t} \rho' u + b \int_{t_0}^{t} \rho' u .$$

Substitute this into (14), and use (15) on $b \int_{t_0}^{t} \rho' u$ but not on $a \int_{t_0}^{t} \rho' u$, to get

$$(16)\quad \rho u \le C + a\int_{t_0}^{t} \rho' u + b\,\rho' G(y) - b\int_{t_0}^{t} \rho'' G(y) - \int_{t_0}^{t} \rho g'(y) u^2 - \int_{t_0}^{t} \rho f$$

($C = \rho_0 u_0 - b\,\rho_0{}' G(y_o)$). Let $h(t)$ be defined by

$$2\,\rho g'(y) h = a\,\rho' .$$

Then

$$a\int_{t_0}^{t} \rho' u - \int_{t_0}^{t} \rho g'(y) u^2 = \int_{t_0}^{t} \frac{a^2 \rho'^2}{4\rho g'(y)} - \int_{t_0}^{t} \rho g'(y)(u-h)^2 .$$

Using this in (16) gives

$$\rho u \le C - \int_{t_0}^{t} \rho f + b\,\rho' G(y) - \int_{t_0}^{t} \rho g'(y)(u-h)^2 + \int_{t_0}^{t}\left(\frac{a^2 \rho'^2}{4\rho g'(y)} - b\rho'' G(y)\right),$$

which leads to

$$(17)\qquad \rho u \le C - \int_{t_0}^{t} \rho f$$

if

$$(18)\qquad (1-a)\rho' \le 0$$

and

$$(19)\qquad \frac{a^2 \rho'^2}{4\rho g'(y)} \le (1-a)\,\rho'' G(y).$$

Now we consider the cases $a = 1$, $a < 1$, and $a > 1$. If $a = 1$, (19) implies that $\frac{1}{4} < 0$, so the choice $a = 1$ cannot exploit this method.

If $a < 1$, then (19) can be written

$$(20)\qquad \frac{a^2}{4(1-a)} \le \frac{\rho''(t)\rho(t)}{\rho'^2(t)} \cdot G(y) g'(y)$$

The inequality (20) which least restricts the right-hand side is for $a = 0$, so

$$(21) \qquad 0 \leq \frac{\rho''(t)\,\rho(t)}{\rho'^2(t)}\, G(y)g'(y)$$

implies (13) for any $a < 1$. But (21) holds for all ρ and g of Theorem 1, and so, since (18) also holds, the hypotheses of Theorem 1 imply (17).

If $a > 1$, (19) can be written

$$(22) \qquad \frac{a^2}{4(1-a)} \geq \frac{\rho''(t)\,\rho(t)}{\rho'^2(t)} \cdot G(y)\, g'(y)\,.$$

The maximum left-hand side is -1 (for a = 2), so (22) is implied for each $a > 1$ by

$$(23) \qquad -1 \geq \frac{\rho''(t)\,\rho(t)}{\rho'^2(t)} \cdot G(y)g'(y)\,.$$

Since (23) involves the solution $y(t)$, it is not very helpful as it stands. What are wanted are hypotheses on ρ and g seperately which, when taken together, imply (23). Such hypotheses are furnished in Theorem 2. Since

$$G(y) = \int_0^y \frac{1}{g}, \quad G'(y) = \frac{1}{g(y)}, \quad \text{and} \quad G''(y) = -\frac{g''(y)}{g^2(y)}\,.$$

the hypotheses

$$\frac{G''(y)\,G(y)}{G'^2(y)} \leq -\frac{1}{c}, \quad G(y) \geq 0, \quad g'(y) > 0$$

imply $0 \leq G(y)\, g'(y) \leq \frac{1}{c}$.

This, together with

$$\frac{\rho''(t)\,\rho(t)}{\rho'^2(t)} \leq -c\,,$$

gives (23). Thus (17) holds under the hypotheses of Theorem 2.

Proceeding now from (17), if $\int^{\infty} \rho f = \infty$, then for large t (say $t \geq t_1$) we get

$$\int_{t_0}^{t} \rho f \geq 2C,$$

so (17) implies that

$$\rho u \leq -\frac{1}{2}\int_{t_0}^{t} \rho f$$

or

$$u \leq -\frac{1}{2}\frac{1}{\rho}\int_{t_0}^{t} \rho f.$$

Integration gives

$$G(y(t)) - G(u(t_1)) \leq -\frac{1}{2}\int_{t_1}^{t} \frac{1}{\rho(s)} \int_{t_0}^{t} \rho(r) f(r) dr\, ds \rightarrow -\infty,$$

which, since $G(y) \geq 0$, is a contradiction.

This completes the proofs of Theorems 1 and 2.

3. Remarks.

1. The results demonstrate that without the condition $f \geq 0$ no condition (9) can be both necessary and sufficient for oscillation of (1).

2. The hypothesis (11) of Theorem 1 is implied by

$$\int^{\infty} \rho f = \int^{\infty} \frac{1}{\rho} = \infty,$$

which probably is easier to check.

3. A necessary and sufficient condition that ρ satisfy (13) with equality is (neglecting multiplication by positive constants) that ρ be of the form $(t + k)^{\beta}$, $\beta = \frac{1}{c+1}$, k constant. It follows that if g satisfies the hypotheses of Theorem 2, then (1) is oscillatory if there are constants k and β $(0 \leq \beta \leq \frac{1}{1+c})$ such that

$$\int^{\infty} (t+k)^{\beta} f(t)\, dt = \infty.$$

Since $(t+k)^{\beta}$ is not a constant multiple of t^{β}, taking $k \neq 0$ gives a nontrivial extension of Belohorec's result.

4. When (6) is specialized to the sublinear case of (3), the appropriate weight function ρ in (9) is g, and, if $f \geq 0$, this is the only possible weight function. More generally, one might ask what relationships exist between g and possible weight functions; the hypotheses in Theorem 2 are a step in this direction. If, in addition to (13), there is a t_0 such that $c\,\rho_0' \int^{t_0} \frac{1}{\rho} \leq 1$, then $c\,\rho' \int^{t} \frac{1}{\rho} \leq 1$ for $t \geq t_0$. Thus $g = p$ and (12) imply equality in (12) so that, except for a multiplicative constant which can be absorbed in f, $g(y) = |y|^{\alpha} \operatorname{sgn} y$, $\alpha = \dfrac{1}{1+c}$. In sum, under these conditions ρ can be taken to be g only in the sublinear case of (3).

On the other hand, if, in addition to (12), $c \circ g' \circ G\text{-}1$ is nonincreasing, then the conditions on ρ in Theorem 2 are fulfilled by g, so ρ can be taken to be g. Let us conclude by stating this fact as a theorem.

THEOREM 3. Let g satisfy (10). Let there be a positive constant c such that $c \circ g' \circ G\text{-}1$ is nonnegative and nonincreasing. Then (1) is oscillatory provided

$$\int^{\infty} f(t)\, g(t)\, dt = \int^{\infty} \frac{1}{g(t)} \int^{t} f(s)\, g(s)\, ds = \infty.$$

REFERENCES

[1] F. V. Atkinson, On second-order non-linear oscillations, Pac. J. Math. 5(1955), 643-647.

[2] N. P. Bhatia, Some oscillation theorems for second order differential equations, J. Math. Anal. Appl. 15(1966), 442-446.

[3] Š. Belohorec, Oscilatorické riešenia istej nelineárnej diferenciálnej rovnice druhého rádee, Mat.-fyz. časop. 11/1961/, 250-255.

[4] Š. Belohorec, Two remarks on the properties of solutions of a nonlinear differential equation, Acta. F. R. N. Univ. Comen.-Mathematica XXII-1969, 19-26.

[5] I. V. Kamenev, Oscillation of solutions of second-order nonlinear equations, Trans. Moscow Electrical-Machine-Design Institute No. 5 (1969-70), 125-136.

[6] I. V. Kamenev, A criterion for the oscillation of solutions of second-order ordinary differential equations, Moscow Electronic-Machine-Design Institute (Transl. from Mat. Zametki, Vol. 8, No. 6 (1970), 773-776).

[7] I. V. Kamenev, Some specifically nonlinear oscillation theorems, Moscow Electrical-Machine-Design Institute (Transl. from Mat. Zametki, Vol. 10, No. 2 (1971), 129-134).

[8] P. Waltman, Oscillation of solutions of a nonlinear equation, SIAM Rev. 5(1963), 128-130.

[9] P. Waltman, An oscillation criterion for a nonlinear second order equation, J. Math. Anal. Appl. 10(1965), 439-441.

[10] A. Wintner, A criterion of oscillatory stability, Quart. Appl. Math. 7(1949), 115-117.

[11] J. S. W. Wong, On two theorems of Waltman, SIAM Jour. 14(1966), 724-728.

[12] J. S. W. Wong, A note on second order nonlinear oscillations, SIAM Rev. 10(1968), 88-91.

[13] J. S. W. Wong, On second order nonlinear oscillation, Funkcialaj Ekvcioj 11(1968), 207-234.

[14] J. S. W. Wong, A second order nonlinear oscillation theorem,PAMS 40(1973), 487-491.

University of Utah, Salt Lake City, Utah

ON GLOBAL CONTROLLABILITY
by
R. Conti

1- "Controllability", like "stability", "regularity, etc. is one of the many valued terms of mathematical language. By what follows it should appear that this statement is true even if we confine ourselves to control systems represented by the classical ODE

$$(U) \qquad dx / dt - A(t)\, x = B(t)\, u(t)$$

and to "global controllability". By this we mean, roughly speaking, the possibility of transferring x from an arbitrary state to zero, or viceversa. Within this context we shall meet definitions of controllability which are of common use since long and classical by now, as well as others which might become of use later.

To proceed further, more details about (U) are necessary. As usual, $x = x(t)$ is an n-vector (the state of the system), $u(t)$ an m-vector (the control), $A(t)$ and $B(t)$ are real matrices of type $n \times n$, $n \times m$ respectively, all depending on t (the time) which varies on some interval $J =]\alpha, \omega[$, $-\infty \leq \alpha < \omega \leq +\infty$. Both $A: t \to A(t)$ and $B: t \to B(t)$ are Lebesgue measurable and locally integrable on J.

For a given control system, controllability, whatever this means, depends in an essential way on the duration of the performance, a subinterval Δ of J .

Δ can be compact with either fixed or variable end. points, or it can be a subinterval with one endpoint in common with J , the other one being either fixed or variable.

Apart from that, a further distinction is represented by the nature of the set U of controls to be used. Anyway U will always be a subset of the space $L^1_{loc}(J, E^m)$, E^m, the euclidean m-dimensional space. As a matter of notations, for any $\theta \in J$ and $\chi \in E^n$ the solution $t \to x(t, \theta, \chi, u)$ of (U) such that

$$x(\theta, \theta, \chi, u) = \chi$$

will be represented by

$$(1.1) \qquad x(t, \theta, \chi, u) = G(t, \theta)\chi + \int_\theta^t G(t, s) B(s)u(s)dy$$

where G is the transition matrix of

$$(E_0) \qquad dx/dt - A(t) x = 0.$$

2. We shall start with controllability on a fixed compact interval ;

Def. A.

Given $\tau, T \in J$, $\tau < T$, we say that (U) is $[\tau, T]$-controllable to zero by means of U iff for every $v \in E^n$ there exists $u_v \in U$ such that

$$(2.1) \qquad x(T, \tau, u_v, v) = 0$$

As shown by (1.1), this requires that U be unbounded.

The classical condition that (U) be a proper system, on $[\tau, T]$, i.e.

(2.2) $$y^* G(\tau, t)\, B(t) = 0, \text{ a.e. } t \in [\tau, T] \Rightarrow y = 0,$$

serves to characterize $[\tau, T]$-controllability in terms of A(via G) and B, when U is the whole space $L^1_{\ell oc}(J, E^m)$. This is a space too large for applications, which generally require the use of controls as simple as possible. Fortunately (2.2) still characterizes $[\tau, T]$- controllability when U is restricted to anyone of the spaces $p'_{\ell oc}$, $1 < p' \leq +\infty$ and even to the space PC of piecewise constant functions.

A further step towards "simplicity", namely the use of constant controls, would lead to replacing (2.2) by the more stringent (and obvious) condition

(2.3) $$\text{rank} \int_{\tau}^{T} G(\tau, t)\, B(t)\, dt = n.$$

When A and B are constant (2.2) and (2.3) can be written

(2.4) $$y^* e^{-tA} B = 0, \ t \in \mathbb{R} \Rightarrow y = 0$$

(2.5) $$\text{rank} \left(\int_0^{T-\tau} e^{-tA}\, dt \right) B = n$$

respectively. While the first is independent of τ, T, the validity of (2.5) may actually depend on τ, T.

It is well known that (2.4) is also equivalent to

(2.6) $$\text{rank} \ \mathcal{C} = n$$

where the $n \times mn$ controllability matrix $\mathcal{C}$ is defined by

(2.7) $$\mathcal{C} = [B, AB, \ldots, A^{n-1}B].$$

This shows that $[\tau, T]$-controllability by means of $U = PC$ can be characterized directly in terms of A and B, avoiding the calculation of e^{-tA}.

Question 2.1.

Is a similar characterization possible when A and B are only piecewise constant?

(For a partial answer see L. E. Zabello [16]).

It is obviously desirable that the dependence of u_v on v in Def. A be a linear one. Whether this is always possible it is not clear. However, a very general condition is known, namely

$$BB^* \in L^1_{\ell oc} .$$

In this case we can define the Kalman matrix

(2.8) $$K(\tau, T) = \int_{\tau}^{T} G(\tau, t) B(t) B^*(t) G^*(\tau, t) dt$$

and (2.2) can be written

(2.9) $$K(\tau, T) > 0$$

so that $K^{-1}(\tau, T)$ exists. Then u_v of (2.1) can be defined by

(2.10) $$u_v(t) = - B^*(t) G^*(\tau, t) K^{-1}(\tau, T) v.$$

Let us now recall, for later reference, a much stronger condition than (2.9) also due to R. E. Kalman [9].

Def. AA

Let $\omega = +\infty$ and let $BB^* \in L^1_{\ell oc}$. We say that (U) is uniformly completely controllable (u.c.c.) by means of $U = L^1_{\ell oc}$ iff there exist constants $\sigma > 0$, $0 < \gamma_1 \leq \gamma_2$, $0 < \gamma_3 \leq \gamma_4$, such that for all $\tau > \alpha$ we have (I_n, the n x n identity matrix):

(2.11) $$\gamma_1 I_n \leq K(\tau, \tau + \sigma) \leq \gamma_2 I_n$$

(2.12) $$\gamma_3 I_n \leq G(\tau + \sigma, \tau) K(\tau, \tau + \sigma) G^*(\tau + \sigma, \tau) \leq \gamma_4 I_n.$$

It can be seen that u.c. controllability implies a severe restriction on G, hence on A. In fact if (2.11) (2.12) hold then there exist two functions $\rho \to \gamma(\rho)$, $\rho \to \Gamma(\rho)$ of $\rho > 0$ such that

$$0 < \gamma(|t-s|) \leq |G(t,s)| \leq \Gamma(|t-s|)$$

for $|t-s| > \sigma$.

Clearly (2.11) implies $[\tau, \tau + \sigma]$ - controllability on every interval $[\tau, \tau + \sigma]$, $\tau > \alpha$.

When A and B are constant u.c. controllability is in fact equivalent to $[\tau, T]$-controllability.

3. We shall now consider various kinds of controllability to zero in a variable time. This means that the initial and/or final instants are not given in advance and the time needed to transfer the state from v to zero depends on v. In this case the set U of controls, may be "bounded" in some sense (which is quite reasonable for the applications) and the transfer is done at the expenses of

time (which gives rise to minimum time control problems).

We start with:

Def. B

Given $\tau \in J$, (U) is $[\tau, T_v]$-controllable to zero by means of U iff for every $v \in E^n$ there exists $u_v \in U$ and $T_v \in J$, $T_v > \tau$ such that

$$x(T_v, \tau, u_v, v) = 0$$

Let us denote by

$$R(t, \theta, \chi, U) = \{x(t, \theta, \chi, u) : u \in U\}$$

the set of points which can be reached at t, by means of U, from θ, χ, either forwards ($\theta < t$) or backwards ($t < \theta$). Also, let

$$S^-(\tau, \chi, U) = \bigcup_{\theta \in [\tau, \omega[} R(\tau, \theta, \chi, U)$$

denote the set of points v from which, starting at τ and using $u \in U$, χ can be reached, soon or later.

When $\chi = 0$, the set $S^-(\tau, 0, U)$ is known as the domain of null controllability.

Clearly, (U) is $[\tau, T_v]$-controllable to zero by means of U, iff

$$S^-(\tau, 0, U) = E^n.$$

This kind of controllability was first considered by J. P. LaSalle [10] for a control set U which we shall denote by $B_{\infty,\infty}$. More generally let us define

$$(3.1)\quad B_{p',r'} = \begin{cases} \{u \in L^1_{loc} : \int_J |u(t)|_{r'}^{p'} dt \leq 1\}, 1 \leq p' < +\infty \\ \{u \in L^1_{loc} : \operatorname{ess\,sup}_{t \in J} |u(t)|_{r'} \leq 1\}, \ p' = +\infty \end{cases}$$

where

$$|u(t)|_{r'} = \begin{cases} (\sum_1^m {}_j |u_j(t)|^{r'})^{1/r'}, & 1 \leq r' < +\infty \\ \max_j |u_j(t)|, & r' = +\infty. \end{cases}$$

In our terminology LaSalle's result is the following

<u>Theorem 3.1</u>

(U) is $[\tau, T_v]$-controllable to zero by means of $U = B_{\infty,\infty}$ if

$$(3.2)\qquad y \neq 0 \Rightarrow \int_\tau^\omega |y^*G(\tau, t)B(t)|_1 \, dt = +\infty$$

In LaSalle's terminology (U) is <u>asymptotically proper</u> when (3.2) holds. More generally we have (<u>R. Conti</u> [3] p.161)

<u>Theorem 3.2</u>

(U) is $[\tau, T_v]$-controllable to zero by means of $U = B_{p',r'}$ $(1 < p' \leq +\infty)$ iff

$$(3.3) \qquad y \neq 0 \Rightarrow \int_{\tau}^{\omega} |y^*G(\tau, t)B(t)|_r^p \, dt = +\infty$$

$(1/p + 1/p' = 1, \ 1/r + 1/r' = 1)$.

It is easy to see that this does depend on p, i.e., on p'. As a matter of fact (3.3) is equivalent to $[\tau, T_V]$ - controllability to zero by means of a wider class of control sets, namely

Theorem 3.3

(3.3) is equivalent to $[\tau, T_V]$- controllability to zero by means of any control set U such that

$$(3.4) \qquad \rho B_{p', r'} \subset U \subset \sigma B_{p', r'}$$

for some $\rho > 0$ and $\sigma \geq \rho$.

In particular, from Theorem 3.3 we have

Theorem 3.4

Let

$$(3.5) \qquad U_\Omega = \{ u \in L^1_{loc} : u(t) \in \Omega, \text{ a.e. } t \in J \}$$

where Ω is any bounded set of E^m such that

$$(3.6) \qquad 0 \in \text{int } \Omega .$$

Then (3.2) is equivalent to $[\tau, T_V]$ - controllability to zero by means of $U = U_\Omega$.

It should also be noted that from (3.3) it follows

Theorem 3.5

If (U) is $[\tau, T_V]$-controllable to zero by means of U such that (3.4) holds, then there exists $T > \tau$ such that

(U) is $[\tau, T]$-controllable to zero by means of $U = PC$.
Since the converse is not true we could ask

Question 3.1

What is another necessary condition C), independent of (2.2), such that C) + (2.2) is equivalent to $[\tau, T_v]$-controllability to zero by means of $B_{p', r'}$?

An answer does not seem impossible, in view of what is known in the case of constant A and B.

When A and B are constant $[\tau, T_v]$-controllability to zero by means of $B_{p', r'}$ is independent both of τ and p' and can be characterized in terms of A and B, directly.

In fact (3.3) becomes

$$y \neq 0 \Rightarrow \int_0^{+\infty} |y^* e^{-tA} B|_r^p \, dt = +\infty$$

which is equivalent to (J. P. LaSalle [10], also R. Conti [3], p. 168)

a) the validity of

(2.6) $\qquad$ rank $C = n$

plus b) all the eigenvalues of A have real parts ≤ 0.

The class of control sets U satisfying (3.4) is fairly general, but it is not large enough to cover such problems as the following, considered by S. H. Saperstone J. A. Yorke [14] (see also R. Gabasov- F. Kirillova [6] pp. 51-55) "Can the motion of a simple pendulum be brought to rest in a finite time by the application of a unit force acting only in one direction?"

Such problems require that the origin lie on the boundary of U or even that it does not belong to U at all.

A recent result in this direction is due to R. F. Brammer [2] who replaced (3.6) by a much weaker assumption, namely (co Ω = convex hull of Ω)

$$\text{int co}\,\Omega \neq \phi$$
$$B\omega = 0 \text{ for some } \omega \in \Omega .$$

and proved that a set of conditions apt to characterize $[\tau, T_v]$-controllability to zero by means of U_Ω is given by conditions a) and b), above plus

c) _there is no real eigenvector_ η _of_ A^* _such that_

$$\eta^* B\omega \leq 0, \text{ for all } \omega \in \Omega .$$

It might be of some interest to consider a similar situation for _non autonomous_ control systems, for instance

Question 3.2

To characterize in terms of G and B control systems (U) which are $[\tau, T_v]$-controllable to zero by means of U when U is a translate set of $B_{p', r'}$ passing through the origin of $L^{p'}_{loc}$. □

4.

As we shall see in a moment the symmetry between the following definition and Def. B is only apparent.

Def. C.

Given $T \in J$, (U) is $[\tau_V, T]$-_controllable to zero by means_ of U iff for every $v \in E^n$ there exist $u_v \in U$ and $\tau_v \in J$, $\tau_v < T$ such that

$$x(T, \tau_v, u_v, v) = 0$$

If we define the set

$$P^-(T, x, U) = \bigcup_{t \in]\alpha, T]} R(t, T, x, U)$$

then (U) is $[\tau_v, T]$-controllable to zero by means of U iff

$$P^-(T, 0, U) = E^n$$

Let us now consider the case $U = B_{p', r'}$, defined by (3.1). It is easy to show that if $1 < p' \leq +\infty$, (U) is $[\tau_v, T]$-controllable to zero by means of $B_{p', r'}$ iff for every $v \in E^n$ there exists $\tau_v \in J$, $\tau_v < T$, such that

$$(4.1) \qquad |y^*v|^p \leq \int_{\tau_v}^{T} |y^*G(\tau_v, t)B(t)|_r^p \, dt, \quad y \in E^n$$

However this does not imply $[\tau, T]$-controllability for some $\tau < T$ as is shown by taking

$$A = \begin{pmatrix} 0 & 1 \\ -1 & 0 \end{pmatrix}, \quad B(t) = \begin{pmatrix} \cos t \\ -\sin t \end{pmatrix}.$$

Therefore it makes sense to ask

Question 4.1

How can $[\tau_v, T]$-controllability to zero by means of $U = B_{p',r'}$ be characterized in terms of G and B?

It is not difficult to see that

$$y \neq 0 \Rightarrow \lim_{\tau \to \alpha} \sup \int_{\tau}^{T} |y^* G(\tau, t) B(t)|_r^p \, dt = +\infty$$

is a necessary condition and we could ask whether this is also a sufficient one.

When A and B are constant there is no problem since $[\tau_v, T]$- and $[\tau, T_v]$-controllability to zero by means of $B_{p',r'}$ are just the same. This is also true for any control set U which, like $B_{p',r'}$, is translation invariant, i.e., such that $u \in U$ implies $u_\theta \in U$ for all $\theta \in \mathbb{R}$, where $u_\theta(t) = u(t + \theta)$.

5.

When neither the initial time nor the final one are prescribed, we have

Def. D.

We say that (U) is $[\tau_v, T_v]$ -controllable to zero by means of U iff for every $v \in E^n$ there exist $\tau_v, T_v \in J$, $\tau_v < T_v$ and $u_v \in U$ such that

$$x(T_v, \tau_v, u_v, v) = 0$$

If A and B are constant and if $U = B_{p', r'}$ (or, more generally, U is translation invariant) then Def. D, C and B are all equivalent.

In general, however, Def. D describes a kind of controllability weaker than both $[\tau_v, T]$- and $[\tau, T_v]$-controllability by means of the same set U.

Therefore, for non constant A, B we can ask

Question 5.1

To characterize $[\tau_v, T_v]$-controllability to zero by means of $U = B_{p', r'}$ in terms of G and B.

6.

Let us now consider some kinds of asymptotic controllability: this means that at least one endpoint of the time interval is "at infinity".

First comes an asymptotic version of Def. A, corresponding to $T = \omega$, namely

Def. E

Given $\tau \in J =]\alpha, \omega[$, (U) is $[\tau, \omega[$-controllable to zero by means of U, iff for every $v \in E^n$ there exists $u_v \in U$

(6.1) $$\lim_{t \to \omega} x(t, \tau, u_v, v) = 0$$

If U is the space L^∞_{loc} or the space PC of piecewise constant functions, it is clear that (U) is $[\tau, \omega[$-controllable if it is $[\tau, T]$-controllable for some $T > \tau$.

However $[\tau, \omega[$-controllability is substantially different from $[\tau, T]$-controllability.

For instance if equation (E_0) is asymptotically stable for $t \to \omega$ then (U) is $[\tau, \omega[$-controllable to zero, no matter what B and U are, provided that $0 \in U$. Therefore, even if we assume that U is a definite subspace of $L^1_{\ell oc}$, there is not much hope to answer

Question 6.1

Characterize $[\tau, \omega[$-controllability to zero by means of U in terms of G and B.

Recalling (1.1) we see that (6.1) is satisfied when there exists an $m \times n$ matrix function $\phi : t \to \phi(t)$ such that $B\phi \in L^1_{\ell oc}$ and

$$(6.2) \qquad \lim_{t \to \omega} \left[G(t,\tau) + \int_\tau^t G(t,s)B(s)\phi(s)ds\right] = 0.$$

so that we can take

$$u_v(t) = \phi(t)\, v.$$

In this case we can say that (U) is linearly $[\tau, \omega[$ -controllable to zero by means of $U = L^1_{\ell oc}$. Similar to what happens to $[\tau, T]$-controllability (Sec. 2) we can ask:

Question 6.2

What conditions (if any) are needed to insure the equivalence between linear and "plain" $[\tau, \omega[$-controllability by means of $U = L^1_{\ell oc}$.

Anyway, the best known and most used condition to

obtain linear $[\tau, \omega[$-controllability is stabilizability. Let us denote by $F: t \to F(t)$ an $m \times n$ matrix function such that $BF \in L^1_{loc}$. We say that (U) is stabilizable iff there is such and F so that

$$(F_0) \qquad dy/dt - [A(t) + B(t)\ F(t)]\ y = 0$$

is asymptotically stable for $t \to \omega$.

If we denote by y_v the solution of (F_0) such that $y_v(\tau) = v$ and G_F is the transition matrix of (F_0) we have

$$y_v(t) = G_F(t, \tau)\ v.$$

On the other hand

$$y_v(t) = G(t,\tau)v + \int_\tau^t G(t, s)\, B(s)F(s)y_v(s)\, ds$$

and it follows

$$(6.3) \qquad \lim_{t \to \omega} \left[G(t,\tau) + \int_\tau^t G(t, s)B(s)F(s)G_F(s,\tau)ds\right] = 0$$

if (F_0) is asymptotically stable for $t \to \omega$. Therefore stabilizability implies linear $[\tau, \omega[$-controllability.

Question 6.3

Prove or disprove the converse implication, i.e., if (6.2) holds for some ϕ, does (6.3) hold for some F ?

Stabilizability is a property well known to engineers and applied mathematicians, because of its connection with the so called "regulator problem". This connection leads also to consider $\omega = +\infty$ and exponential stabilizability (rather then plain stabilizability), which means the existence of an F such that (F_0) is exponentially asymptotically stable for $t \to +\infty$.

When A and B are constant, stabilizability is necessarily exponential and can be characterized by the existence of a positive solution R of a certain quadratic matrix equation (see. D. L. Lukes [12]).

For non-constant A and B the role of that equation is taken over by a Riccati differential matrix equation. It can be shown that a sufficient condition for exponential stabilizability is represented by the existence of a solution R(t) such that

$$\rho I_n \leq R(t) \leq \sigma I_n$$

with $0 < \rho \leq \sigma$.

This will happen, in particular, when (U) is u.c.c. (Recall Def AA of Sec. 2). In this case (F_0) turns to be exponentially asymptotically stable with any prescribed exponent (Sec. M. Ikeda, H. Maeda, S. Kodama, [8]). This results parallels the case of constant A, B where $[\tau, T]$- controllability is equivalent to an arbitrary "pole assignment", i.e., to the existence of a constant F such that the set of eigenvalues of A + BF is any set of complex numbers given in advance (see. W.M. Wonham [15]).

7. Contrary to $[\tau, \omega[$-controllability, the (apparently) "symmetric" notion of $]\alpha, T]$-controllability (see next Def. F) does not look like a familiar one in control theory. The same remark applies to the other kinds of asymptotic

controllability to be considered later (see Definitions G, H, I). So the first question is: what is their applicability?

Anyway, their theoretical interest does not seem negligeable as they differ substantially from $[\tau, \omega[$ - controllability and each other.

Def. F.

Given $T \in J =]\alpha, \omega[$, (U) is $]\alpha, T]$-controllable to zero by means of U, iff for every $v \in E^n$ there exist $u_v \in U$ such that

$$(7.1) \qquad \lim_{t \to \alpha} x(t, T, u_v, 0) = v$$

Probably at least as much hopeless as Question 6.1 is

Question 7.1

To characterize in terms of G and B, $]\alpha, T]$-controllability to zero by means of $U = L^1_{loc}(J, E^m)$.

We note that $[\tau, T]$-controllability for some $\tau < T$ does not imply $]\alpha, T]$-controllability as simple examples show.

We can say that (U) is linearly $]\alpha, T]$-controllable to zero by means of $U = L^1_{loc}$ when there exists an $m \times n$ matrix function $\psi = t \to \psi(t)$ such that $B\psi \in L^1_{loc}$ and

$$(7.2) \qquad \lim_{t \to \alpha} \int_T^t G(t, s)B(s)\psi(s)ds = I_n$$

so that (7.1) holds with

$$u_v(t) = \psi(t) v.$$

In particular, (U) will be linearly $]\alpha, T]$-controllable if we assume the validity of

Condition 7.1

Let there exist $F: t \to F(t)$ such that $BF \in L^1_{loc}$ and

$$(7.3) \qquad \lim_{t \to \alpha} [G_F(t, T) - G(t, T)] = I_n.$$

In fact
$$\int_T^t G(t, s)B(s)F(s)\, G_F(s, T)ds$$
$$= G_F(t, T) - G(t, T)$$

so that (7.2) holds with $\psi(t) = F(t)\, G_F(t, T)$.

A particular case of Cond. 7.1 is

Condition 7.2

Let $m = n$, let $B^{-1}(t)$ exist and let (E_0) be asymptotically stable for $t \to \alpha$.

Take $F(t) = - B^{-1}(t)\, A(t)$, whence $A(t) + B(t)F(t) = 0$

so that
$$G_F(t, T) - G(t, T) = I_n - G(t, T).$$

Another condition (also implying $m = n$ and the existence of $B^{-1}(t)$) for linear $]\alpha, T]$-controllability is

Condition 7.3.

Let $m = n$ and let B be a non singular matrix solution of (E_0).

In fact, if B is a matrix solution of (E_0) we have

$$(7.4) \qquad \int_T^t G(t,s)B(s)\psi(s)\,ds = B(t)\int_T^t \psi(s)ds.$$

Take any $n \times n$ matrix function $Y : t \to Y(t)$ locally absolutely continuous on $]\alpha, T]$, such that

$$\lim_{t \to \alpha} Y(t) = I_n. \qquad Y(T) = 0$$

and take

$$\psi(t) = -B^{-1}(t)[A(t)Y(t) - dY(t)| dt]$$

Then

$$B(t)\int_T^t Y(s)ds = Y(t)$$

and (7.4) yields (7.2).

8.

For the sake of completeness we shall list three more definitions of asymptotic controllability. Apart from their applicability, the common problem, as usual, is that of characterizing in terms of G and B (or, directly, in terms of A and B, when these are constant) each one of these types of controllability.

First comes a "totally" asymptotic version of $[\tau, T]$-controllability, namely

Def. G

We say that (U) is J-controllable to zero by means of U iff for every $v \in E^n$ there exist $u_v \in U$, $\theta_v \in J$, $\chi_v \in E^n$ such that

$$\lim_{t \to \alpha} x(t, \theta_v, u_v, \chi_v) = v$$

$$\lim_{t \to \omega} x(t, \theta_v, u_v, \chi_v) = 0$$

Two more definitions arise if we consider the asymptotic versions corresponding to $\tau = \alpha$ in Def. B and to $T = \omega$ in Def. C, i.e., respectively.

Def. H.

(U) is $]\alpha, T_v]$-controllable to zero by means of U iff for every $v \in E^n$ there exist $T_v \in J$, $u_v \in U$ such that

$$\lim_{t \to \alpha} x(t, T_v, u_v, 0) = v$$

Def. I

(U) is $[\tau_v, \omega[$-controllable to zero by means of U iff for every $v \in E^n$ there exist $T_v \in J$, $u_v \in U$, such that

$$\lim_{t \to \omega} x(t, \tau_v, u_v, 0) = v$$

It should be noted that, contrary to what happens, to Definitions B and C, Definitions I and H are not equivalent when A and B are constant.

In this case, however, if U is translation invariant Def. H is equivalent to Def. F and Def. I is equivalent to Def. E.

9. So far, all the definitions dealt with controllability to zero.

It is immediate to see that $[\tau, T]$-controllability to zero is equivalent to $[\tau, T]$-controllability from zero, i.e., that Def. A is the same as

Def. A′

Given $\tau, T \in J$, $\tau < T$, (U) is $[\tau, T]$-controllable from zero by means of U iff for every $w \in E^n$ there exist $u_w \in U$ such that

$$x(T, \tau, u_w, 0) = w$$

On the contrary, each one of the remaining Definitions (B to I) yields a new type of controllability when we replace the initial state v by 0 and the final state 0 by an abitrary vector w.

Therefore we can add 8 more definitions (B′ to I′) to our list.

Further, Def. A (and A′) is apparently more general than

Def. A″

Given $\tau, T \in J$, $\tau < T$, we say that (U) is $[\tau, T]$-controllable by means of U iff for every pair $v, w \in E^n$ there exist $u_{v,w} \in U$ such that

$$x(T, \tau, u_{v,w}, v) = w$$

Actually, Def. A'' is equivalent to Def. A and Def. A′. In other words, transferability in a fixed compact interval $[\tau, T]$ to zero or from zero or from any state to any other state are all just the same. On the contrary the corresponding Definitions B'', C'', etc. are actually more restrictive, in general, than both Def. B and B′, C′ and C′, etc, respectively. All in all, the list of different kinds of controllability (apart from Def. AA) adds up to $1 + 3.8 = 25$ distinct items.

10.

From a control system represented by

$$(U) \qquad dx/dt - A(t)\, x = B(t)u(t)$$

a new one, represented by

$$(U_c) \qquad dx/dt - A(t)\, x = B(t)u(t) + c(t, x, u(t))$$

can be obtained by adding the perturbing term c. When (U) possesses some kind of controllability it is of practical importance to know whether this is preserved for (U_c).

To clear the ground from accessory complications one has to assume that to each $\theta \in J =]\alpha, \omega[$, $\chi \in E^n$ and $u \in L^1_{\ell oc}(J, E^m)$ there corresponds a unique (Caratheodory) solution $t \Rightarrow x(t, \theta, \chi, u)$ of (U_c) passing through θ, χ and defined on the whole interval J. Such a solution also verifies

$$x(t, \theta, x, u) = G(t, \theta)x + \int_{\theta}^{t} G(t, s)B(s)u(s)ds +$$

$$+ \int_{\theta}^{t} G(t, s)c(s, x(s, \theta, x, u), u(s))\, ds$$

From this follows immediately that if c is independent of x and u, i.e., if

$$c(t, x, u) = c(t)$$

its presence does not affect at all the $[\tau, T]$-controllability of (U) (Defs. A, A', A'').

This is no longer true in general for asymptotic controllability.

Next, let c depend linearly on x and u, i.e., let

$$c(t, x, u) = A_0(t)x + B_0(t)\, u$$

where A_0 and B_0, like A and B, are measurable and locally integrable on J. Then it can be proved (see. E. B. Lee - L. Markus [11]. p-100: J. P. Dauer [4]) that $[\tau, T]$-controllability is a "generic property", in the sense expressed by

Theorem 10.1

If (U) is $[\tau, T]$-controllable by means of $U = L^{\infty}_{\ell oc}(J, E^m)$ there exists $\rho > 0$ such that if

$$\int_{\tau}^{T} |A_0(t)|\, dt + \int_{\tau}^{T} |B_0(t)|\, dt < \rho$$

(U_c) is also $[\tau, T]$-controllable.

Further, independently of whether (U) is $[\tau, T]$-controllable or not, for every $\varepsilon > 0$ there exist A_0 and B_0 such that

$$\int_\tau^T |A_0(t)| \, dt + \int_\tau^T |B_0(t)| \, dt < \varepsilon$$

and (U_c) is $[\tau, T]$ - controllable.

Question 10.1

Are there similar results available for the other kinds of controllability?

The linearity of c with respect to u seems to be more relevant than linearity with respect to x in order to preserve $[\tau, T]$-controllability (H. Hermes [7]. E. J. Davison - E. G. Kunze [5]). Anyway this kind of controllability is preserved as long as c is "almost" linear with respect to both x and u (D. L. Lukes [13]. G. Aronsson [1]).

Question 10.2

What can be said about the "resistance" of the other kinds of controllability against nonlinear perturbations ?

REFERENCES

[1] G. Aronsson - Global controllability and bang-bang steering of certain non-linear systems. SIAM J. Control 11 (1973), 607-619;

[2] R. F. Brammer- Controllability in linear autonomous systems with positive controllers . SIAM J. Control

10 (1972), 339-353:

[3] R. Conti - Problemi di controllo e di controllo ottimale. UTET, Torino, 1974:

[4] J. P. Dauer- Perturbations of linear control systems, SIAM J. Control 9 (1971), 393-400; 11 (1973), 395.

[5] E. J. Davison - E. G. Kunze - Some sufficient conditions for the global and local controllability of nonlinear time varying systems. SIAM J. Control 8 (1970), 489-497;

[6] R. Gabasov - F. Kirillova - Qualitative Theory of optimal processes (in Russian), Nauka, Moscow, 1971;

[7] H. Hermes - Controllability and the singular problems. J. SIAM Control, Ser. A, 2(1965), 241-259:

[8] M. Ikeda - H. Maeda. S. Kodama, - Stabilization of linear systems. SIAM J. Control 10 (1972), 716-729:

[9] R. E. Kalman - Contributions to the theory of optimal control Bol. Soc Mat. Mexicana, (2) 5(1960), 102-119.

[10] J. P. LaSalle- The time optimal control problem. Contributions to the theory of nonl. oscillations. vol. 5(1960), 1-24.

[11] E. B. Lee - L. Markus- Foundations of optimal control Theory, J. Wiley and Sons, N. York, 1967.

[12] D. L. Lukes- Stabiliability and optimal control. Funkcialj Ekvatory , 11(1968), 39-50.

[13] D. L. Lukes- Global controllability of nonlinear systems SIAM J. Control 10(1972), 112-126; 11(1973), 186.

[14] S. H. Saperstone- J. A. Yorke- Controllability of linear oscillatory systems using positive controls. SIAM J. Control 9 (1971) 253-262.

[15] W. M. Wonham. On pole assignment in multi-input controllable systems. IEEE Trans. Automatic Control. AC-12, (1967) 660-665.

[16] L. E. Zabello- On the controllability of linear non-stationary systems (in Russian) Autom i Telem. 8(1973), 13-19.

Via G B Amici 14 A, Florence, Italy 50131

MATRICES OF RATIONAL FUNCTIONS

W. A. Coppel

The properties of the degree of a matrix of rational functions are obtained in a simplified way, which enables them to be generalized to matrices whose elements are not necessarily rational functions. On the basis of these results a theory of realizations is developed, which similarly generalizes the theory of state space realizations of a matrix of rational functions.

A complete account of this work has appeared in Bull. Asutral. Math. Soc. 11 (1974), 89-113.

Australian National University

SUPPRESSION OF OSCILLATIONS IN NONLINEAR SYSTEMS*

Jane Cronin

1. Introduction.

Outside stimuli sometimes suppress biological oscillations in the sense that while the stimulus is applied, the biological oscillation ceases. However if the stimulus is withdrawn, the oscillation resumes. See Oatley Goodwin [16, p. 15], Hastings [11], Njus, Sulzman, and Hastings [15, p. 2]. The purpose of this note is to propose mathematical models (systems of nonlinear ordinary differential equations) which might be used to describe such phenomena. That is, we study some nonlinear systems of ordinary differential equations which may have stable almost periodic solutions and show that if certain parameters are changed, no biologically significant solutions exist and there is a unique critical point toward which all biologically significant solutions tend. If the values of the parameters were returned to their initial values, then the periodic

*Supported by the U. S. Army Research Office (Durham) Grant No. DA-ARO-D-31-124-72-G69.

solutions would again exist. We also show that the addition of a periodic term may obliterate a given periodic solution.

The systems of differential equations for which we obtain these results are particularly designed for the study of biological systems in two respects. First, we consider n-dimensional systems where there is no restriction on n. The restriction to systems of dimension two or three is often quite satisfactory in describing problems in physics (see for example the applications described by Stoker [19] of Minorsky [13]). But biological systems are far more complicated and in many cases a realistic description requires consideration of a larger set of variables. Secondly, we consider only solutions all of whose components are nonnegative and we impose the hypothesis that any solution all of whose components are nonnegative at a given time t_o has the property that for all $t \geq t_o$, all the components are nonnegative. This is a realistic assumption if the components represent concentrations of biochemicals or light intensity or any physical variable for which negative values are meaningless.

After each theorem we cite examples of biological oscillations for which the system of ordinary differential equations might serve as an approximate description or we point out how the theorem suggests possible directions for experimentation. The majority of the examples described here occur in the work of Hastings [2, 11, 15] and our results may be regarded as an attempt to use mathematical language to describe the aspect of the behavior of biological clocks considered in [11].

This discussion has biological significance only if it is assumed that biological systems can be described by systems of ordinary differential equations. For example, we disregard the possibility of delay terms and stochastic terms in the equations. Moreover we use stability concepts which are closely related to the standard Lyapunov definitions of stability and the practical significance of these definitions is questionable. (See Hahn [10, p. 8, p. 278]). This is especially true in the case of biological problems. Finally, whether any continuum method can be used to obtain realistic descriptions of biological phenomena is questionable. See, for example, Gann, Seif, and Schoeffler [9] and Seif and Gann [18].

2. Suppression of periodic solutions by addition of constant terms

We consider the n-dimensional autonomous system

$$\dot{x} = Ax + f(x) \tag{2.1}$$

where $\dot{x}$ denotes differentiation of x with respect to t.

In all of our discussion we will assume that the solutions

of differential equations studied have maximal domains, i.e., no solution can be extended. Also without saying so explicitly we will in several places use the fact that if a solution of a differential equation stays in a closed bounded set in the domain of the functions in the differential equation, then there is a value $\bar{t}$ such that the solution is defined for all $t \geq \bar{t}$. (See Nemytskii and Stepanov [14, Theorem 1.21, page 8].) Finally the norm of an n-vector $x = (x_1, \ldots, x_n)$, denoted by $|x|$, is

$$|x| = \sum_{i=1}^{n} |x_i|$$

and

$$|x| \leq n^{\frac{1}{2}} \left(\sum_{i=1}^{n} x_i^2\right)^{\frac{1}{2}}$$

We assume that equation (2.1) satisfies the following hypotheses:

H-1) $A = (a_{ij})$ is a constant matrix such that if

$$z^n + A_1 z^{n-1} + \cdots + A_{n-1} z + A_n = 0$$

is the characteristic equation of A, then the determinants $\delta_1 = A_1$ and

$$\delta_k = \begin{vmatrix} A_1 & A_3 & A_5 & \cdots & A_{2k-1} \\ 1 & A_2 & A_4 & \cdots & A_{2k-2} \\ 0 & A_1 & A_3 & \cdots & A_{2k-3} \\ 0 & 1 & A_2 & \cdots & A_{2k-4} \\ & & \cdot & \cdot & \cdot \\ 0 & 0 & 0 & \cdots & A_k \end{vmatrix}, \quad (k = 2, \ldots, n),$$

where $A_j = 0$ for $j > n$, are all positive. Then by the Routh-Hurwitz criterion there is a positive number σ such that for each eigenvalue λ of A,

$$\mathcal{R}(\lambda) \leq -3\sigma$$

where $\mathcal{R}(\lambda)$ denotes the real part of λ. See Marden [12, p. 141]. It follows by standard arguments that there exists a nonsingular real matrix P such that $PAP^{-1} = B$ and such that if the change of variables $y = Px$ is introduced, then for all y,

$$yBy \leq -\sigma \sum_{i=1}^{n} y_i^2 .$$

(The matrix B is described in Coddington and Levinson [4, pp. 340-341].) We shall assume that this change of variables has been made and hence that for x,

$$\text{(L)} \qquad xAx \leq -\sigma \sum_{i=1}^{n} x_i^2 .$$

H-2) Let

$$\mathcal{P} = \{x = (x_1, \ldots, x_n) / x_i \geq 0 \text{ for } i = 1, \ldots, n\}.$$

Let $\mathcal{O}$ be an open set in R^n (real Euclidean n-space) such that $\mathcal{P} \subset \mathcal{O}$. We assume that f is defined and continuous on $\mathcal{O}$ and that there exists a positive number M such that if $x \in \mathcal{P}$ then

$$|f(x)| < M.$$

H-3) f has continuous first derivatives at each point of $\mathcal{O}$ and given $\varepsilon > 0$, then there is a positive number $B(\varepsilon)$ such that if $x \in \mathcal{P}$ and $|x| \geq B(\varepsilon)$, then

$$\left| \frac{\partial f_i}{\partial x_j}(x) \right| < \varepsilon \quad (i, j = 1, \ldots, n).$$

That is, the derivatives go to zero as $|x| \to \infty$ (for $x \in \mathcal{O}$).

H-4) Suppose there exists $m \in (0,1)$ such that if

$$\mathcal{K} = \{(x_1, \ldots, x_n) \in \mathcal{O} / x_i \geq m \sum_{j=1, j\neq i}^{n} x_j, \ i=1, \ldots, n\},$$

then if $x(t)$ is a solution of (2.1) such that for some t_o, $x(t_o) \in \partial\mathcal{K}$ (where $\partial\mathcal{K}$ denotes the point set boundary of $\mathcal{K}$) then for all $t \geq t_o$ for which $x(t)$ is defined, $x(t) \in \mathcal{K}$. That is, no solution "escapes" the set $\mathcal{K}$.

It is reasonable to impose a hypothesis like H-4) on a system of equations which describes some biological system because very often the variables $x_1, \ldots, x_n$ denote concentrations of biochemicals or other quantities which have physical meaning only if they are nonnegative. Hence if for some value t_o a solution $x(t)$ is such that $x(t_o) \in \mathcal{O}$, then in order for solution $x(t)$ to have physical significance, it is necessary that for all $t \geq t_o$, $x(t) \in \mathcal{O}$. For mathematical reasons, we impose the slightly stronger condition that $x(t)$ remain in $\mathcal{K}$: Hypotheses H-1) and H-2) are also reasonable to impose if (2.1) describes a biological system because if the variables $x_1, \ldots, x_n$ denote, for example, concentrations of biochemicals then in order for

$x_1, \ldots, x_n$ to have physical meaning, they must be nonnegative and also must remain, in some sense, bounded. That is, we must exclude the possibility that there are a solution x(t) and values t_1 and t_2 where $x(t_1) \in \mathcal{K}$, $t_2 > t_1$ (t_2 may be ∞) and $\lim_{t \uparrow t_2} |x(t)| = \infty$. Hypotheses H-1), H-2) and H-4) exclude this possibility as the following lemma shows.

<u>Lemma 1.</u> There is a positive number R_o such that if $R \geq R_o$ and if B_R is the ball in R^n with center 0 and radius R, i.e.,

$$B_R = \{(x_1, \ldots, x_n) / \sum_{i=1}^{n} x_i^2 \leq R^2\},$$

then if x(t) is a solution of (2.1) such that for some t_o, $x(t_o) \in \partial(B_R \cap \mathcal{K})$ then for all $t \geq t_o$, x(t) is defined and $x(t) \in B_R \cap \mathcal{K}$.

<u>Proof:</u> Because of (H-4), it is sufficient to show that if $x(t_1) \in \partial B_R \cap \partial(B_R \cap \mathcal{K})$, then there is a positive number δ such that if $t \in (t_1, t_1 + \delta)$, then $|x(t)| < R$ and the tangent to x(t) at t_1 is not tangent to ∂B_R. But at $t = t_o$

$$r \dot{r} = \sum_{i=1}^{n} x_i \dot{x}_i = x A x + x f(x).$$

From condition (L) in hypothesis H-1), we have:

$$x A x \leq - \sigma \sum_{i=1}^{n} x_i^2 .$$

Suppose $R > n^{\frac{1}{2}} \left(\frac{M+\varepsilon_o}{\sigma}\right)$ where ε_o is positive and M is the constant in H-2). Then

$$\begin{aligned} x\,Ax + x\,f(x) &\leq -\sigma \sum_{i=1}^{n} x_i^2 + M\,|x| \\ &\leq -\sigma R \left(\sum_{i=1}^{n} x_i^2\right)^{\frac{1}{2}} + M\,|x| \\ &\leq -\sigma\, n^{\frac{1}{2}} \left(\frac{M+\varepsilon_o}{\sigma}\right) n^{-\frac{1}{2}}\, |x| + M\,|x| \\ &\leq -\varepsilon_o\, |x| \end{aligned}$$

This completes the proof of Lemma 1.

Definition 1. Solution $x(t)$ of (2.1) is uniformly stable if there exists $K > 0$ such that $\varepsilon > 0$ implies there is a positive $\delta(\varepsilon)$ so that if $u(t)$ is a solution of (2.1) and if there exist numbers t_1, t_2 such that $t_2 \geq K$ and such that

(C) $$|u(t_1) - x(t_2)| < \delta(\varepsilon)$$

then for all $t \geq 0$

$$|u(t+t_1) - x(t+t_2)| < \varepsilon\ .$$

Solution $x(t)$ of (2.1) is asymptotically stable if $x(t)$ is uniformly stable and if, in addition, whenever condition (C) holds, then there exists a number t_3 such that

$$\lim_{t\to\infty} |u(t) - x(t_3 + t)| = 0.$$

(This definition of asymptotic stability is closely related to the standard Lyapunov definitions of stability and asymptotic stability. However it is stronger than the condition of asymptotic orbital stability which is the usual asymptotic stability condition imposed on solutions of autonomous systems.)

Definition. A critical point p of equation (2.1) is globally asymptotically stable in a set $E \subset R^n$ if $p \in R^n$, if p is asymptotically stable, and if for each solution $x(t)$ of (2.1) such that for some t_o, the point $x(t_o) \in E$, it is true that $\lim_{t\to\infty} x(t) = p$.

Theorem 1. Let $R \geq R_o$ where R_o is the constant obtained in Lemma 1. If $B_R \cap \mathcal{K}$ contains no stable critical point and if there is a uniformly stable solution $x(t)$ of (2.1) such that for some t_o, $x(t_o) \in B_R \cap \mathcal{K}$, then the Ω-set of $x(t)$ contains the underlying point set of a uniformly stable almost periodic solution $\bar{x}(t)$ (which is not a critical point) of (2.1) and for all t, $\bar{x}(t) \in B_R \cap \mathcal{K}$. If $B_R \cap \mathcal{K}$ contains no asymptotically stable critical point and if there is an asymptotically stable solution $x(t)$ such that for some t_o, $x(t_o) \in B_R \cap \mathcal{K}$, then the Ω-set of $x(t)$ is the underlying point set of an asymptotically stable periodic solution $\bar{x}(t)$ (which is not a critical point) of (2.1) and for all t, $\bar{x}(t) \in B_R \cap \mathcal{K}$. If there exists an asymptotically stable solution $x(t)$ of (2.1) such that for some t_o, $x(t_o) \in B_R \cap \mathcal{K}$ and if for each asymptotically stable critical point p of (2.1) such that $p \in B_R \cap \mathcal{K}$ there is a neighborhood $N(p)$ and a number t_p such that for $t \geq t_p$, $x(t) \notin N(p)$, then the Ω-set of x(t) is the underlying point set of an asymptotically stable periodic solution $\bar{x}(t)$ (which is not a critical point) of (2.1) and for all t, $\bar{x}(t) \in B_R \cap \mathcal{K}$.

Proof: See Cronin [7]. This theorem is also a special case of a theorem of Sell [20, Theorem 1].

Theorem 1 shows that equation (2.1) may have a periodic or almost periodic solution. The problem we consider is the following: what happens to the solutions in $\mathcal{K}$ of (2.1) if a constant vector $K = (k_1, \ldots, k_n)$, where $K \in \mathcal{K}$, is added to the right side of (2.1). That is, we study the equation

(2.2) $$\dot{x} = Ax + f(x) + K.$$

If $|K|$ is sufficiently small, then Theorem 1 may be applicable and thus (2.2) may have an almost periodic or periodic solution. However if $|K|$ is sufficiently large, it turns out that (2.2) has no almost periodic solutions in K. t hat is, we prove the following theorem.

Theorem 2. Suppose that equation (2.1) satisfies hypotheses H-1), H-2), H-3), H-4). Then there is a positive number B such that if $K \in \mathcal{K}$ and $|K| > B$, then the equation (2.2) has a unique critical point x_o in $\mathcal{K}$ and x_o is globally asymptotically stable in $\mathcal{K}$.

Proof: The proof is a consequence of the following lemmas.

Lemma 2. Given a vector $K \in \mathcal{K}$, then there is a number $R_1(K)$ such that if a solution x(t) of (2.2) intersects the set $\partial B_R \cap \mathcal{K}$, where $R \geq R_1(K)$, at $t = t_o$, i.e., if $|x(t_o)| = R$

and if $x(t_o) \in \mathcal{K}$, then the tangent to $x(t)$ at t_o is not tangent to ∂B_R and for all $t \geq t_o$, $x(t) \in B_R \cap \mathcal{K}$.

Proof: It is sufficient to show that no solution of (2.3) escapes $\mathcal{K}$. The remainder of the proof is essentially the same as the proof for Lemma 1.

From hypothesis H-4), it follows that if $\bar{x}(t)$ is a solution of (2.1) such that for some t_o, $\bar{x}(t_o) \in \partial\mathcal{K}$, then the tangent to $\bar{x}(t)$ at t_o is contained in $\partial\mathcal{K}$ or is directed into the interior of $\mathcal{K}$. Hence if $x(t)$ is a solution of (2.2) such that $x(t_1) = \bar{x}(t_o)$, then the tangent to $x(t)$ at t_1 is

$$A[x(t_1)] + f[x(t_1)] + K = A[\bar{x}(t_o)] + f[x(t_o)] + K.$$

This is the sum of two vectors which are in $\partial\mathcal{K}$ or directed into the interior of $\mathcal{K}$ and hence is directed into the interior of $\mathcal{K}$. This completes the proof of Lemma 2.

Remark: Notice that the solution $x(t)$ of (2.2) in Lemma 2 is defined for all $t \geq t_o$.

Lemma 3 Given $B > 0$ then there exists $R_2(B) > 0$ such that if $|K| > R_2(B)$ and if x_o is a critical point of (2.2) then $|x_o| > B$.

Proof: If

$$A x_o + f(x_o) + K = 0$$

then since, by H-1), the eigenvalues of A are all non-zero, there exists A^{-1} and hence

$$x_o = -A^{-1} f(x) - A^{-1}K,$$

and

$$|x_o| \geq |A^{-1}K| - |A^{-1} f(x_o)|.$$

Then by H-2)

$$|x_o| \geq |A^{-1} K| - |A^{-1}| M.$$

Hence if K is sufficiently large, $|x_o| > B$.

Lemma 4. There exists $\beta > 0$ such that if x_o is a critical point of (2.2) and $|x_o| > \beta$, then x_o is asymptotically stable.

Proof: If x_o is a critical point then to determine if x_o is asymptotically stable, we study the eigenvalues of the matrix

$$\mathcal{a} = (a_{ij} + \frac{\partial f_i}{\partial x_j}(x_o)).$$

By hypotheses H-1), H-3) and the Routh-Hurwitz criterion, it follows that if $|x_o|$ is sufficiently large, the eigenvalues of $\mathcal{a}$ all have negative real parts and hence that x_o is an asymptotically stable critical point.

Lemma 5. There is a positive number K_o such that if $|\bar{K}| > K_o$ then there exist positive numbers R_3 and R_4 such that $R_3 < R_4$ and there exists a critical point x_o, asymptotically stable, of equation (2.2) with $K = \bar{K}$, i.e.,

$$\dot{x} = A x + f(x) + \bar{K} \tag{2.3}$$

such that x_o is an element of the interior of $\overline{B_{R_4} \cap (\mathcal{K} - B_{R_3})}$

Proof: Let $K_o^{(1)} = R_2(\beta)$ where β where β is the number given by Lemma 4 and $R_2(\cdot)$ is the function given by Lemma 3. Take $R_3 > \mathcal{B}$. Also for use in Lemma 6, we apply Hypothesis H-3) and choose R_3 large enough so that if $\mathcal{C}(\mathcal{K} - B_{R_3})$ denotes the convex hull of $\mathcal{K} - B_{R_3}$ and if $\bar{x} \in \mathcal{C}(\mathcal{K} - B_{R_3})$, then

$$|f_x(\bar{x})| < \varepsilon_o(3\sigma)$$

where $\sigma > 0$ is such that for all the eigenvalues λ_i of matrix A, it is true that $R(\lambda_i) < -3\sigma$ and $\varepsilon_o(3\sigma)$ is a positive value chosen so that Gronwall's Lemma can be used. Now we show there is a positive number $K_o^{(2)}$ such that if $\bar{x} \in \partial[\mathcal{K} - B_{R_3}]$ and if $\bar{K}$ is a fixed point in $\mathcal{K}$ such that $\bar{K} > K_o^{(2)}$, the solution of (2.3) which passes through $\bar{x}$ is either tangent of $\partial[\mathcal{K} - B_{R_3}]$ or is such that its tangent at $\bar{x}$ is directed into the interior of $\mathcal{K} - B_{R_3}$. This follows for $\bar{x} \in \partial\mathcal{K}$ by the same argument as used for the proof of Lemma 2. If $\bar{x} \in \mathcal{K} \cap (\partial B_{R_3})$, we have:

$$\begin{aligned} x\mathring{x} &= x A x + x f(x) + x K \\ &\geq x K - |A|\,|x|^2 - M|x| \end{aligned} \tag{2.4}$$

Since $x \in \mathcal{K}$, then by H-4), for $i = 1, \ldots, n$,

$$x_i \geq m \sum_{\substack{j=1 \\ j \neq i}}^{n} x_j$$

$$(1+m)\,x_i \geq m \sum_{j=1}^{n} x_j = m|x|$$

or

$$x_i \geq \frac{m\,|x|}{1+m} \quad .$$

Hence since $K \in \mathcal{K}$,

$$(2.5) \qquad \begin{aligned} x K = x_1 K_1 + \cdots + x_n K_n &\geq K_1 \frac{m\,|x|}{1+m} + \cdots + K_n \frac{m\,|x|}{1+m} \\ &= |K| \frac{x}{1+m} \end{aligned}$$

Substituting from (2.5) into (2.4), we obtain:

$$\begin{aligned} x\dot{x} &\geq |K| \frac{x}{1+m} - |A|\,|x|^2 - M\,|x| \\ &= \left(\frac{|K|}{1+m} - |A|\,|x| - M\right)|x| \end{aligned}$$

$$(2.6) \qquad \begin{aligned} &= \left(\frac{|K|}{1+m} - |A|\sqrt{n}\left(\sum_{i=1}^{n} x_i^2\right)^{\frac{1}{2}} - M\right)|x| \\ &= \left(\frac{|K|}{1+m} - |A|\sqrt{n}\,R_3 - M\right)|x| \,. \end{aligned}$$

Hence there is a positive number $K_o^{(2)}$ such that if $|K| > K_o^{(2)}$, $x\dot{x} > 0$ at each point $\bar{x}$ $\cap (\partial B_{R_3})$. Hence if $|\bar{K}| > K_o^{(2)}$ then a solution of (2.3) which passes through $\bar{x} \in \mathcal{K} \cap (\partial B_{R_3})$ is such that its tangent at $\bar{x}$ is directed to the interior of $\mathcal{K} - \partial B_{R_3}$. Now let $\bar{K}$ be a fixed element of $\mathcal{K}$ such that

$$|\bar{K}| > \max (K_o^{(1)}, K_o^{(2)})$$

and let $R_4 > R_1(\bar{K})$ where $R_1(\cdot)$ is the function given by Lemma 2. Then by Lemma 2, if $\bar{x} \in \mathcal{K} \cap \partial B_{R_4}$, the solution of (2.3) which passes through $\bar{x}$ is such that its tangent is directed to the interior of $\mathcal{K} \cap B_{R_4}$. Thus if $x(t)$ is a solution of (2.4) such that for some t_o,

$$x(t_o) \in \partial[\mathcal{K} \cap (B_{R_4} - B_{R_3})]$$

then the tangent to $x(t)$ at t_o is either directed into the interior of $\mathcal{K} \cap (B_{R_4} - B_{R_3})$ or is tangent to the surface $\partial[\mathcal{K} \cap (B_{R_4} - B_{R_3})]$ at $x(t_o)$. Hence it follows by Satz 1, p. 479, of Alexandroff-Hopf [1] that there is a critical point x_o of (2.4) such that

$$x_o \in \text{Int}\,[\mathcal{K} \cap (B_{R_4} - B_{R_3})].$$

Since $R_3 > \mathcal{B}$, then by Lemma 4, x_o is asymptotically stable. This completes the proof of Lemma 5.

Lemma 6. The critical point obtained in Lemma 5 is globally asymptotically stable in $\mathcal{K}$.

Proof: We must show that if x(t) is a solution of (2.3) such that there exists a number t_o so that $x(t_o) \in \mathcal{K}$ then $\lim_{t \to \infty} x(t) = x_o$. By hypothesis H-4) and the argument used in the proof of Lemma 2, if $t > t_o$ and x(t) is defined, then $x(t) \in \mathcal{K}$. By inequality (2.6) if for some t,

$$\left(\sum_{i=1}^{n} [x_i(t)]^2\right)^{\frac{1}{2}} \leq R_3 ,$$

there is a number $\tilde{t}$ such that for all $t \geq \tilde{t}$,

$$x(t) \in \mathcal{K} - B_{R_3}.$$

By Lemma 2, there is a value $\bar{t}$ such that for all $t \geq \bar{t}$, $x(t) \in \mathcal{K} \cap B_{R_4}$. Hence there is a value τ such that for all $t \geq \tau$,

$$x(t) \in \overline{(B_{R_4} \cap \mathcal{K}) - B_{R_3}}$$

(and hence x(t) is defined for all $t \geq \tau$).

Now for $t \geq \tau$, we have

$$\dot{x}(t) = A\,x(t) + f[x(t)] + \bar{K}, \tag{2.7}$$

where

$$x(t) \in \overline{(B_{R_4} \cap \mathcal{K}) - B_{R_3}}$$

and

$$0 = A\,x_o + f(x_o) + \bar{K}. \tag{2.8}$$

Subtracting (2.8) from (2.7) and using the Mean Value Theorem, we obtain:

$$\begin{aligned}(x(t) - x_o)^{\cdot} &= A[x(t) - x_o] + f[x(t)] - f(x_o) \\ &= A[x(t) - x_o] + \\ &\qquad \{f_x[x_o + \theta(x(t) - x_o)]\}\,\{x(t) - x_o\}\end{aligned}$$

where $\{f_x[x_o + \theta\,(x(t) - x_o)]\}$ denotes the matrix $\left\{\dfrac{\partial f_i}{\partial x_j}\right\}$ and f_i is the ith component of f, x_j is the jth component of x and the elements of the ith row of the matrix are evaluated at a point $x_o + \theta_i(x(t) - x_o)$ where $\theta_i \in (0,1)$.

Since x_o and $x(t)$ are in $\overline{(B_{R_4} \cap \mathcal{K}) - B_{R_3}}$, then $x_o + \theta_i(x(t) - x_o)$ is in the convex hull of $\mathcal{K} - B_{R_3}$. Hence by the condition placed on the selection of R_3 in the proof of Lemma 5, it follows that

$$|f_x[x_o + \theta(x(t) - x_o)]| < \varepsilon_o(3\sigma)$$

where $\varepsilon_o(3\sigma)$ is chosen so that

$$c_2[\varepsilon_o(3\sigma)] - a < 0$$

where c_2, a are such that if $\Phi(t)$ is a fundamental matrix of $\dot{x} = Ax$ such that $\Phi(0)$ is the identity then

$$|\Phi(t)| \leq c_2 e^{-at}.$$

By a standard argument using Gronwall's Inequality (see Cesari [3, p. 35]) it follows that

$$\lim_{t \to \infty} |x(t) - x_o| = 0.$$

This completes the proof of Lemma 6.

Since x_o is globally asymptotically stable in $\mathcal{K}$, it is the only critical point in $\mathcal{K}$. This completes the proof of Theorem 2.

Remarks

1. The set $\mathcal{K}$ described in hypothesis H-4) is chosen because it is particularly simple. Other sets could be used.

2. One generalization of Theorem 2 might possibly be obtained by replacing Ax with a more general function F(x) and imposing the hypothesis that 0 is a globally

asymptotically stable critical point of the equation

$$\dot{x} = F(x)$$

in the set $\mathcal{O}$. However it seems that to carry through an analog of the proof of Theorem 1 for this more general case would require very explicit hypotheses on $F(x)$.

3. Hypothesis H-1) can be replaced by the hypothesis: each eigenvalue λ of matrix A is such that $\mathcal{R}(\lambda) < 0$. The only change in the proof of Theorem 2 is a slight change in the proof of Lemma 4. Hypothesis H-1) as given has the advantage that it is somewhat more concrete.

There are a number of biological oscillators for which equations (2.1) and (2.2) might serve as approximate descriptions. Experiments with Gonyaulax show that constant bright light may cause oscillations to cease (Hastings [11], Njus, Sulzman and Hastings [15, p. 2]). Also Oatley and Goodwin [16, p. 15] point out that oscillators are sometimes blocked by large stimuli. The administration of thyroxin to patients suffering from periodic catatonic schizophrenia sometimes produces a stabilization of the patient's condition. A mathematical analysis suggesting an explanation for this has been given by Danziger and Elmergreen [8]. See also Rashevsky [17] and Cronin [6]. That analysis is a special case of the discussion in this section.

3. Suppression of periodic solutions by addition of linear terms

We assume that the n-dimensional system

(3.1) $$\dot{x} = A x + f(x)$$

satisfies the following hypotheses.

$\mathcal{H}$-1) A is a constant matrix such that for each eigenvalue λ of A, it is true that $\mathcal{R}(\lambda) < 0$.

$\mathcal{H}$-2) f is defined and continuous on $\mathcal{O}$ an open set in R^n such that $\mathcal{P} \subset \mathcal{O}$, and there exists a positive number M such that if $x \in \mathcal{P}$, then

$$|f(x)| < M.$$

$\mathcal{H}$-3) f has continuous first derivatives at each point of $\mathcal{O}$, and there is a positive number $\mathcal{M}$ such that if $x \in \mathcal{O}$, then

$$\left| \frac{\partial f_i}{\partial x_j}(x) \right| < \mathcal{M} .$$

$\mathcal{H}$-4) If x(t) is a solution of (3.1) such that for some t_o, $x(t_o) \in \partial\mathcal{P}$, then for all $t \geq t_o$ for which x(t) is defined, $x(t) \in \mathcal{P}$. That is, no solution "escapes" the set $\mathcal{P}$.

By application of Theorem 1, it follows that equation (3.1) may have an almost periodic or periodic solution. The problem is to determine what kind of changes in the linear terms in x in equation (3.1) may cause such an almost periodic or periodic solution to disappear. To this end, we impose the following additional hypotheses.

$\mathcal{H}$-5) Suppose that for all nonnegative μ, there is an $n \times n$ matrix $L(\mu)$ such that each element of $L(\mu)$ is a continuous function of μ for all nonnegative

μ and such that if $\lambda_1(\mu), \ldots, \lambda_n(\mu)$ denote the eigenvalues of

(3.2) $$A + \mu L(\mu)$$

and $\mathcal{R}\lambda_1(\mu), \ldots, \mathcal{R}\lambda_n(\mu)$ denote the real parts of the eigenvalues $\lambda_1(\mu), \ldots, \lambda_n(\mu)$, then

$$\max_{i=1,\ldots,n} \mathcal{R}\lambda_i(\mu)$$

decreases without bound as $\mu \to \infty$. Also hypothesis $\mathcal{H}$-4) holds for the equation

(3.3) $$\dot{x} = Ax + f(x) + \mu[L(\mu)]x$$

for all $\mu > 0$.

Theorem 3. If equation (3.1) satisfies hypotheses $\mathcal{H}$-1), $\mathcal{H}$-2), $\mathcal{H}$-3), $\mathcal{H}$-4), and if $L(\mu)$ satisfies hypothesis $\mathcal{H}$-5), then there is a positive number μ_1 such that if $\mu > \mu_1$, then the equation (3.3) has a critical point x_o in $\mathcal{O}$ and x_o is globally asymptotically stable in $\mathcal{O}$.

Proof: The proof is a consequence of the following lemmas.

Lemma 7. There is a positive number μ_o such that if $\mu \geq \mu_o$ then there is a positive number R_μ such that if $R \geq R_\mu$, and if $x(t)$ is a solution of equation (3.3) which intersects ∂B_R, the point set boundary of the ball in R^n with center 0 and radius r, say at t_o, i.e., if $x(t_o) \in \partial B_R$, then for all $t \geq t_o$, $x(t) \in B_R$.

Proof: This follows by standard arguments.

Lemma 8. Given $\mu \geq \mu_o$, then if $R \geq R_\mu$, equation (3.3) has a critical point in the interior of the set $\mathcal{O} \cap B_R$.

Proof: Follows from hypothesis $\aleph$-5), Lemma 7 and Satz 1, p. 479 of Alexandroff-Hopf [1].

Lemma 9. If μ is sufficiently large and if $R \geq R_\mu$, then the critical point x_o obtained in Lemma 8 is globally asymptotically stable in $\mathcal{O} \cap B_R$.

Proof: Follows by an argument similar to the part of the proof of Lemma 6 in which Gronwall's Inequality in used.

Exactly parallel arguments may be used to prove the following theorem.

Theorem 4. If equation (3.1) satisfies $\aleph$-1), $\aleph$-2), $\aleph$-3) and if $L(\mu)$ satisfies hypothesis $\aleph$-5) (except for the last sentence of $\aleph$-5)), then equation (3.3) has a unique critical point x_o in R^n and x_o is globally asymptotically stable in R^n.

A study of a biological oscillation to which equations (3.1) and (3.3) and Theorems 3 or 4 might be applicable is the analysis of the effect of light or rhythms made by Njus, Sulzman and Hastings [15, p. 2] who use a mathematical description based on the membrance model introduced in [15].

4. Suppression of periodic solution by addition of periodic terms

The theorems stated in this section are only mathematical remarks. However it seems worthwhile to state them explicitly because they are suggestive of directions for experimental work.

Theorem 5. Given the n-dimensional system

(4.1) $$\dot{x} = f(t, x)$$

where f has continuous first derivatives with respect to t and x at each point of an open set U in (t, x)-space such that $R \times \{0\} \subset U$ where R is the real line and suppose f has (minimal) period $T > 0$ as a function of t, Let $g(t, x)$ have continuous first derivatives in U and suppose that for each x, function $g(t, x)$ has minimal period $T_1 > 0$ where $T_1 \neq \frac{T}{m}$ and m is any positive integer. Then the equation

(4.2) $$\dot{x} = f(t, x) + g(t, x)$$

does not have a solution of period T.

Proof: Suppose that $x(t)$ is a solution of (4.2) which has period T. Then for all t,

(4.3) $$\dot{x}(t + T) = f(t + T, x(t + t)) + g(t + T, x(t + T))$$

and

(4.4) $$\dot{x}(t) = f(t, x(t)) + g(t, x(t)).$$

Since $\dot{x}(t)$, $x(t)$ and f, as a function of t, all have period T then subtracting (4.4) from (4.3), we obtain: for all t,

$$g(t + T, x(t)) = g(t, x(t)).$$

This contradicts the periodicity hypothesis on g as a function of t.

Corollary 5. Given the n-dimensional system

(4.5) $$\dot{x} = f(x)$$

where f has continuous first derivatives with respect to x at each point of an open set V in R^n, suppose that (4.5) has a solution $x(t)$ of (minimal) period $T > 0$. Let $g(t, x)$

have continuous first derivatives in U and suppose that for each x, $g(t, x)$ has minimal period $T_1 > 0$ where $T_1 \neq \frac{T}{m}$, where m is any positive integer. Then the equation

$$\dot{x} = f(x) + g(t, x) \tag{4.6}$$

does not have a solution of period T.

Finally we consider the equation

$$x = A(t)\, x + \mu f(t, x, \mu) \tag{4.7}$$

where the matrix $A(t)$ is continuous for all real t and $A(t)$ has minimal period $T > 0$. Suppose that the linear space of solutions of period T of the transposed equation

$$\dot{x} = -[A(t)]^T\, x$$

has dimension $k > 0$ and let $y^{(1)}(t), \ldots, y^{(k)}(t)$ be a basis for this space. Suppose that f has continuous first derivatives with respect to t, x and μ at each point of an open set U in (t, x, μ)-space such that $R \times \{0\} \times \{0\} \subset U$ and that f has period T as a function of t. Suppose finally that there is a nontrivial periodic solution of

$$\dot{x} = A(t)\, x$$

from which there branches a solution of (4.7) of period T.

Theorem 6. Let $g(t)$ be continuous and have period T. If there is a j such that

$$\int_0^T g(t) y^{(j)}(t)\, dt \neq 0,$$

then for all sufficiently small $|\mu|$ the equation

$$\dot{x} = A(t)\, x + \mu f(t, x, \mu) + g(t) \tag{4.8}$$

has no solution of period T.

Proof: This remark follows from standard considerations. See, e.g., [5, pp. 64-69].

Theorem 5, Corollary 5 and Theorem 6 suggest possible directions for experiments. For example, suppose that equation (4.5) describes a biological system which oscillates. Then (4.5) has a solution $x(t)$ of period T. Then Corollary 5 shows that if the periodic term $g(t, x)$ is added to the equation, the periodic solution $x(t)$ no longer exists. Notice that $g(y, x)$ may be small. Theorem 6 gives a similar conclusion. Thus if a small forcing term (Zeitgeber) of the right frequency is imposed on the biological system, the oscillations may cease.

REFERENCES

[1]. Paul Alexandroff and Heinz Hopf, Topologie, Springer, Berlin, 1935.

[2] Frank A. Brown, Jr., J. Woodland Hastings, John D. Palmer, The Biological Clock, Two Views, Academic Press, New York, 1970.

[3] Lamberto Cesari, Asymptotic Behavior and Stability Problems in Ordinary Differential Equations, 3rd edition, Springer-Verlag, Berlin, 1971.

[4] Earl A. Coddington and Norman Levinson, Theory of Ordinary Differential Equations, McGraw-Hill, New York, 1955.

[5] Jane Cronin, Fixed Points and Topological Degree in Nonlinear Analysis, Mathematical Surveys, Number 11, American Mathematical Society,

Providence, Rhode Island, 1964.

[6] Jane Cronin, "The Danziger-Elmergreen Theory of Periodic Catatonic Schizophrenia," Bulletin of Mathematical Biology 35(1973) 689-707.

[7] Jane Cronin, "Periodic Solutions in n Dimensions and Volterra Equations," to appear in Journal of Diff. Eqns.

[8] Lewis Danziger and George L. Elmergreen, "The Thyroidpituitary Homeostatic Mechnanism," Bulletin of Mathematical Biophysics 18 (1956) 1-13.

[9] Donald S. Gann, Fritz J. Seif and James D. Schoeffler, "A Quantized Variable Approach to Description of Biological and Medical Systems," Application of Walsh Functions, 1972 Proceedings, edited by R. W. Zeek and A. E. Showalter, Naltional Technical Information Service AD 744650, Springfield VA.

[10] Wolfgang Hahn, Stability of Motion, Springer-Verlag, New York, 1967.

[11] J. W. Hastings, "Are Circadian Rhythms Conditional ?" Abstract, Proceedings of the XI Conference of the International Society for Chronobiology, Hannover, Germany, July 25-28, 1973.

[12] Morris Marden, The Geometry of the Zeros of a Polynomial in a Complex Variable, Mathematical Surveys Number 3, American Mathematical Society, Providence, Rhode Island, 1949.

[13] Nicholas Minorsky, Nonlinear Oscillations, D. Van Nostrand Company, Inc. Princeton, New Jersey, 1962.

[14] V. V. Nemytskii and V. V. Stepanov, Qualitative Theory of Differential Equations, Princeton University Press, Princeton, New Jersey, 1960.

[15] David Njus, Frank M. Sulzman and J. W. Hastings, "Membrane Model for the Circadian Clock," Nature 248 (1974) 116-119.

[16] Keith Oatley and B. C. Goodwin, "The Explanation and Investigation of Biological Rhythms," Biological Rhythms and Human Performance, edited by W. P. Colquhoun, Academic Press, New York, 1971, Chapter 1, pp. 1-38.

[17] N. Rashevsky, Some Medical Aspects of Mathematical Biology, Charles C. Thomas, Springfield, Illinois, 1964.

[18] Fritz J. Seif and Donald S. Gann, "An Orthogonal Transform Approach to the Description of Biological and Medical Systems, "Applications of Walsh Functions, 1972 Proceedings, edited by R. W. Zeek and A. E. Showalter, National Technical Information Service Ad 744650, Springfield VA.

[19] J. J. Stoker, Nonlinear Vibrations in Mechanical and Electrical Systems, Interscience Publishers, Inc., New York, 1950.

[20] George R. Sell, "Periodic solutions and asymptotic stability," J. Differential Equations 2(1966), 143-157.

Rutgers University, New Brunswick, New Jersey

DUALITY OF VARIOUS NOTIONS OF CONTROLLABILITY AND OBSERVABILITY

Szymon Dolecki

1. Introduction.

After the lecture of Professor Roberto Conti [1] we have an idea of great variety of controllability concepts in the finite-dimensional control theory.

What we shall do is to take one of those concepts, namely controllability on an interval, say [0, T], and to split it into three different notions in the case when a state space is infinite-dimensional.

Actually we shall begin by writing down three definitions of observability. Then the three notions of controllability will appear in a natural way, as corresponding dual properities.

Let us consider a process given by a strongly continuous semigroup of bounded linear operators $S(t)$, $t \geq 0$, defined on a Banach space X. Every initial state $x_0 \in X$ determines the trajectory $S(\cdot)x_0$. We have to choose a topological space in which we shall examine the trajectories. Let us pick up $L^2(0, T;X)$ for example.

Supported in part by the Office of Naval Research under contract NR 041-404.

Now it is vehement human curiosity that pushes us to observe the trajectory with the aid of a linear operator B:

$$B: X \to \tilde{X},$$

such that the map:

$$x_0 \overset{C}{\to} BS(\cdot)x_0 \in L^2(0, T; \tilde{X})$$

is bounded. We have

$$C : X \to Y \overset{def}{=} L^2(0, T, \tilde{X}).$$

<u>Definition</u> 1.

An observed process $\{S(\cdot), B\}$ is said to be

(1a) <u>distinguishable</u>, if

(1a) $$\text{Ker } C = \{0\}$$

(1b) <u>observable</u>, if the operator (defined on the closed span of all observations)

(1b) $$BS(\cdot)\, x_0 \to x_0$$

is bounded. In other words, if there is a k, such that $\|x_0\| \leq k \|Cx_0\|$ for each x_0.

(1c) <u>finally observable</u>, if (1b) is valid with x_0 replaced by the final state $x_T = S(T)x_0$, that is when the following formula holds:

(1c) $$\|S(T)x_0\| \leq k \|Cx_0\|.$$

Remark 1.

S. Rolewicz [11] introduces a definition of observability (stronger than (c)). The additional requirement is the possibility of extension of C^{-1} to the whole space Y.

Following Mizel and Seidman [9] and Luxemburg and Korevaar [8] I studied the problem of final observability (1c) for diffusion processes [2] [3]. This notion of observability was chosen, because distinguishability in this case is rather trivial and (initial) observability does not occur.

If the process is observable and the semigroup $S(t)$ is one-to-one then by a theorem due to D. L. Russell (private communication) $S(t)$, $t \geq 0$ should be a group of operators. But $S(t)$ is compact in the above case.

It turned out that there exists a critical time T_0 (such a number that for T less than T_0 the system is not finally observable, but for T greater than T_0 the system is finally observable) given by the formula

$$T_0 = -\liminf_n \frac{\log \| B\psi_n \|}{\lambda_n} \tag{2}$$

where ψ_n are the normalized eigenvectors of the infinitesimal generator and $-\lambda_n$ are its eigenvalues.

It we consider all continuous linear functionals $B \in X^*$, then the set:

$$\{ B: T_0(B) = \infty \}$$

is of second (residual) Baire category.

However we can choose a non-degenerate Gaussian measure μ on X^* so that

$$\mu\{B: T_0(B) = 0\} = 1.$$

These were results concerning particular systems, namely one dimensional diffusion processes. However they proved to be very useful in a general perturbation theory of observed systems [4].

2. Duality theorem.

Let us consider a dual system $\{B^*, X^*(t),\ 0 \leq t \leq T\}$ (to have strong continuity of $S^*(\cdot)$ assume that X is reflexive). Then the dual operator C^* (after a change of a scale) is given by

$$C^*u = \int_0^T S^*(T-t)B^*(t)u(t)dt \tag{3}$$

where $u \in L^2(0, T, \tilde{X}^*) = Y^*$. The formula

$$x(T) = S^*(T)\, x_0 + \int_0^T S^*(T-t)B^*u(t)dt \tag{4}$$

describes a control process, that starting fom $x(0) = x_0$ is transferred to $x(T)$ at time T by control u.

It is well-known [6] [7] that distinguishability of $\{S(\cdot), B\}$ is equivalent to approximate controllability at $\{B^*, S^*(\cdot)\}$ a. i. to the condition that

$$\{\overline{x(T)}\} = X^*.$$

It is also well-known [6] [7] that observability of the original system is equivalent to exact controllability of its dual system:

$$\{x(T)\} = X^*.$$

Final observability turns out to be dual with respect to null-controllability [1].

Null-controllability means that for each initial state x_0 we are able to pick out a control $u(\cdot)$ in such a way that the corresponding $x(T)$ is null. From (4) and (3) it follows that the range of $S^*(T)$ is then included in the range of C^*. We introduce the special notation for $S(T) \overset{\text{def}}{=} F$ and we express the necessary and sufficient condition for null-controllability as follows:

$$\mathcal{R}(F^*) \subset \mathcal{R}(C^*) \tag{5}$$

where $\mathcal{R}$ stands for the range of an operator.

We shall formulate the problem in a more abstract way relaxing essentially our previous assumptions.

We are concerned with three Banach spaces X, Y and Z, with linear densely defined operator C and with a bounded linear operator F interconnected as in the following graph:

$$\begin{array}{ccc} X & \supset \operatorname{dom} C & \xrightarrow{C} Y \\ \searrow F & & \\ Z & & \end{array} \tag{6}$$

The dual graph is defined in a natural way:

(6*) $$X^* \overset{C^*}{\leftarrow} \operatorname{dom} C^* \subset Y^*$$
$$\nwarrow F^* \quad Z^*$$

Now we introduce the weakest topology in X for which F is continuous and we obtain the space X_F (not necessarily metric or complete). Define all linear continuous functionals on X_F by X^F and define operator C^F dual to C in respect to F-topology. The F-dual graph is the following:

(6^F) $$X^F \overset{C^F}{\leftarrow} \operatorname{dom} C^F \subset Y^*$$

Theorem.

The following statements are equivalent

(i) C is F-observable:

$$\| Fx \| \leq k \| Cx \| \quad x \in \operatorname{dom} C$$

(ii) $\mathcal{R}(F^*) \subset \mathcal{R}(C^*)$

(5)

(iii) C^F is exactly controllable:

$$X^F = \mathcal{R}(C^F).$$

An application of the theorem combined with controllability results by D. L. Russell [12] gives the answer to a problem posed in [10]:

We consider the heat equation

$$\frac{\partial u}{\partial t} = \Delta u$$

in domain Ω with smooth boundary $\partial\Omega$ along with the homogeneous boundary condition

$$\alpha u \,|_{\partial\Omega \times (0,\, T)} + \beta \frac{\partial u}{\partial \nu}\,|_{\partial\Omega \times (0,\, T)} = 0.$$

Then the following maps are continuous (and linear) [4]:

$$\begin{array}{lcr} u\,|_{\partial\Omega \times (0,\, T)} & & \text{if } \beta \neq 0 \\ & \searrow & \\ & & u(\cdot\,, T) \\ & \nearrow & \\ \frac{\partial u}{\partial \nu}\,|_{\partial\Omega \times (0,\, T)} & & \text{if } \alpha \neq 0 \end{array}$$

Proof of the theorem (more detailed proof is given in [4]).

$\mathcal{R}(F^*)$ is equal X^F and is isometrically isomorphic to $(X/\ker F)^*$. On the other hand the intersection of $\mathcal{R}(C^*)$ with X^F is equal to $\mathcal{R}(C^F)$. (See [4], [6], [7]).

(ii) $\Rightarrow$ (iii)

If $\mathcal{R}(F^*) \subset \mathcal{R}(C^*)$, then $X^F \subset \mathcal{R}(C^*)$ and

$$X^F \subset \mathcal{R}(C^*) \cap X^F = \mathcal{R}(C^F)$$

(iii) $\Rightarrow$ (ii)

If $X^F \cap \mathcal{R}(C^*) = X^F$, then $\mathcal{R}(F^*) = X^F$ is included in $\mathcal{R}(C^*)$.

(ii) $\Leftrightarrow$ (iii) $\Rightarrow$ (i)

For each F-continuous functional ξ there is an element $\eta \in Y^*$ such that

$$\xi = C^* \eta .$$

We estimate:

$$|\xi(x)| = |(C^*\eta)x| = |\eta(Cx)| \leq \|\eta\|_{Y^*} \|Cx\|_Y .$$

We define a linear functional on $(X/\ker F)^* = X^F$:

If $Cx \neq 0$

$$\psi_x(\xi) = \frac{\xi(x)}{\|Cx\|_Y} . \tag{7}$$

If $Cx = 0$, then $\xi(x) = 0$ and we put

$$\psi_x(\xi) = 0. \tag{8}$$

For each x we have

$$|\psi_x(\xi)| \leq \|\eta\|_{Y^*} .$$

Thus by the Banach-Steinhaus theorem on uniform boundedness [5] there is a k such that

$$\|\psi_x(\xi)\| \leq k \|\xi\|_{X^F} \tag{9}$$

or

$$|\xi(x)| \leq k \|\xi\|_{\overline{(X/\ker F)^*}} \|Cx\|_Y$$

hence $\|x\|_{X/\ker F} = \|Fx\| \leq k \|Cx\|_Y$.

(i) $\Rightarrow$ (ii) $\Leftrightarrow$ (iii)

If (i) holds we define for each F-continuous functional ξ a functional η in the following way:

$$\eta(Cx) = \xi(x). \tag{10}$$

The definition of η is correct for if $Cx_1 = Cx_2$, then

$$0 = \|C(x_1 - x_2)\| \geq \frac{1}{k} \|F(x_1 - x_2)\|$$

and therefore $\xi x_1 = \xi x_2$ because of F-continuity of ξ.

We have to prove that η is continuous and to extend it to a functional $\hat{\eta}$ on X^*. Then from definitions

$$\xi = C^* \hat{\eta} \tag{11}$$

which is equivalent to (ii) and (iii).

The proof completed, we want to emphasize that the very mild assumptions on operator C make the theorem easy to use in various applications.

REFERENCES

[1] R. Conti, On global controllability, (L. W. Neustadt Memorial Lecture), these proceedings.

[2] S. Dolecki, Observability for the one-dimensional heat equation, St. Math. 48(1973), 291-305.

[3] __________, Observability for regular processes (to appear).

[4] S. Dolecki and D. L. Russell, A general theory of observation and control: duality, perturbation and reconstruction (to appear in SIAM J. on Control).

[5] N. Dunford and Y. T. Schwartz, Linear operators, Part I, Interscience 1959.

[6] S. Goldberg, Unbounded linear operators, McGraw Hill 1966.

[7] S. G. Krein, Linear equations in Banach space, Nauka 1971 (in Russian).

[8] W. A. J. Luxemburg and J. Korevaar, Entire functions and Muntz-Szasz type approximation, Trans. Am. Math. Soc. 57 (1971), 23-37.

[9] V. Y. Mizeland and T. I. Seidman, Observation and predictions for the heat equation, J. Math. Anal. Appl. 28 (1969), 303-312.

[10] ____________, II, J. Math. Anal. Appl. 38 (1972), 149-166.

[11] S. Rolewicz, On optimal observability of linear systems in infinite-dimensional states, St. Math. 44 (1972), 411-416.

[12] D. L. Russell, A unified boundary value controllability theory for hyperbolic and parabolic partial differential equations, St. Appl. Math. (3) Sept. 73.

Institute of Mathematics, Polish Academy of Sciences, Warsaw, Poland. This paper was prepared while the author was Visiting Assistant Professor at the University of Wisconsin-Madison, Madison, Wisconsin 53706

A CASE HISTORY IN SINGULAR PERTURBATIONS

A. Erdélyi

1. We shall consider singular perturbation problems P_ε consisting of a nonlinear ordinary differential equation of the second order,

$$\varepsilon y'' + f(t, y, y', \varepsilon) = 0 \quad 0 \leq t \leq 1 \tag{1.1a}$$

containing a small positive parameter ε, and the boundary conditions

$$y(0, \varepsilon) = \alpha(\varepsilon), \quad y(1, \varepsilon) = \beta(\varepsilon). \tag{1.1b}$$

Here t is the independent variable, $y = y(t, \varepsilon)$, and primes denote differentiation with respect to t.

As $\varepsilon \to 0+$, one would expect the solution of P_ε to approach a solution of the limiting differential equation,

$$f(t, y, y', 0) = 0 \quad 0 \leq t \leq 1. \tag{1.2a}$$

Now, this differential equation is one of the first order, and it cannot be expected that its solution will satisfy both of the limiting boundary conditions

$$y(0, 0) = \alpha(0), \quad y(1, 0) = \beta(0). \tag{1.2b}$$

Thus, the very nature of the limiting problem P_o is at question here.

In problems of applied mathematics the nature of P_o

often suggests itself by physical reasoning. For instance, problems of this character arise in mathematical models of viscous flow when one of the boundary conditions expresses the condition of zero velocity at the wall of the channel. The limiting differential equation is that of non-viscous flow, and clearly the boundary condition expressing the adherence to the walls of a viscous fluid must be suppressed in formulating P_0. However, the character of P_0 is not always as obvious as in this instance.

Analytical investigations will naturally start with a formulation of P_0 and investigation of the relationship between the solution of P_ε and the solution, y_0, of P_0. Clearly $y(t,\varepsilon) \to y_0(t)$ as $\varepsilon \to 0+$ must not be expected in the whole interval $0 \leq t \leq 1$: in general one of the boundary conditions (1, 2b) will be violated, or else y_0 will be a discontinuous solution of (1. 2a). In either case, the approach of $y(t,\varepsilon)$ to $y_0(t)$ will not be uniform, and the nature of y in the neighborhood of the exceptional points (in the boundary layer in the case of viscous flow) will be of especial interest. The applied mathematician will also seek approximations for small ε to $y(t,\varepsilon)$, perhaps starting with y_0. The differential equation (1. 1a) being nonlinear, the existence or uniqueness of solutions of P_ε must not be taken for granted. Since (1. 2a) is an implicit differential equation, it is not at all clear that it will have a solution over the whole interval. Another natural question to ask is: will the existence of a solution to P_0 guarantee the existence of one to P_ε? And if so, what uniqueness properties will the latter solution have?

The problem described here would seem to be very special, and not many genuine problems of applied mathematics can be cast in this form. Nevertheless, Friedrichs [15] used a simple linear problem of the form (1.1) to illustrate Prandtl's boundary layer theory, and since then problems of the form (1.1) have often been used as model problems to explore the nature of singular perturbations, develop and test techniques for the construction of approximations, or prove the validity of such approximations: see for instance [7], (21]. Because of this, P_ε deserves some attention.

The sequence of investigations to be described here has two interesting features.

At first analysts and applied mathematicians worked at different aspects of the problem, and there was very little connection between the efforts of the two groups. Analysts asked, and in part answered the questions most natural from their point of view, but they produced no useful approximations to the solution of P_ε. Applied mathematicians on their part constructed approximations (see for instance [7], [20]) whose validity they were unable to test - except in cases where an exact solution or a comparison with known facts was available. In the course of time, however, a situation developed in which analysts can both provide significant pointers towards the construction of approximations and test the validity of tentative approximations, while applied mathematicians on their part are able to construct approximate solutions which serve to prove the existence of a solution to P_ε as well as provide an approximation to that solution. To this extent, the history of P_ε is a success story.

Another interesting feature of the history of P_ε is by no means peculiar to this problem but it manifests itself especially clearly here. More than once significant progress was made almost by accident by an investigator who set out to achieve one aim and then, almost unnoticed by himself, brought about an improvement in another direction.

2. For orientation we consider a very special linear case of (1.1), namely

$$\varepsilon y'' + cy' = h'(t) \tag{2.1a}$$

$$y(0,\varepsilon) = \alpha, \quad y(1,\varepsilon) = \beta \tag{2.1b}$$

with given constants c, α, β and a given function h. In these exploratory calculations we assume that all continuity and differentiability requirements are satisfies.

The solution of (2.1) may be written down explicitly as

$$y(t,\varepsilon) = (1-e^{-c/\varepsilon})^{-1}\{[\alpha + \varepsilon^{-1}\int_0^t e^{cu/\varepsilon}h(u)du](e^{-ct/\varepsilon})$$

$$+ [\beta - \varepsilon^{-1}\int_t^1 e^{c(u-1)/\varepsilon} h(u)du](1-e^{-ct/\varepsilon})\}, \tag{2.2}$$

with an abvious modification (2.7) in case $c = 0$.

In $c > 0$ and $0 < t \leq 1$, then

$$y(t,\varepsilon) \to \beta + c^{-1}[h(t)-h(1)] \quad \text{as } \varepsilon \to 0+. \tag{2.3}$$

As expected, the right-hand side of (2.3) satisfies the limiting differential equation

$$cy' = h'(t); \tag{2.4}$$

it also satisfies the boundary condition at $t = 1$ but not at $t = 0$. Moreover, (2.3) holds uniformly in $[\delta,1]$ for each $\delta > 0$ but does not hold uniformly in $]0,1]$.

To investigate the behavior of (2.2) near $t = 0$, we set $t = \varepsilon\tau$ and make $\varepsilon \to 0+$ keeping $\tau \geq 0$ fixed. The result is

$$(2.5)\quad y(\varepsilon\tau,\varepsilon) \to \alpha e^{-c\tau} + \{\beta + c^{-1}[h(0) - h(1)]\}(1 - e^{-c\tau}) \text{ as } \varepsilon \to 0+.$$

The differential equation (2.1a) may be written in terms of τ as

$$(2.6)\qquad \ddot{y} + c\dot{y} = \varepsilon h'(\varepsilon\tau),$$

where dots denote differentiation with respect to τ. The righthand side of (2.5) satisfies the limiting form of (2.6), and it satisfies the boundary condition at $t = 0$ (i.e., $\tau = 0$) but not at $t = 1$ (i.e., $\tau = 1/\varepsilon$).

Similar circumstances, with the role of $t = 0$ and $t = 1$ interchanged and τ being defined by $t = 1 - \varepsilon\tau$, arise when $c < 0$. The case $c = 0$ is more difficult. In this case

$$(2.7)\quad y(t,\varepsilon) = [\alpha + \varepsilon^{-1}\int_0^t h(u)du](1-t) + [\beta - \varepsilon^{-1}\int_t^1 h(u)du]t$$

does not approach a finite limit as $\varepsilon \to 0+$, unless h is a constant. Indeed in this case (2.4) demands $h'(t) = 0$ but (2.7) does not satisfy (2.4).

Approximations to any order (in case $c > 0$) may be developed systematically by first constructing an "outer expansion"

$$\Sigma f_n(t)\varepsilon^n$$

and an "inner expansion"

$$\Sigma g_n(t/\varepsilon)\varepsilon^n$$

by substitution in (2.1a), and then matching these two expansions. Ideas developed by Lagerstrom and his associates, in particular Kaplun, allow also the construction of a uniformly valid "composite expansion". This construction has been carried out for (2.1) in [10]; and the special case $h'(t) = a + 2bt$ has been used more recently [21] to illustrate basic concepts and techniques in "layertype" singular perturbation problems. Matched expansions have been investigated in a roader context with great precision by Fraenkel [14] and in even greater generality but with less detail by Eckhaus [9].

An alternative approach, employing two-variable expansions has been used for (2.1) in [13].

3. We return to P_ε and assume that $f_{y'}(= \partial f/\partial y')$ is positive. The case $f_{y'} < 0$ is similar (interchange t and $1 - t$); those cases in which $f_{y'}$ is allowed to vanish or change sign are more difficult, have not as yet been fully investigated, and will only be briefly mentioned here.

From (2.1) and other examples one would expect that if $f_{y'} > 0$, $0 < t \leq 1$, and $\varepsilon \to 0+$, then y approaches a solution of the limiting equation (1.2a); and this solution of (1.2a) would be expected to satisfy the second boundary condition (1.2b). The behavior of y in the neighborhood of $t = 0$ could then be investigated by means of a "stretching transformation" corresponding to $t = \varepsilon\tau$ in the case of (2.1), and one would expect that systematic approximation procedures can be based on matching techniques or two-variable expansions.

Coddington and Levinson [4] showed that such facile generalizations have severe limitations. Specifically, they pointed out that these generalizations fail in the simple case

$$(3.1) \qquad f(t, y, y', \varepsilon) = y' + (y')^3$$

In this case P_ε has a solution only if $|\alpha(\varepsilon) - \beta(\varepsilon)| < \varepsilon\pi/2$. Thus starting with a value of $\varepsilon > 0$ for which a solution exists, the solution may cease to exist as ε decreases, may reappear and disappear again, and will ultimately fail to exist for all sufficiently small values of ε if $\alpha(0) \neq \beta(0)$ ($\alpha(\varepsilon)$ and $\beta(\varepsilon)$ being assumed continuous at $\varepsilon = 0$). Other strange behavior of such problems was exemplified by Wasow [34].

A closer look at the example (3.1) suggests that its peculiar nature is due to the rapid increase of f with increasing y'. Accordingly, Coddington and Levinson (and several later investigators) restricted themselves to the quasilinear case, assuming that f is a linear function of y' with coefficients which depend of t and y. (The absence of ε in f, and also in α and β is not an important restriction.) In this particular case a very satisfactory result can be established [4]. Under reasonably liberal conditions P_ε possesses a unique solution and as $\varepsilon \to 0+$, this solution approaches, uniformly in any closed subinterval of $]0,1]$, that solution of (1.2a) satisfying the second boundary condition (1.2b).

4. At this stage there was a considerable gap between the work of analysts on the one hand and applied mathematicians

on the other hand. The analysts did not describe the structure of the solution near $t = 0$ (in the "boundary layer") and they did not provide explicit approximations to the solution even outside the boundary layer. Applied mathematicians were able to construct approximate solutions by boundary layer techniques but they were unable to prove the existence of an exact solution of P_ε and they were unable to show that their approximate solutions were valid approximations to the exact solution when the latter existed.

This gap was considerably narrowed by Wasow [33] who assumed f linear in y' and analytic in y and ε and was able to construct series expansions which proved the existence of a unique solution to P_ε, provided approximations, and also revealed the structure of the solution near $t = 0$. For the proof of his results Wasow developed a pattern which was followed, with slight changes, by a succession of investigators.

Wasow starts with the solution $u(t)$ of the problem P_o:

(4.1) $$f(t, u, u', 0) = 0, \qquad u(1) = \beta(0)$$

and shows first the existence of a solution $y^* = u + v$ of (1.1a) with $y^*(1,\varepsilon) = \beta(\varepsilon)$ and

(4.2) $$v = 0(\varepsilon), \quad v' = 0(\varepsilon) \text{ uniformly for } 0 \le t \le 1.$$

Of course, $y^*(0,\varepsilon) \neq \alpha(\varepsilon)$ in general. Wasow then proves that if $|y^*(0,\varepsilon) - \alpha(\varepsilon)|$ is sufficiently small, then P_ε has a unique solution close to u. This solution is of the form

(4.3) $$y = u + v + w,$$

where the boundary layer correction w vanishes exponentially outside the immediate neighborhood of $t = 0$. More precisely,

$$w = O(e^{-\phi/\varepsilon}), \quad w' = O(\varepsilon^{-1} e^{-\phi/\varepsilon}) \text{ uniformly for } 0 \leq t \leq 1, \tag{4.4}$$

where

$$\phi(t) = \int_0^t f_{y'}(s, u(s), u'(s), 0)\,ds . \tag{4.5}$$

Wasow's construction of v and w involves "linearization" of (1.1), around u for the construction of v and then around $y^* = u + v$ for the construction of w. The differential equation for v is written in the form

$$\varepsilon v'' + p(t,\varepsilon)v' + q(t,\varepsilon)v = g(t, v, v', \varepsilon), \tag{4.6}$$

where

$$\begin{aligned} p(t,\varepsilon) &= f_{y'}(t, u(t), u'(t), \varepsilon) \\ q(t,\varepsilon) &= f_{y}(t, u(t), u'(t), \varepsilon) \\ g(t, v, v', \varepsilon) &= -f(t, u(t) + v, u'(t) + v', \varepsilon) - \varepsilon u''(t) \\ &\quad + p(t,\varepsilon)v' + q(t,\varepsilon)v; \end{aligned} \tag{4.7}$$

and there is a corresponding form for the differential equation for w.

This two-step construction of the solution in the form (4.3) has remained the pattern followed until very recently.

5. Next comes one of those instances when an investigator found something he was not seeking.

In 1960, I attempted to reproduce Wasow's results by real variable methods, without using power series, hoping to reduce Wasow's analyticity assumptions to modest differentiability assumptions. In this I was successful (although my conditions were not quite as liberal as those of Coddington and Levinson) but in the process I also discovered, somewhat to my surprise and much to my pleasure, that with my approach the assumption of quasilinearity became unnecessary: instead of assuming that f in (1.1a) be a linear function of y', it was sufficient to assume that $f_{y'y'} = 0(\varepsilon)$.

The result [11] is perhaps worth recalling in some detail. With a fixed $\delta > 0$, let D_δ be the set of those quadruplets (t, y, y', ε) satisfying

$$0 \leq t \leq 1, \ 0 < \varepsilon < \varepsilon_o, \ |y - u(t)| < \delta,$$
$$|y' - u'(t)| < \delta(1 + \varepsilon^{-1} e^{-\phi(t)/\varepsilon}).$$

The basic assumptions are as follows: -

(I) For some $\delta > 0$, f possesses in D_δ partial derivatives of order one and two with respect to y and y'; and f and these partial derivatives are continuous functions of t, y, y'.

(II) There is $u \in C^2[0,1]$ satisfying (4.1).

(III) With p and ϕ as in (4.7) and (4.5), we have $\phi'(t) > 0$ and $p(t,\varepsilon) = \phi(t) + \varepsilon p_1(t,\varepsilon)$; and $\varepsilon^{-1} f(t,u(t),u'(t),\varepsilon)$, $p_1, q, f_{yy}, f_{yy'}$, $\varepsilon^{-1} f_{y'y'}$ are uniformly bounded.

(IV) $\beta(\varepsilon) - \beta(0) = 0(\varepsilon)$.

Under these assumptions there exist positive numbers ε_1 and μ_1 so that whenever $0 < \varepsilon < \varepsilon_1$ and $|u(0) - \alpha(\varepsilon)| < \mu_1$, then P_ε possesses a solution (4.3) for which (4.2) and (4.4) hold.

Under the further assumption

(V) f_y uniformly bounded, $f_{y'} \geq \delta_1 > 0$

this solution is unique.

The existence of and estimates for v are established by converting (4.6) together with the boundary conditions $v'(0,\varepsilon) = 0$, $v(1,\varepsilon) = \beta(\varepsilon) - \beta(0)$ into a nonlinear integral equation and applying the Banach fixed point theorem (the contraction mapping principle). Most of the effort goes into the study of the linear differential equation.

$$\varepsilon V'' + p(t,\varepsilon)V' + q(t,\varepsilon)V = h(t,\varepsilon) . \tag{5.1}$$

Once the properties of solutions of (5.1) have been ascertained, v can be constructed easily enough; and the construction of w is achieved by a similar process of linearization of (1.1a) around $y^* = u + v$.

In comparison with these results, both [4] and [33] are restricted to the quasilinear case, $f_{y'y'} = 0$. In this more restricted context, the assumptions of [11] are more stringent than those of Coddington and Levinson (who do not assume the existence of f_{yy} and $f_{yy'}$) but the results are somewhat more detailed and take in the entire interval [0,1]. In contradistinction, the assumptions of [11] are more liberal than those of Wasow (who assumes analytic dependence on y) but they do not allow the duplication of all of Wasow's results, not even in the quasilinear case.

6. The problem (1.1) was reconsidered by Willett [36], who noted that the condition $f_{y'}(t, u(t), u'(t), 0) > 0$ could be relaxed somewhat and wished to replace the restriction imposed on $u(0) - \alpha(\varepsilon)$ by a more easily verifiable condition. With this aim he chose an approach which, he noted, "includes the possibility of determining to any order

of ε the asymptotic nature of the solutions".

It is perhaps not quite obvious from Willett's presentation that apart from the improvements in existing results, he achieved a break-through by radically changing the point of view. Up to this time all investigators take (4.1) as their point of departure and compare solutions of (1.1) with those of (4.1). Willett does not refer to (4.1) at all. He assumes that a tentative approximate solution of (1.1) is available and uses this both to establish the existence of an exact solution and to validate the approximation. His assumptions provide guidelines for the construction of approximate solutions but do not prescribe the construction. This is left to the ingenuity of applied mathematicians who have developed powerful methods: see [7], [9], [21], [30] and similar books. In comparison with older work, Willett's results are at the same time more useful in applied mathematics and more dependent on applied mathematical skill in the construction of approximate solutions.

Instead of describing these results in the form in which they are presented in [36], I shall recast them in a form [12] which allows a direct comparison with section 5.

We assume given: a function $\eta(\varepsilon)$ measuring the accuracy of the approximation, a function $\phi(t,\varepsilon)$ describing decay in the boundary layer, and an approximate solution $u(t,\varepsilon)$ for P_ε. The functions p, p_1, q are defined as in sections 4 and 5 except that now u and ϕ depend on ε; and we also define

$$\Phi(t,\varepsilon) = 1 + \varepsilon^{-1} \exp(-\phi(t,\varepsilon)/\varepsilon) . \tag{6.1}$$

Some of the assumptions of section 5 must be modified and others are omitted.

(II′) For each ε, $0 < \varepsilon < \varepsilon_o$, we have $u \in C^2[0,1]$.

(III′) $\phi \in C^2[0,1]$ for fixed ε, $0 < \varepsilon < \varepsilon_o$, $\phi(0,\varepsilon) = 0$, $\phi'(t,\varepsilon) > 0$; ϕ', $(\phi')^{-1}$, ϕ'', p_1, q, f_{yy}, $f_{yy'}$, $\varepsilon^{-1} f_{y'y'}$ and uniformly bounded.

We add one assumption which expresses and u is an approximate solution.

(VI) We have $\eta = O(\varepsilon)$,

(6.2) $$\varepsilon u'' + f(t, u, u', \varepsilon) = O(\eta[1 + \varepsilon^{-1} e^{-m\phi/\varepsilon}])$$

with $m > 1$, and

(6.3) $$u(0,\varepsilon) - \alpha(\varepsilon) = O(\eta), \quad u(1,\varepsilon) - \beta(\varepsilon) = O(\eta).$$

The conclusion is that under assumptions (I), (II′), (III′), (VI), P_ε has, for each sufficiently small positive ε, a unique solution y such that

(6.4) $$y - u = (\eta), \quad y' - u' = O(\eta\Phi).$$

It will be noted that (6.4) differs from the estimates arising from (4.2) and (4.4) with $\eta(\varepsilon) = \varepsilon$. This happens because the solution of (4.1) is not an approximate solution of P_ε in the sense of (VI). A boundary layer correction would have to be added to the solution of (4.1) before it can satisfy the first relation of (6.3) and qualify as an approximate solution.

It may also be mentioned that (VI) could be relaxed. Corresponding results are known under the less restrictive conditions $\eta = o(1)$, $m > 0$.

The construction of $y = u + v + w$ and the proof of (6.4)

are quite similar to the work described in section 5.

7. The next step forward is a purely technical one, but it has not only introduced improvements in the proof of available results but has already led to new results of considerable significance.

It has been mentioned in section 5 that the most laborious part of the work described there is the investigation of (5.1). K. W. Chang [2] applied an ingenious transformation which simplifies the argument considerably.

In (5.1) assume

(VII) p, q continuous in t, $p > 0$; p, q, p^{-1} uniformly bounded. Under these assumptions Chang proves that for sufficiently small ε,

$$\varepsilon r' + (pr + q + \varepsilon r^2) = 0, \quad r(0) = 0 \tag{7.1}$$

determines a uniformly bounded function r, and with this function

$$\varepsilon s' = (p + 2\varepsilon r)s - 1, \quad s(1) = 0 \tag{7.2}$$

determines a uniformly bounded function s for which $(s(0))^{-1}$ is uniformly bounded.

Using these functions, set

$$A = (1 - \varepsilon rs)V + \varepsilon sV', \quad B = -rV + V'. \tag{7.3}$$

Then (5.1) is equivalent to

$$A' - rA = sh, \quad \varepsilon B' + (p + \varepsilon r)B = h. \tag{7.4}$$

It is now easy enough to express A and B in terms of p, r, s, h, and estimate A and B. Once this is done, V and

V' can be recovered;

(7.5) $$V = A - \varepsilon s B, \quad V' = rA + (1 - \varepsilon r s)B .$$

The same transformation applied to (4.6) leads to a system of nonlinear integral equations of which the linear part is in diagonal form and which is much easier to handle than the integral equation used in [11].

Chang developed his method in the hope that it will enable him to avoid breaking up $y - u$ in two parts, v and w, as has been customary ever since [33]. In this he was successful in the quasilinear case [written communication, 1973] but it seems that at present no such one-stage proof using the Banach fixed point theorem is known.

8. A one-stage proof in the nonlinear case was finally achieved by Habets [16] who applied Schauder's (rather than Banach's) fixed point theorem to the system of nonlinear integral equations derived from (4.6) by Chang's transformation.

Habets uses the set-up described in section 6 except that he assumes $p_1 = 0$, i.e., $p = \phi'$, and expresses his other assumptions in terms of p, q, g rather than f. As in section 6, u is allowed to depend on ε. He assumes $\eta = O(\varepsilon)$, $m > 1$, (6.3), (II'), (VII), and in addition

(VIII) g is a continuous function of t, v, v' satisfying

$$g(t, v, v', \varepsilon) = O(|v|^2 + |vv'| + \varepsilon|v'|^2 + (1 + \varepsilon^{-1} e^{-m\phi/\varepsilon})\eta) .$$

Under these conditions he proves the existence of a solution y to P_ε for which (6.4) holds. Under the additional assumption

$$\text{(IX)} \qquad |g(t, v_1, v_1', \varepsilon) - g(t, v_2, v_2', \varepsilon)| \\ \leq K\{|v_1 - v_2| \max(|v_1|, |v_2|, |v_1'|, |v_2'|) \\ + |v_1' - v_2'| \max(|v_1|, |v_2|, \varepsilon|v_1'|, \varepsilon|v_2'|)\}$$

he also proves uniqueness of y in some neighborhood of u.

It should be noted that (I) and (III′) are sufficient but they are not necessary for the validity of (VIII) and (IX).

9. We have reached the end of the line. As far as the comparison of the solution of (1.1) with that of the limiting (degenerate) problem (4.1) goes, the results outlined in section 5 are probably the most useful ones. For approximate solutions and their validation, section 7 presents the most powerful results; those outlined in section 6 are more restrictive but the conditions on which they are based are somewhat easier to verify than those of section 7.

There remains to indicate some ramifications of this work.

For the quasilinear problem, O'Malley [25] has produced a simpler proof. More general boundary conditions than (1.1b) have been considered by Macki [23], Searl [31], Cohen [6]. Systems of two nonlinear (in some cases quasilinear) differential equations (in place of a single second order equation) have been investigated by Harris [18], Macki [23], O'Malley [26]. A further extension to differential equations involving vector-valued functions has been considered by Harris [19], by Chang [3] and by Habets [17].

Throughout this presentation we have assumed that $f_{y'}$ has a constant sign. In this case the boundary layer develops at one of the end points (at t = 0 if $f_{y'} > 0$). When $f_{y'}$ is allowed to vanish complications arise which have not yet been fully considered. If $f_{y'}$ vanishes at t = 0 (and is positive otherwise), the nature of the boundary layer changes. Such a situation (for a linear equation) was investigated by Watts [35]. If $f_{y'}$ changes its sign in the interior of the interval, then in general the limiting solution will be discontinuous. Of newer investigations we may mention O'Malley [27], [28]. Under exceptional circumstances "resonance" occurs and the limiting solution is continuous. This phenomenon seems to have been described first by Ackerberg and O'Malley [1]. Watts [35], Lakin [22], Zauderer [37], Cook and Eckhaus [8], and Matkowski [24] contributed to the understanding of it.

Lastly, in certain cases the limiting problem has multiple solutions. Such problems have been discussed for instance by Cohen [5] and O'Malley [29].

All these investigations are more or less in the spirit of the work described in earlier sections. For an entirely different approach to boundary value problems containing a small parameter it will be sufficient to refer to the survey article by Vasiléva [32] who gives references to earlier literature.

REFERENCES

[1] Ackerberg, R. C. and O'Malley, R. E., Jr. Boundary layer problems exhibiting resonance. Studies in Appl. Math. 49 (1970) 277-295.

[2] Chang, K. W. Approximate solutions of nonlinear

boundary value problems involving a small parameter. SIAM J. Appl. Math.

[3] Chang, K. W. Diagonalization method for a vector boundary value problem of singular perturbation type.

[4] Coddington, E. A. and Levinson, N. A boundary value problem for a nonlinear differential equation with a small parameter. Proc. Amer. Math. Soc. 3 (1952) 73-81.

[5] Cohen, D. S. Multiple solutions of singular perturbation problems. SIAM J. Math. Anal. 3 (1972) 72-82.

[6] Cohen, D. S. Singular perturbations of nonlinear two-point boundary value problems. J. Math. Anal. Appl. 43 (1973) 151-160.

[7] Cole, J. Perturbation techniques in applied mathematics. Blaisdell, Waltham, Mass., 1968.

[8] Cook, L. P. and Eckhaus, W. Resonance in a boundary value problem of singular perturbation type. Studies in Appl. Math. 52 (1973) 129-139.

[9] Eckhaus, W. Matched asymptotic expansions and singular perturbations. North Holland/American Elsevier, 1973.

[10] Erdélyi, A. An expansion procedure for singular perturbations. Atti Accad. Sci. Torino, Cl. Fis. Mat. Nat. 95 (1961) 651-672.

[11] Erdélyi, A. On a nonlinear boundary value problem involving a small parameter. J. Australian Math. Soc. 2 (1962) 425-439.

[12] Erdélyi, A. Approximate solutions of a nonlinear

boundary value problem. Arch. Rat. Mech. Anal. 29 (1968) 1-17.

[13] Erdélyi, A. Two-variable expansions for singular perturbations. J. Inst. Math. Applic. 4 (1968) 113-119.

[14] Fraenkel, L. E. On the method of matched asymptotic expansions, I, II, III. Proc. Cambridge Phil. Soc. 65 (1969) 209-284.

[15] Friedrichs, K. O. Special topics in fluid dynamics. New York University Press, 1953.

[16] Habets, P. Singular perturbations of a nonlinear boundary value problem.

[17] Habets, P. Singular perturbations of a vector boundary value problem.

[18] Harris, W. A., Jr. Singular perturbations of a boundary value problem for a nonlinear system of differential equations. Duke Math. J. 29 (1962) 429-446.

[19] Harris, W. A., Jr. Singularly perturbed boundary value problems revisited. Symposium on ordinary differential equations, W. A. Harris and Y. Sibuya eds., 54-64. Springer Lecture Notes in Math. no. 312, 1973.

[20] Kaplun, S. (Edited by P. A. Lagerstrom, L. N. Howard, and C. S. Liu). Fluid mechanics and singular perturbations. Academic Press, 1967.

[21] Lagerstrom, P. A. and Casten, R. G. Basic concepts underlying singular perturbation techniques. SIAM Review 14 (1972) 63-120.

[22] Lakin, W. D. Boundary value problems with a turning point. Studies in Appl. Math. 51 (1972) 261-275.

[23] Macki, J. W. Singular perturbations of a boundary value problem for a system of nonlinear ordinary differential equations. Arch. Rat. Mech. Anal. 24 No. 3 (1967) 219-232.

[24] Matkowsku, B. J. On boundary layer problems exhibiting resonance.

[25] O'Malley, R. E., Jr. On a boundary value problem for a nonlinear differential equation with a small parameter. SIAM J. of Appl. Math. 17 (1969) 569-581.

[26] O'Malley, R. E., Jr. Singular perturbations of a boundary value problem for a system of nonlinear differential equations. J. of Diff. Equat. 8 (1970) 431-447.

[27] O'Malley, R. E., Jr. On boundary value problems for a singularly perturbed differential equation with a turning point. SIAM J. Math. Anal. 1 (1970) 479-490.

[28] O'Malley, R. E., Jr. On nonlinear singular perturbation problems with internal nonuniformities. J. Math. Mech. 19 (1970) 1103-1112.

[29] O'Malley, R. E., Jr. On multiple solutions of a singular perturbation problem. Arch. Rat. Mech. Anal. 49 (1972) 89-98.

[30] O'Malley, R. E., Jr. Introduction to singular perturbations. Academic Press, New York, 1974.

[31] Searl, J. W. Singular perturbations. Ph. D. Thesis, Edinburgh, 1969.

[32] Vasiléva, A. B. Asymptotic behavior of solutions to certain problems involving nonlinear differential equations containing a small parameter multiplying the

highest derivatives. Uspehi Mat. Nauk. (Russian Math. Surveys) 18 no. 3 (1963) 13-84.

[33] Wasow, W. Singular perturbations of boundary value problems for nonlinear differential equations of the second order. Comm. Pure Appl. Math. 9 (1956) 93-113.

[34] Wasow, W. The capriciousness of singular perturbations. Niew Arch. Wisk. (3) 18 (1970) 190-210.

[35] Watts, A. M. A singular perturbation problem with a turning point. Bull. Australian Math. Soc. 5 (1971) 61-73.

[36] Willett, D. On a nonlinear boundary value problem with a small parameter multiplying the highest derivative. Arch. Rat. Mech. Anal. 23 (1966) 276-287.

[37] Zauderer, E. Boundary value problems for a second order differential equation with a turning point. Studies in Appl. Math. 51 (1972) 411-413.

University of Edinburgh, Eidenburgh EH1 1HZ

ON THE STRONG LIMIT-POINT CONDITION OF SECOND-ORDER DIFFERENTIAL EXPRESSIONS

W. N. Everitt

1. This paper is concerned with some properties of the formally symmetric, second-order differential expression M given by, for suitably differentiable complex-valued functions f,

(1.1) $$M[f] = -(pf')' + qf \quad \text{on } [0,\infty) \quad (' \equiv d/dx).$$

The coefficients p and q are real-valued on $[0,\infty)$ and satisfy the basic conditions

(1.2)
(i) p is locally absolutely continuous on $[0,\infty)$, and $p(x) > 0$ $(x \in [0,\infty))$

(ii) q is locally Lebesgue integrable on $[0,\infty)$.

A property is 'local' on $[0,\infty)$ if it is satisfied on all compact sub-intervals of $[0,\infty)$. $L(0,\infty)$ and $L^2(0,\infty)$ denote the classical Lebesgue complex function spaces on $[0,\infty)$; $L^2(0,\infty)$ is identified with the Hilbert function space of equivalence classes.

With the basic conditions (1.2) satisfied the formally symmetric differential expression M is <u>regular</u> at all points of $[0,\infty)$ but is <u>singular</u> at the end-point infinity; for this terminology see [16, section 15.1]. At the singular point

∞ M may be classified as either limit-point (LP) or limit-circle (LC); for this standard terminology see [4, page 1306], [16, section 17.5] or [17, sections 2.1 and 2.19]; this classification is due originally to Weyl, see his now classic memoir [18].

Let the linear manifold $\Delta = \Delta(p, q)$ of $L^2(0, \infty)$ (Δ depends on the coefficients p and q) be defined by: $f \in \Delta$ if (i) $f \in L^2(0, \infty)$, (ii) f and f' are locally absolutely continuous on $[0, \infty)$, and (iii) $M[f] \in L^2(0, \infty)$.

When $f, g \in \Delta$ it is known, from Green's formula, that the limit

$$(1.3) \qquad \lim_{\infty} p(fg' - f'g) = \lim_{x \to \infty} p(x)(f(x)g'(x) - f'(x)g(x))$$

exists and is finite. A necessary and sufficient condition for M to be LP at ∞ is that the limit in (1.3) should be zero for all $f, g \in \Delta$; for this result see [5] or [16, section 17.4].

M is said to be strong limit-point (SLP) at ∞ if

$$(1.4) \qquad \lim_{\infty} pfg' = 0 \quad (f, g \in \Delta);$$

clearly SLP at ∞ implies LP at ∞ but it is known that the converse result is false, see section 2 below.

M is said to have the Dirichlet (D) property at ∞ if

$$(1.5) \qquad p^{\frac{1}{2}}f' \text{ and } |q|^{\frac{1}{2}}f \in L^2(0, \infty) \quad (f \in \Delta)$$

and the conditional Dirichlet (CD) property at ∞ if

$$(1.6) \qquad p^{\frac{1}{2}}f' \in L^2(0, \infty) \text{ and } \lim_{X \to \infty} \int_0^X qfg \text{ exists and is finite} \quad (f, g \in \Delta).$$

Clearly D implies CD but the converse is false, see section 2 below.

It is known that with M given by (1.1) and the unweighted space $L^2(0,\infty)$, then M is D at ∞ implies that M is SLP at ∞, see section 2 below; it is not known if this statement holds good when D is replaced by CD.

If M is D at ∞, or if M is CD and SLP at ∞ then the Dirichlet formula (DF) is valid on $[0,\infty)$, i.e.

$$(1.7)\qquad \int_0^\infty pfg + \int_0^{\to\infty} qfg = -p(0)f(0)g'(0) + \int_0^\infty fM[g] \quad (f,g \in \Delta).$$

Finally in these definitions let $\Delta_0 \subset \Delta$ denote the linear manifold of $L^2(0,\infty)$ determined by: $f \in \Delta_0$ if $f \in \Delta$ and $f(0) = 0$; define the linear, unbounded operator T : $D(T) \subset L^2(0,\infty) \to L^2(0,\infty)$ by

$$(1.8)\qquad D(T) = \Delta_0 \quad \text{and} \quad Tf = M[f] \qquad (f \in D(T)).$$

It is known that T is self-adjoint in $L^2(0,\infty)$ if and only if M is LP at ∞, see [16, section 18.3]. The operator T is said to be bounded below by αI, where α is a real number and I is the identity operator in $L^2(0,\infty)$, if

$$(1.9)\qquad (Tf,f) \geq \alpha(f,f) \qquad (f \in D(T))$$

where $(\cdot\,,\cdot)$ denotes the inner-product in $L^2(0,\infty)$. When (1.9) holds we write $T \geq \alpha I$.

This paper is concerned with establishing conditions on the coefficients p and q which ensure that M satisfies one or more of the above properties.

We outline the contents of the paper. In section 2 there is a short review of some of the known results concerning the above properties of the differential expression M, including references to the proofs of the statements made above. In section 3 we state two theorems which establish some of these properties for M when q is allowed to take large negative values. Sections 4 and 5 contain the proof of these two theorems. In section 6 examples are given to relate the result of the second theorem to an earlier result of Brinck. There is a list of references.

Acknowledgement The author thanks Professor F. V. Atkinson, University of Toronto, for several helpful discussions during his visit to the University of Dundee, in the spring and summer of 1974, under the auspices of the Science Research Council of the United Kingdom.

2. The original concepts of the LP and LC classification are due to Weyl, see [18].

The term SLP was introduced by Everitt, Giertz and Weidmann in [8], but see the earlier result of Everitt in [6] in which the question was posed as to whether or not the LP and SLP conditions were equivalent; this question was answered in the negative by McLeod, see [10, section 2] and [15, section 5].

The term D was also introduced in [8] but the idea of the Dirichlet integral has a much earlier usage; see the work of Wintner in [19] and particularly the opening remarks and references in the recent paper [14] by Kalf.

The term CD is not yet a standard terminology. This property seems to have been fully considered for the first time by Brinck in [2], although again Kalf in [14] comments upon earlier results.

In his very recent paper [14] Kalf has shown, in wider circumstances than are considered in this paper, that if M is D at ∞ then M is SLP at ∞; however the converse of this result is false, see Everitt, Giertz, McLeod [10, section 2] and McLeod [15, sections 4 and 5]. It is not known at the present time if CD implies SLP; certainly the converse of this statement is false, see [14, sections 4 and 5].

We comment now on some of the known conditions on the coefficients p and q, in addition to the basic conditions (1.2), which ensure that M has some of the above properties:

(i) for some real number k let $q(x) \geq k > -\infty (x \in [0,\infty))$; then for any coefficient p, M is LP at ∞, a result due to Weyl, see [18, page 238]; M is also D at ∞, a result which may be due to Wintner, see [19], but there could be an earlier reference; the fact that M is SLP at ∞ seems to have been first noted by Everitt in [6]; see also Kalf [14, example 1]; the operator T satisfies $T \geq kI$, see (1.9), a result which follows from the DF given in (1.7).

(ii) let $p = 1$, i.e. $p(x) = 1$ $(x \in [0,\infty))$, and for some real number $r \geq 1$ let $q \in L^r(0,\infty)$; then M is D and SLP at ∞; this result is due to Everitt, Giertz and Weidmann in [8]; however it was proved earlier that M is LP at ∞ by Hartman in [12]; see also Kalf [14, example 2]; $T \geq \alpha I$ where α is given in terms of the norm of q in $L^r(0,\infty)$, see

[7, section 1].

(iii) <u>let</u> $p = 1$, <u>and for some non-negative number</u> K <u>let</u>

$$\int_x^{x+h} q \geq -K \quad (x \in [0,1]\ ,\ h \in [0,1]); \tag{2.1}$$

then M is SLP and CD at ∞; these results may be found essentially in the paper [2] by Brinck, who discusses an example to show that M may not be D at ∞, see [2, page 230]; in this case $T \geq -2K^2 I$, see [2, page 231]

(iv) <u>let</u> $p = 1$, <u>and for some real number</u> k <u>let</u>

$$q(x) \geq kx^2 \qquad (x \in [0,\infty)); \tag{2.2}$$

then M is LP at ∞, see [17, section 2.20]; however examples exist to show that M may not be SLP of D at ∞, see [10, sections 2 and 3] and [15, sections 4, 5 and 6]; on the other hand when (2.2) is satisfied there is a 'weighted' D result for M at ∞, see [10] and [15]

(v) <u>let</u> $p = 1$, <u>and for some non-negative number</u> K <u>let</u>

$$\int_0^X q_- \leq KX^3 \qquad (X \in [0,\infty))$$

<u>where</u> $q_-(x) = -\min\{0, q(x)\}$ $(x \in [0,\infty))$; then M is LP at ∞, see Hartman and Wintner [13]; however in general M is not SLP or D at ∞, nor does M seem to satisfy any 'weighted' D result at ∞ unless something like exponentially small weights are introduced.

The most comprehensive test for M to be D, and SLP, at ∞ is to be found in the recent paper [14] by Kalf; this covers both finite and infinite singular points for second order differential expressions. In particular it contains an earlier result of Everitt and Giertz [9] on M with p = 1, giving both the D and SLP conditions at a finite singular point.

Additional recent work on SLP, D and CD classification of second and higher order differential expressions may be found in the work of Atkinson [1], see in particular [1, sections 8, 9 and 10]; of Brown and Evans [3]; and of Everitt, Hinton and Wong [11].

3. In this section we state the two theorems to be proved in this paper. All the notations of section 1 apply.

Theorem 1. Let the coefficients p and q of M satisfy the basic conditions (1.2); let the real numbers A, B, δ and k be chosen so that A, B are non-negative, and δ is positive; let the coefficient q be written in the form

$$q = q_1 - q_2 \quad \text{on} \quad [0,\infty) \tag{3.1}$$

where both q_1 and q_2 are real-valued on $[0,\infty)$; let p, q_1 and q_2 satisfy

(i) $\int_0^\infty \{p(x)\}^{-\frac{1}{2}} dx = \infty \qquad \int_0^\infty x\,\{p(x)\}^{-1} dx = \infty$

(ii) $\int_0^x q_2 \leq \frac{1}{2}\, p'(x) + Ax + B$ (almost all $x \in [0,\infty)$)

(iii) $q_1(x) \geq (1+\delta)\,\{p(x)\}^{-1}. \left[\int_0^x q_2\right]^2 - k \quad (x \in [0,\infty))$;

then (a) $p^{\frac{1}{2}} f' \in L^2(0,\infty)$ and $|q_1|^{\frac{1}{2}} f \in L^2(0,\infty)$ $(f \in \Delta)$

(b) M is LP at ∞.

Proof. This is given in section 4 below.

Theorem 2. Let the coefficients p and q of M satisfy the basic conditions (1.2); let the real numbers A, B, K, L, δ, ϵ and k be chosen so that A, B, K, L are non-negative, and δ, ϵ are both positive; let $q = q_1 - q_2$ as in (3.1); let p, q_1 and q_2 satisfy

(i) $|p'(x)| \leq Kx + L$ (almost all $x \in [0, \infty)$)

(ii) $\left| \int_0^x q_2 \right| \leq Ax + B \quad (x \in [0, \infty))$

(iii) $q_1(x) \geq (1 + \delta)\{p(x)\}^{-1} [\int_0^x q_2]^2 + \epsilon \{p(x)\}^{-1} x^2 - k \; (x \in [0, \infty))$

then (a) all of $p^{\frac{1}{2}} f'$, $|q_1|^{\frac{1}{2}} f$, $xp^{-\frac{1}{2}} f$, $p^{-\frac{1}{2}} |\int_0^x q_2| f \in L^2(0, \infty) (f \in \Delta)$

(b) M is SLP at ∞

(c) M is CD at ∞, i.e. $\lim_{X \to \infty} \int_0^X q_2 fg$ exists and is finite for all $f, g \in \Delta$

(d) $T \geq -kI$.

Proof. This is given in section 5 below.

Remarks 1. In both theorems if q_2 becomes large and positive then q_1 must also be positive to compensate; but note that the compensation takes place between q_1 and the integral of q_2, and not between q_1 and q_2 directly; it would seem from the results stated in (iv) and (v) of section 2 above that some compensation of this kind is essential if M is to satisfy something like the D condition at ∞.

2. It does not seem possible to obtain further conclusions from the hypotheses of theorem 1 and this will be brought out in the proof below; indeed it seems likely that, in general, M is not SLP, and not D or even CD at ∞ under these conditions, but no example has been found to be definite in this respect.

3. When $p = 1$ in theorem 2 note that $q_1(x) \geq \epsilon x^2$ $(x \in [0,\infty))$ so that in this case $\lim q_1 = \infty$ at ∞; however it may well be the case that $\lim\inf(q_1 - q_2) = -\infty$; see example (ii) of section 6 below.

4. In section 6 we shall produce examples to show that neither theorem 2 (with $p = 1$) nor the Brinck theorem is included in the other, although both theorems have similar conclusions.

5. Note that $p = 1$ satisfies condition (i) of both theorems.

4. In this section we give proof of Theorem 1.

The proof depends on the criterion (1.3) for the LP case at ∞. We note that it is sufficient to argue with the linear manifold Δ_o, rather than Δ since the proof is only concerned with behaviour at ∞; also in considering, an element $f \in \Delta_o$ we may assume, without loss of generality, that f is realvalued on $[0,\infty)$ since otherwise we work separately with the real and imaginary parts of f.

The following straight-forward results are required, which are stated without proof:

(i) let f be locally Lebesgue integrable on $[0,\infty)$; then for all $X > 0$

$$(4.1) \qquad \int_0^X dx \int_0^x f(t)dt = \int_0^X (X-x)f(x)dx$$

(ii) let f be a non-negative locally Lebesgue integrable function on $[0,\infty)$; then

$$(4.2) \qquad \lim_{X\to\infty} \int_0^X f(x)dx < \infty \text{ if and only if } \lim_{X\to\infty} \int_0^X \left(1-\frac{x}{X}\right)f(x)dx < \infty$$

(iii) let $f \in L^2(0,\infty)$; then

$$(4.3) \qquad \lim_{X\to\infty} X^{-1} \int_0^X x\, f(x)^2 dx = 0$$

(iv) for any real numbers a, b and any $\tau > 0$

$$(4.4) \qquad 2ab \leq \tau a^2 + \tau^{-1} b^2 .$$

Now let $f \in \Delta_0$, so that $f(0) = 0$; then we have the following identity, valid for all $X > 0$ (on using (3.1))

$$\int_0^X \{pf'^2 + (q_1 + k)f^2\} = (pff')(X) + \int_0^X q_2 f^2 + \int_0^X f\{M[f]+kf\}$$

(4.4a)

$$= (pff')(X) + f(X)^2 \int_0^X q_2$$

$$-2\int_0^X ff'[\int_0^X q_2]dx + \int_0^X f\{M[f] + kf\}$$

on integration by parts to obtain the second line. On using the Cauchy-Schwarz inequality and (4.4) this yields

$$\int_0^X \{pf'^2 + (q_1 + k)f^2\} \leq (pff')(X) + f(X)^2 \int_0^X q_2 + \tau \int_0^X pf'^2$$

$$+ \tau^{-1} \int_0^X p^{-1} [\int_0^x q_2]^2 f^2 + \int_0^X f\{M[f]+kf\}$$

and on rearrangement

$$(1 - \tau) \int_0^X pf'^2 + \int_0^X \{q_1 + k - \tau^{-1} p^{-1} [\int_0^x q_2]^2\} f^2$$

(4.5)

$$\leq (pff')(X) + f(X)^2 \int_0^X q_2 + \int_0^X f\{M[f] + kf\}.$$

Now let $\tau = (1 + \delta)^{-\frac{1}{2}}$ where δ is the positive number in condition (iii) of theorem 1; thus $0 < \tau < 1$. With this choice of τ and using again (iii) of theorem 1 it may be verified that (4.5) implies the inequality

$$(4.6)\qquad (1-\tau)\int_0^X \{pf'^2 + (q_1+k)f^2\} \le (pff')(X) + f(X)^2\int_0^X q_2 + \int_0^X f\{M[f]+kf\}.$$

At this stage we note that $q_1 + k \ge 0$ on $[0,\infty)$, from (iii) of theorem 1.

Now integrate (4.6) over $0 \le X \le Y$, replace Y by X, use (4.1), divide through by $X > 0$, to obtain

$$(4.7)\qquad (1-\tau)\int_0^X \left(1-\frac{x}{X}\right)[pf'^2 + (q_1+k)f^2] \le X^{-1}\int_0^X pff' + X^{-1}\int_0^X f(x)^2\left[\int_0^x q_2\right]dx + \int_0^X \left(1-\frac{x}{X}\right)f\{M[f]+kf\}$$

which is valid for all $X > 0$.

Integration by parts gives

$$\int_0^X pff' = \tfrac{1}{2}(pf^2)(X) - \tfrac{1}{2}\int_0^X p'f^2$$

so that the right-hand side of (4.7) becomes, on using (ii) of theorem 1

$$(4.8)\qquad \tfrac{1}{2}X^{-1}p(X)f(X)^2 + X^{-1}\int_0^X \left\{\int_0^x q_2 - \tfrac{1}{2}p'(x)\right\} f(x)^2\,dx + \int_0^X \left(1-\frac{x}{X}\right)\{M[f]+kf\}$$

$$\leq \tfrac{1}{2}X^{-1}p(X)f(X)^2 + X^{-1}\int_0^X (Ax+B)f(x)^2dx + \int_0^X (1-\frac{x}{X})| f\{M[f]+kf\}| .$$

The second term in this last expression tends to zero as X tends to infinity, on using (4.3). The third term tends to a finite limit on using (4.2) and recalling that $f \varepsilon \Delta_o$ implies f and $M[f] \varepsilon L^2(0,\infty)$.

Consider now the first term of (4.8); we state that there is a sequence $\{X_n : n = 1, 2, 3, \ldots\}$ satisfying $0 < X_n < X_{n+1}$ $(n = 1, 2, 3, \ldots)$ and $\lim X_n = \infty$ as $n \to \infty$, such that

(4.9) $$\lim_{n\to\infty} X_n^{-1} p(X_n)f(X_n)^2 = 0.$$

For otherwise there is a positive η such that $X^{-1}p(X)f(X)^2 \geq \eta > 0$ for all large X which gives a contradiction on using the second part of (i) of theorem 1 and $f \varepsilon L^2(0,\infty)$. Thus (4.9) holds.

Using now (4.7) and (4.8), with $X = X_n$ (recall that the integrand on the left-hand side of (4.6) is non-negative) we see that, since $0 < 1 - \tau < 1$

$$\lim_{n\Rightarrow\infty} \tfrac{1}{2}\int_0^{\frac{1}{2}X_n} [pf'^2 + (q_1+k)f^2] \leq \lim_{n\Rightarrow\infty} \int_0^{X_n} (1-\frac{x}{X_n}) \times [pf'^2 + (q_1+k)f^2]$$

$$\leq (1-\tau)^{-1} \lim_{X\to\infty} \int_0^{X_n} (1-\frac{x}{X_n}) \times | f\{M[f]+kf\} |$$

$$< \infty \quad (\text{on using } (4.2)).$$

This completes the proof of the first part (a) of theorem 1.

Suppose now that M is not LP at ∞ ; then from the criterion (1.3) we can find $f, g \in \Delta$ such that

$$\lim_{\infty} p(fg' - f'g) = \mu \neq 0,$$

(4.10)

i.e. $$p^{\frac{1}{2}} | fg' - f'g| \geq \tfrac{1}{2}|\mu| \, p^{-\frac{1}{2}}$$

on some interval $[X_o, \infty)$. Now from (a) theorem 1 the left-hand side of (4.10) is $L(X_o, \infty)$ and this gives a contradiction to the first part of condition (i) of theorem 1.

This completes the proof of the theorem.

To return to remark 2 following the statements of theorems 1 and 2 in section 3 above. It seems different to obtain further information about M at ∞ on the basis of the hypotheses of theorem 1. When (a) has been obtained it follows from (4.4a) that

$$\lim_{X\to\infty} \{ p(X)f(X)f'(X) + f(X)^2 \int_0^X q_2 \}$$

exists and is finite but seemingly no conclusion can be drawn about the limit of pff' itself at ∞. Even the special case $p = 1$ does not seem to yield further results about ff' at ∞. It follows from this that nothing can be said about

$$\lim_{X\to\infty} \int_0^X q_2 fg$$

so that both SLP and CD at ∞ for M remain in doubt. It seems likely that condition (ii) of theorem 1 is not strong enough and (ii) of theorem 2 should be seen in this light since this does yield SLP and CD for M. No example has yet emerged to settle this difference between the conclusions of theorems 1 and 2.

5. In this section we give the proof of theorem 2.

We note that condition (i) of the theorem implies

$$0 < p(x) \le \tfrac{1}{2} Kx^2 + Lx + p(0) \quad (x \in (0,\infty)) \tag{5.1}$$

and from this it follows that p satisfies condition (i) of theorem 1.

We now follow in part of the proof of theorem 1. We have the identity (4.4a) and the inequality (4.5). With the same choice of τ but using now condition (iii) of theorem 2 we obtain in place of (4.6)

$$\begin{aligned}(1-\tau)\int_0^X \{pf'^2 + (q_1 + k)f^2\} + \tau\epsilon \int_0^X p^{-1}x^2 f^2 \\ \le (pff')(X) + f(X)^2 \int_0^X q_2 + \int_0^X f\{M[f] + kf\}.\end{aligned} \tag{5.2}$$

Again we note that $q_1 + k \geq 0$ on $[0, \infty)$ from condition (iii) of the theorem.

Now integrate again as before to obtain a result equivalent to (4.7) but with the following additional term on the left-hand side

$$\tau\epsilon \int_0^X (1 - \frac{x}{X}) p^{-1} x^2 f^2 \ .$$

In place pf (4.8) we have

$$\tfrac{1}{2} X^{-1} p(X) f(X)^2 + \tfrac{1}{2} X^{-1} \int_0^X |p'| f^2 + X^{-1} \int_0^X \left| \int_0^x q_2 \right| f^2$$

$$+ \int_0^X (1 - \frac{x}{X}) \left| f\{M[f] + kf\} \right| .$$

In this expression the second and third terms tend to zero as $X \to \infty$ on using (4.3) and conditions (i) and (ii) of the theorem. Since we have seen that condition (i) of theorem 1 holds for p the same argument as given in section 4 shows that there is a sequence $\{X_n : n = 1, 2, 3 \ldots\}$ for which (4.9) holds. Proceeding as before we now find that the proof of result (a) of the theorem is complete. Note that it is essential to have $\epsilon > 0$ to obtain $xp^{-\frac{1}{2}} f \in L^2(0, \infty)$.

From the identity

$$\int_0^X f^2 = Xf(X)^2 - 2 \int_0^X xp^{-\frac{1}{2}} f \, p^{\frac{1}{2}} f' \qquad (X \in [0, \infty))$$

it follows that $\lim X f(X)^2$ $(X \to \infty)$ exists and is finite; this limit must be zero since $f \in L^2(0, \infty)$. From this result and condition (ii) of the theorem it now follows that

$$\lim_{X\to\infty} f(X)^2 \int_0^X q_2 = \lim_{X\to\infty} f(X)g(X)\int_0^X q_2 = 0 \quad (f,\ g \ \varepsilon\ \Delta).$$

Returning to the second line of the identity (4.4a) and replacing the quadratic terms in f with the corresponding terms in two elements $f, g \ \varepsilon\ \Delta_0$, it now follows on letting X tend to infinity that $\lim pfg'$ $(X \to \infty)$ exists and is finite. If this limit is not zero then for some $\mu > 0$ and $X_0 > 0$

$$p^{\frac{1}{2}}|fg'| \geq \mu p^{-\frac{1}{2}} \quad \text{on} \quad [X_0, \infty).$$

The left-hand side is $L(X_0, \infty)$, from result (a) of the theorem, and this contradicts the result, given above, that $p^{-\frac{1}{2}} \not\in L(X_0, \infty)$. Thus the limit is zero and M is SLP at ∞. This completes the proof of (b) of the theorem.

Returning to the first line of the identity (4.4a), with one f replaced by g, we now see that

$$\lim_{X\to\infty} \int_0^X q_2 fg$$

exists and is finite for all $f, g \ \varepsilon\ \Delta$. Since we have seen in (a) that $|q_1|^{\frac{1}{2}} f \ \varepsilon\ L^2(0,\infty)$ it now follows that M is CD at ∞ and the proof of result (c) is complete.

Finally from the inequality (5.2), with $f \ \varepsilon\ \Delta_0$, we obtain, when $X \to \infty$,

$$\int_0^\infty \bar{f}\,\{M[f] + kf\} \geq 0 \quad (f \ \varepsilon\ \Delta_0)$$

and this is equivalent to stating that

$$(Tf,\ f) \geq -k(f,f) \quad (f \ \varepsilon\ D(T)),$$

i.e. $T \geq -kI$ which is result (d) of the theorem.

This completes the proof of theorem 2.

6. In this last section we discuss examples which relate to the result of theorem 2 above and the Brinck theorem given in (iii) of section 2.

(i) If we take $q(x) = -K$ $(x \in [0,\infty))$, where K is a positive number, then the Brinck condition (2.1) is satisfied; on the other hand, taking $p = 1$, it is not possible to write $q = q_1 - q_2$ in such a way as to satisfy both conditions (ii) and (iii) of theorem 2.

(ii) Let $p = 1$ and let $q = q_1 - q_2$ on $[0,\infty)$ be defined by

$$q_1(x) = 2x^2 \qquad (x \in [0,\infty))$$

$$q_2(x) = -x e^x \cos(e^x) - \sin(e^x) \qquad (x \in [0,\infty)),$$

i.e.

$$\int_0^x q_2 = -x \sin(e^x) \qquad (x \in [0,\infty)).$$

Let $A = 1$, $B = 0$, $\delta = \varepsilon = \frac{1}{2}$. Then it may be seen that all the conditions of theorem 2 are satisfied.

Take, for $n = 1, 2, 3, \ldots$

$$x_n = \log[(2n + \tfrac{1}{2})\pi] \text{ and } x_n + h_n = \log[(2n + 1)\pi]$$

so that $h_n = 0(n^{-1})$ $(n \to \infty)$.

Then, as $n \to \infty$,

$$\int_{x_n}^{x_n+h_n} q_1 = \frac{2}{3}[(x_n + h_n)^3 - x_n^3] \le 4x_n^2 h_n = \underline{0}\,(1)$$

and

$$-\int_{x_n}^{x_n+h_n} q_2 = (x_n+h_n)\sin(\exp[x_n+h_n]) - x_n \sin(\exp[x_n])$$

$$= -x_n \to -\infty .$$

Thus

$$\lim_{n \to \infty} \int_{x_n}^{x_n + h_n} q = -\infty$$

and the Brinck condition cannot be satisfied, in spite of the positive contribution from the term q_1.

REFERENCES

[1] Atkinson, F. V. "Limit-n criteria of integral type", (to appear in Proc. Royal Soc. Edinburgh).

[2] Brinck, I. "Self-adjointness and spectra of Sturm-Liouville operators" Math. Scand. 7 (1959), 219-239.

[3] Brown, B. M. and Evans, W. D. "On the limit-point and strong limit-point classification of 2n order differential expressions with widely oscillating coefficients", (to appear in Math. Zeit.).

[4] Dunford, N. and Schwartz, J. T. Linear operators: Part II (Interscience, New York, 1963).

[5] Everitt, W. N. "A note on the self-adjoint domains of second-order differential equations", Quart. J. Math. (Oxford) (2) 14 (1963) 41-45.

[6] Everitt, W. N. "On the limit-point classification of second-order differential operators", Journ. Lond. Math. Soc. 41 (1966) 531-544.

[7] Everitt, W. N. "On the spectrum of a second-order linear differential equations with a p-integrable coefficient", Applicable Anal. 2 (1972) 143-160.

[8] Everitt, W. N., Giertz, M. and Weidmann, J. "Some remarks on a separation and limit-point criterion of

second-order, ordinary differential expressions", Math. Ann. 200 (1973) 335-346.

[9] Everitt, W. N. and Giertz, M. "A Dirichlet type result for ordinary differential operators". Math. Ann. 203 (1973) 119-128.

[10] Everitt, W. N., Giertz, M. and McLeod, J. B. "On the strong and weak limit-point classification of second-order differential expressions" Proc. Lond. Math. Soc. (3) 29 (1974) 142-158.

[11] Everitt, W. N., Hinton, D. B. and Wong, J. S. W. "On the strong limit-n classification of linear ordinary differential expressions of order 2n', (to appear in Proc. Lond. Math. Soc.).

[12] Hartman, P. "Differential equations with non-oscillatory eigenfunctions", Duke Math. J. 15 (1948) 697-709.

[13] Hartman, P. and Wintner, A. "A criterion for the non-degeneracy of the wave equation", Amer. J. Math. 71 (1949) 206-213.

[14] Kalf, H. "Remarks on some Dirichlet type results for semi-bounded Sturm-Liouville operators", (to be published.).

[15] McLeod, J. B. "The limit-point classification of differential expressions", Spectral theory and asymptotics of differential equations (Mathematics Studies 13, North-Holland, Amsterdam, 1974; edited by E M de Jager).

[16] Naǐmark, M. A. Linear differential operators: Part II (Ungar, New York, 1968).

[17] Titchmarsh, E. C. Eigenfunction expansions: Part I (Oxford University Press, 1962).

[18] Weyl, H. "Über gewöhnliche Differentialgleichungen mit Singularitäten and die zugehörigen Entwicklungen willkürlicher Funktionen", Math. Ann. 68 (1910) 220-269.

[19] Wintner, A. "(L^2)-connections between the potential and kinetic energies of linear systems", Amer. J. Math. 69 (1947) 5-13.

University of Dundee, Dundee, Scotland

A NONLINEAR PARABOLIC PARTIAL DIFFERENTIAL EQUATION IN POPULATION GENETICS

By

W. H. Fleming

A selection-migration model is considered in which the frequency of a gene type depends on time t and place x in a habitat R. The frequency $u(t,x)$ obeys a nonlinear parabolic partial differential equation, with zero normal derivative boundary conditions. Results concern existence, stability or instability, and bifurcation of equilibrium solutions.

Brown University, Providence, Rhode Island

FIXED POINTS AND UNIQUENESS

Rubén Flores and Carlos Imaz

In all that follows let X be a connected and complete metric space, with some metric ρ. Let $T:X \to X$ be any transformation, and T^n, $n = 1, 2, 3, \ldots$ denote the iterates of T. For any point $x \in X$, $\{T^n x\}$ stands for the sequence of iterates of x under T.

For a given $x \in X$, $C(x)$ denotes the set

$$C(x) = \{y \in X \mid \{T^n y\} \to x\}$$

and N will denote the set

$$N = \{y \in X \mid \{T^n y\} \text{ is not Cauchy}\},$$

then clearly

$$X = \bigcup_x C(x) \cup N.$$

Definition. We say that $T:X \to X$ is a local contraction if for all $x \in X$ there exists a neighborhood $V(x)$ such that $y \in V(x)$ implies $\rho(T^n x, T^n y) \to 0$ as $n \to \infty$.

Observation 1. The property of local contraction does not imply continuity.

Proposition 1. If $T:X \to X$ is a local contraction then the sets $C(x)$ and N are open.

The proof is straightforward and will be omitted.

Proposition 2. It $T:X \to X$ is a local contraction then it has at most one fixed point.

Proof. By Proposition 1 the sets $C(x)$ and N are open, therefore (since X is connected) either $X = C(x)$ for some $x \in X$ or $X = N$. If $X = N$ then there is no fixed point. If $X = C(x)$ there is at most one.

Observation 2. Both cases, $X = N$ or $X = C(x)$ do occur.

Observation 3. If T is continuous and $X = C(x)$ then x is a (unique) fixed point.

Proposition 3. If $T:X \to X$ is a continuous local contraction and for some $y \in X$ the sequence $\{T^n y\}$ is Cauchy, then T has a unique fixed point x, and for all $y \in X$ the sequence $\{T^n y\} \to x$.

Proof. Direct consequence of Proposition 2 and Observation 3.

Observation 4. The contraction principle of Banach (for connected spaces) follows from Proposition 3, since it is simple to prove that a contraction satisfies the hypothesis of Proposition 3.

As an application let us consider an initial value problem

$$\dot{x}(t) = f(t, x(t))$$
$$x(0) = x_0.$$

With such a problem we associate a space X of functions (a space where the solutions should be), and a mapping

$T:X \to X$ given as usual by

$$Tx(t) = x_0 + \int_0^t f(s, x(s)ds$$

Proposition 4. Suppose $f(t, x)$ satisfies Kamke's uniqueness condition with a function $w(t, r)$, then the associated map T is a local contraction.

Sketch of proof.

Let $d(t) = \limsup_{n\to\infty} \|T^n x(t) - T^n y(t)\|$, then

$$d(t) \leq \limsup_{n\to\infty} \int_0^t w(s, \|T^{n-1}x(s) - T^{n-1}y(s)\|)ds$$

For a fixed t_0 pick a subsequence n_k such that

$$\|T^{n_k-1}x(t_0) - T^{n_k-1}y(t_0)\| \to d(t_0) .$$

One can do this in such a way that

$$\lim_{n\to\infty} \|T^{n_k-1}x(t) - T^{n_k-1}y(t)\| = d^*(t)$$

exists uniformly. Clearly $d^*(t) \leq d(t)$.

Then

$$d(t) \leq \int_0^t w(s, d^*(s))ds \leq \int_0^t w(s, d(s))ds$$

and, as usual, it follows that $d(t) \equiv 0$.

In other words, in the same manner that contraction is related to lipschitz condition for uniqueness, local contraction is at least related with Kamke's condition for uniqueness.

Universidad de Sonora and
Centro de Investigación del IPN, México

PERIODIC SOLUTIONS OF PERTURBED LOTKA-VOLTERRA SYSTEMS

H. I. Freedman and Paul Waltman

The Lotka-Volterra system of differential equations

$$(1)\qquad \begin{aligned} x' &= \alpha x - \beta xy \\ y' &= -\gamma x + \delta xy, \quad \alpha, \beta, \gamma, \delta > 0 \end{aligned}$$

is one of the early attempts in ecology to model a predator-prey relationship. (For readable accounts of the use of mathematics in ecology, see the recent books of May [7] and Smith [9].) While the derivation of these simple equations has logical appeal, the fact that all solutions with initial conditions in the first quadrant are periodic seems to be biologically unacceptable. It has recently been suggested [6], [8] that as appropriate model for a predator-prey systems should yield an unstable equilibrium point and a surrounding stable limit cycle. The system

$$(2)\qquad \begin{aligned} x' &= \alpha x - \beta xy - \epsilon f_1(x, y) \\ y' &= -\gamma x + \delta xy - \epsilon f_2(x, y) \end{aligned}$$

is studied as a perturbation of the Lotka-Volterra system (1). We wish to determine conditions on the functions f_1 and f_2 that are sufficient to produce a periodic solution. The basic idea is to find conditions which make the critical point in the first quadrant an unstable spiral and then find conditions which guarantee that trajectories cross a certain closed curve containing the critical point from "outside" to "inside". We sketch the principal results; details can be found in [2], [3].

The only critical point of (1) interior to the first quadrant is $\left(\frac{\gamma}{\delta}, \frac{\alpha}{\beta}\right)$. The implicit function theorem can be applied to the right hand side of (2), a function of x, y and ϵ, to assure the existence of a critical point (x^*, y^*) of (2) for sufficiently small ϵ and to determine that

$$x^* = \frac{\gamma}{\delta} + \frac{\beta}{\alpha\delta} f_2\left(\frac{\gamma}{\delta}, \frac{\alpha}{\beta}\right)\epsilon + o(\epsilon)$$

$$y^* = \frac{\alpha}{\beta} - \frac{\delta}{\beta\gamma} f_1\left(\frac{\gamma}{\delta}, \frac{\alpha}{\beta}\right)\epsilon + o(\epsilon) .$$

A translation of the critical point to the origin by $u = x - x^*$, $y = y - y^*$ produces a system

$$\begin{aligned} u' &= a_{11}u + a_{12}v - \beta uv - \epsilon g_1(u, v) \\ v' &= a_{21}u + a_{22}v + \delta uv - \epsilon g_2(u, v) \end{aligned} \qquad (3)$$

where

$$g_1(u, v) = f_1(u+x^*, v+v^*) - f(x^*, y^*) - \kappa u - \lambda v$$

$$g_2(u, v) = f_2(v+x^*, v+y^*) - f_2(x^*, y^*) - \mu u - \nu v$$

and $\kappa = f_{1u}(x^*, y^*)$, $\lambda = f_{1v}(x^*, y^*)$, $\mu = f_{2u}(x^*, y^*)$, $\nu = f_{2v}(x^*, y^*)$. It is the system (3) that we analyze. The stability of the origin is determined by the quantity

$$[\cdot]_1 = [\frac{\delta}{\gamma} f_1(\frac{\gamma}{\delta}, \frac{\alpha}{\beta}) + \frac{\beta}{\alpha} f_2(\frac{\gamma}{\delta}, \frac{\alpha}{\beta}) - f_{1x}(\frac{\gamma}{\delta}, \frac{\alpha}{\beta}) - f_{2y}(\frac{\gamma}{\delta}, \frac{\alpha}{\beta})].$$

<u>Lemma.</u> If $[\cdot]_1 > 0$, the origin for the system (3).

If $[\cdot]_1 < 0$, the origin would be a stable spiral and if $[\cdot]_1 = 0$, the test fails. In the latter case, by determining x^*, y^* to order ϵ^2 one can determine a new condition $[\cdot]_2 > 0$ which is sufficient for instability. (The process can, in principle, be continued to determine a sequence of conditions $[\ ]_1 = \cdots [\ ]_i = 0$, $[\ ]_{i+1} > 0$, but computations beyond order ϵ^2 are tedious.)

To study behavior away from the critical point we use the curve family

$$\Gamma_C = \{(u, v) \mid \varphi_1(u) + \varphi_2(v) = C\}$$

$$\varphi_1(y) = \frac{1}{\beta}\left(y + \frac{a_{12}}{\beta}\log\left|\frac{a_{12} - \beta y}{a_{12}}\right|\right)$$

$$\varphi_2(y) = \frac{1}{\delta}\left(y - \frac{a_{21}}{\delta}\log\left|\frac{a_{21} + \delta y}{a_{21}}\right|\right)$$

These curves are the trajectories of the unperturbed system. If we define

$$V(u, v) = \varphi_1(u) + \varphi_2(v),$$

(with the aid of several preliminary computations) it is possible to determine the behavior of $\frac{dV}{dt}$ along solutions curves. The derivative takes the form

$$\frac{d}{dt} V(u,(t),\ v(t)) = h_1(u, v)\epsilon + h_2(u, v)\epsilon + o(\epsilon)$$

where $h_i(u, v)$ involve only the knowledge of the f_i, $i = 1, 2$ and of their derivatives evaluated at the critical point of the unperturbed system. The principal result is:

<u>Theorem:</u> <u>Let</u> f_i, $i = 1, 2$ <u>be such that</u> $[\ ]_1 > 0$ <u>and that</u> $h_1(u, v) \leq 0$, $h_2(u, v) \leq 0$, <u>and they are not simultaneously</u>

zero, on Γ_C for some sufficiently large C. Then there exists a periodic solution of the system (3) for sufficiently small $\epsilon > 0$.

In the case that $f_i\left(\frac{\gamma}{\delta}, \frac{\alpha}{\beta}\right) = 0$, $i = 1, 2$, the critical point does not depend on ε. Under the further assumption that

$$(4) \qquad \kappa^2 + \lambda^2 + \mu^2 + \nu^2 \neq 0$$

several simplifications result. First of all,

$$[\cdot]_1 = -\kappa - \nu, \; [\cdot]_i = 0, \quad i = 1, 2, \cdots .$$

Further,

$$h_1(u, v) = \lambda uv - f_1(u+\frac{\gamma}{\delta}, \; v+\frac{\alpha}{\beta})u$$

$$h_2(u, v) = \mu uv - f_2(u+\frac{\gamma}{\delta}, \; v+\frac{\alpha}{\beta})v .$$

If (4) is not satisfies, that is, the perturbation term and its first partial derivatives vanish at the point $\left(\frac{\alpha}{\beta}, \frac{\gamma}{\delta}\right)$, the foregoing type analysis does not apply. It is possible in this case, however, to modify the implicit function technique of W. S. Loud [4], [5], and find conditions for the existence of a periodic solution. These results, which make use of "critical cases" of the implicit function theorem, [1], can be found in [3].

REFERENCES

[1] H. I. Freedman, The implicit function theorem in the scalar case, Canad. Math. Bull. 12 (1969), 721-732.

[2] H. I. Freedman and P. Waltman, Perturbations of two dimensional predator-prey equations, SIAM J.

Appl. Math. To appear.

[3] H. I. Freedman and P. Waltman, Perturbations of two dimensional predator-prey equations with an unperturbed critical point, SIAM J. Appl. Math. To appear.

[4] W. S. Loud, Periodic solutions of perturbed second order autonomous equations, Mem. Amer. Math. Soc. #47, 1964.

[5] W. S. Loud, Periodic solutions of $x'' + cx' + g(x) = \epsilon f(t)$, Mem. Amer. Math. Soc. # 31, 1959.

[6] Robert May, Limit cycles in predator-prey communities, Science 177 (1972), 900-902.

[7] Robert May, Stability and Complexity in Model Ecosystems, Princeton University Press, 1973.

[8] P. Samuelson, Generalized predator-prey oscillations in ecological and economic equilibrium, Proc. Nat. Acad. Sci. USA 68 (1971), 980-983.

[9] J. Maynard Smith, Models in Ecology, Cambridge University Press, Cambridge, 1974.

University of Alberta, Alberta, Canada and
University of Iowa, Iowa City, Iowa

CONTINUOUS DEPENDENCE ON A PARAMETER IN ORDINARY AND FUNCTIONAL DIFFERENTIAL EQUATIONS

G. Gonzalez
and
Z. Vorel

It was shown in [1] that a functional differential equation of retarded type in R^n (cf. [5]) is equivalent to an ordinary differential equation in L_p, $p \geq 1$. This result is used in the presented paper to obtain some basic properties (as existence, uniqueness and continuous dependence on a parameter) of solutions of functional differential equations from the corresponding results for ordinary differential equations in a real Banach space. Initial data for functional equations are an element of L_p in the initial interval $[-h, 0]$, $h > 0$, and the value of the solution at 0 assuming we consider the solution in an interval $[0, a]$, $a > 0$.

Let $\tilde{K}$ be a set in $L_p[-h, a, R^n]$, $p \geq 1$, with the property that $x \in \tilde{K}$ implies: x is absolutely continuous in $[0, a]$ and the truncation x_t defined for $0 \leq t \leq a$ by $x_t(s) = x(s)$ if $-h \leq s \leq t$ and $x_t(s) = x(t)$ if $t \leq s \leq a$, also belongs to $\tilde{K}$.

As usually, we shall identify a class $\mathfrak{x} \in \tilde{K}$ with its representative x whose restriction to $[0, a]$ is absolutely continuous.

Consider a function $g: \tilde{K}\ [0, a] \to R^n$ with the property that if $x \in \tilde{K}$, $y \in \tilde{K}$, and if the restrictions to $[-h, t]$ of x and y coincide $(x_{[-h, t]} = y_{[-h, t]})$ for some $t \in [0, a]$ then $g(x, t) = g(y, t)$. To stress the causal character of g we shall write $g(x_t, t)$ instead of $g(x, t)$. Fix an element $y \in \tilde{K}$ and let y_0 be its truncation at 0.

Instead of a differential equation with right side g and initial condition given by y_0 we shall consider the corresponding integral equation

(1) $$x(t) = y_0(0) + \int_0^t g(x_s, s)\, ds \qquad (0 \leq t \leq a)$$

$$x(t) = y_0(t) \quad \text{for} \quad -h \leq t < 0$$

For $x \in \tilde{K}$, $t \in [0, a]$ define a function

$$G: \tilde{K} \times [0, a] \to L_p[-h, a, R^n] \text{ by}$$

(2) $$G(x, t)(s) = \begin{cases} 0 & \text{if } -h \leq s < t \\ g(x_t, t) & \text{if } t \leq s \leq a \end{cases}$$

The following theorem was proved in [1].

<u>Theorem 1.</u> Let $g(x, \cdot) \in L_p[0, a, R^n]$ for $x \in \tilde{K}$.

Then $G(x, \cdot)$ defined by (2) is in $L_p[0, a, L_p[-h, a, R^n]]$ and $\int_0^t G(x, s)\, ds$, which is a class in $L_p[-h, a, R^n]$, contains a continuous function φ, $\varphi(s) = \begin{cases} 0 \text{ for } -h \le s < 0 \\ \int^{\min(s,t)} g(x_u, u)\, du \end{cases}$

Moreover, if $y \in \tilde{K}$ is a solution of (1) with the initial condition y_0, then the function $z : [0, a] \to \tilde{K}$ defined by $z(t) = y_t$ is a solution of the ordinary differential equation (in integral form)

$$(3) \qquad z(t) = y_0 + \int_0^t G(z(s), s)\, ds \qquad (0 \le t \le a)$$

and vice versa, if z is a solution of (3) then $z(a) \in \tilde{K}$ is a solution of (1). Also $z(a)_t = z(t)$ $(0 \le t \le a)$.

An ordinary differential equation in R^n is a special case of (1). Theorem 1 shows that (1) is a special case of the ordinary equation (3). This suggests that exixtence, uniqueness and continuous dependence theorems for equation (3) may yield corresponding results for (1). In the following an existence theorem of Carathéodory type is proved as a basis for subsequent study of continuous dependence of solutions of (3) and of (1) on parameters. Note that the distance of solutions of (1) is measured by an L_p-norm in Theorem 5 and Remark 4.

Let K be a convex and compact subset of a real Banach space B, $f: K \times [0, a] \to B$, $f(x, \cdot)$ measurable for $x \in K$, $f(\cdot, t)$ continuous for almost all $t \in [0, a]$, $|f(x, t)| \le m(t)$ for $(x, t) \in K \times [0, a]$. Let $\Omega \subset C[0, a, B]$ be a subset of the space of continuous functions from $[0, a]$ into B with sup norm and with the property that $x \in \Omega$ implies

$x(t) \in K$ for $t \in [0, a]$.

Lemma 1. $f(x(\cdot), \cdot)$ is (Bochner) integrable for every $x \in \Omega$.

Proof: As x is uniformly continuous in $[0, a]$ there exists a sequence of step functions converging uniformly to x. The dominated convergence theorem [2] yields the results.

Theorem 2. Let $x^o \in K$ be fixed and assume that the set Ω defined above has these additional properties: Ω is convex and closed, $v \in \Omega$ implies $v(0) = x^o$, and if $T: \Omega \to C[0, a, B]$ is defined by $(Tv)(t) = x^o + \int_0^t f(v(s), s)ds$, then $T\Omega \subset \Omega$. Under these hypotheses T has a fixed point.

Proof: If $v_i \to v$, $v_i \in \Omega$ for $i = 1, 2, 3, \ldots$, then

$$\sup_{0 \leq t \leq a} |(T v_i - Tv)(t)| \leq \int_0^a |f(v_i(s), s) - f(v(s)(s)| ds \to 0$$

as $i \to \infty$ by the continuity of $f(\cdot, t)$ and by the dominated convergence theorem, which proves that T is continuous. Schauder's theorem may be used to finish the proof if the closure of $T\Omega$ is compact. But $|(Tv)(t_2) - Tv(t_1)| \leq \int_{t_1}^{t_2} m(s)ds$ for $v \in \Omega$, $0 \leq t_1 < t_2 \leq a$ and $(Tv)(t) \in K$ for $t \in [0, a]$ shows that Ascoli's theorem applies.

Remark 1. If $B = R^n$, $K = \{x \in R^n \mid |x-x^o| \leq \int_0^a m(t)dt\}$,

$\Omega = \{ v \in C[0, a, R^n] : |v(t) - x^o| \le \int_0^t m(s)\,ds \}$, the hypotheses of Theorem 2 are satisfied and the Caratheodory existence theorem for ordinary equations in R^n is obtained.

Now Theorem 2 will be used to obtain an existence theorem for (1). Let $\tilde{y}_0 \in L_p[-h, a, R^n]$ be fixed, the restriction of $\tilde{y}_0$ to $[0, a]$ constant.

Theorem 3. Let g be the function from Theorem. 1, $\tilde{K} = \{x \in L_p[-h, a, R^n]:\ x(t) = \tilde{y}_0(t)$ a.e in $[-h, 0)$, $x(0) = \tilde{y}_0(0)$, $|x(t_2) - x(t_1)| \le \int_{t_1}^{t_2} m_1(t)\,dt$ for $0 \le t_1 < t_2 \le a \}$ where m_1 is an integrable function. Assume $g(x_t, t) \le m_1(t)$ for $x \in \tilde{K}$, $0 \le t \le a$, and $|g((x^j)_t, t) - g(x_t, t)| \to 0$ whenever $|x^j - x|_{L_p} \to 0$ with $j \to \infty$ for $x^j \in \tilde{K}$ and for almost all $t \in [0, a]$. Then (1) has a solution with the initial condition $\tilde{y}_0$.

Proof: By Theorem 1 and (2) $G(x, \cdot)$ is measurable, $G(\cdot, t)$ is continuous for almost all $t \in [0, a]$ and $|G(x, t)| \le m_1(t)a^{1/p}$ for $x \in \tilde{K}$, $0 \le t \le a$. Define $\Omega = \{ Z \in C[0, a^{1/p}[-h, a, R^n]] :\ z(0) = \tilde{y}_0,\ z(t) \in \tilde{K}$ for $0 \le t \le a$, $|z(t_2) - z(t_1)| \le \int_{t_1}^{t_2} m_1(s)ds.\ a^{1/p}$, $z(t) = y_t$ for some $y \in \tilde{K}\}$. As $\tilde{K}$ is convex and compact Ω is convex and closed. To prove the last assertion consider $z^n(t) = (y^n)_t \in \tilde{K}$ converging uniformly to $z(t)$. We shall show that $(z(a))_t = z(t)$ $(0 \le t \le a)$. For every $\varepsilon > 0$

$|(z(a))_t - z(t)| \leq |(z(a) - y^n)_t| + |(y^n)_t - z(t)| < \varepsilon$ for all $t \in [0,a]$ and for n sufficiently large, as $y^n \to z(a)$ implies $|(z(a) - y^n)_t| \to 0$ uniformly in $[0,a]$ in view of the fact that the L_p-convergence in $\tilde{K}$ implies the uniform convergence of continuous restrictions to $[0,a]$. Now we shall show that if $z \in \Omega$, $(Tz)(t) = \varphi_0 + \int_0^t G(z(s),s)ds$ for $0 \leq t \leq a$ then $Tz \in \Omega$. By Theorem 1

$$[(Tz)(t)](s_2) - [(Tz)(t)](s_1)| \leq \int_{\min(t,s_1)}^{\min(t,s_2)} |g(y_u, u)| du$$

$\leq \int_{s_1}^{s_2} m_1(u)du$ for $0 \leq s_1 < s_2 \leq a$, $y = z(a)$. Thus $(Tz)(t) \in \tilde{K}$ for $0 \leq t \leq a$. Further,

$$|(Tz)(t_2) - (Tz)(t_1)| = |\int_{t_1}^{t_2} G(z(s),s)ds|_{L_p} =$$

$$= \left(\int_0^a |\int_{\min(t_1,s)}^{\min(t_2,s)} g(y_u, u)\, du|^p ds\right)^{1/p}$$

$$\leq a^{1/p} \int_{t_1}^{t_2} m_1(u)\, du.$$

To check the last property of Ω, i.e. there is a $y \in \tilde{K}$ such that $(Tz)(t) = y_t$ $(0 \leq t \leq a)$, define $y(s) = \varphi_0 + \int_0^s g(z(a)_u, u)du$ for $0 \leq s \leq a$ and use the first part of Theorem 1.

Thus $G = f$ satisfies the hypotheses of Theorem 2 with $\tilde{K} = K$ and Theorem 3 follows from Theorem 1.

The following discussion of continuous dependence shows that the usual two different ways to measure the distance of right sides of differential equations, i. e. that $|\int_0^t [f_1(x, s) - f_2(x, s)]\, ds|$ for all x from a subset of R^n and $|\int_0^t [f_1(\mathfrak{x}(s), s) - f_2(\mathfrak{x}(s), s)]\, ds|$ for a set of continuous functions $\mathfrak{x}$ with values in R^n, have a common basis.

Let Λ be a metric space with a limit point $\lambda_0 \in \Lambda$, $\lambda_j \neq \lambda_0$, $d(\lambda_j, \lambda_0) \to 0$ as $j \to \infty$. Let D be a non-empty set in a real Banach space B and consider a function $F: D \times [0, a] \times \Lambda \to B$ satisfying

$|F(x, t, \lambda_0) - F(y, t, \lambda_0)| \leq L(t)\,|x-y|$ for some integrable function L and for (x, t), (y, t) from a subset $S \subset D \times [0, a]$.

<u>Lemma 2.</u> Let $x: [0, a] \times \Lambda \to D$ satisfy

(4) $x(t, \lambda) = x(0, \lambda) + \int_0^t F(x(s, \lambda), s, \lambda)\, ds$, $(x(t, \lambda), t) \in S$

for $(t, \lambda) \in [0, a] \times \Lambda \cdot \forall$ $|x(t, \lambda) - x(t, \lambda_0)| \leq$

$\{ |x(0, \lambda) - x(0, \lambda_0)| + \sup_{0 \leq t \leq a} |\int_0^t [F(x(s, \lambda), s, \lambda) - F(x(s, \lambda), s, \lambda_0)]\, ds| \}. \exp \int_0^t L(s) ds.$ Proof follows from Gronwell's inequality.

Remark 2. For a $\lambda \neq \lambda_0$ there may be more than one solution of (4) satisfying the initial condition $x(0, \lambda)$.

In Lemma 2, for each $\lambda \in \Lambda$ one of these solutions was selected and denoted by $x(\cdot, \lambda)$. The integral in (4) is assumed to exist in the Bochner sense.

<u>Theorem 4.</u> In addition to the hypotheses of Lemma 2 assume that D is bounded, there exist integrable functions m_2, m_3 in $[0, a]$ and a continuous increasing ω on $[0, \operatorname{diam} D]$, $\omega(0) = 0$, such that

$$|F(x, t, \lambda_0)| \leq m_2(t), \quad (x, y \in D,\ 0 \leq t \leq a),$$

$$|F(x, t, \lambda) - F(y, t, \lambda)| \leq m_3(t)\, \omega(|x-y|),$$

$$\left| \int_{t_1}^{t_2} [F(x, s, \lambda) - F(x, s, \lambda_0)]\, ds \right| \to 0 \text{ as } \lambda \to \lambda_0 \text{ uniformly}$$

in $D \times [0, a]$. Then $x(t, \lambda) \to x(t, \lambda_0)$ uniformly on $[0, a]$.

Proof: We prove first that for $\varepsilon > 0$ there exist positive numbers δ_1 and δ_2 such that $d(\lambda, \lambda_0) < \delta_1$ and $0 \leq t_1 < t_2 \leq a$, $t_2 - t_1 < \delta_2$ imply $|x(t_2, \lambda) - x(t_1, \lambda)| < \varepsilon$. Choose δ_1 and δ_2 so small that

$$\left| \int_{t_1}^{t_2} [F(x(t_1, \lambda), t, \lambda) - F(x(t_1, \lambda), t, \lambda_0)]\, dt \right| < \frac{\varepsilon}{2} \text{ for}$$

$$d(\lambda, \lambda_0) < \delta_1, \quad \omega(\operatorname{diam} D) \int_{t_1}^{t_2} m_3(t)dt + \int_{t_1}^{t_2} m_2(t)\, dt < \frac{\varepsilon}{2}$$

for $t_2 - t_1 < \delta_2$. Now

$$|x(t_2, \lambda) - x(t_1, \lambda)| \leq \int_{t_1}^{t_2} |F(x(t, \lambda), t, \lambda) - F(x(t_1, \lambda), t, \lambda)|dt + \left| \int_{t_1}^{t_2} [F(x(t_1, \lambda), t, \lambda) - F(x(t_1, \lambda), t, \lambda_0)]\, dt \right| + \int_{t_1}^{t_2} |F(x(t_1, \lambda), t, \lambda_0)|dt$$

$< \varepsilon$ for $t_2 - t_1 < \delta_2$, $d(\lambda, \lambda_0) < \delta_1$.

Next we prove that for $\eta > 0$ there exists a $\delta > 0$ such that $d(\lambda, \lambda_0) < \delta$ implies

$$\sup_{0 \leq t \leq a} \left| \int_0^t [F(x(s,\lambda), s, \lambda) - F(x(s,\lambda), s, \lambda_0)] \, ds \right| < \eta .$$

Fix η and choose ε such that

$\omega(\varepsilon) \int_0^a m_3(s) ds + \varepsilon (1 + \int_0^a L(s) ds) < \eta$. Let $t \in (0, a]$

and let $0 = s_0 < s_1 < \cdots s_k = t$, $s_j - s_{j-1} = \frac{t}{k} < \delta_2$. Then

$$\left| \int_0^t [F(x(s,\lambda), s, \lambda) - F(x(s,\lambda, s, \lambda_0)] \, ds \right|$$

$$\leq \sum_{j=1}^{k} \Big\{ \int_{s_{j-1}}^{s_j} |F(x(s,\lambda), s, \lambda) - F(x(s_{j-1}, \lambda), s, \lambda)| \, ds$$

$$+ \left| \int_{s_{j-1}}^{s_j} [F(x(s_{j-1}, \lambda), s, \lambda) - F(x(s_{j-1}, \lambda) s, \lambda_0)] \, ds \right|$$

$$+ \int_{s_{j-1}}^{s_j} |F(x(s_{j-1}, \lambda), s, \lambda_0) - F(x(s,\lambda), s, \lambda_0)| \, ds \Big\}$$

$$\leq \omega(\varepsilon) \int_0^t m_3(s) \, ds + \frac{\varepsilon}{k} \cdot k + \varepsilon \int_0^t L(s) ds < \eta$$

for $d(\lambda, \lambda_0) < \delta \leq \delta_1$ and for all $t \in [0, a]$.

Now Theorem 4 follows from Lemma 2.

Remark 3. If $B = R^n$ and $F(\cdot, \cdot \lambda)$ satisfies existence conditions (see Remark 1) for each $\lambda \in \Lambda$, Theorem 4 can be strengthened in the usual way (cfr. [3], without assuming that solutions x exist on $[0, a]$.

For functional differential equations (1) ($B = L_p[-h, a, R^n]$) the assumptions of Theorem 4 can be weakened due to the fact that solutions z of (3) in Theorem 1 are generated by functions from $\tilde{K}$, i.e. $\int_0^t G(z(s), s)ds = \int_0^t G(z(a), s)ds$.

Theorem 5. Let $H \subset \{x \in L_p[-h, a, R^n] : x$ is absolutely continuous in $[0, a]\}$, $x \in H \Rightarrow x_t \in H$, $\tilde{h} : H \times [0, a] \times \Lambda \to R^n$, $\tilde{h}(x, \cdot, \lambda) \in L_p[0, a, R^n]$ for $x \in H$, $\lambda \in \Lambda$. Define $F : H \times [0, a] \times \Lambda \to L_p[-h, a, R^n]$,

$$F(x, t, \lambda)(s) = \begin{cases} 0 & \text{for } -h \leq s < t \\ \tilde{h}(x_t, t, \lambda) & \text{for } t \leq s \leq a. \end{cases}$$

Let $x : [0, a] \times \Lambda \to H$ satisfy $x(t, \lambda) =$

$$x(t, \lambda) = x(0, \lambda) + \int_0^t F(x(s, \lambda), s, \lambda)\, ds \text{ in } [0, a] \times \Lambda,$$

$x(0, \lambda) \to x(0, \lambda_0$ as $\lambda \to \lambda_0$. If there is an integrable function L such that (5) $|F(y, s, \lambda_0) - F(z, s, \lambda_0)| \leq L(s)\, |y_s - z_s|$ for $s \in [0, a]$, y and z from H, and if

$$(6) \quad \sup_{0 \leq t \leq a} \left| \int_0^t [F(y, s, \lambda) - F(y, s, \lambda_0)]\, ds \right| \to 0$$

uniformly with respect to $y \in H$ as $\lambda \to \lambda_0$ then

$$\sup_{0 \leq t \leq a} | x(t, \lambda) - x(t, \lambda_0) | \to 0 \text{ as } \lambda \to \lambda_0.$$

Proof. By Theorem 1 $x(t,\lambda) = x(t,\lambda)_t = x(a,\lambda)_t$ and by (2) $F(x(a,\lambda),t,\lambda) = F(x(a,\lambda)_t, t,\lambda)$ for $(t,\lambda) \in [0,a] \times \Lambda$. Define $S = \{(z,t) \in H \times [0,a] : z = z_t\}$. Now (5) holds on S and the result follows from (6) and from Lemma 2.

Corollary : Let $\tilde{h}$ be as in Theorem 5, let y satisfy the functional differential equation

$$(7)\quad y(t,\lambda) = y(0,\lambda) + \int_0^t \tilde{h}(y(\cdot,\lambda)_s, s,\lambda)ds$$

for $0 \le t \le a$, $\lambda \in \Lambda$, with the initial conditions $y(\cdot,\lambda) \to y(\cdot,\lambda_0)$ in $L_p(-h, 0, R^n]$, $y(0,\lambda) \to y(0,\lambda_0)$ as $\lambda \Rightarrow \lambda_0$. If

$$(8)\quad |\tilde{h}(y_s, s,\lambda_0) - \tilde{h}(z_s, s,\lambda_0)|(a-s)^{1/k} \le L(s)|y_s - z_s|$$

for almost all $s \in [0,a]$ and y, z from H, and if

$$(9)\quad \sup_{0 \le t \le a} \left(\int_0^a \left| \int_0^{\min(\tau,t)} [\tilde{h}(y_s, s,\lambda) - \tilde{h}(y_s, s,\lambda_0)]\, ds \right|^p d\tau \right)^{1/p}$$

$\to 0$ as $\lambda \to \lambda_0$ uniformly with respect to $y \in H$ then

$$(10)\quad \sup_{0 \le t \le a} \left(\int_{-h}^t |y(s,\lambda) - y(s,\lambda_0)|^p ds + (a-t)|y(t,\lambda) - y(t,\lambda_0)|^p \right)^{1/p} \to 0 \text{ as } \lambda \to \lambda_0.$$

Remark 4. Condition (9) implies but is not implied by the uniform convergence of

$$(11)\quad \int_0^t [\tilde{h}(y_s, s,\lambda) - \tilde{h}(y_s, s,\lambda_0)]\, ds \text{ to } 0 \text{ on } H \times [0, a-\varepsilon]$$

for every ε in $(0,a)$. Similarly, (10) implies uniform convergence $y(\cdot,\lambda) \to y(\cdot,\lambda_0)$ on every interval $[0, a-\varepsilon]$, ε

in $(0, a)$. It is also easy to see that (9) is implied by uniform convergence of (11) on $H \times [0, a]$ and that (10) does not imply uniform convergence $y(\cdot, \lambda) \to y(\cdot, \lambda_0)$ on $[0, a]$.

Example. Let $h = 1$, $a > 0$, $H = \{x \in L_1[-1, a, R^1]:$

$|x| \leq 1$, x absolutely continuous in $[0, a]\}$,

$\Lambda = [0, 1]$, $\lambda_0 = 0$, $|x(\cdot, \lambda)|_{L_1[-1,0]} \to 0$, $x(0, \lambda) \to 0$ as $\lambda \to 0$.

Let (7) have the form

$$(12)\quad x(t, \lambda) = x(0, \lambda) + \int_0^t [\lambda^{-\alpha}(\gamma\, x(s-1) + x(s)) \sin \frac{s}{\lambda} + \lambda^{-\beta} \cos \frac{s}{\lambda}]\, ds \text{ for } \lambda \in (0, 1],\ t \in [0, a],\ \alpha, \beta, \gamma \text{ real},$$

$x(t, 0) = 0$. In this case condition (8) is trivally satisfied and (9) holds if $\alpha < 0$, $\beta < 1$, so that (10) implies that the solutions $x(\cdot, \lambda)$ of (12) converge to 0 uniformly on every bounded subinterval of $[0, \infty)$. If $\gamma = 0$ (12) is an ordinary differential equation with initial condition $x(0, \lambda)$ and the uniform convergence of solutions to 0 follows from Theorem 4 for $\alpha \leq 0$, $\beta < 1$, if we put $B = R^1$ (cfr. [4]). If we use Theorem 5 instead of Theorem 4 on the case $\gamma = 0$ we obtain a weaker result, i.e. continuous dependence of solutions is guaranteed for $\alpha \leq 0$ $\beta < 1$. Results in [4] and [5] can be applied to the case $\gamma = 0$ only and give convergence for $\alpha < \frac{1}{2}$, $\beta < 1$, $\alpha + \beta < 1$, although the solutions of this ordinary differential equation converge to zero if $\alpha < 1$, $\beta < 1$, $\alpha + \beta < 1$, as it is easy to verify.

REFERENCE

[1] Gonzalez, J. and Imaz, C. and Vorel Z., Functional and ordinary differential equations. Bol. Soc. Mat. Mex., to appear.

[2] Dunford, N. and Schwartz, J. T., Linear Operators, Part I. New York. Interscience 1958.

[3] Kurzweil, J. and Vorel, Z., Continuous dependence of solutions of differential equations on a parameter Czechoslovak Math. J. 7(82), 568-583 (1957) (in Russian).

[4] Kurzweil, J. Generalized Ordinary Differential Equations and Continuous Dependence on a parameter. Czechoslovak Math. J., 7(82), 418-446, (1957).

[5] Neustadt, L. W., On the Solutions of Certain Integral- like Operator Equations. Arch. Rat. Mech. Anal., Vol. 38, 2, 131-160, (1970).

*Escuela Superior De Fisica y Matematicas
Del Instituto Politecnico Nacional, Mexico

**University of Southern California, Los Angeles, California

THE SOLUTION OPERATOR WITH INFINITE DELAYS

Jack K. Hale

For a retarded functional differential equation for which the solution at time t depends only on the past history over the interval $[t-r, t]$, the solution operator takes the initial data on an interval $[\sigma-r, \sigma]$ into the restriction of the solution to $[t-r, t]$ and is extremely simple. In fact, under very weak hypotheses, this map is completely continuous for $t - \sigma \geq r$. Because of this property, complete information about the spectral sets for linear operators is easily obtained. If the complete continuity is exploited to obtain results on the existence of periodic solutions for nonlinear equations, one must generally suppose the period ω is $\geq r$. To eliminate this latter restriction is not easy and one approach taken in [6] is to discuss additional properties of the solution operator in $0 \leq t - \sigma \leq r$. In fact, it is observed that the operator is an α-contraction in this interval.

If the solution at time t of a retarded functional differential equation depends on the past history over $(-\infty, t]$, then the operator never becomes completely continuous so that spectral properties are not immediate and crude methods for obtaining periodic solutions will not work. It is the

This research was supported in part by the Office of Army Research under DA-ARO-D-31-124-73-G-130, and in part by the National Science Foundation under GP 28931X2.

purpose of this lecture to show how some very simple ideas in [5] can be of assistance in solving these problems.

Let $\mathcal{B}$ be a Banach space of functions mapping $(-\infty, 0]$ into E^n, an n-dimensional normed linear vector space. We suppose constant functions are in $\mathcal{B}$. If x is a function defined on $[-\infty, A)$, $A > 0$, and t is fixed in $(-\infty, A)$ let x_t: $(-\infty, 0] \to E^n$ be defined by $x_t(\theta) = x(t+\theta)$, $-\infty < \theta \leq 0$. The class of functions to be considered will be supposed to satisfy $x_t \in \mathcal{B}$, $t \in (-\infty, A)$. Let $f: R \times \mathcal{B} \to E^n$ be completely continuous and consider the retarded functional differential equation

$$(1) \qquad \dot{x}(t) = f(t, x_t) .$$

For any $(\sigma, \phi) \in R \times \mathcal{B}$, we suppose there is a solution $x(\sigma, \phi)$ through (σ, ϕ) defined for $t \geq \sigma - r$ and, furthermore, $x(\sigma, \phi)(t)$ is continuous in (σ, ϕ, t).

The solution operator $T(t, \sigma): \mathcal{B} \to \mathcal{B}$, $t \geq \sigma$, is defined by $T(t,\sigma)\phi = x_t(\sigma, \phi)$. Let $S(t): \mathcal{B} \to \mathcal{B}$, $t \geq 0$, $U(t,\sigma): \mathcal{B} \to \mathcal{B}$, $t \geq \sigma$ be defined by

$$(2) \quad \text{(a)} \quad S(t)\phi(\theta) = \begin{cases} 0 & \text{for } t + \theta \geq 0 \\ \theta(t+\theta) - \phi(0) & \text{for } t + \theta < 0 \end{cases}$$

$$\text{(b)} \quad U(t,\sigma)\phi(\theta) = \begin{cases} \phi(0) + \int_\sigma^t f(s, x_s(\sigma, \phi))ds & \text{for } t + \theta \geq 0 \\ \phi(0) & \text{for } t + \theta < 0 \end{cases}$$

With these definitions, the solution operator can be written as

$$(3) \qquad T(t,\sigma) = S(t-\sigma) + U(t,\sigma) .$$

Hypothesis (H_1): The norm in $\mathcal{B}$ is such that the operator $U(t,\sigma)$ is weak completely continuous; that is, for any bounded set $B \subset \mathcal{B}$ for which $\{T(s,\sigma)B, \sigma \leq s \leq t\}$ is bounded, it follows that $U(t,\sigma)B$ is precompact.

This latter condition is satisfied for example by all $\mathcal{B}$ with fading memory [3] and for $\mathcal{B}$ consisting of all continuous bounded functions ϕ with $\phi(\theta)$ approaching a limit as $\theta \to -\infty$ with the topology of uniform convergence.

The properties of the operator $S(t)$ are more sensitive to the particular type of space $\mathcal{B}$ being considered. To facilitate the discussion, we make

<u>Hypothesis (H_2)</u>: There is a function $\gamma(t): [0,\infty) \to [0,\infty)$ continuous and bounded such that

(4) $$|S(t)\phi| \leq \gamma(t)|\phi| \quad \text{for } t \geq 0, \ \phi \in \mathcal{B}_0$$

where $\mathcal{B}_0 = \{\phi \in \mathcal{B}: \phi(0) = 0\}$.

Let $\alpha(B) = \inf\{d: B$ has a finite cover of diameter $< d\}$ be the Kuratowski measure of compactness of any bounded set $B \subset \mathcal{B}$. If $Q: \mathcal{B} \to \mathcal{B}$ is a bounded linear operator, then $\alpha(Q) = \inf\{k: \alpha(QB) \leq k\alpha(B)$ for any bounded $B \subset \mathcal{B}\}$.

Since $\mathcal{B}$ is essentially the same as $\mathcal{B}_0 \times E^n$ and E^n is locally compact, hypothesis (H_2) implies that $\alpha(S(t)) \leq \gamma(t)$ for all t.

Using the above observations together with the fact that $\alpha(B+C) \leq \alpha(B) + \alpha(C)$ for bounded sets B, C one obtains the following

<u>Proposition 1.</u> If hypotheses (H_1), (H_2) are satisfied, then $T(t,\sigma)$ is a <u>weak $\gamma(t)$-set contraction</u>; that is, for any bounded set $B \subset \mathcal{B}$ with $\{T(s,\sigma)B, \sigma \leq s \leq t\}$ bounded, one has

$$\alpha(T(t,\sigma)B) < \gamma(t)\,\alpha\,(B)\,.$$

As an application of Proposition 1 and Theorem 9 of [6], we have

Theorem 1. If hypotheses (H_1), (H_2) are satisfies, if there is an $\omega > 0$ such that $f(t+\omega,\phi) = f(t,\phi)$ for all t, ϕ, $\gamma(\omega) < 1$, and if there is a bounded set B in $\mathcal{B}$ which attracts orbits of compact sets of (1), then equation (1) has an ω-periodic solution. If $f(t,\phi)$ is independent of t and the other hypotheses are satisfied, then there is a constant solution of (1).

As another application of Proposition 1, consider the linear system

$$\dot{x}(t) = Lx_t \tag{5}$$

where $L: \mathcal{B} \to E^n$ is continuous and linear. In this case, we may take $\sigma = 0$ and the solution operator $T(t)$, $t \geq 0$, is a strongly continuous semigroup of linear operators on $\mathcal{B}$ with infinitesimal generator A given by

$$A\phi(\theta) = \dot{\phi}(\theta),\ -r \leq \theta \leq 0$$

$$D(A) = \{\phi \in \mathcal{B}: \dot{\phi} \in \mathcal{B},\ \dot{\phi}(0) = L\phi\}.$$

The point spectrum of A is given by

$$\{\lambda : \det[\lambda I - L(e^{\lambda \cdot} I)] = 0\}\,. \tag{6}$$

For a proof of these results, see [9].

Under hypothesis (H_2), $\alpha(T(t)) \leq \gamma(t)$. Using known results in [1], [4], [7], one can easily prove

Theorem 2. Under hypothesis (H_2), for any $t > 0$, $\varepsilon > 0$, there are only a finite number of points $\rho(t)$ in the spectrum of $T(t)$ with modulus $>\gamma(t) + \varepsilon$, each such $\rho(t)$ must be of the form $\rho(t) = e^{\lambda t}$ for some λ satisfying (6). Also, the generalized eigenspace of $\rho(t)$ is finite dimensional.

As an immediate corollary, we have

Corollary 1. Suppose there is a $\delta \in R$ such that $\gamma(\tau) = e^{\delta\tau}$ for some $\tau > 0$ and let $\alpha = \max(\delta, \{\mathrm{Re}\,\lambda: \lambda$ satisfies (6)$\})$. Then for any $\varepsilon > 0$ there is an $M_\varepsilon > 0$ such that

$$\|T(t)\| \le M_\varepsilon \exp(\alpha+\varepsilon)t, \quad t \ge 0. \tag{7}$$

In particular, if

$$\{\lambda \in \mathbb{C}: \mathrm{Re}\,\lambda > \delta,\ \lambda \text{ satisfies } (6)\} = \emptyset \tag{8}$$

then

$$\|T(t)\| \le M_\varepsilon \exp(\delta+\varepsilon)t, \quad t \ge 0. \tag{9}$$

The special case $\gamma(t) \equiv 1$ $(\delta = 0)$ and condition (8) imply small exponential growth of the semigroup and the case $\gamma(t) = \exp(-\mu t)$, $\mu > 0$, and condition (8) imply exponential decay of the semigroup (compare [2], [8]).

To conclude this lecture, we give two types of Banach spaces and the corresponding $\gamma(t)$. Following [3], consider the space of all functions which are locally L_1 on $(-\infty, 0]$ with

$$|\varphi| = |\phi(0)| + \int_{-\infty}^{0} p(\theta)|\phi(\theta)|\,d\theta$$

where p is continuous, $\int_{-\infty}^{0} p < \infty$. From (4), one easily obtains

$$\gamma(t) \le \sup_{-\infty<\theta<0} p(\theta-t)/p(t).$$

Therefore, $\gamma(t) \le 1$ always. Furthermore, if

$p(\theta) = \exp -\mu t$, $\mu > 0$, then $\gamma(t) = \exp(-\mu t)$.

For the second illustration, suppose $\mathcal{B}$ is the space of bounded continuous functions which approach a limit as $\theta \to \infty$ with the topology of uniform convergence. Then $\gamma(t) = 1$.

REFERENCES

[1] Ambrosetti, A., Propieta spettrali di certi operatori lineari non compatti. Rend. Sem. Mat. Univ. Padova 42(1969), 189-200.

[2] Barbu, V. and S. Grossmann, Asymptotic behavior of linear integral differential equations. Trans. Am. Math. Soc. 173(1972), 277-289.

[3] Coleman, B. D. and V. J. Mizel, On the stability of the solutions of functional differential equations. Arch. Rat. Mech. Anal 30(1968), 173-196.

[4] Gokberg, T. C. and M. G. Krein, Introduction to the Theory of Linear Nonself-Adjoint Operators. Transl. Math. Monographs, Am. Math. Soc. 18(1969).

[5] Hale, J. K., Functional differential equations with infinite delays. J. Math. Ana. Appl. 47(1974).

[6] Hale, J. K. and O. Lopes, Fixed point theorems and dissipative processes. J. Differential Equations, 13(1973), 391-402.

[7] Hille, E. and R. S. Phillips, Functional Analysis and Semigroups. Am. Math. Soc. Colloquium Publ. 31(1957).

[8] MacCamy, R. C., Exponential stability for a class of functional differential equations. Arch. Rat. Mech.

Ana., 40(1971), 120-138.

[9] Naito, T., On autonomous linear functional differential equations with infinite retardations. J. Differential Equations. To appear.

Lefschetz Center for Dynamical Systems
Brown University, Providence, Rhode Island

BROUWER FIXED POINT THEOREM VERSUS CONTRACTION MAPPING THEOREM IN OPTIMAL CONTROL THEORY

by

Hubert Halkin

In order to derive meaningful necessary conditions in nonlinear optimal control one had, until recently, to prove and to use a result of the following type:

Proposition I. If U is an open subset of R^n, if $x_0 \in U$ and if φ is a continuous function from U into R^m which admits as derivative at the point x_0 a linear map A from R^n onto R^m, then $\varphi(x_0) \in \text{int}\ \varphi(U)$.

The proof of Proposition I, which will be repeated in this paper, is an easy consequence of the Brouwer Fixed Point Theorem but, apparently, cannot be obtained without using this important (but non constructive) result or one of its algebraic topological equivalents.

It is important to remark that in Proposition I we assume that φ is differentiable at x_0 but we do not assume that φ is continuously differentiable in a neighborbood of x_0. As a matter of fact we do not even assume, in Proposition I, that φ is differentiable in a neighborhood of x_0. With the stronger assumtpion of continuous differentiability in a neighborhood of x_0 the preceeding result becomes.

<u>Proposition I*</u> If U is an open subset of R^n, if $x_0 \in U$ and if φ is a continuously differentiable function from U into R^m which admits as derivative at the point x_0 a linear map A from R^n onto R^m then $\varphi(x_0) \in \operatorname{int} \varphi(U)$

Proposition I* is a particular case of Proposition I and is a lot easier to prove than Proposition I. Indeed Proposition I* is nothing but a corollary of the classical Implicit and Inverse Function Theorems of advanced calculus. It should be recalled here those theorems of advanced calculus are based (directly or indirectly) on the Contraction Mapping Theorem, a weak but constructive cousin of the Brouwer Fixed Point Theorem.

The easy Proposition I* was quite adequate in classical calculus of variations where weak (bilateral) variations lead to continuously differentiable functions φ. In the theory of nonlinear optimal control with its strong (unilateral) variations there was never any hope of obtaining continuously differentiable functions φ; this led to the use of the harder and non constructive Proposition I.

In the last few years I have finally succeeded in deriving necessary conditions in nonlinear optimal control without using the Brouwer Fixed Point Theorem. The crux of this new proof was the realization (Halkin [2]) that although the function φ encountered in nonlinear optimal control was not continuously differentiable in a neighborhood of x_0 its derivative A at the point x_0 did not only satisfy the usual requirement of differentiability:

For all $\varepsilon > 0$ there exists a $\delta > 0$ such that $|\varphi(x) - \varphi(x_0) - A(x - x_0)| \leq \varepsilon |x - x_0|$ whenever $x \in U$ and $|x - x_0| \leq \delta$.

but also the following stronger requirement:

For all $\varepsilon > 0$ there exists a $\delta > 0$ such that $|\varphi(x_2) - \varphi(x_1) - A(x_2 - x_1)| \leq \varepsilon |x_2 - x_1|$ whenever $x_1, x_2 \in U$ and $|x_1 - x_2|, |x_2 - x_0| \leq \delta$.

With this stronger requirement I was then able to move (see Proposition II given below) that $\varphi(x_0) \in \text{int } \varphi(U)$ without using the Brouwer Fixed Point Theorem.

At the occasion of a recent lecture at the University of Grenoble I was fortunate to have a few numerical analysts among my listeners who pointed out to me that the stronger requirement mentionned above was nothing else than the concept of strong differentiability which they had been using for many years. For that reason I shall say that the linear map A is the <u>strong derivative</u> of the function φ at the point x_0 if it satisfies the stronger requirement mentionned above. I shall also say that the function φ is <u>strongly differentiable</u> at the point x_0 if there exists a linear map A which is the strong derivative of φ at x_0.

<u>Proposition II.</u> If U is an open subset of R^n, if $x_0 \in U$ and if φ is a function from U into R^m which admits as strong derivative at the point x_0 a linear map A from R^n <u>onto</u> R^m then $\varphi(x_0) \in \text{int } \varphi(U)$.

It is easy to see that if φ is continuously differentiable in a neighborhood of x_0 and admits A as its derivative at x_0 then φ admits A as its strong derivative at x_0. Consequently the (classical) Proposition I* is nothing but a particular case of Proposition II. Moreover since strong differentiability at x_0 implies differentiability at x_0 and continuity in a neighborhood of x_0 it follows that Proposition II is a particular case of Proposition I. Examples of differentiability without strong differentiability are easily obtained. Consider for instance the function $\varphi(x) = x^2 \sin \frac{1}{x}$ with $\varphi(0) = 0$; the function φ is differentiable but not strongly differentiable at the point $x = 0$. Examples of strong differentiability without continuous differentiability are also easily obtained. The continuous function φ defined on $[-1, +1]$ such that $\varphi(0) = 0$ and $\varphi'(t) = \frac{1}{k}$ whenever $\frac{1}{k+1} < |t| < \frac{1}{k}$ for any positive integer k is strongly differentiable at 0 but is not continuously differentiable on any neighborhood of 0.

As I mentionned above, Proposition I* is a particular case of Proposition II which is itself a particular case of Proposition I. From a purely traditional point of view it would then be sufficient to give a proof of Proposition I. However since the proof of Proposition I requires the use of the Brouwer Fixed Point Theorem whereas Proposition II can be proved using the simpler (and more constructive) Contraction Mapping Theorem, I will also give this simpler and direct proof of Proposition II. On the other hand since Proposition I* is a particular case of Proposition II and since I do not know a proof of Proposition I* which would be simpler than the proof of Proposition II, I do not see the interest of making any further mention of Proposition I* in this paper. The only reason to mention Proposition I* in the earlier part of this introduction was motivational: Proposition I* has a structure similar to the structure of the results in which we are interested here (Proposition I and II) and is easily recognized as a corollary of a classical result (the Implicit Function Theorem).

Let me now recall the particular form of the two basic theorems which shall be used later.

Brouwer Fixed Point Theorem. If f is a continuous function from a compact convex subset K of R^m into itself then for some $x^* \in K$ we have $f(x^*) = x^*$.

Contraction Mapping Theorem. If f is a function from a closed subset K of R^m into itself such that for some $k < 1$ we have $|f(x_2) - f(x_1)| \leq k\, |x_2 - x_1|$ for all $x_1, x_2 \in K$ then for some $x^* \in K$ we have $f(x^*) = x^*$.

Instead of giving separately the proof of Proposition I and the direct proof of Proposition II I shall combine those two proofs using a procedure described in the following remark.

Essential guidelines for the reader. The remainder of this paper contains three types of messages: the normal message, the bracketed message and the braced message. The reader should split his personality into two parts Reader I and Reader II. Reader I should read the normal and the braced messages, whereas Reader II should read the normal and the bracketed message. For instance, if the text reads: "when the function is [strongly] differentiable we shall establish our result by using the {Brouwer Fixed Point} [Contraction Mapping] Theorem", Reader I should read: "when the function is differentiable we shall establish our result by using the Brouwer Fixed Point Theorem" and Reader II should read: "when the function is strongly differentiable we shall establish our result by using the Contraction Mapping Theorem". Reader I will be presented with the proof of Proposition I and Reader II will be presented with the direct proof of Proposition II.

<u>[Direct] Proof of Proposition {I} [II].</u>

There is no loss of generality by assuming that $x_0 = 0$ and that $\varphi(x_0) = 0$. Let A^{-1} be a linear mapping from R^m into R^n such that $AA^{-1}y = y$ for all $y \in R^m$. This is possible since A is a linear mapping from R^n <u>onto</u> R^m. Let $\eta > 0$ be such that $A^{-1}y \in U$ whenever $|y| \leq \eta$. For all $y \in R^m$ with $|y| \leq \eta$ let $\psi(y) = \varphi(A^{-1}y)$. Since φ admits A as its [strong] derivative at 0 and since A^{-1} admits (trivially) A^{-1} as its [strong] derivative at 0 it follows (Chain Rule for [strong] differentiability) that ψ admits AA^{-1}, the identity by construction, as its [strong] derivative at x_0. Let $\delta^* \in (0, \eta]$ be such that

$$|\psi(y) - y| \leq \frac{1}{2}|y| \text{ whenever } |y| \leq \delta^*$$

$$\left[\begin{array}{l}\text{and even} \\ |\psi(y_2) - \psi(y_1) - (y_2 - y_1)| \leq \frac{1}{2} |y_2 - y_1| \\ \text{whenever } |y_1| \text{ and } |y_2| \leq \delta^* \end{array}\right].$$

For every y_* with $|y_*| \leq \frac{1}{2}\delta^*$ let ψ_{y_*} be a function defined over $\{y: |y| \leq \delta^*\}$ by $\psi_{y_*}(y) = y_* + y - \psi(y)$. The function ψ_{y_*} is continuous and maps $\{y: |y| \leq \delta^*\}$ into itself since $|\psi_{y_*}(y)| \leq |y_*| + |y - \psi(y)| \leq \frac{1}{2}\delta^* + \frac{1}{2}\delta^* = \delta^*$.

[Moreover for all y' and $y'' \in \{y: |y| \leq \delta^*\}$ we have $|\psi_{y_*}(y'') - \psi_{y_*}(y')| \leq |\psi(y'') - \psi(y') - (y'' - y')| \leq \frac{1}{2} |y'' - y'|$.]

Let y_{**} be the fixed point of ψ_{y_*}. {The existence of y_{**} is guaranteed by the Brouwer Fixed Point Theorem}. [The existence of y_{**} is guaranteed by the Contraction Mapping Theorem]. We have then $y_* + y_{**} - \psi(y_{**}) = y_{**}$ and hence $\psi(y_{**}) = y_*$. Let $x_{**} = A^{-1} y_{**}$. We have then $x_{**} \in U$ and $\varphi(x_{**}) = y_*$. In other words we have proved that $y_* \in \varphi(U)$ whenever $|y_*| \leq \frac{1}{2}\delta^*$. This concludes the proof of Proposition {I} [II].

REFERENCES

[1] Halkin, H, "Implicit Functions and Optimization Problems without Continuous Differentiability of the

Data," SIAM Journal on Control, 12, 1974 pp. 229-236.

[2] Halkin, H, "Necessary Conditions in Mathematical Programming and Optimal Control Theory" Proceedings of the 14th Biennial Seminar of the Canadian Mathematical Congress, (to appear) Springer Verlag.

University of California, San Diego
La Jolla, California 92037

RECENT RESULTS IN THE ASYMPTOTIC INTEGRATION OF LINEAR DIFFERENTIAL SYSTEMS

by

W. A. Harris, Jr.(1)

and

D. A. Lutz(2)

The stability and asymptotic behavior of solutions of an autonomous linear differential system $x' = Ax$ are determined by the spectrum of the constant matrix A. If $B(t)$ is a suitably small perturbation, then the stability and asymptotic behavior of solutions of $x' = [A + B(t)]x$ are still determined by the limiting system $x' = Ax$. For example, if $P^{-1} AP = \Lambda$ is a diagonal matrix, $B(t)$ continuous for $t \geq t_0$, and $\int_{t_0}^{\infty} \|B(t)\| \, dt < \infty$, then $x' = [A + B(t)]x$ has a fundamental matrix $X(t)$ satisfying, as $t \to \infty$,

$$X(t) = [P + o(1)] \exp \int_{t_0}^{t} \Lambda \, ds.$$

The classical theorems of Levinson [12], Hartman-Wintner [11] and Wintner [16] are deeper results of this

(1) Supported in part by the United States Army under contract DAHC04-74-G0013

(2) Supported in part by the National Science Foundation under Grant GP-28149

nature which describe the asymptotic behavior of solutions of the nonautonomous linear differential system $x' = A(t)x$ in terms of the eigenvalues of the matrix $A(t)$. Extensions of these results, which have been employed by Devinatz [4, 5] and Fedoryuk [6] to determine the deficiency index of certain differential operators, have been discussed in a systematic manner through preparatory lemmas by Harris-Lutz [7]. These methods are not directly applicable to the physically important case of the adiabatic oscillator, as typified by the scalar equation $x'' + [1 + g(t)]\,x = 0$, and special ad hoc methods have been employed, eg. Atkinson [1, 3], Wintner [16]. However, Harris-Lutz [8] have shown how the asymptotic integration of the adiabatic oscillator may also be effected through preparatory lemmas.

The essence of the method employed by Harris-Lutz [7, 8] is the transformation of the original differential system into one to which the fundamental result of Levinson [12] or Hartman-Wintner [11] applies. The germ of this method has been utilized recently by Harris-Lutz [9] to give a unified treatment of asymptotic integration of linear differential systems through transformation into L-diagonal canonical form and the fundamental result of Levinson [12]. (See eg. Raporport [13] and Cesari [2] for other applications of this canonical form.)

We are concerned with the linear differential system

$$y'(t) = [\Lambda + V(t)]\,y(t) \tag{1}$$

where Λ is a constant diagonal matrix with distinct eigenvalues (or $\Lambda = 0$) and $V(t) \to 0$ as $t \to \infty$. Utilizing the transformation

$$(2) \qquad y(t) = [I + Q(t)]\, u(t)$$

where $Q(t) \to 0$ as $t \to \infty$ with the normalization diag $Q \equiv 0$, we obtain the linear differential system

$$(3) \qquad u'(t) = [\Lambda + \hat{V}(t)]\, u(t)$$

where $\hat{V}(t) \to 0$ as $t \to \infty$. If $\hat{V}$-diag $\hat{V}$ is sufficiently regular in the sense of integrability and Λ + diag $\hat{V}$ satisfies Levinson's dicotomy conditions, then the asymptotic integration of the linear differential system (3), and hence also (1), can be effected.

We require, in the spirit of the original results, that (i) diag $\hat{V}$ must be computable, and (ii) in at most a finite number of repeated applications, $\int_{t_0}^{\infty} \| \hat{V}(s) - \text{diag } \hat{V}(s) \| ds < \infty$.

The matrices Q and $\hat{V}$ satisfy the equation

$$(4) \qquad (I + Q)\hat{V} = \Lambda Q - Q\Lambda + V + VQ - Q'$$

The diagonal terms contain no derivatives due to our assumption, diag $Q \equiv 0$, and we have

$$\text{diag } \hat{V} = \text{diag } V + 0(\|Q\| \; \|V\|).$$

If we can choose Q in an appropriate manner so that $\int_{t_0}^{\infty} \|\hat{V}(s) - \text{diag } \hat{V}(s)\| \, ds < \infty$, we will have achieved our goal.

Three specific choices of Q have proved useful:

I. $\Lambda Q - Q\Lambda + V + VQ - \text{diag } (V + VQ) = 0.$

II. $V - \text{diag } V - Q' = 0.$

III. $\Lambda Q - Q\Lambda + V - \text{diag} V - Q' = 0.$

Case I. This is Levinson's basic result in which $\Lambda + \text{diag } \hat{V}$ represents the eigenvalues of the matrix $\Lambda + V$, and $\hat{V} - \text{diag } \hat{V} = 0(\|Q'\|) = 0(\|V'\|)$ which is integrable due to the assumption $\int_{t_0}^{\infty} \|V'(s)\| ds < \infty$, see eg. Harris-Lutz [7].

Case II. If $\Lambda = 0$, this case covers the results of Wintner [16]. Preliminary transformations also allow the adiabatic oscillator in both resonance and nonresonance to be treated in this manner, see Harris-Lutz [8].

Case III. The results of Hartman-Wintner [11], are representative of this case. One can show that if $\int_{t_0}^{\infty} \|V(s)\|^2 ds < \infty$, then $\int_{t_0}^{\infty} \|Q(s)\|^2 ds < \infty$, and hence $\hat{V}\text{-diag } \hat{V} = 0(\|V\| \ \|Q\|)$, and $\int_{t_0}^{\infty} \|\hat{V}(s)\text{-diag } \hat{V}(s)\| ds < \infty$, see Harris-Lutz [9].

REFERENCES

[1] F. V. Atkinson, "The asymptotic solution of second order differential equations, Ann. Mat. Para Appl. 37 (1954), 347-378.

[2] L. Cesari, "Asymptotic Behavior and Stability Problems in Ordinary Differential Equations," New York, Academic Press, 1963.

[3] W. Coppel, "Stability and Asymptotic Behavior of Differential Equations, Boston, Heath, 1965.

[4] A. Devinatz, "An asymptotic theory for systems of linear differential equations," Trans. Amer. Math Soc. 160 (1971) 353-363.

[5] __________, "The deficiency index of a certain class of ordinary self-adjoint differential operators, Adv. in Math. 8 (1972) 434-473.

[6] M. Fedoryuk, "Asymptotic methods in the theory of one-dimensional singular differential operators," Trans. Moskow Math. Soc. (1966) 333-386.

[7] W. A. Harris, Jr. and D. A. Lutz, "On the asymptotic integration of linear differential systems," J. Math. Anal. Appl. 48 (1974) 1-16.

[8] __________, "Asymptotic integration of adiabatic oscillators," J. Math. Anal. Appl. (to appear).

[9] __________, "A unified theory of asymptotic integration," (to appear).

[10] P. Hartman, "On a class of perturbations of the harmonic oscillator," Proc. Amer. Math. Soc. 19(1968) 533-540.

[11] P. Hartman and A. Wintner, "Asymptotic integration of linear differential equations," Amer. J. Math. 77(1955) 45-86 and 932.

[12] N. Levinson, "The asymptotic nature of solutions of linear differential equations," Duke Math J. 15(1948)

[13] I. M. Rapoport, "On some asymptotic methods in the theory of differential equations," Kiev, 1954.

[14] A. Wintner, "The adiabatic linear oscillator," Amer. J. Math 68(1946) 385-397.

[15] __________, "Asymptotic integration of the adiabatic oscillator," Amer. J. Math. 69(1947) 251-272.

[16] __________, "On a theorem of Bôcher in the theory of ordinary linear differential equations," Amer. J. Math. 76(1954) 183-190, and Addenda 78(1956) 895-897

University of Southern California, Los Angeles, CA 90007

and

University of Wisconsin-Milwaukee, Milwaukee, WI 53201

INTEGRATION OF LINEAR DIFFERENTIAL EQUATIONS BY LAPLACE-STIELTJES TRANSFORMS

Philip Hartman[1]

1. Introduction. This is a report on some of the results in the papers Hartman [4], D'Archangelo [1] and D'Archangelo-Hartman [2]. These papers deal with N-th order equations and with systems of first order equations, both homogeneous and inhomogeneous. For the sake of brevity, we shall mainly describe the general results only in the case for homogeneous, linear, first order system of dimension N, say

(1.1) $$y' = [A+g(t)]y,$$

where A is a constant matrix and g(t) is a matrix representable as a Laplace-Stieltjes transform

(1.2) $$g(t) = \int_0^\infty e^{-st}\, dG(s), \quad G(0) = G(+0) = 0,$$

absolutely convergent for $\operatorname{Re} t > 0$, where G(s) has entries which are complex-valued and locally of bounded variation on $s \geqq 0$. (In particular, the last part of (1.2) implies that $g(t) \to 0$ as $t \to \infty$.)

[1] This report was partially supported by NSF Grant GP- 30483A2.

The problems considered are the existence of solutions of (1.1) representable as Laplace-Stieltjes transforms or linear combinations of such transforms with coefficients of the form $e^{\lambda t}t^{\mu}$. First, our results can be interpreted as an extension of the theory of regular singular points, since the case where $g(t)$ is a power series in $z = e^{-t}$ (i.e., dG is a discrete measure with support on the positive integers) corresponds to a regular singular point at $z = 0$ (and solutions are linear cominations of power series in $z = e^{-t}$ with coefficients of the form $z^{-\lambda}(\log z)^{\mu} = e^{\lambda t}(-t)^{\mu}$). Second, our results provide sharpened variants of known results in the theory of asymptotic integrations. Finally, they can be considered as a contribution to the theory of the representations of solutions in terms of definite integrals, a theory important to the study of solutions "in the large".

2. <u>An existence theorem.</u> In the theory of regular singular points, we deal with sets of eigenvalues of A differing by an integer. Here we have to consider sets of eigenvalues differing by a real number, i.e., eigenvalues with the same imaginary part, say $\operatorname{Im} \lambda = \tau$. In the theory of asymptotic integrations, we impose conditions on the perturbations $g(t)$ of the form $\int^{\infty} t^{-\gamma} |g(t)| dt < \infty$. Correspondingly, these are replaced here by conditions of the form

$$\int_{+0} s^{-\gamma} |dG(s)| < \infty . \tag{2.1}$$

We list some assumptions which we make below.

Assumption (A1). The constant NxN matrix A is similar to the Jordan matrix

(2.2) $$J = \text{diag}\,[J(\lambda(1), e(1)), \ldots, J(\lambda(M), e(M))],$$

where $J(\lambda, e)$ is an exe Jordan block with λ on the diagonal and 1 on the superdiagonal and $N = e(1) + \cdots + e(M)$. For a fixed real number τ, let

$$\{\lambda(1), \ldots, \lambda(K)\} = \text{subset of } \{\lambda(k)\}: \text{Im} = \tau$$

enumerated as follows

$$\text{Re}\,\lambda(1) = \cdots = \text{Re}\,\lambda(K_1) < \text{Re}\,\lambda(K_1+1) = \cdots = \text{Re}\,\lambda(K_2) < \cdots$$
$$= \text{Re}\,\lambda(K_h),$$

$$e(K_i+1) \geqq e(K_i+2) \geqq \cdots \geqq e(K_{i+1}) \text{ for } i = 0, \ldots, h\text{-}1,$$

where $K_0 = 0 < K_1 < \cdots < K_h = K$. Put

$$E(0) = 0 \text{ and } E(i) = \sum_{j=0}^{i=1} e(K_{j+1}) \text{ for } i = 1, \ldots, h.$$

Assumption (A2). Let $\delta_i > 0$ be arbitrary if $i = 1$ and

$$0 < \delta_i < \min\,[\lambda(K_{j+1}) - \lambda(K_j): j = 1, \ldots, i\text{-}1] \text{ if } i > 1.$$

Assumption (A3). The matrix $g(t)$ is representable as a Laplace-Stieltjes transform (1.2), absolutely convergent for $\text{Re}\, t > 0$, where $G(s)$ is locally of bounded variation.

Theorem 2.1. [2]. Assume (A1) - (A3). For some fixed i, $1 \leqq i \leqq h$, assume that (2.1) holds with

$$(2.3) \qquad \gamma \geq m(i) = \max\,[e(K_{j+1}) \;:\; j = 0, \ldots, i-1].$$

Then, for $K_{i-1} < n \leq K_i$ and $\nu = 0, \ldots, e(n) - 1$, (1.1) has a solution representable in the form

$$(2.4) \qquad y = e^{\lambda(n)t}\{\xi_\nu^n(t) + t^\nu w_\nu^n(t) + t^{\nu+E(i-1)} w_\nu^{n0}(t)\},$$

where $\xi^n(t)$ is a vector with components which are polynomials in t of degree $\leq \nu$ and $= \nu$ for at least one component, w_ν^n and w_ν^{n0} are vectors representable as Laplace-Stieltjes transforms of the form

$$(2.5) \qquad w_\nu^n(t) = \int_0^{\delta_i} e^{-st}\, dW_\nu^n(s), \quad W_\nu^n(0) = W_\nu^n(+0) = 0,$$

$$(2.6) \qquad w_\nu^{n0}(t) = \int_{\delta_i}^{\infty} e^{-st}\, dW_\nu^{n0}(s),$$

absolutely convergent for $\operatorname{Re}\, t > 0$ and

$$(2.7) \qquad \int_{+0} s^{-\gamma+m(i)}\, |dW_\nu^n(s)| < \infty .$$

The simple case $A = 0$ and certain cases of $N = 2$ go back to Wintner; cf. [8]- [10]. The exponent $\nu + E(i-1)$ in (2.4) is suggested by the theory of regular singular points.

3. Discussion of proof and consequences. The proof is quite technical and leads to a refined theorem in which assumption (2.1), (2.3) is weakened and assertion (2.7) is strengthened. For reasons of brevity, we shall not state

this refined theorem but refer to it as "Theorem 2.1R" (cf. Theorem 2.1 and the Remark following it in [2]). For example, a consequence of Theorem 2.1R (but not of Theorem 2.1) is the following concerning an N-th order equation, where $D = d/dt$.

Theorem 3.1[4]. _In the N-th order equation_

$$D^k P(-D)\, x + \sum_{j=0}^{N-1} (-1)^{k+j} b_j(t) D^j x = 0,$$

let $0 < k \leqq N$, $P(\lambda) = \lambda^{N-k} + \cdots$ _a polynomial with constant coefficients such that_ $P(\lambda) \neq 0$ _for_ $\lambda \geqq 0$, $b_j(t)$ _a Laplace-Stieltjes transform_

$$b_j(t) = \int_0^\infty e^{-st} dB_j(s), \quad B_j(0) = B_j(+0) = 0,$$

absolutely convergent for $\operatorname{Re} t > 0$, _such that_

$$\int_0 s^{-k+j} |dB_j(s)| < \infty \text{ for } 0 \leq j < k.$$

Then there exist k _solutions_ $x = x_\nu(t)$, $\nu = 0, \ldots, k-1$, _of the form_

$$x_\nu(t) = t^\nu [1 + \int_0^\infty e^{-st} dX_\nu(s)], \quad X_\nu(0) = X_\nu(+0) = 0,$$

absolutely convergent for $\operatorname{Re} t > 0$. _If, in addition,_

$$\int_0 s^{-k+j-\gamma} |dB_j(s)| < \infty \quad \text{for} \quad 0 \leq j < \nu$$ _and some constant_ $\gamma > 0$, _then_ $\int_0 s^{-\gamma} |dX_\nu(s)| < \infty$.

The proof of Theorem 2.1R involves an induction on the order N of the system (1.1). The first step of the induction requires a generalization of Theorem 3.1, $\nu = 0$, to systems (cf. Theorem 1.1 of D'Archangelo [1]) to give K_1 particular solutions with suitable properties [corresponding to $n = 0, \ldots, K_1$ and $\nu = 0$ in (2.4)], say, solutions

$$(3.1) \qquad y_n(t) = e^{\lambda(n)t}[1_0^n + w_0^n(t)],$$

when $1_0^n \neq 0$ is a specified eigenvector of A, $A1_0^n = \lambda(n)1_0^n$,

$$(3.2) \qquad w_0^n(t) = \int_0^\infty e^{-st}dW_0^n(s), \; W_0^n(0) = W_0^n(+0) = 0,$$

absolutely convergent for $\operatorname{Re} t > 0$, and the j-th component W_0^{nj} of W_0^n satisfies

$$\int_0 s^{-\gamma(j)} |dW_0^{nj}(s)| < \infty$$

for a specified constant $\gamma(j) \geqq 0$. After obtaining these K_1 solutions, the order N of the system is reduced by a variation of constants.

In the proof of the existence of the solutions (3.1), we can suppose that $A = J$ is in the Jordan normal form and that $\lambda(1) = \cdots = \lambda(K_1) = 0$. Fix n and write W, W^j for W_0^n, W_0^{nj}. The existence of the solution (3.1) is essentially reduced to the existence of a solution W of a functional equation

$$(3.3) \qquad W = H + TW$$

in a Banach space V and T : $V \to V$ is a linear operator. Here H is a certain column of the matrix $G(t) = (G_{jk}(t))$, V is the Banach space of N-vectors W(t) of bounded variation on $[0, \infty)$, continuous from the right, $W(0) = W(+0) = 0$, with norm

$$\| W \| = \sum_{j=1}^{N} \{ \int_0^1 s^{-c(j)} |dW^j(s)| + \int_1^\infty |dW^j(s)| \} < \infty ,$$

where $c(j) \geqq 0$ is a specified constant, and T is of the form

$$(TW)^j(t) = \sum_i \sum_{k=1}^{N} \int_0^t (s+\lambda_j)^{-c(i,j)} d\{ \int_0^s G_{ik}(s-u) dW^k(u) \},$$

λ_j is the j-th diagonal element in A = J, the first sum is over certain row indices i depending on the Jordan structure of A and on the index j, and $c(i,j) \geqq 0$ is a specified constant. It turns out that there exists a constant C such that

$$\| T \| \leq C \sum_{j=1}^{N} \sum_{k=1}^{N} (\int_0^1 s^{-\gamma(j,k)} |dG_{jk}(s)| + \int_1^\infty |dG_{jk}(s)|) \leq \infty,$$

and again $\gamma(j, k)$ is a specified constant.

If $U > 0$ is a sufficiently large number and the independent variable t is replaced by $t + U$, so that $dG_{jk}(s)$ is replaced by $e^{-sU} dG_{jk}(s)$, then $\| T \| < 1$, and we obtain the existence of W by the contraction principle. This gives the existence of the solution (3.1) but only with the absolute convergence of the integral in (3.2) for $\operatorname{Re} t \geqq U$. In order to obtain the absolute convergence for $\operatorname{Re} t > 0$, we appeal to the following Lemma 3.1. (It can be mentioned that if V is replaced by a space of functions of locally bounded variation, then the functional equation (3.3) can also be solved by successive approximations $W_0 = H$ and $W_{n+1} = H + TW_n$ without the change of variables $t \Rightarrow t + U$ and the use of Lemma 3.1, but the arguments are much more complicated; cf. [1].)

Lemma 3.1. Let B be a constant NxN matrix, h(t) an NxN matrix function representable as a Laplace-Stieltjes transform

$$h(t) = \int_0^\infty e^{-st} dH(s), \quad H(0) = H(+0) = 0, \tag{3.4}$$

absolutely convergent for $\operatorname{Re} t > 0$. Let z(t) be a solution of

$$z' = [B+h(t)]z \tag{3.5}$$

representable as a Laplace-Stieltjes transform

$$z(t) = z(\infty) + \int_0^\infty e^{-st} dZ(s), \quad Z(0) = Z(+0) = 0, \tag{3.6}$$

absolutely convergent for large Re t, where $z(\infty)$ is a constant vector. Then the integral in (3.6) is absolutely convergent for $\operatorname{Re} t > 0$.

This lemma is also used at other points in the remainder for the proof. It can be used, for example, to show that if (2.4) is a solution (1.1) where the Laplace-Stieltjes transforms (2.5), (2.6) are absolutely convergent for large Re t, then they are in fact absolutely convergent for $\operatorname{Re} t > 0$.

After having obtained the solutions (3.1), the proof of Theorem 2.1R is completed by a variation of constants which again involves a number of propositions. Some are simple technical results concerning representations, in terms of Laplace-Stieltjes transforms, of certain functions, for example,

$$\int_T^t (t-s)^k e^{\Delta s} x(s)ds \quad \text{or} \quad \int_T^t s^k e^{\Delta s} x(s)ds ,$$

where $x(s)$ is a given Laplace-Stieltjes transform. Another states that if $Y = I + w$, where $w = (w_{jk}(t))$ is a matrix of Laplace-Stieltjes transforms, absolutely convergent for large Re t,

$$w_{jk}(t) = \int_0^\infty e^{-st} dW_{jk}(s),\ W_{jk}(0) = W_{jk}(+0),\ \int_0 s^{-\sigma(j)} |dW_{jk}(s)| < \infty$$

for $j \neq k$ and $\sigma(j) \geqq 0$, then $Y^{-1} = I + v$, where v has the same properties (with the same $\sigma(j)$) as w.

4. <u>A second order equation.</u> As an application of our results to the study of solutions "in the large," we shall mention some results of [4] on a scalar second order equation

$$(4.1) \qquad y'' + [1 + q(t)]y = 0 ,$$

especially since our original motivation for these papers came from the following: If $J_\mu(t)$, $K_\mu(t)$ are the standard Bessel functions of order μ, then $w = \pi t(J_\mu^2 + K_\mu^2)/2$ has a representation as an explicit definite integral (the "Nicholson integral"; cf. [7], p. 444) for $\operatorname{Re} t > 0$. From this integral, it is easily seen that w is completely monotone for $t > 0$ if $\mu \geqq 1/2$ and, hence, by the Hausdorff-Bernstein theorem, has a representation of the form

$$w(t) = 1 + \int_0^{\infty} e^{-st} dW(s), \quad W(0) = W(+0) = 0, \tag{4.2}$$

$$dW(s) \geqq 0. \tag{4.3}$$

The functions $t^{1/2}J_{\mu}(t)$, $t^{1/2}K_{\mu}(t)$ are linearly independent solutions of the differential equation (4.1), where $q(t) = (1/4-\mu^2)/t^2$. This result was generalized in [3] to show that whenever $q(t)$ has a completely monotone derivative $q'(t)$ and $q(\infty) = 0$, then (4.1) has a pair of linearly independent solutions $y_1(t)$, $y_2(t)$ such that $w = y_1^2 + y_2^2$ has a representation of the type (4.2)-(4.3). The theorems in [3] are actually more general in that they only impose conditions of monotony on a finite number of derivations of q and assert that a corresponding certain number of inequalities $(-1)^n D^n w \geqq 0$ hold. The proof of $(-1)^n D^n w \geqq 0$ for $n = 0, \ldots, m$ involved an existence theorem for each m. Theorem 3.1 above implies that if $q(t)$ has a representation

$$q(t) = \int_0^{\infty} e^{-st} dQ(s), \quad Q(0) = Q(+0) = 0,$$

absolutely convergent for $\mathrm{Re}\, t > 0$ (without any hypothesis on the sign of $dQ(s)$), then (4.1) has a pair of linearly independent solutions $y_1(t)$, $y_2(t)$ such that $w(t) = y_1^2 + y_2^2$ has a representation of the form (4.2). In fact, $W(t)$ is determined in [3] as the unique solution of the Volterra integral equation

$$t(t^2+1)dW(t) = -\ 2t dQ(t) - 2d\{\int_0^t (t+s)Q(t-s)dW(s)\}, \tag{4.4}$$

$W(0) = W(+0) = 0$, and the existence proof shows that $dQ \leqq 0 \Rightarrow dW \geqq 0$ (as in the Bessel case $\mu \geqq 1/2$). If Q has a continuous derivative Q' for $s > 0$, then W also has a continuous derivative for $s > 0$, and (4.4) becomes an integral equation for $W'(t)$, $t > 0$,

$$t(t^2+1)W'(t) = -2tQ'(t) - 2\int_0^t (t+s)Q'(t-s)W'(s)ds.$$

There is an extensive discussion of conditions on Q which permit the implication $Q' \geqq 0 \Rightarrow W' \leqq 0$ (as in the Bessel case $0 < \mu < 1/2$). The paper [4] does not, of course, yield results when we only have information about a finite number of derivatives of q; cf. [3], [6]. It does however yield, for example, a new representation of $J_\mu^2 + K_\mu^2$ as a definite integral for real or purely imaginary μ, simpler than that of Nicholson; namely,

$$\pi(J_\mu^2 + K_\mu^2)/2 = \int_0^\infty e^{-st}P(1+s^2/2)ds \text{ for } \operatorname{Re} t > 0$$

and

$$\pi t(J_\mu^2 + K_\mu^2)/2 = 1 + \int_0^\infty e^{-st}sP^1(1+s^2/2)ds \text{ for } \operatorname{Re} t > 0,$$

where $y = P(\sigma) = P_{\mu-1/2}(\sigma)$ is the unique solution of the Legendre equation

$$D[(1-\sigma^2)Dy] + (\mu^2-1/4)y = 0, \quad D = d/d\sigma,$$

regular at $\sigma = 1$ and satisfying $P(1) = 1$,

$$P(\sigma) = 1 - (1/4-\mu^2)(\sigma-1)/2 + \cdots \text{ and}$$

$$P^1(\sigma) = DP(\sigma).$$

Furthermore, for $\sigma \geqq 1$, we have that $P^1(\sigma) > 0$ if $\mu > 1/2$, $P^1(\sigma) < 0$ if $0 < \mu < 1/2$, and $P^1(\sigma)$ is oscillatory at $\sigma = \infty$ if μ $(\neq 0)$ is purely imaginary.

There is also a similar discussion in [4] of the cases (4.1) with $q(t) = \beta/t^{\gamma}, \gamma > 0$, which is not complete however if $1 < \gamma < 2$.

REFERENCES

[1] J. D'Archangelo, Linear ordinary differential equations with Laplace-Stieltjes transforms as coefficients, Trans. Amer. Math. Soc. 195(1974) 115-145.

[2] J. D'Archangelo and P. Hartman, Integration of ordinary linear differential equations by Laplace-Stieltjes transforms, ibid.

[3] P. Hartman, On differential equations and the function $J_{\mu}^2 + K_{\mu}^2$, Amer. J. Math. 83(1961) 154-188.

[4] P. Hartman, On differential equations, Volterra equations and the functions $J_{\mu}^2 + Y_{\mu}^2$, ibid. 95(1973) 553-593.

[5] P. Hartman, Ordinary Differential Equations, Baltimore (1973).

[6] P. Hartman, Positive and monotone solutions of linear ordinary differential equations, J. Differential Equations.

[7] G. N. Watson, A Treatise on the Theory of Bessel Functions, 2nd ed., Cambridge (1958).

[8] A. Wintner, On the existence of Laplace solutions for linear differential equations of second order, Amer. J. Math. 72 (1950) 442-450.

[9] A. Wintner, On the small divisors in integrations by Laplace transforms, ibid. 73 (1951) 173-180.

[10] A. Wintner, Ordinary differential equations and Laplace transforms, ibid. 79 (1957) 265-294.

The Johns Hopkins University, Baltimore, Maryland

SOLUTIONS NEAR BIFURCATED STEADY STATES

by

F. Hoppensteadt*

Multi-time expansion solutions can be constructed for a class of nonlinear differential equations near bifurcated steady states. The solution is constructed as the sum of expansions for the bifurcated state and one each for slow and fast time transients near the state. The expansion for the transient gives an approximation to the domain of attraction of stable bifurcated states and a description of the short and intermediate time behavior of the solutions.

1. <u>Introduction</u>. An elementary example is presented here which illustrate a procedure for constructing solutions to systems of ordinary differential equations near bifurcated steady states. This method was derived in [1] for studying instabilities in nonlinear diffusion processes and certain fluid flow problems.

Consider a system of two scalar differential equations

$$dx/dt = \mathcal{F}(x, y, \lambda), \quad dy/dt = \mathcal{G}(x, y, \lambda)$$

---

*This research was partially supported by U.S. Army Research Office, Durham, under Grant No. DA-ARO-D-31-124-72-G47.

where $x, y \in R^1$ and λ is a real parameter. Suppose that for some value of λ, say $\lambda = \lambda_0$, this system has a steady state $x = x_0$, $y = y_0$:

$$\mathcal{F}(x_0, y_0, \lambda_0) = 0, \quad \mathcal{G}(x_0, y_0, \lambda_0) = 0.$$

The functions $\mathcal{F}$ and $\mathcal{G}$ are assumed to be smooth (i.e. having derivatives of any order ≥ 2) in the vicinity of the point (x_0, y_0, λ_0).

New variables,

$$u = x - x_0, \quad v = y - y_0, \quad \varepsilon = \lambda - \lambda_0$$

are introduced into the problem with the result that

$$\mathcal{F}(x, y, \lambda) = Au + Bv + F(u, v, \varepsilon)$$

$$\mathcal{G}(x, y, \lambda) = Cu + Dv + G(u, v, \varepsilon)$$

where $A = (\partial \mathcal{F}/\partial x)(x_0, y_0, \lambda_0)$, etc., and $F(u, v, 0) = \mathcal{O}(|u| + |v|)(|u| + |v|)$, etc.

The linear part of the system is determined by the matrix

$$\begin{pmatrix} A & B \\ C & D \end{pmatrix} .$$

This is assumed to be similar to

$$\begin{pmatrix} 0 & 0 \\ 0 & a \end{pmatrix}$$

Thus, an invertible linear transformation $\binom{u}{v} \rightarrow P\binom{u}{v}$ takes the system into an equivalent one whose linear part is determined by Diag $(0, a)$. Without loss of generality, we suppose that the coefficient matrix is already is canonical form (i.e., $P = I$), so

$$du/dt = F(u, v, \varepsilon)$$

(1)

$$dv/dt = av + G(u, v, \varepsilon) .$$

In [1], $\binom{u}{v}$ was taken to be an element of a space of functions and a, F, G were taken as unbounded operators acting in that space.

The solutions of this system which begin near $u = 0$, $v = 0$, for ε near zero will be analyzed. In particular, an initial state is prescribed,

$$u(0) = \xi(\varepsilon), \quad v(0) = \eta(\varepsilon) ,$$

and then its evolution in time is studied, This is done by the following perturbation scheme. First, the steady state problem is studied, and the method of Newton polygons is used to determine all steady states lying near $u = 0$, $v = 0$ for ε near zero. This step determines the correct asymptotic scaling for the problem. Once the problem is rescaled, it appears in a form which is amenable to the method of matched asymptotic expansions, and the theory can be applied to deduce the structure of solutions, their long time behavior, and other qualitative features such as approximate domains of stability.

2. Steady State Problems.

The steady state system corresponding to (1) is

(2) $$F(u, v, \varepsilon) = 0, \qquad av + G(u, v, \varepsilon) = 0.$$

Because of the conditions on $\mathcal{J}$ and $\mathcal{L}$, it follows that

$$F(0,0,0) = 0, \quad G(0,0,0) = 0.$$

The second equation in (2) can be solved for v as a function of u and ε by means of the implicit function theorem.

$$v = v^*(u, \varepsilon)$$

This is a smooth function near u = 0, ε = 0, which satisfies $v^*(0,0) = 0$. Substituting this into the first equation in (2) gives

$$0 = F(u, v^*(u,\varepsilon), \varepsilon) = \sum_{k,j} F_{k,j}\, u^{k}\varepsilon^{j} .$$

The solutions of this equation which satisfy u = 0(1) as ε → 0, can be determined in a constructive way by the method of Newton polygons. Two special, but typical, cases are described:

Suppose that $F_{k,0} = 0$ for k < K, but $F_{K,0} \neq 0$, And either $F_{0,1} \neq 0$ (forced case) or $F_{1,1} \neq 0$ (unforced). In these cases, the solutions can be constructed as power series in the parameter $\mu = \varepsilon^{1/K}$ in the forced case and $\mu = \varepsilon^{1/K}$ in the unforced case. The parameter μ determines the correct asymptotic scale for the problem.

3. Rescaling the Problem.

Suppose now that the initial state is restricted so that

$$\xi(\varepsilon) = \mu\ \Xi(\mu) \quad \text{and} \quad \eta(\varepsilon) = V^*(\xi(\varepsilon),\varepsilon) + \varepsilon H(\mu)$$

where Ξ and H are smooth functions of μ at μ = 0.

This makes precise the restriction that the initial data lie near the bifurcated state.

Next, the problem is rescaled by introducing the new variables

$$u = \mu U, \quad v = v^*(u, \varepsilon) + \varepsilon V$$

where $\varepsilon = \mu^K$ in the forced case and $\varepsilon = \mu^{K-1}$ in the unforced case. The problem (1) becomes

$$(3) \qquad dU/dt = \mu^{K-1} f(U, V, \mu), \quad dV/dt = aV + g(U, V, \mu)$$

where

$$\mu^K f(U, V, \mu) = F(\mu U, v^*(\mu U, \varepsilon) + \varepsilon V, \varepsilon), \text{ etc.}$$

f is a smooth function at $\mu = 0$.

4. Multi-time Analysis. Problems in the form (3) are amenable to the method of matched asymptotic expansions. For example, the following convergence result is proved in [2].

Theorem : Let f, g and μ be as described above. Suppose that the initial value problem

$$dU_0/d\sigma = f(U_0 0, 0), \; U_0(0) = \Xi(0),$$

has a solution $U_0(\sigma)$ for $0 \leq \sigma < \infty$ which approaches a finite limit, U^*, at an exponential rate ($f_U(U^*, 0, 0) < 0$). Also, suppose that the initial value problem

$$dV_0/dt = aV_0 + g(\Xi(0), V_0, 0), \; V_0(0) = H(0),$$

has a solution $V_0(t)$ defined for $0 \leq t < \infty$ which approaches zero at an exponential rate $(a < 0)$. Then the problem (3) has a unique solution, $U(t,\mu)$, $V(t,\mu)$, which is defined for $0 \leq t < \infty$ and satisfies

$$U(0,\mu) = \Xi(\mu), \quad V(0,\mu) = H(\mu).$$

Also,

$$U(t,\mu) = U_0(\mu^{k-1}t) + O(\mu), \; V(t,\mu) = V_0(t) + O(\mu)$$

where the error terms $O(\mu)$ hold uniformly for $0 \leq t < \infty$.

The equation for U_0 has been referred to as the Landau equation for the system, in analogy to its occurrence in the stability theory of fluid flow. It is useful to observe that the theorem determines the domain of attraction of the stable steady state of (3), $U = U^* + O(\mu)$, $V = O(\mu)$, by the zero order problems for U_0 and V_0.

In terms of the original system (1), the theorem shows that

$$u = \mu U^* + \mu[U_0(\mu^{K-1}t) - U^*] + O(\mu^2)$$

$$v = v^*(\mu U^*, \varepsilon) + [v^*(\mu U_0(\mu^{K-1}t), \varepsilon) - v^*(\mu U^*, \varepsilon)]$$

$$+ \varepsilon V_0(t) + O(\mu(\mu + \varepsilon)).$$

Thus, the solution can be written to leading order as the sum of the bifurcated steady state, a term which decays on the slow time scale, $\sigma = \mu^{K-1}t$, and a term which decays on the fast time scale t.

REFERENCES

[1] F. Hoppensteadt, N. Gordon, Asymptotic solution of nonlinear partial differential equations near bifurcated states, (in press)

[2] F. Hoppensteadt, Asymptotic stability in singular perturbation problems II. J. Diff. Eqns. 15(1974) 510-521.

RECENT ADVANCES IN THE ANALYTIC THEORY OF NONLINEAR DIFFERENTIAL EQUATIONS WITH AN IRREGULAR TYPE SINGULARITY

Po-Fang Hsieh

1. Introduction.

Consider a system of nonlinear differential equations:

$$xw' = h(x, w) \tag{1.1}$$

where x is a complex variable, w and h are n-dimensional vectors, h is holomorphic in a neighborhood of $(0, 0)$ and satisfies $h(0, 0) = 0$. Then $x = 0$ is called a Briot-Bouquet type singularity, after C. A. Briot and J. C. Bouquet who proved in [1] the following

Theorem 0. *If (1.1) has a formal solution*

$$w \sim xd_1 + x^2 d_2 + \ldots + x^k d_k + \ldots$$

where d_k are constant n-column vectors, then this power series is convergent in a neighborhood of $x = 0$.

Since then, many mathematicians, such as H. Poincaré, H. Dulac, E. Picard, J. Malmquist, W. J. Trjitzinsky, H. L. Turrittin, M. Hukuhara devoted to the study of the solution of (1.1).

This work is partially supported by a Faculty Research Fellowship, Western Michigan University.

In order to study (1.1) in the case that the Jacobian matrix $h_w(0,0)$ is a singular matrix of the form

$$h_w(0,0) = \begin{pmatrix} 0 & 0 \\ 0 & H \end{pmatrix}$$

where H is a non-singular matrix of lower dimension, it is necessary to study a system of equation of the form of (cf [8]):

(1.2) $$x^{\sigma+1}y' = f(x,y,z), \quad xz' = g(x,y,z).$$

Here we assume that

1) σ is a positive integer;
2) $y = \text{col}(y_1, \ldots, y_m)$, $z = \text{col}(z_1, \ldots, z_n)$;
3) $f = \text{col}(f_1, \ldots, f_m)$, $g = \text{col}(g_1, \ldots, g_n)$ with f_j and g_k being holomorphic in

(1.3) $$|x| < a, \quad \|y\| < b, \quad \|z\| < c,$$

(a, b, c: positive constants, $\|y\|$ denotes $\max\{|y_j|\}$),
and we have

$$f(0,0,0) = 0, \quad g(0,0,0) = 0$$

4) The Jacobian matrix $A = f_y(0,0,0)$ is nonsingular and has a Jordan canonical form in lower triangular form;
5) The eigenvalues $\{\mu_1, \ldots, \mu_n\}$ of the matrix $B = g_z(0,0,0)$ have positive real parts and B has a canonical form; namely

(1.4) $$\text{Re}\,\mu_k > 0 \quad (k = 1, 2, \ldots, n).$$

A system of equations (1.2) satisfying Assumptions 1)

and 3) along with the fact that the matrices A and B are nonsingular is said to have an irregular type singularity at $x = 0$. This paper will survey recent results in the analytic study of the system (1.2) under Assumption 1) through 5).

2. Distinct eigenvalues of A.

M. Hukuhara [5] in 1940 studies the formal solutions of a system of nonlinear equations. Since then, many attempts were made to give analytic meanings to the formal solutions. Recently, M. Iwano contributed a series of study [7, 8, 9, 10, 11, 12] assuring decisive progress in analytic theory of nonlinear differential equations. The author is indebt to Professor Masahiro Iwano for many discussions and sharing his preprints.

For a row vector $p = (p_1, \ldots, p_m)$ of non-negative integers $\{p_j\}$, we shall denote by $|p| = p_1 + \cdots + p_m$, $y^p = y_1^{p_1} y_2^{p_2} \cdots y_m^{p_m}$ and $p \cdot y = p_1 y_1 + \cdots + p_m y_m$. Also denote $1_m(y) = \operatorname{diag}(y_1, \ldots, y_m)$, an m by m diagonal matrix.

An m-vectorial function $f(x, y, z)$ is said to have Property-$\mathfrak{A}$ with respect to y and z for

$$(2.1) \qquad 0 < |x| < a, \quad \underline{\theta} < \arg x < \overline{\theta}, \quad \|y\| < b, \quad \|z\| < c$$

if it is holomorphic in (2.1) and can be expressed as a uniformly convergent power series

$$(2.2) \qquad f(x, y, z) = f_o(x, z) + F(x, z)y + \sum_{|p|=2}^{\infty} y^p f_p(x, z)$$

where f_o and f_p are m-vectors, F is an m by m matrix holomorphic in

(2.3) $$0 < |x| < a, \underline{\theta} < \arg x < \overline{\theta}, \quad \|z\| \leq c$$

expressible in uniformly convergent power series of z with coefficients admitting asymptotic power series expansions in powers of x as x tends to 0 in $\underline{\theta} < \arg x < \overline{\theta}$.

For a given set of complex constants $\{\nu_1, \nu_2, \ldots, \nu_m\}$, let

(2.4) $$\Lambda_j(x) = -\frac{\nu_j}{\sigma x^{\sigma}}, \quad (j = 1, 2, \ldots, m).$$

A sector $\underline{\theta} < \arg x < \overline{\theta}$ is said to have Property-$\mathcal{J}$ with respect to the monomials $\{\Lambda_j(x) \mid j = 1, 2, \ldots, m\}$ if, for each j, there is a direction in this sector on which $\arg \Lambda_j(x) > 0$, and

$$|\arg \Lambda_j(x)| < \frac{3\pi}{2} \pmod{2\pi}$$

for $j = 1, 2, \ldots, 3$ when $\underline{\theta} < \arg x < \overline{\theta}$. It is noteworthy that finite set of monomials, such a sector always exists. As a matter of fact, for a given set of finite monomials (2.4), we can choose $\arg \nu_j (j = 1, \ldots, m)$ suitably on a sheet of Riemann surface that any fixed direction in the x-plane can be contained in a sector which have Property-$\mathcal{J}$ with respect to $\{\Lambda_j(x) \mid j = 1, \ldots, m\}$.

Iwano did the analytic study of (1.2) first in [7] under the following assumptions.

I. $B = 1_n(\mu)$, where $\mu = \text{col}(\mu_1, \mu_2, \ldots, \mu_n)$;

II. For each (1+n) - row vector (ℓ, q) of non-negative integers such that $\ell + |q| \geq 2$, we have

(2.5) $$\mu_k \neq \ell + q \cdot \mu \qquad k = 1, 2, \ldots, n;$$

III. The eigenvalues $\nu_1, \nu_2, \ldots, \nu_m$ of A satisfy

(2.6) $$\mathrm{Re}\nu_1 \geq \mathrm{Re}\nu_2 \geq \cdots \geq \mathrm{Re}\nu_s > 0;$$

IV. The eigenvalues $\nu_1, \nu_2, \ldots, \nu_m$ are mutually distinct;

V. For each m-row vector p of non-negative integers such that $|p| \geq 2$, we have

(2.7) $$\nu_j \neq p \cdot \nu, \qquad j = 1, 2, \ldots, m,$$

where $\nu = \mathrm{col}(\nu_1, \nu_2, \ldots, \nu_m)$.

Theorem 1. (Iwano [7]) Under the Assumptions I ~ V, there exists a sector $\underline{\theta}_1 < \arg x < \overline{\theta}_1$, which has Property-$\mathcal{J}$ with respect to $\{\Lambda_j(x) \mid j = 1, 2, \ldots, m\}$, and a transformation

(2.8) $$y = P_1(x, u, v), \qquad z = Q_1(x, u, v)$$

such that

(i) $P_1(x, u, v)$ and $Q_1(x, u, v)$ are m- and n-columer vectors having Property-$\mathfrak{A}$ with respect to u and v for

(2.9) $$0 < |x| < a_1,\ \underline{\theta}_1 < \arg < \overline{\theta}_1 \quad \|u\| \leq b_1,\ \|v\| \leq c_1,$$

where $0 < a_1 \leq a$, $0 < b_1 \leq b$, $0 < c_1 \leq b$, and satisfy

(2.10) $$\frac{\partial P_1}{\partial u}(0, 0, 0) = 1_m, \qquad \frac{\partial Q_1}{\partial v}(0, 0, 0) = 1_n;$$

(ii) The equations (1.2) are transformed into

(2.11) $x^{\sigma+1}u' = \mathrm{diag}(\lambda_1(x, v), \ldots, \lambda_m(x, v))u, \quad xv' = 1_n(\mu)v$

__where__ $\lambda_j(x, v)$ $(j=1, \ldots, m)$ __are polynomials of degree__ σ __in__ x __with coefficients holomorphic in__ v.

Let

(2.12) $$V(x) = \mathrm{col}\,(x^{\mu_1}C_1, x^{\mu_2}C_2, \ldots, x^{\mu_n}C_n)$$

be a general solution of the second equation (2.11). Here $C_I, \ldots, C_n$ are arbitrary constants. Denote $C = \mathrm{col}(C_1, \ldots, C_n)$. By the fact that

(2.13) $$\int \frac{V(x)^q}{x^{s+1}}\,dx = \begin{cases} \dfrac{V(x)^q}{x^s(q\cdot\mu - s)}, & \text{if } q\cdot\mu \neq s, \\ C^q \log x, & \text{if } q\cdot\mu = s. \end{cases}$$

Then, the first equations of (2.11) has a general solution

(2.14) $$U(x, V(x)) = \mathrm{Diag}\,(u_1, u_2, \ldots, u_n)$$

where

$$u_j = e^{\Omega_j(x,\, V(x))}\, e^{\hat{\lambda}_j(C)}\, C_j$$

with $\Omega_j(x, v)$ polynomials of degree σ in powers of x having coefficients holomorphic in v, $\hat{\lambda}_j(C)$ polynomials of C.

Thus, under the Assumptions I ~ V, the equations (1.2) have a general solution

$$Y = P_1(x, U(x, V(x)), V(x)), \quad z = Q_1(x, U(x, V(x)), V(x))$$

with P_1, Q_1 having the properties described in (i) of Theorem 1.

3. A non-distinct eigenvalue case of A.

In order to investigate the case when the eigenvalues of A are not distinct, instead of III and IV, we shall assume;

III′. Let $\nu_1, \nu_2, \ldots, \nu_s$ be distinct eigenvalues of A with multiplicities $m_1, m_2, \ldots, m_s$ $(m_1+m_2+\cdots+m_s = m)$ respectively, then

$$\text{(3.1)} \qquad \operatorname{Re}\nu_1 > \operatorname{Re}\nu_2 \geq \cdots \geq \operatorname{Re}\nu_s > 0.$$

In this case we need some more assumptions on the Jacobian matrix $f_y(x, y, z)$ in order to obtain a result comparable to Theorem 1. We shall assume:

IV′. When the matrix $f_y(x, 0, z)$ is expanded into power series of x, the doefficients up to the powers of $\sigma - 1$ and that of x^σ at $z = 0$ are all diagonalizable.

In another word, $f_y(x, 0, z)$ can be reduced by a non-singular transformation to an m by m matrix $F(x, z)$ of the form;

$$\text{(3.2)} \qquad F(x, z) = \sum_{j=1}^{s} \oplus \lambda_j(x, z) 1_{m_j} + x^\sigma F_\sigma(z) + x^{\sigma+1} G(x, z)$$

where $\lambda_j(x, z)$ are scalar functions of polynomial in x of degree σ with constant coefficients for x^σ, and $F_\sigma(0) = 0$.

For a given row vector $p = (p_1, p_2, \ldots, p_m)$ of non-negative integers, denote according to the multiplicities of ν_j,

$$(3.2)\quad \begin{cases} \hat{p}_1 = (p_1, \ldots, p_{m_1}),\ \hat{p}_2 = (p_{m_1+1}, \ldots, p_{m_1+m_2}), \ldots, \\ \hat{p}_s = (p_{m-m_s+1}, \ldots, p_m) \end{cases}$$

Let

$$(3.3)\qquad R = \{p \mid \text{there exists an index } j \text{ such that } \nu_j = \sum_{j=1}^{s} |\hat{p}_i| \nu_i,\ |p| \geq 2\}$$

Then, by (3.1), R is a finite set.

Consider the monomials of the form

$$(3.4)\qquad \{\Lambda_{jp}(x) = \Lambda_j(x) - \sum_{i=1}^{s} |\hat{p}_i| \Lambda_i(x) \mid j=1, 2, \ldots, s;\ 2 \leq |p| \leq M\}$$

where M is a sufficiently large positive integer. Since all of these monomials have the same degree with respect to x^{-1}, it is easy to see that a sector which has Property-$\mathcal{T}$ with respect to $\{\Lambda_{jp}(x) \mid j=1, \ldots, s;\ 2 \leq |p| \leq M\}$ when M is large, also has Property-$\mathcal{T}$ with respect to $\{\Lambda_{jp}(x) \mid j=1, \ldots, j;\ |p| \geq 2\}$.

Theorem 2. (Hsieh [3]). Under the Assumptions I, II, III′ and IV′, there exists a sector $\underline{\theta}_2 < \arg x < \overline{\theta}_2$, which has Property-$\mathcal{T}$ with respect to $\{\Lambda_{jp}(x) \mid j=1, \ldots, s;\ |p| \geq 2\}$, and a transformation

$$(3.5)\qquad y = P_2(x, u, v), \quad z = Q_2(x, u, v)$$

such that

(i) $P_2(x, u, v)$ and $Q_2(x, u, v)$ are m- and n-column vectors having Property-$\mathfrak{A}$ with respect to u and v for

$$(3.6)\qquad 0 < |x| < a_2,\ \underline{\theta}_2 < \arg x < \overline{\theta}_2,\ \|u\| \leq b_2,\ \|v\| \leq c_2,$$

where $0 < a_2 \leq a$, $0 < b_2 \leq b$, $0 < c_2 \leq c$, and satisfy

$$\frac{\partial P_2}{\partial u}(0, 0, 0) = 1_m, \quad \frac{\partial Q_2}{\partial v}(0, 0, 0) = 1_n;$$

(ii) The equations (1.2) are transformed into

$$(3.7) \quad \begin{cases} x^{\sigma+1}u' = (\sum_{j=1}^{s} \oplus \lambda_j(x, v)1_{m_j})u + \sum_{p \in R} u^p H_p(x, v), \\ \\ xv' = 1_n(\mu)v \end{cases}$$

where $H_p(x, v)$ are m- column vectors and have Property-$\mathfrak{A}$ with respect to v in

$$(3.8) \quad 0 < |x| < a_2, \ \underline{\theta}_2 < \arg x < \overline{\theta}_2, \ \|v\| < c_2.$$

Furthermore, if for all p in R, the following condition is satisfied:

$$(3.9) \quad \nu_j = \sum_{i=1}^{s} |\hat{p}_i| \nu_i \ \text{implies} \ \lambda_j(x, v) \equiv \sum_{i=1}^{s} |\hat{p}_i| \lambda_i(x, v),$$

then $H_p(x, v)$ are polynomials in x in degree $\leq \sigma$ whose coefficients are holomorphic in $\|v\| < c_2$ for all p in R.

An example that the condition (3.9) being satisfied is:

$$(3.10) \quad \lambda_j(x, v) = \nu_j(1 + x + \cdots + x^\sigma)h(v), \ j=1, 2, \ldots, s)$$

where $h(v)$ is a function holomorphic in $\|v\| \leq c_3$. When (3.9) is satisfies, the system (3.7) can be solved by quadrature. For this purpose, in addition to (2.12) and (2.13), we need

$$(3.11) \quad \int \frac{\log x}{x^{s+1}} dx = -\frac{1}{sx^s}[\log x + \frac{1}{s}], \quad s \neq 0$$

and

$$
(3.12) \quad \int \frac{V(x)^q \log x}{x^{s+1}} dx = \begin{cases} \dfrac{V(x)^q}{(q \cdot \mu - s)x^s} \left[\log x - \dfrac{1}{q \cdot \mu - s}\right], & \text{if } q \cdot \mu \neq s, \\ \dfrac{1}{2} C^q (\log x)^2, & \text{if } q \cdot \mu = s. \end{cases}
$$

Then the equations (3.7)' have a general solution $\{U(x, v(x)), V(x)\}$ where

$$
U(x, v) = \mathrm{col}(\hat{U}_1(x, v), \hat{U}_2(x, v), \ldots, \hat{U}_s(x, v))
$$

with $\hat{U}_j(x, v)$ m_j-column vectors, determined successively from s to 1, in the following form:

$$
\hat{U}_s(x, v) = e^{\Omega_{s-1}(x, v)} x^{\hat{\lambda}_{s-1}(C)} C_s
$$

$$
\hat{U}_s(x, v) = e^{\Omega_1(x, v)} x^{\hat{\lambda}_{s-1}(C)} \{C_{s-1} + \varphi_{s-1}(x; C, C_s, \log x, v)\}
$$

$$
\hat{U}_1(x, v) = e^{\Omega_1(x, v)} x^{\hat{\lambda}_1(C)} \{C_1 + \varphi_1(x; C, C_2, \ldots, C_s, \log x, v)\}.
$$

Here $\Omega_j(x, v)$ are polynomials of degree σ in powers of x having coefficients holomorphic in v for $\|v\| \leq c_2$, $\hat{\lambda}_j(C)$ are polynomials of C. C_j are constant m_j-column vectors, and φ_j are m_j-column vectors whose components are polynomials of x^{-1} with coefficients polynomials of C, $C_{j+1}, \ldots, C_s$, $\log x$ and holomorphic in v for $\|v\| \leq c_2$.

<u>4. A general case.</u> For a more general case, we shall assume III′ and

I′. The eigenvalues $\{\mu_k\}$ of B satisfy

(4.1) $0 < \mathrm{Re}\mu_1 \leq \mathrm{Re}\mu_2 \leq \cdots \leq \mathrm{Re}\mu_n$.

This is just a more specific condition than (1.4).

Without loss of generality, we shall assume that A and B are in lower triangular Jordan canonical forms; namely;

(4.2) $$A = \sum_{i=1}^{s} \oplus \, [\nu_i 1_{m_i} + D_i]$$

and

(4.3) $$B = 1_n(\mu) + E .$$

where

(4.4) $$D_i = \begin{pmatrix} 0 & & & & \\ \delta_{i2} & 0 & & \mathbf{0} & \\ & \cdot & \cdot & & \\ \mathbf{0} & & \cdot & \cdot & \\ & & & \cdot & \cdot \\ & & & \delta_{im_i} & 0 \end{pmatrix}, \quad \delta_{ij} = 1 \text{ or } 0,$$

and

(4.5) $$E = \begin{pmatrix} 0 & & & & \\ \delta_2 & 0 & & \mathbf{0} & \\ & \cdot & \cdot & & \\ \mathbf{0} & & \cdot & \cdot & \\ & & & \cdot & \cdot \\ & & & \delta_n & 0 \end{pmatrix}$$

with $\delta_k = 1$ or 0 and $\delta_k = 1$ only if $\mu_k = \mu_{k-1}$.

Let S be a set of (1+n) - row vectors (ℓ, q) of non-negative integers such that, for some k,

$$(4.6)\quad \begin{cases} q = q_1, \ldots, q_{k-1}, \quad 0, \ldots, 0), \\ \mu_k = \ell + \mu_1 q_1 + \ldots + \mu_{k-1} q_{k-1} . \end{cases}$$

By (4.1), we know that there are only finitely many elements in S. For each (ℓ, q) in S, we shall denote by $b_{\ell q}$ a $b_{\ell q, k} \neq 0$ only if (4.6) is satisfied.

We have

Theorem 3. (Hsieh [4]). Under the Assumptions I′ and III′, for a sector $\underline{\theta}_2 < \arg < \bar{\theta}_2$, given in Theorem 2, which has Property-$\mathcal{T}$ with respect to $\{\Lambda_{jp}(x) \mid j=1, \ldots, s; |p| \geq 2\}$, there exists a transformation

$$(4.7)\qquad y = P_3(x, u, v), \quad z = Q_3(x, u, v)$$

such that

(i) $P_3(x, u, v)$ and $Q_3(x, u, v)$ are m- and n-column vectors having Property-$\mathfrak{U}$ with respect to u and v for

$$(4.8)\qquad 0 < |x| < a_3, \ \underline{\theta}_2 < \arg x < \bar{\theta}_2, \ \|u\| \leq b_3, \ \|v\| \leq c_3$$

where $0 < a_3 \leq a$, $0 < b_3 \leq b$, $0 < c_3 \leq c$ and satisfy

$$(4.9)\qquad \frac{\partial P_3}{\partial u}(0, 0, 0) = 1_m, \quad \frac{\partial Q_3}{\partial v}(0, 0, 0) = 1_n ;$$

(ii) The transformation (4.7) reduces (1.2) into

$$(4.10)\quad \begin{cases} x^{\sigma+1} u' = \Sigma \oplus (\nu_i 1_{m_i} + D_i + C_i(x, v))u + \sum_{p \in R} u^p H_p(x, v) \\ xv' = (1_n(\mu) + E)v + \sum_{(\ell, q) \in S} x^\ell v^q b_{\ell q} \end{cases}$$

where $C_i(x, v)$ are m_i by m_i matrices, $H_p(x, v)$ are m-column vectors having Property-$\mathfrak{A}$ with respect to v in

$$(4.11) \qquad 0 < |x| < a_3, \ \underline{\theta}_2 < \arg x < \overline{\theta}_2, \ \|v\| \le c_3,$$

and, in particular,

$$(4.12) \qquad C_i(0, 0) = 0 \qquad (i = 1, 2, \ldots, s).$$

The detail of the proof of these theorems are in the respective references. Even the domains, such as (4.8), are given in nicely shaped forms, actually in the process of proof we need to modify the circular boundaries in order to assume the validity of the problem and analyticity of the transformation. Such a modified domain is called a Stable domain. The idea of Stable domains goes back to A. M. Liapunov, M. Hukuhara, M. Nagumo, P. Hartman et. al for the study of existence theorems and stability conditions. For our problem, M. Iwano [9, 10, 11] modifies Hukuhara's method [6] to construct the suitable stable domains.

Once the stable domain is found, these theorems can be proved by a Tychonoff type fixed point theorem as in [7, 11, 12] or by a successive approximation method as in [2, 3, 4].

5. Open problem.

As pointed out in §2, the location of a sector which has Property-$\mathcal{J}$ gives no serious restriction on the problem. Thus, the condition (3.1) on III′ poses no difficulty. However, the relaxation of (4.1) remains a challenge to all interested persons.

REFERENCES

[1] C. A. A. Briot and J. C. Bouquet, Recherches sur les proprietes des fonctions definies par des equations differentielles, J. Ecole Poly. Tome 21, Cahier 36 (1856).

[2] P. F. Hsieh, Successive approximations method for solutions of nonlinear differential equations at an irregular type singular point, Comment. Math. Univ. St. Pauli, 20 (1971), 27-53.

[3] P. F. Hsieh, A general solution of a system of nonlinear differential equations at an irregular type singularity, Funk. Ekv. 16 (1973), 103-136.

[4] P. F. Hsieh, On a system of nonlinear differential equations at an irregular type singularity, (to appear).

[5] M. Hukuhara, Integration formelle dún systѐme déquations differentielles nonlineaires dans le voisinage dun point singulier, Ann. Mat. Pura Appl. (4) 19 (1940), 35-44.

[6] M. Hukuhara, Sur les points singuliers des equations differentielles lineaires, III, Mem. Fac. Sei. Kyusyu Imp. Univ., Ser. A, 2 (1941), 125-137.

[7] M. Iwano, Analytic expression for bounded solutions of non-linear ordinary differential equations with an irregular type singular point, Ann. Mat. Pura Appl. (4) 82 (1969), 189-256.

[8] M. Iwano, A general solution of a system of nonlinear ordinary differential equations $xy' = f(x, y)$ in the case

when $f_y(0, 0)$ is the zero matrix, Ann. Mat. Pura Appl. (4) 83 (1969), 1-42.

[9] M. Iwano, Determination of stable domains for bounded solutions of simplified equations, Funk. Ekv. 12 (1969), 251-268.

[10] M. Iwano, Bounded solutions and stable domains of nonlinear ordinary differential equations, Analytic Theory of Differential Equations, Lecture notes in Math. 183, (1971), 59-127, Springer-Verlag.

[11] M. Iwano, Analytic integration if a system of nonlinear ordinary differential equations with an irregular type singularity, Ann. Mat. Pura Appl. (4) 94 (1972).

[12] M. Iwano, Analytic integration of a system of nonlinear ordinary differential equations with an irregular type singularity, II., (to appear).

Western Michigan University, Kalamazoo, Michigan

NAGUMO CONDITIONS FOR ORDINARY DIFFERENTIAL EQUATIONS

Joan E. Innes and Lloyd K. Jackson*

1. Introduction.

The classical Nagumo condition for the second order differential equation

$$(1) \qquad y'' = f(x, y, y')$$

is a growth condition on $f(x, y, y')$ which guarantees that solutions of (1) become unbounded on their maximal intervals of existence. If (1) is a scalar equation with $f(x, y, y')$ continuous on $(a, b) \times R^2$ and if for each $M > 0$ and each compact interval $[c, d] \subset (a, b)$ there is a corresponding positive continuous function $\phi(s)$ such that $|f(x, y, y')| \leq \phi(|y'|)$ for $c \leq x \leq d$ and $|y| \leq M$ and such that $\int_0^\infty \frac{s\,ds}{\phi(s)} = +\infty$, then solutions of (1) either extend to (a, b) or become unbounded on their maximal intervals of existence.

This property of solutions along with the assumed existence of solutions of certain types of differential inequalities plays an important role in demonstrating the existence of solutions of boundary value problems not only

* Partially supported by National Science Foundation Grant GP-34532.

for second order equations but for higher order equations as well, see for example, [1], [2], [3].

In Section 2 of this paper we consider the third order equation

$$(2) \qquad y''' = f(x, y, y', y'')$$

in which $f(x, y, y', y'')$ is continuous on $(a, b) \times R^3$ and we determine a growth condition on the function $f(x, y, y', y'')$ which insures that solutions of (2) either extend to (a, b) or become unbounded on maximal intervals of existence.

In Sections 3 and 4 we illustrate the use of this property of solutions in dealing with boundary value problems and in establishing a compactness condition for bounded sets of solutions of (2).

2. A Nagumo condition for third order equations.

Assume that in the differential equation

$$(2) \qquad y''' = f(x, y, y', y'')$$

the function $f(x, y, y', y'')$ is continuous on $(a, b) \times R^3$. For each real number $p \geq 0$ define the function $\phi_p(s)$ on $0 \leq s < +\infty$ by $\phi_p(s) = \text{Max}\{1, s^p\}$. The following theorem establishes a Nagumo condition for equation (2).

Theorem 1. Assume that corresponding to each compact interval $[c, d] \subset (a, b)$ and each positive number M there is an $h > 0$ such that

$$(3) \qquad |f(x, y, y', y'')| \leq h\, \phi_p(|y'|)\phi_q(|y''|)$$

for $c \leq x \leq d$, $|y| \leq M$ where $0 \leq p \leq 1$, $q \geq 0$, and

$p + 2q \leq 3$. Then every solution of (2) either extends to (a, b) or becomes unbounded on its maximal interval of existence.

Proof. Assume that the conditions of the theorem are satisfied and that (2) has a solution $y(x)$ which does not extend to (a, b) but is bounded on its maximal interval of existence. In particular assume that $y(x)$ is a solution on $[c, d)$ where $a < c < d < b$, that $[c, d)$ is right maximal, and that there is an $M > 0$ such that $|y(x)| \leq M$ on $[c, d)$. It suffices to consider only this case since a left maximal interval can be treated in a similar way. Let $h > 0$ be a constant such that (3) is satisfied for $c \leq x \leq d$ and $|y| \leq M$.

Since $[c, d)$ is right maximal, $y''(x)$ cannot be bounded on $[c, d)$. We will consider separately the two possible cases: $\lim_{x \to d} |y''(x)| = +\infty$ and $\lim_{x \to d} |y''(x)|$ doesn't exist. Our arguments are based on the fact that the solution satisfies

$$(4) \qquad |y'''(x)| \leq h\, \phi_p(|y'(x)|)\, \phi_q(|y''(x)|)$$

on $[c, d)$. Consequently, since $-y(x)$ also satisfies (4) on $[c, d)$, it suffices to consider the two cases

$$\lim_{x \to d} y''(x) = +\infty$$

and

$$-\infty \leq \liminf_{x \to d} y''(x) < \limsup_{x \to d} y''(x) = +\infty .$$

First assume $\lim_{x \to d} y''(x) = +\infty$. Then there is an x_o, $c \leq x_o < d$, such that $y''(x) \geq 1$ on $[x_o, d)$. Hence, it follows from (4) that

$$(y''(x))^{1-q} y'''(x) \leq h\, \phi_p(|y'(x)|) y''(x)$$

on $[x_o, d)$ which implies

$$\int_{x_o}^{x} (y''(s))^{1-q} y'''(s)ds \leq h \int_{x_o}^{x} \phi_p(|y'(s)|)y''(s)ds$$

for $x_o \leq x < d$. If $0 \leq q \leq 2$, the integral on the left becomes positively infinite as x approaches d. It follows that $y'(x)$ cannot be bounded on $[x_o, d)$, and, since $y''(x) > 0$ on $[x_o, d)$, we conclude that $y'(x) \to +\infty$ as $x \to d$. Thus there is an x_1, $x_o \leq x_1 < d$, such that $y'(x) \geq 1$ on $[x_1, d)$. Then for $x_1 \leq x < d$ we have

$$\int_{x_1}^{x} (y''(s))^{1-q} y'''(s)ds \leq h \int_{x_1}^{x} (y'(s))^p y''(s)ds$$

which, if $0 \leq q < 2$, yields

$$y''(x) \leq c_1 |y'(x)|^k + c_2$$

where c_1 and c_2 are positive constants and $k = \dfrac{p+1}{2-q}$. Finally from this last inequality we obtain

$$\int_{x_1}^{x} \frac{y'(s)y''(s)ds}{c_1|y'(s)|^k + c_2} \leq y(x) - y(x_1)$$

for $x_1 \leq x \leq d$. If $k \leq 2$, that is, $p + 2q \leq 3$, this contradicts the boundedness of $y(x)$ on $[c, d)$ since the integral on the left then becomes positively infinite as $x \to d$. If $q = 2$, we do not obtain a contradiction in this way.

Now assume $\liminf_{x \to d} y''(x) < \limsup_{x \to d} y''(x) = +\infty$. In this case we can choose an $\alpha \geq 1$ and sequences $\{x_n\}$, $\{t_n\}$ in $[c, d)$ such that $x_n < t_n < x_{n+1}$ for each n,

$\lim x_n = d$, $y''(x_n) = \alpha$ for each n, $y''(x) > \alpha$ on (x_n, t_n) for each n, and $y''(t_n) \to +\infty$ as $n \to \infty$. Then, since $[c, d)$ is maximal and $|y(x)| \leq M$ on $[c, d)$, it follows that

$|y'(x_n)| \to +\infty$ as $n \to \infty$. Since we can pick subsequences and renumber, if necessary, we can assume that $y'(x_n) \to +\infty$ or $y'(x_n) \to -\infty$. We will consider these two subcases separately.

Assume that $y'(x_n) \to +\infty$ in which case with out loss of generality we can assume $y'(x) \geq 1$ on $[x_n, t_n]$ for each n. We then have on integrating (4)

$$\int_{\alpha}^{y''(x)} s^{1-q} ds \leq h \int_{y'(x_n)}^{y'(x)} s^p \, ds$$

on $[x_n, t_n]$ for each $n \geq 1$ from which it follows that

$$\frac{1}{2-q}\{(y''(x))^{2-q} - \alpha^{2-q}\} \leq \frac{h}{p+1}\{(y'(x))^{p+1} - (y'(x_n))^{p+1}\}$$

on $[x_n, t_n]$ for each $n \geq 1$. Since $0 \leq q < 2$, we conclude that $[y'(t_n)]^{p+1} - [y'(x_n)]^{p+1} = v_n \to +\infty$ as $n \to +\infty$. Solving the above inequality for $y''(x)$ we obtain

$$y''(x) \leq \{\frac{(2-q)h}{p+1}[(y'(x))^{p+1} - (y'(x_n))^{p+1}] + \alpha^{2-q}\}^{\frac{1}{2-q}}.$$

From this it follows that

$$\int_{y'(x_n)}^{y'(t_n)} \frac{s \, ds}{\{\frac{(2-q)h}{p+1}[s^{p+1} - (y'(x_n))^{p+1}] + \alpha^{2-q}\}^{\frac{1}{2-q}}} \leq y(t_n) - y(x_n) \leq 2M$$

for each n. Setting $u = s^{p+1} - (y'(x_n))^{p+1}$, we convert this inequality to

$$\int_0^{v_n} \frac{[u + (y'(x_n))^{p+1}]^{\frac{1-p}{1+p}} du}{(p+1)\{\frac{(2-q)h}{p+1}u + \alpha^{2-q}\}^{\frac{1}{2-q}}} \leq 2M$$

for each $n \geq 1$. If $p \leq 1$ and $\frac{1-p}{1+p} - \frac{1}{2-q} \geq -1$, this leads to a contradiction for sufficiently large n since $v_n \to +\infty$ as $n \to \infty$. Hence, if $0 \le q < 2$, $0 \le p \le 1$, and $p + 2q \le 3$, this case is impossible.

Now consider the remaining sub-case in which $y'(x_n) \to -\infty$ as $n \to \infty$. We can assume $y'(x_n) < -1$ for all n. For each n let $s_n = t_n$ if $y'(t_n) \le -1$, and, if $y'(t_n) > -1$, let s_n be chosen such that $x_n < s_n < t_n$, $y'(s_n) = -1$, and $y'(x) < -1$ on $[x_n, s_n)$. Then on $[x_n, s_n]$ we have

$$\int_{\alpha}^{y''(x)} s^{1-q} ds \le h \int_{y'(x_n)}^{y'(x)} (-s)^p ds$$

which implies

$$(y''(x))^{2-q} - \alpha^{2-q} \le \frac{(2-q)h}{p+1} \{(-y'(x_n))^{p+1} - (-y'(x))^{p+1}\}$$

on $[x_n, s_n]$. From this inequality and the definition of the sequence $\{s_n\}$ we conclude that $w_n = |y'(x_n)|^{p+1} - |y'(s_n)|^{p+1} \to +\infty$ as $n \to \infty$. Solving the above inequality for $y''(x)$, multiplying by $-y'(x)$, and then integrating from x_n to s_n and making suitable changes of variable, we obtain

$$\int_0^{w_n} \frac{[|y'(x_n)|^{p+1} - u]^{\frac{1-p}{1+p}} du}{(p+1)\{\frac{h(2-q)}{p+1} u + \alpha^{2-q}\}^{\frac{1}{2-q}}} \le 2M$$

for each n. Since $0 \le u \le |y'(x_n)|^{p+1} - u$ on $0 \le u \le \frac{1}{2} w_n$, it follows that, if $p \le 1$, then

$$\int_0^{\frac{1}{2}w_n} \frac{u^{\frac{1-p}{1+p}}du}{p+1\{\frac{h(2-q)}{p+1}u+\alpha^{2-q}\}^{\frac{1}{2-q}}} \leq 2M$$

for each n. As before this is contradictory for sufficiently large n if $\frac{1-p}{1+p} - \frac{1}{2-q} \geq -1$. Thus the second subcase is also impossible if $0 \leq q < 2$, $0 \leq p \leq 1$, and $p + 2q \leq 3$.

This completes the proof of the theorem.

The restriction $p \leq 1$ is probably the result of the cumbersome methods used in the proof of the theorem. It is likely that the theorem is true if p and q satisfy the inequalities $p \geq 0$, $q \geq 0$, and $p + 2q \leq 3$. The theorem is not true if $p \geq 0$, $0 \leq q < 2$, and $p + 2q > 3$. In this case the equation $y''' = +|y'|^P |y''|^q$ has a solution $y'(x) = \alpha(d-x)^{-r}$ where α is a suitable constant and $r = \frac{2-q}{p+q-1}$ satisfies $0 < r < 1$.

3. Compactness of bounded collections of solutions.

In recent years there has been considerable research concerned with the question of when the uniqueness of solutions of boundary value problems for nonlinear ordinary differential equations implies the existence of solutions of such problems, for example, [4], [5], [6], [7], [8], and [9]. In much of this work it is either assumed or proven that solutions of the differential equation satisfy the following compactness condition which we are formulating for an equation of order n: (K) If $[c, d]$ is a compact interval and if $\{y_k(x)\}$ is a sequence of solutions such that $|y_k(x)| \leq M$ on $[c, d]$ for some $M > 0$ and all $k \geq 1$, then there is a subsequence $\{y_{k_j}(x)\}$ such that $\{y_{k_j}^{(i)}(x)\}$ converges uniformly on $[c, d]$ for each $0 \le i \le n-1$.

In [5] the following theorem is proven concerning this compactness condition for equations of order 3.

Theorem 2. Suppose that the differential equation

$$y''' = f(x, y, y', y'') \tag{2}$$

satisfies the following conditions:

(A) $f(x, y, y', y'')$ is continuous on $(a, b) \times R^3$,

(B) All solutions of (2) extend to (a, b), and

(C) If $y(x)$ and $z(x)$ are any solutions of (2) such that $y(x_i) = z(x_i)$ for $i = 1, 2, 3$ for some $a < x_1 < x_2 < x_3 < b$, then $y(x) \equiv z(x)$ on $[x_1, x_3]$.

Then any bounded sequence of solutions of (2) on any compact subinterval of (a, b) satisfies the compactness condition (K).

The condition (C) asserts that 3-point boundary value problems for (2) have at most one solution.

We consider now a weakened form of Theorem 2.

Theorem 3. Assume that equation (2) satisfies the conditions:

(A) $f(x, y, y', y'')$ is continuous on $(a, b) \times R^3$,

(B)* Each solution of (2) either extends to (a, b) or is unbounded on its maximal interval of existence, and

(C)* Each point of (a, b) is contained in an open interval on which 3-point boundary value problems for (2) have at most one solution.

Then any bounded sequence of solutions of (2) on any compact subinterval of (a, b) satisfies the compactness condition (K).

Proof. Let $[c, d]$ be a compact subinterval of (a, b) and assume that $\{y_k(x)\}$ is a sequence of solutions on $[c, d]$ which is uniformly bounded on $[c, d]$. It suffices to show that there is a subsequence $\{y_{k_j}(x)\}$, a sequence $\{x_j\} \subset [c, d]$, and an $N > 0$ such that $|y'_{k_j}(x_j)| + |y''_{k_j}(x_j)| \leq N$ for all $j = 1, 2, \ldots$. For in this case the Kamke Theorem can be applied to obtain a subsequence of $\{y_{k_j}(x)\}$ which converges uniformly along with its first and second derivative sequence on compact subintervals of the maximal interval existence of the limit solution. Because of the assumption $(B)^*$ and the boundedness of the sequence $\{y_k(x)\}$, the convergence would take place on $[c, d]$.

Thus to prove the theorem it suffices to show that the uniform convergence of $|y'_k(x)| + |y''_k(x)|$ to $+\infty$ on $[c, d]$ as $k \to \infty$ is not possible. In the proof of Theorem 2 in [5] this was done in two steps. First it was shown that assumptions (B) and (C) imply that 2-point boundary value problems for (2) have at most one solution. If one examines the argument used in [5], it is easily seen that $(B)^*$ and (C) imply that solutions of 2-point boundary value problems for (2) on (a, b) have at most one solution. The second step in [5] is to show that, if 2-point boundary value problems on (a, b) have at most one solution, then $|y'_k(x)| + |y''_k(x)| \to +\infty$ uniformly on $[c, d]$ as $k \to \infty$ is not possible. If (C) is replaced by $(C)^*$, let $x_0 \in (c, d)$

and let (α, β) be an open interval containing x_o on which 3-point boundary value problems have at most one solution. Then as remarked above in the presence of $(B)^*$ we can conclude that 2-point boundary value problems on (α, β) have at most one solution. It follows that, if $[c_1, d_1]$ is a compact subinterval of $[c,d]$ such that $[c_1, d_1] \subset (\alpha, \beta)$, then $|y_k'(x)| + |y_k''(x)| \to +\infty$ uniformly on $[c_1, d_1]$ as $k \to \infty$ is not possible. As remarked at the beginning this suffices for the proof of the theorem.

Corollary: If equation (2) satisfies (A) and $(C)^*$ of Theorem 3 and the Nagumo condition of Theorem 1, then solutions satisfy the compactness condition (K) on compact subintervals of (a, b).

As an example the hypotheses of the corollary are satisfied by the equation

$$y''' = x^2 y' + y'|y''|. \tag{5}$$

First it is clear that condition (A) and the Nagumo condition are satisfied. The second order equation

$$z'' = x^2 z + z|z'| = f(x, z, z') \tag{6}$$

is such that f is non-decreasing in z for fixed x and z' and f satisfies a Lipschitz condition in z' on compact subsets. It is shown in [1] that this implies that 2-point boundary value problems for (6) with boundary data $z(x_1) = z_1$, $z(x_2) = z_2$ have at most one solution. It is easy to see that this implies that 3-point boundary value problems for (5) have at most one solution.

Theorem 3 is false if the condition concerning the

uniqueness of solutions of 3-point boundary value problems is omitted. All solutions of the differential equation

$$(7) \qquad y''' = -(y')^3$$

extend to the whole real line. The sequence of solutions $\{y_k(x)\}$ of (7) in which $y_k(x)$ satisfies initial conditions

$$y_k(0) = 0, \quad y_k'(0) = 0, \quad y_k''(0) = k$$

is uniformly bounded.

4. A boundary value problem.

As remarked earlier the assumption that solutions of the differential equation either extend or become unbounded plays an important role in existence theorems for boundary value problems. In this section we give an illustration of such an existence theorem and an example of a particular boundary value problem in which the hypotheses of the theorem are satisfied. The following theorem is one of the results proven in [3].

Theorem 4. Assume that $f(x, y, y', y'')$ is continuous on $[a, b] \times R^3$, that $f(x, y, y', y'')$ is nonincreasing in y for each fixed x, y', y'', and that each solution of

$$(2) \qquad y''' = f(x, y, y', y'')$$

either extends to $[a, b]$ or become unbounded. Assume that there are functions $\phi, \psi \in C^{(3)}[a, b]$ such that $\phi(x) \leq \psi(x)$, $\phi'(x) \leq \psi'(x)$, $\phi'''(x) \geq f(x, \phi(x), \phi'(x), \phi''(x))$, and $\psi'''(x) \leq f(x, \psi(x), \psi'(x), \psi''(x))$ on $[a, b]$. Then for α, β, γ with $\phi(a) \leq \alpha \leq \psi(a)$, $\phi'(a) \leq \beta \leq \psi'(a)$, and $\phi'(b) \leq \gamma \leq \psi'(b)$, there is a solution $y(x)$ of (2) such that $y(a) = \alpha$, $y'(a) = \beta$, and

$y'(b) = \gamma$. Furthermore, $\phi(x) \le y(x) \le \psi(x)$ and $\phi'(x) \le y'(x) \le \psi'(x)$ on $[a, b]$.

The theorem is proven by constructing a sequence of functions $\{f_k(x, y, y', y'')\}$ such that each function is uniformly bounded on $[a, b] \times R^3$ and such that the sequence converges uniformly to $f(x, y, y', y'')$ on compact subsets of $[a, b] \times R^3$. Then it is shown that for each $k = 1, 2, \ldots,$ there is a solution $y_k(x)$ of the boundary value problem

$$y''' = f_k(x, y, y', y'')$$

$$y(a) = \alpha,\ y'(a) = \beta,\ y'(b) = \gamma$$

with $\phi(x) \le y_k(x) \le \psi(x)$ and $\phi'(x) \le y_k'(x) \le \psi'(x)$ on $[a, b]$. The Kamke Theorem can then be applied to the sequence $\{y_k(x)\}$ to obtain the result.

The differential equation

(8) $$y''' = -y|y'| + y'y''$$

satisfies

$$|-y|y'| + y'y''| \le (M+1)\phi_1(|y'|)\phi_1(|y''|)$$

for $c \le x \le d$ and $|y| \le M$ where $[c, d]$ is any compact interval. Hence, by Theorem 1 each solution extends to $(-\infty, +\infty)$ or becomes unbounded on its maximal interval of existence. Thus (8) satisfies the hypotheses of Theorem 4. Obviously, $\phi(x) \equiv 0$ is a solution of (8) and $\psi(x) = e^{x^2}$ satisfies $\psi'''(x) \le -\psi(x)|\psi'(x)| + \psi'(x)\psi''(x)$ on $[1, \infty)$. Hence, for any $[a, b] \subset [1, \infty)$ and any α, β, γ satisfying $0 \le \alpha \le e^{a^2}$, $0 \le \beta \le 2ae^{a^2}$, and $0 \le \gamma \le 2be^{b^2}$ the equation (8) has a solution $y(x)$ satisfying $y(a) = \alpha$, $y'(a) = \beta$, $y'(b) = \gamma$,

$0 \leq y(x) \leq e^{x^2}$ on $[a, b]$, and $0 \leq y'(x) \leq 2xe^{x^2}$ on $[a, b]$.

REFERENCES

[1] L. K. Jackson, Subfunctions and second order differential inequalities, Advances in Math. 2(1968), 307-363.

[2] K. Schrader, Existence theorems for second order boundary value problems, Journal of Diff. Equs. 5(1969), 572-584.

[3] G. Klaasen, Differential inequalities and existence theorems for second and third order boundary value problems, Journal of Diff. Equs. 10(1971), 529-537.

[4] A. Lasota and Z. Opial, On the existence and uniqueness of solutions of a boundary value problem for an ordinary second order differential equation, Colloq. Math. 18(1967), 1-5.

[5] L. K. Jackson and K. Schrader, Existence and uniqueness of solutions of boundary value problems for third order differential equations, Journal of Diff. Equs. 9(1971), 46-54.

[6] K. Schrader and P. Waltman, An existence theorem for nonlinear boundary value problems, Proc. Amer. Math. Soc. 21(1969), 653-656.

[7] P. Hartman, On n-parameter families and interpolation problems for nonlinear ordinary differential equations, Trans, Amer. Math. Soc. 154(1971), 201-226.

[8] G. Klaasen, Existence theorems for boundary value problems for nth order ordinary differential equations,

Rocky Mount. J. Math. 3(1973), 457-473.

[9] L. K. Jackson, Existence and uniqueness of solutions of boundary value problems for third order differential equations, Journal of Diff. Equs. 13(1973), 432-437.

University of Nebraska, Lincoln, Nebraska

PERIODIC SOLUTIONS OF ORDINARY DIFFERENTIAL EQUATIONS AS FIXED POINTS OF THE TRANSLATION MAPPING NEAR CERTAIN CRITICAL CASES

Michel Jean

Introduction

Let $\begin{array}{l} \mathbb{R} \to \mathbb{R} \\ t \mapsto q(t,z,Z) \end{array}$ be the solution of the differential equation $\dot{x}(t) = F(t, x(t))$, $x(t) \in \mathbb{R}^n$, $F(t+T,Y) = F(t,y)$, with initial conditions (z, Z) (conditions insuring the existence and uniqueness are assumed). Several constructive methods have been used to exhibit the T-periodic solutions of E. One first method is the Galerkin's method where a periodic function is decomposed on a bases. Usually the Fourier coefficients are exhibited as the roots of an algebraic system by the Newton's method. This method has been developed by CESARI [1], URABE [2]. In other methods fixed points are sought directly in a space of periodic functions by a Newton's method, ANTOSIEWICZ [3], D'HONDT [5]. It is known that $t \mapsto q(t, z, Z)$ is a T-periodic solution if and only if Z is a fixed point of the so called translation mapping $\begin{array}{l} \Theta: \mathbb{R}^n \to \mathbb{R}^n \\ Y \mapsto q(z+T, z,Y) \end{array}$. In a previous paper [6], we have used a Newton's method to locate the fixed points of the translation mapping and given existence and non existence theorems. Let us denote by $\Theta'(Z)$ the derivative of Θ at a fixed point Z, by j the identity mapping. The Newton's method applied only if $j - \Theta'(Z)$ is a one-to-one mapping. If $\Theta'(Z)$ has one eigenvalue equal to 1, the Newton's method fails (It should be noted that when this critical case arises, similar critical cases arise also in the methods mentioned above). We develop here a method allowing us to assert the existence of a fixed point of Θ near a critical case, we state an existence theorem, and we give a numerical example. One should mention other successful and more or less constructive methods in critical cases. One is based on the existence of a pseudo inverse, see MAWHIN [7], STRASBERG [8]. The projection method developed by ANTOSIEWICZ [4] can yield results in certain critical cases. SCHMITT [9] has used for $n = 2$ a method based on the index of Seifert which happens to be quite constructive.

Notations.

$\mathbb{R}$ will be the real line. The set of continuous linear mappings from $\mathbb{R}^n$ into $\mathbb{R}^n$ will be denoted $\mathcal{L}(\mathbb{R}^n, \mathbb{R}^n)$, or more briefly by $\mathcal{L}$. j will be the identity from $\mathbb{R}^n$ onto $\mathbb{R}^n$. We do not specify the norm $\|\cdot\|$ in $\mathbb{R}^n$. The norm in $(\mathbb{R}^n, \mathbb{R}^n)$ is the one induced by $\| \ \|$, and will be denoted with the same symbol. The symbol $\circ$ of composition of mappings will be often omitted. Also we omit writing canoncial immersions.

Let $F : \mathbb{R} \times \mathbb{R}^n \Rightarrow \mathbb{R}^n$ be a mapping locally bounded.

$$(t,y) \to R(t,y)$$

Furthermore, for every $f : I \subseteq \mathbb{R} \to \mathbb{R}^n$ continuous, $\begin{matrix} \mathbb{R} \to \mathbb{R}^n \\ t \to F(t, f(t)) \end{matrix}$ is regulated. We say a continuous mapping $f: I \subseteq \mathbb{R} \to \mathbb{R}^n$ is a solution of E

$$\mathring{x}(t) = F(t, x(t))$$

if f is differentiable almost everywhere and, $\dot{f}(t)$ being the derivative at t, if

$$\dot{f}(t) = F(t, f(t))$$

or z being a point of I

$$\forall t \quad f(t) = f(z) + \int_z^t F(s, f(s))\, ds \,.$$

We shall suppose that F satisfies conditions insuring that for every $(z, Y) \in \mathbb{R} \times \mathbb{R}^n$ there exists a unique solution of E, defined on $\mathbb{R}$, with value Y at z. This solution will be denoted $q(\ \ , z, Y)$ and its value at time t, $q(t, z, Y)$. Of course $q(z, z, Y) = Y$. We introduce now the variational equation of E. We suppose that for every t, the mapping $\begin{matrix} \mathbb{R}^n \to \mathbb{R}^n \\ Y \to F(t,Y) \end{matrix}$ has a Frechet derivative. We denote by $J(t,Y) \in \mathcal{L}$ the derivative at Y. We suppose that for every continuous $f : I \subseteq \mathbb{R} \to \mathbb{R}^n$, $\begin{matrix} I \to \mathcal{L} \\ t \to J(t, f(t)) \end{matrix}$ is regulated. Furthermore J will be locally bounded and locally Lipschitzian in Y. Then, there exists one, and only one, continuous mapping from $\mathbb{R}$ into $\mathcal{L}$, that we denote by $A(\ \ , z, Y)$, the value at t being $A(t, z, Y)$, differentiable almost everywhere and satisfying

$$\overset{\circ}{A}(t, z, Y) = J(t, q(t, zY)) \circ A(t, z, Y)$$

$$A(z, z, Y) = j$$

or equivalently

$$\forall t \quad A(t, z, Y) = j + \int_z^t J(s, q(s, z, Y)) \circ A(s, z, Y)ds$$

$A(\ , z, Y)$ is called the fundamental solution of the variational equation of E, along $q(\ , z, Y)$. For every t, z, $A(t, z, Y)$ is a linear homeomorphism from $\mathbb{R}^n$ onto $\mathbb{R}^n$.

We have the property that for every t, z the mapping

$$\begin{aligned} \mathbb{R}^n &\to \mathcal{L} \\ Y &\to A(t,x,Y) \end{aligned}$$ is locally Lipschitzian.

The translation mapping.

We suppose from now on, that the mapping F is T-periodic, i. e.

$$\forall t \in \mathbb{R}, \ Y \in \mathbb{R}^n, \quad F(t+T, Y) = F(t, Y).$$

It is known that a necessary and sufficient condition for the solution $q(\ , z, Z)$ to be T-periodic is that Z be a fixed point of the so called translation mapping.

$$\begin{aligned} \Theta : \mathbb{R}^n &\to \mathbb{R}^n \\ Y &\to q(z+T, z, Y) \end{aligned}$$

With the assumptions stated in the previous paragraph, Θ is a homeomorphism from $\mathbb{R}^n$ onto $\mathbb{R}^n$. Furthermore Θ is differentiable and its derivative at Y is

$$\Theta'(Y) = A(z+T, z, Y)$$

Our purpose is to seek the periodic solutions of E, as fixed points of the translation mapping near a point $Z^o \in \mathbb{R}^n$. First we shall state various theorems. Later we shall give comments about the way to apply them in practical cases.

Theorem 1 (existence).

Let $Z^o \in \mathbb{R}^n$ and let S be the closed ball with center Z^o and radius η_o. Let be $A \in \mathcal{L}$. Suppose that $j-A$ has an inverse $(j-A)^{-1}$. Let $\underline{e}$, $\underline{m}$, σ, δ, be positive numbers such that

$$\|(j-A)^{-1}(\Theta(Z^o) - Z^o)\| \leq \underline{e}$$

$$\|(j-A)^{-1}\| \leq \underline{m}$$

$$\|\Theta'(Z^o) - A\| \leq \sigma$$

$$\sup_{Y \in S} \|\Theta'(Y) - \Theta'(Z^o)\| \leq \delta$$

If the following inequalities are true

$$\underline{m}(\sigma + \delta) < 1$$

$$\eta_o \geq \eta'' \quad \text{where} \quad \eta'' \geq \frac{\underline{e}}{1 - \underline{m}(\sigma + \delta)}$$

then Θ has a fixed point in the closed ball with center Z^o and radius η'' and it is unique in S.

Theorem 2 (non existence).

Let Z^o, Z, S be as in theorem 1. Let $\overline{e}$, $\overline{m}$, σ, δ, be positive numbers such that

$$\bar{e} \leq \| \Theta(Z^o) - Z^o \|$$

$$\| j-A \| \leq \bar{m}$$

$$\| \Theta'(Z^o) - A \| \leq \sigma$$

$$\sup_{Y \in S} \| \Theta'(Y) - \Theta'(Z^o) \| \leq \delta$$

If the following inequalities are true

$$\sigma + \delta + \bar{m} > 0$$

$$0 < \eta' \leq \eta_o \quad \text{where} \quad \eta' \leq \frac{\bar{e}}{\sigma + \delta + \bar{m}}$$

then Θ has no fixed point in the closed ball with center Z^o and radius η'.

The proof of these theorems is straightforward and can be found in [6]. About theorem 1, it is proved that the mapping $\Phi : \mathbb{R}^n \rightarrow \mathbb{R}^n$

$$\Phi(Y) = Z^o + (j-A)^{-1}(\Theta(Y) - \Theta(Z^o)-A(Y-Z^o) + \Theta(Z^o) - Z^o)$$

is contractive in S.

We shall develop now the case when there is one (actually only one) eigenvalue of A equal to 1, or close to 1, the others being "different enough from 1". If this eigenvalue is 1, j-A has no inverse, and the theorem 1 cannot be applied. If this eigenvalue is different from 1, but close to 1, j-A does have an inverse but $\| (j-A)^{-1} \|$ is large and the condition $\eta_o \geq \frac{\underline{e}}{1 - \underline{m}(\delta + \sigma)}$ would require small $\underline{e}$, δ, σ. It may not be possible to fulfill this requirement.

We shall suppose that $\mathbb{R}^n$ is the direct algebraic sum of two supplementary spaces H_1, H_2, $\mathbb{R}^n = H_1 \oplus H_2$ where H_1 and H_2 are invariant for A; $A(H_1) = H_1$, $A(H_2) = H_2$.
We shall use the following mappings

$$A_1 : H_1 \to H_1 \quad , \quad A_2 : H_2 \to H_2$$
$$Y \mapsto A(Y) \qquad\qquad Y \mapsto A(Y)$$

$$j_1 : H_1 \to H_1 \quad , \quad j_2 : H_2 \to H_2$$
$$Y \mapsto Y \qquad\qquad Y \mapsto Y$$

We shall denote respectively $pr_1 : \mathbb{R}^n \to H_1$, $pr_2 : \mathbb{R}^n \to H_2$ the projection mappings on H_1, H_2, i.e. $Y \in \mathbb{R}^n$ admits a unique decomposition $Y = Y_1 + Y_2$ where $Y_1 = pr_1 Y \in H_1$, $Y_2 = pr_2\, Y \in H_2$. We have the obvious identities

$$A = A_1 pr_1 + A_2 pr_2$$

The norm in H_1, H_2 will be the norm induced by the norm in $\mathbb{R}^n$ (still denoted $\| \ \|$).

We shall suppose that $j_1 - A_1$ is a bijection. The aim of the following theorem is to exhibit a set which is a part of a continuous manifold with dimension equal to dim H_2 where

$$pr_1(\Theta(Y) - Y) = 0$$

for every Y in this set.

<u>Theorem 3.</u>

Let $Z^o \in \mathbb{R}^n$, A be as defined before. We suppose that $j_1 - A_1$ is a one-to-one mapping. Let $P_1 \in H_1$ be

$$P_1 = (j_1 - A_1)^{-1}\, pr_1\, (\Theta\,(Z^o) - Z^o)$$

Let $\eta_1, \eta_2 > 0$ and let $\Sigma_1 \subseteq H_1$, $\Sigma_2 \subseteq H_2$ be

$$\Sigma_1 = \{\, Y \in H_1 : \ \|Y\| \leq \eta_1 \}$$

$$\Sigma_2 = \{\, Y \in H_2 : \ \|Y\| \leq \eta_2 \}$$

respectively, the closed ball in H_1 with center 0 and radius η_1, and the closed ball in H_2 with center 0 and radius η_2. Suppose that there exists positive numbers m_1, δ, e_1 such that

$$\|P_1\| \leq e_1$$

$$\|(j_1 - A_1)^{-1}\| \leq m_1$$

$$\sup_{X \in \Sigma_1 + \Sigma_2} \|\Theta'(Z^o + X) - \Theta'(Z^o)\| \leq \delta$$

$$\|\Theta'(Z^o) - A\| \leq \sigma$$

If the following relations are satisfied,

$$m_1(\delta + \sigma) < 1$$

$$e_1 + m_1(\delta + \sigma)\,\eta_2 \leq (1 - m_1(\delta + \sigma))\,\eta_1$$

then there exists a continuous mapping $g : \Sigma_2 \rightarrow \Sigma_1$ such that

$$\forall\, Y_2 \in \Sigma_2 \quad pr_1\, \Theta(Z^o + Y_2 + g(Y_2)) = pr_1(Z^o + Y_2 + g(Y_2))$$

Proof.

Suppose $M \in \mathcal{L}$ with $M = M_1\, pr_1 + M_2\, pr_2$ where M_1 is a linear bijection from H_1 onto H_1 and M_2 is a linear mapping from H_2 into H_2. Let

$$\psi : \mathbb{R}^n \to H_1$$
$$\xi \mapsto \psi(\xi)$$

$$\psi(\xi) = pr_1(-M\,\Theta\,(Z^o + \xi) + M\,\Theta\,(Z^o) + M\,\xi)$$

$$P_1 = M_1\, pr_1\, (\Theta\,(Z^o) - Z^o)$$

then

$$pr_1\,(\Theta(Z^o + \xi) = pr_1(Z^o + \xi)$$

if and only if

$$\psi(\xi) = P_1 .$$

We choose

$$M_1 = (j_1 - A_1)^{-1}\, pr_1 \;, \quad M_2 = 0$$

so that

$$M = (j_1 - A_1)^{-1}\, pr_1$$

Consider the mapping

$$\varphi : \mathbb{R}^n \to H_1 \qquad \varphi(\xi) = -\psi(\xi) + pr_1\xi$$
$$\xi \mapsto \varphi(\xi)$$

φ is differentiable and its derivative at the point ξ is

$$\varphi'(\xi) = pr_1\, M(\Theta'(Z^o + \xi) - j) + pr_1$$

$$= pr_1\, M(\Theta'(Z^o + \xi) - \Theta'(Z^o)) + pr_1 M(\Theta'(Z^o) - A)$$

$$+ pr_1(M(A-j) + j).$$

We note that with the choice $M = (j_1 - A_1)^{-1}\, pr_1$

$$pr_1(M(A-j) + j) = 0 .$$

We may write

$$\|\varphi'(\xi)\| \leq \|M\| \; \|\Theta'(Z^o + \xi) - \Theta'(Z^o)\|$$

$$+ \|M\| \; \|\Theta'(Z^o) - A\| \quad \sup_{X \in \Sigma_1 + \Sigma_2} \|\varphi'(X)\| \leq (\delta + \sigma) m_1$$

As $\Sigma_1 + \Sigma_2$ is a convex set, we have the following estimations with the mean value theorem

$$\forall \xi, \xi' \in \Sigma_1 + \Sigma_2 \quad \|\varphi(\xi) - \varphi(\xi')\| \leq \|\xi - \xi'\| \sup_{X \in \Sigma_1 + \Sigma_2} \|\varphi'(X)\|$$

(1)

$$\|\varphi(\xi) - \varphi(\xi')\| \leq \|\xi - \xi'\| \; m_1(\delta + \sigma)$$

Let us recall the well known Banach theorem: let X be a normed space, Y a Banach space, $U = \{x \in X: \|x\| \leq a\}$, $V = \{y \in Y: \|y\| \leq b\}$. Let $f: U \times V \to Y$ be continuous and suppose there is a $\lambda \in [0,1[$ such that

$$\forall x \in U, \forall y, y' \in V \quad \|f(x, y) - f(x, y')\| \leq \lambda \|y - y'\|$$

If

$$\forall x \in U \qquad \|f(x, 0)\| \leq b(1 - \lambda)$$

then there exists a unique continuous mapping $g: U \to V$ such that

$$\forall x \in U \qquad g(x) = f(x, g(x))$$

We apply this theorem to the continuous mapping

$$\Sigma_2 \times \Sigma_1 \to H_1$$

$$(x_2, x_1) \mapsto f(x_2, x_1)$$

$$f(x_2, x_1) = \varphi(x_2+x_1) + P_1$$

The inequality (1) and the hypotheses imply that the conditions of Banach's fixed point theorem are satisfied and there exists a continuous mapping $g: \Sigma_2 \to \Sigma_1$ such that

$$\forall Y_2 \in \Sigma_2 \qquad \varphi(Y_2 + g(Y_2)) + P_1 = g(Y_2)$$

$$\psi(Y_2 + g(Y_2)) = P_1$$

$$pr_1 \,\Theta(Z^o + Y_2 + g(Y_2)) = pr_1(Z^o + Y_2 + g(Y_2))$$

In the second theorem we shall give sufficient conditions for the existence of a fixed point in the case where dim H_1 equals 1. The proof is based on the following argument: if the conditions of the previous theorem are satisfied there exists an arc and for every Y belonging to this arc

$$pr_1(\Theta(Y) - Y) = 0$$

If we may exhibit two points Y^1, Y^2 on the arc where $pr_2(\Theta(Y^1)-Y^1)$, $pr_2(\Theta(Y^2)-Y^2)$ have "opposite sign", there is a point $\hat{Y}$ on the arc where $pr_2(\Theta(\hat{Y}) - \hat{Y}) = 0$. Then

$\Theta(\hat{Y}) = \hat{Y}$.

Theorem 4.

We assume that $\dim H_2 = 1$. We suppose that H_2 has for basis a unit vector h. If $Y \in H_2$, we denote by RY the component of Y, i.e, the unique real number r such that $Y = rh$. We have $\|Y\| = |RY|$. We assume that the conditions of theorem 1 are satisfied. Let $Z^1, Z^2 \in Z^o + \Sigma_1 + \Sigma_2$, $\alpha > 0$ be such that

$$\|pr_1(Z^1 - Z^o)\| \leq \alpha < \eta_1 \qquad \|pr_1(Z^2 - Z^o)\| \leq \alpha < \eta_1$$

$$\|pr_2(Z^1 - Z^o)\| < \eta_2 \qquad \|pr_2(Z^2 - Z^o)\| < \eta_2$$

If

$$\|pr_2(\Theta(Z^1) - Z^1)\| > (\delta + \sigma)(\eta_1 + \alpha) \quad \|pr_2(\Theta(Z^2) - Z^2)\| > (\delta + \sigma)(\eta_1 + \alpha)$$

$R\,pr_2(\Theta(Z^1) - Z^1)$ and $R\,pr_2(\Theta(Z^2) - Z^2)$ have opposite sign then Θ has a fixed point in $\Sigma_1 + \Sigma_2$.

Proof.

We may write the identity for every $\xi_1 \in H_1$

$$R\,pr_2(\Theta(Z^1 + \xi_1) - (Z^1 + \xi_1)) - R(pr_2\,\Theta(Z^1 + \xi_1) - pr_2\,\Theta(Z^1) - pr_2\xi_1)$$
$$= R\,pr_2(\Theta(Z^1) - Z^1)$$

and the inequality

$$\|pr_2\,\Theta(Z^1+\xi_1)-pr_2\,\Theta(Z^1)-pr_2\,\xi_1\| \leq \|pr_2\,\Theta(Z^1+\xi_1)$$

$$-pr_2\,\Theta(Z^1)-pr_2\,\Theta'(Z^o)\xi_1\| + \|pr_2\,\Theta'(Z^o)\,\xi_1-pr_2 A\xi_1\|$$

$$+\|pr_2(A-j)\,\xi_1\|$$

We have $pr_2(A-j)\,\xi_1 = 0$. The mean value theorem yields for every ξ_1

$$\|\Theta(Z^1+\xi_1)-\Theta(Z^1)-\Theta'(Z^o)\,\xi_1\| \leq \|\xi_1\| \sup_{X\in[0,\xi_1]} \|\Theta'(Z^1+X)-\Theta'(Z^o)\|$$

where $[0,\xi_1]$ is the closed segment with extremities $0,\xi_1$;

$$\sup_{X\in[0,\xi_1]} \|\Theta'(Z^1+X)-\Theta'(Z^o)\| =$$

$$\sup_{X\in[0,\xi_1]} \|\Theta'(Z^o+Z^1-Z^o+X)-\Theta'(Z^o)\| =$$

$$\sup_{X\in[Z^1-Z^o,\,Z^1-Z^o+\xi_1]} \|\Theta'(Z^o+Y)-\Theta'(Z^o)\|$$

$Z^1 - Z^o$ is in $\Sigma_1+\Sigma_2$. We have for every ξ_1 with $Z^1 - Z^o + \xi_1 \in \Sigma_1 + \Sigma_2$

$$\sup_{Y\in[Z^1-Z^o,\,Z^1-Z^o+\xi_1]} \|\Theta'(Z^o+Y)-\Theta'(Z^o)\| \leq \delta$$

At last we have for every $\xi_1 \in H_1$ with $Z^1-Z^o+\xi_1 \in \Sigma_1+\Sigma_2$,

$\|\xi_1\| \leq \eta_1 + \alpha$

$$\| pr_2 \Theta(Z^1 + \xi_1) - pr_2 \Theta(Z^1) - pr_2 \xi_1 \| \leq (\delta + \sigma)(\eta_1 + \alpha)$$

(2)

$$| R(pr_2 \Theta(Z^1 + \xi_1) - pr_2 \Theta(Z^1) - pr_2 \xi_1) | \leq (\delta + \sigma)(\eta_1 + \alpha)$$

(1), (2), and our assumptions imply that for every $\xi_1 \in H_1$, with $Z^1 - Z^o + \xi_1 \in \Sigma_1 + \Sigma_2$, $\|\xi_1\| \leq \alpha + \eta_1$, $R\, pr_2(\Theta(Z^1 + \xi_1) - (Z^1 + \xi_1))$ has the same sign as $R\, pr_2(\Theta(Z^1) - Z^1)$. In particular it is true for $\xi_1 = \xi^1 = pr_1(Z^o - Z^1) + g(pr_2(Z^1 - Z^o))$. Indeed as $pr_2(Z^1 - Z^o) \in \Sigma_2$, $g(pr_2(Z^1 - Z^o))$ is defined and in Σ_1 (theorem 1) and we have

$$\|\xi^1\| \leq \|pr_1(Z^o - Z^1)\| + \|g(pr_2(Z^1 - Z^o))\| \leq \alpha + \eta_1$$

$$Z^1 - Z^o + \xi^1 = pr_2(Z^1 - Z^o) + g(pr_2(Z^1 - Z^o)) \in \Sigma_1 + \Sigma_2$$

In the same way we can assert that $R\, pr_2(\Theta(Z^2 + \xi^2) - (Z^2 + \xi^2))$ has the same sign as $R\, pr_2(\Theta(Z^2) - Z^2)$ where $\xi^2 = pr_1(Z^o - Z^2) + g(pr_2(Z^2 - Z^o))$. Now, consider the continuous real function h defined on the interval $[a, b] \subseteq \mathbb{R}$, where $a = R\, pr_2(Z^1 - Z^o)$, $b = R\, pr_2(Z^2 - Z^o)$ (we assume $a < b$ for instance) with values

$$h(x) = R\, pr_2(\Theta(Z^o + x + g(x)) - (Z^o + x + g(x))$$

We have

$$h(a) = R\ pr_2(\Theta(Z^1+\xi^1) - (Z^1+\xi^1))$$

$$h(b) = R\ pr_2(\Theta(Z^2+\xi^2) - (Z^2+\xi^2))$$

As h(a) and h(b) have opposite sign, there is an $\hat{x} \in [a, b] \subseteq \Sigma_2$ such that $h(\hat{x}) = 0$

$$pr_2(\Theta(Z^o + \hat{x} + g(\hat{x})) - (Z^o + \hat{x} + g(\hat{x})) = 0$$

We have also by theorem 1

$$pr_1(\Theta(Z^o + \hat{x} + g(\hat{x})) - (Z^o + \hat{x} + g(\hat{x})) = 0$$

Finally

$$\Theta(Z^o + \hat{x} + g(\hat{x})) = Z^o + \hat{x} + g(\hat{x})$$

Some comments are necessary to explain the way the previous theorems can be applied in practical cases. The estimation of δ is based on the following lemma.

Lemma.

Suppose that there exists a mapping

$$\ell: \mathbb{R} \times \mathbb{R}^n \times \mathbb{R}_+ \to \mathbb{R}_+ \qquad (t, Z, \xi) \mapsto \ell(t, Z, \xi)$$

such that for every continuous mappings $f: \mathbb{R} \to \mathbb{R}^n$, $t \mapsto f(t)$, $r: \mathbb{R} \to \mathbb{R}_+$, $t \mapsto r(t)$, the mapping

$$\mathbb{R} \to \mathbb{R}_+ \qquad t \mapsto \ell(t, f(t), r(t))$$

is regulated and locally bounded.

Furthermore ℓ is monotone, non decreasing with respect to ξ, and

$$\forall t \in \mathbb{R},\ Z,\ Y \in \mathbb{R}_n \quad \|J(t,Y) - J(t,Z)\| \leq \ell(t, Z, \|Y - Z\|)$$

Let S be the closed ball with center Z^o and radius η_o

$$q : \mathbb{R} \to \mathbb{R}^n \qquad q(t) = q(t, z, Z^o)$$
$$t \mapsto q(t)$$

$$a : \mathbb{R} \to \mathcal{L} \qquad a(t) = A(t, z, Z^o)$$
$$t \mapsto a(t)$$

$$\overline{w} : \mathbb{R} \to \mathbb{R}_+ \qquad \overline{w}(t) = \sup_{Y \in S} \| A(t,z,Z^o)^{-1} A(t,z,Y) - j \|$$
$$t \mapsto w(t)$$

$$g : \mathbb{R} \times \mathbb{R}_+ \to \mathbb{R}_+$$
$$(t,y) \mapsto g(t,y)$$

$$g(t,y) = \| a(t)^{-1} \| \, \| a(t) \| (1+y) \, \ell(t, q(t), \eta_o \, \| a(t) \| (1+y))$$

Suppose that the differential equation

$$\dot{x}(t) = g(t, x(t))$$

possesses a maximal solution u_m defined on $[z, z+T]$. Then $\overline{w}$ is dominated by u_m and one has the estimation

$$\sup_{Y \in S} \| \Theta'(Y) - \Theta'(Z^o) \| \le \| \Theta'(Z^o) \| \, u_m(z+T)$$

The proof of this lemma can be found in [6].

Numerical example.

We give now a numerical example for a Duffing equation as an application of theorems 3 and 4. Numerical applications of theorems 1 and 2 are developed in [6]. We consider the Duffing equation

(1) $$\ddot{x}(t) + 0.2\,\dot{x}(t) + 0.25\,x(t)^3 = E \cos(0.5\,t)$$

or the equivalent equation in $\mathbb{R}^2$

(1′) $$\dot{x}_1(t) = x_2(t)$$
$$\dot{x}_2(t) = -0.2\,x_2(t) - 0.25\,x_1(t)^3 + E \cos(0.5\,t)$$

with $E = 2.97\,965$

We seek the periodic solutions with period 4π.

We have $\Theta(Y) = q(4\pi, 0, Y)$. Before giving more detail, some general explanations are needed. Z being a point in $\mathbb{R}^2$, we shall have to compute an approximate value $\overline{\Theta(Z)}$ of $\Theta(Z)$. The method used here is the Runge-Kutta method performed on the equation (1').

We shall need also a number ϵ_1 such that

$$\| \overline{\Theta(Z)} - \Theta(Z) \| \leq \epsilon_1$$

ϵ_1 acts as an upper bound of the errors in computation. It can be estimated by various procedures such as "deferred approach to the limit". In a similar way we shall have to compute an approximate value $\overline{\Theta'(Z)}$ of $\Theta'(Z)$ with a Runge-Kutta method performed on the variational equation of (1').

A number ϵ_2 such that

$$\| \overline{\Theta'(Z)} - \Theta'(Z) \| \leq \epsilon_2$$

will be estimated. S being a closed ball with center 0 and radius η_o, we shall need to exhibit a number δ such that

$$\sup_{X \in S} \| \Theta'(Z_o+X) - \Theta'(Z_o) \| \leq \delta$$

This number δ will be constructed in computing an approximate value $\overline{u_m(T)}$ of $u_m(T)$ with a Runge-Kutta method performed on the equation $\dot{x}(t) = g(t, x(t))$ introduced in the previous lemma. ϵ_3 such that

$$\| \overline{u_m(T)} - u_m(T) \| \leq \epsilon_3$$

will be estimated. Any number greater than

$$\overline{(u_m(T)} + \epsilon_3)\,(\|\Theta'(Z)\| + \epsilon_2)$$

is an admissible δ. The explicit form of g and more detail can be found in [6].

The norm used in $\mathbb{R}^2$ is the Euclidean norm. The computation has been made on UNIVAC 1110 with a 4th order Runge-Kutta method, double precision, steps 2×10^{-2}, 10^{-2}, 0.5×10^{-2}

The procedure proceeds as follows.

We take

$$Z^o = (3.41\,85718\,40,\ 1.14\,23125\,10)$$

A computation gives an approximate value $\overline{\Theta'(Z^o)}$ of $\Theta'(Z^o)$ we only mention the first figures)

$$\overline{\Theta'(Z^o)} = \begin{pmatrix} 0.158560\ldots & 0.086273\ldots \\ 0.754768\ldots & 0.921537\ldots \end{pmatrix}$$

and we estimate

$$\|\Theta'(Z^o) - \overline{\Theta'(Z^o)}\| \leq \epsilon_2 = 10^{-6}$$

We make the choice

$$A = \Theta'(Z^o)$$

so that we have

$$\sigma = \epsilon_2 = 10^{-6}$$

We note that approximate values $\bar{\lambda}_1, \bar{\lambda}_2$ of the eigenvalues λ_1, λ_2 of A are

$$\bar{\lambda}_1 = 0.08108\ldots \qquad \bar{\lambda}_2 = 0.99901\ldots$$

and H_1 has for basis the corresponding unit eigenvector h_1 with approximate value

$$\bar{h}_1 = (0.744015\ldots, -0.668162\ldots)$$

and H_2 has for basis the corresponding unit eigenvector h_2 with approximate value

$$\bar{h}_2 = (0.102114\ldots, 0.994772\ldots)$$

We choose as Σ_1 the closed ball in H_1 with center 0 and radius $\eta_1 = 4 \times 10^{-6}$, and as Σ_2 the closed ball in H_2 with center 0 and radius $\eta_2 = 10^{-4}$. Then $\Sigma_1 + \Sigma_2$ is included in S the closed ball in $\mathbb{R}^2$ with center 0 and radis $\eta_o = 1.1 \times 10^{-4}$. We then proceed to the construction of δ associated to S.

An approximate value $\overline{u_m(T)}$ of $u_m(T)$ is

$$\overline{u_m(T)} = 0.01419$$

and we estimate

$$|u_m(T) - \overline{u_m(T)}| \leq \epsilon_3 = 10^{-5}$$

We have $\|A\| \leq 1.20\,311$. So we find

$$\delta = 0.01709$$

and

$$\delta + \sigma \leq 0.0171$$

We take as upper bound m_1 of $(j_1 - A_1)^{-1}$ a number greater than $(-1-\lambda_1)^{-1}$

$$m_1 = 1.0883$$

so that

$$m_1(\delta + \sigma) \leq 0.0187$$

$$0.9813 \leq 1 - m_1(\delta + \sigma)$$

The computation gives an approximate value $\overline{P}_1$ of $P_1 = (j_1 - A_1)^{-1} pr_1(\Theta(Z^o) - Z^o)$

$$\overline{P}_1 = 0$$

and we estimate

$$\| P_1 - \overline{P}_1 \| \leq \epsilon_4 = 10^{-9}$$

so that

$$\| P_1 \| \leq e = 10^{-9}$$

We have

$$10^{-9} + 0.0187 \times 10^{-4} < 0.0392 \times 10^{-4} < 0.9813.4 \times 10^{-6}$$

i.e.

$$e + m_1(\delta + \sigma)\eta_2 < (1 - m_1(\delta + \sigma))\eta_1$$

The condition of theorem 3 is satisfied.

Now we take for Z^1 a computed value of $Z^o + \Delta \ell\, h_2$, and for Z^2 a computed value of $Z^o - \Delta\ell\, h_2$, where $\Delta \ell = 10^{-4}$, i.e.

$$Z^1 = (3.41858205146 \quad , \quad 1.14241198726)$$

$$Z^2 = (3.41856162854 \quad , \quad 1.14221303274)$$

We estimate

$$| \, pr_1(Z^{-1} - Z^o) \, | \leq \alpha = 10^{-11}$$

$$| \, pr_1(Z^2 - Z^o) | \leq \alpha = 10^{-11}$$

$$| \, pr_2(Z^1 - Z^o) | \leq \Delta 1 + 10^{-11} = 10^{-4} + 10^{-11}$$

$$| \, pr_2(Z^2 - Z^o) | \leq \Delta 1 + 10^{-11} = 10^{-4} + 10^{-11}$$

so that $\underline{Z^1, Z^2 \text{ are in } \Sigma_1 + \Sigma_2}$. We compute an approximate value $\overline{pr_2(\Theta(Z^1) - Z^1)}$ of $pr_2(\Theta(Z^1) - Z^1)$

$$\overline{pr_2(\Theta(Z^1) - Z^1)} = -0.103 \ 10^{-6}$$

and we estimate

$$\overline{| \, pr_2(\Theta(Z^1) - Z^1)} - pr_2(\Theta(Z^1) - Z^1) | \leq \epsilon_5 = 10^{-9}$$

We compute also an approximate value $\overline{pr_2(\Theta(Z^2) - Z^2)}$ of $pr_2(\Theta(Z^2) - Z^2)$

$$pr_2(\Theta(Z^2) - Z^2) = 0.094 \times 10^{-6}$$

and we estimate

$$\overline{| pr_2(\Theta(Z^2)-Z^2) - pr_2(\Theta(Z^2)-Z^2)|} \leq \epsilon_5 = 10^{-9}$$

So we have

$$R\ pr_2(\Theta(Z^1) - Z^1) \leq -0.102\times 10^{-6}$$

$$R\ pr_2(\Theta(Z^2) - Z^2) \geq 0.093\times 10^{-6}$$

$$(\delta + \sigma)(\eta_1 + \alpha) \leq 0.0685$$

It follows from theorems 3 and 4 that provided the estimations of ϵ are right there is a fixed point by Θ in the set $\Sigma_1 + \Sigma_2$.

Remark 1

We find as approximate value $\overline{\Theta(Z^o)}$ of $\Theta(Z^o)$

$$\overline{\Theta(Z^o)} = (3.418571839\ ,\ 1.142312505)$$

and we estimate

$$\| \Theta(Z^o) - \overline{\Theta(Z^o)} \| \leq 10^{-9}$$

We estimate that $\bar{\lambda}_1$ and $\bar{\lambda}_2$ are approximations of the eigenvalues of $\Theta'(Z^o)$ up to 10^{-5}.

Remark 2

It has been observed by SIMOND [10], following UEDA [11], that for E = 2, 975, there exist two stable periodic solutions with even harmonic components and one unstable periodic solution. For E = 3.025 the periodic solutions with even harmonic components have disappeared and one stable periodic solution remains with odd harmonics. SIMOND has established the existence by the Galerkin Urabe procedure. Our case $E = 2.97965$, $\lambda_2 \simeq 0.99901$ is close to the critical case corresponding to the change of situation i.e. the vanishing of even harmonic components.

Remark 3.

To locate Z^o one has made use of a procedure suggested by theorems 3 and 4. A Y^o is given, a is computed, and the corresponding eigenvectors h_1, h_2 are computed. Let us consider the sequence in $H_1 \oplus H_2$ defined by

$$pr_1 Y^{n+1} = pr_1 Y^n + (j_1 - A_1)^{-1} pr_1(\Theta(Y^n) - Y^n)$$

$$pr_2 Y^{n+1} = pr_2 Y^n + \frac{\Delta L}{n+1} \text{ sgn}(pr_2(\Theta(Y^n) - Y^n) \times pr_2(\Theta(Y^o) - Y^o))$$

where the function $x \to \text{sgn } x$ is defined by sgn $x = +1$ if $x > 0$, sgn $x = -1$ if $x < 0$, sgn $x = 0$ if $x = 0$. ΔL is given and the following condition must be fulfilled

$$\text{sgn } pr_2 Y^1 = - \text{sgn } pr_2 Y^o$$

The sequence ($pr_1 Y^n$) is convergent thanks to the contraction principle which implies that $pr_1(\Theta(Y^n) - Y^n)$ tends to zero. The sequence ($pr_2 Y^n$), (actually an improved form of ($pr_2 Y^n$)) is monitored by sgn $pr_2(\Theta(Y^n) - Y^n)$ and $pr_2(\Theta(Y^n) - Y^n)$ approaches zero.

Remark 4

In the case where there is more than one eigenvalue equal to 1, dim $H_2 > 1$, we think a theorem similar to theorem 4 can be written, in using in the subspace H_2 where the contraction fails, the fixed point theorem: let X be a normed space, $K \subseteq X$ a compact convex set, $T : K \to X$ a continuous mapping. If

$$(\forall x \in K)\ (\exists \lambda \in]-1, +1[)(\lambda x + (1-\lambda)\ Tx \in K)$$

then T has a fixed point. This is the degenerate form of the theorem which has been used to prove theorem 4.

REFERENCES

[1] L. Cesari, Functional analysis and periodic solutions of nonlinear differential equations. Contrib. Diff. Eqs 1, 149, (1963).

[2] M. Urabe, Galerkin's procedure for nonlinear periodic systems. Archs ration. Mech. Analysis, 20, 120, (1965).

[3] H. A. Antosiewicz, Newton's method and boundary value problems. J. Comp. Syst. Sci., 2, 177-202, (1968).

[4] H. A. Antosiewicz, A general approach to linear problems for nonlinear ordinary differential equations. Ordinary Differential Equations, Academic Press, (1972).

[5] T. D'Hondt, Newton's method applied to nonlinear periodic differential systems. Equations différentielles et fonctionnelles non linéaires. Actes de la Conférence Internationale Equa. Diff. 1973, Bruxelles et Louvain-La-Neuve. Hermann, Paris, (1973).

[6] M. Jean, Les solutions périodiques et les points fixes de l'application translation dans une équation différentielle ordinaire. Internat. J. Nonlinear Mech. 9(1974), 369-382.

[7] J. Mawhin, Equations fonctionnelles non linéaires et solutions périodiques. Equa-Diff 70, C.N.R.S. Marseille, (1970).

[8] M. Strasberg, Sur un algorithme pour les équations différentielles périodiques faiblement non linéaires, non autonomes et critiques. Publications du Services de Mathématiques Université Libre de Bruxelles, (1971).

[9] B. Schmitt, Index de Seifert et systèmes différentiels à coefficients périodiques. Thèse, Université Louis-Pasteur, Strasbourg I, (1973).

[10] G. Simond, Sur les solutions périodiques comportant des harmoniques pairs de l'équation de Duffing. Thèse 3e Cycle, Université de Provence, Marseille, (1974).

[11] Y. Ueda, Some problems in the theory of nonlinear oscillations. Nippon Printing and Publishing Company, Ltd, Osaka, Japan, (1968).

Laboratoire de Mecanique et d'Acoustique
13 274 Marseille Cedex 2 , Marseille, France

TOWARD A UNIFICATION OF ORDINARY DIFFERENTIAL EQUATIONS WITH NONLINEAR SEMI-GROUP THEORY

James L. Kaplan and James A. Yorke

In this paper we study the question of the existence of solutions, in some weak sense, for the initial value problem

$$\text{(E)} \qquad \frac{dx}{dt}(t) = F(x(t)), \quad x(0) = x_o ,$$

where F is a function whose domain (dom F) and range are subsets of a Banach space B. Let Q denote the closure of the domain of F. If the problem (E) is to have a solution for $t > 0$ it seems reasonable to require that at x_o the vector $F(x_o)$, if defined, should be tangent to Q or point into Q. We do not wish to require, however, that F be defined everywhere on Q. This motivates the following definition.

Definition. We say F is subtangential to Q at x_o if there exists a sequence $\{x_n\} \varepsilon$ dom F with $x_n \to x_o$, and $\tau_n \downarrow 0$ such that

$$\left| \frac{x_n - x_o}{\tau_n} - F(x_n) \right| \to 0 \quad \text{as} \quad n \to \infty .$$

We say F is subtangential to Q on $Q_1 \subset Q$ if F is subtangential to Q at each $x_o \in Q_1$.

The use of the subtangential condition to establish the existence of a solution to the Cauchy problem dates back over thirty years. It was first used by Nagumo in 1942 [7] to prove the following result.

Theorem 1. (Nagumo). *Let* $Q \subset R^n$ *be a closed set and let* $F:Q \to R^n$ *be continuous. Assume* F *is subtangential to* Q. *Then for each* $x_o \in Q$ *there exists* $\eta > 0$ *and a solution* $x(\cdot):[0,\eta) \to Q$ *of* (E).

Theorem 1 was later rediscovered at least four times, by Brezis, Crandall [2], Hartman [5] and Yorke [9]. The simplest proof is still the original one due to Nagumo.

Unfortunately, there is no hope of extending Theorem 1 to the case in which $F:Q \to B$, where B is an arbitrary Banach space. In fact, it is well known that the continuity of $F:B \to B$ is insufficient to guarantee the existence of solutions of (E). Dieudonné [4] gave an elegant example in the Banach space c_o in which a "solution" $x(t)$ exists, although it is in the second dual c_o^{**}. Thus $x(t) \in c_o$ only for $t=0$. S.N. Chow and J. D. Schuur noticed [1] that Dieudonne's F is defined on all of c_o^{**} and is weak * continuous. Such equations always have solutions since all closed bounded subsets of the second dual of a Banach space are weak * compact. Nevertheless, Dieudonne's example shows that solutions need not exist in non-reflexive Banach spaces. Cumbersome examples exist which show that (E) need not have a solution even when B is a Hilbert or L^p space.

The case of reflexive Banach spaces has been discussed

by Yorke [10]. A cumbersome example there shows that F can be continuous and yet there may exist no solution on any non-trivial interval $[0,\varepsilon)$, even if B is a Hilbert space. (It is perhaps worth remarking parenthetically that (E) has a unique solution for most functions F, in the sense of Baire category, even though the set of Lipschitz continuous functions F is only a set of $1^{\underline{st}}$ category.)

In this present work, we establish the existence of a generalized solution to the initial value problem (E). Our principal hypothesis requires that the function -F should be decomposable into the sum of an accretive function and a Lipschitz function. Significantly, we do not require the continuity of F. Our choice of conditions on F, as well as our methods of proof, are directly motivated by an elegant paper of Crandall and Liggett [3]. In that paper, the authors establish an "exponential formula" for the generation of certain nonlinear semi-groups. As a consequence, they obtain sufficient conditions for the existence of a (possibly weak) solution of the Cauchy problem $\frac{du}{dt} + Au = 0$, $u(0) = x_o$, as the limit of piecewise constant functions, where A is an accretive nonlinear operator on a Banach space B. The conditions imposed by Crandall and Liggett on the operator A are more restrictive than our corresponding hypotheses on -F.

In an as yet unpublished paper, also on the generation of nonlinear semi-groups, Takahashi [8] establishes results quite similar in spirit to the ones presented here. He provides conditions for the convergence of difference approximations for the Cauchy problem for quasi-dissipative operators in general Banach spaces. His proofs, however, bear no resemblance to our own.

Let us turn our attention to the statement of our principal results.

Define, for $x \varepsilon Q$,

$$\rho(x) = \liminf_{y \to x} |F(y)|.$$

Let $Q_\rho = \{x \varepsilon Q: \rho(x) < \infty\}$. It can be shown that if F is subtangential to Q at x_o then $|F(x_n)| \to \rho(x_o)$ as $n \to \infty$, where $\{x_n\}$ is the sequence given in the definition of subtangential. Crandall and Liggett assume a much stronger condition that guarantees that $|F(x_n)| < \rho(x_o)$. We will sometimes require an intermediate condition. We will require that the rate of convergence of $|F(x_n)|$ to $\rho(x_o)$ should be bounded above by a Lipschitz rate. Precisely, we assume the following.

(1) There exists a continuous function $K: Q_\rho \to [0, \infty)$ such that if $x_o \varepsilon Q_\rho$, the sequence $\{x_n\}$ in the definition of subtangential can be chosen so that

$$|F(x_n)| < \rho(x_o) + K(x_o)|x_n - x_o|.$$

Hypothesis (1) will enable us to establish a local existence theorem for a "solution" of (E) on $[0, T]$ for some $T > 0$.

<u>Definition.</u> We say a function $G: \text{dom}\, G \to B$, $\text{dom}\, G \subset B$, is <u>dissipative</u> if for every x, y, ε $\text{dom}\, G$

$$d(x + h\, G(x),\ y + h\, G(y))$$

is monotonic non-increasing for $h \varepsilon (-\infty, 0]$. Here $d(\cdot, \cdot)$ is the natural distance function on B. (-G is said to be accretive if G is dissipative.)

We say F is dissipative plus locally Lipschitz if there is a dissipative function G_D and a locally Lipschitzean function G_L with $\operatorname{dom} F \subset \operatorname{dom} G_D \cap \operatorname{dom} G_L$ such that

$$F(x) = G_A(x) + G_L(x)$$

for all $x \varepsilon \operatorname{dom} F$.

Definition. Let $\varepsilon > 0$. We say y is a backward ε-approximate solution on $[0, T]$ for (E) if

(i) there exists a finite sequence $0 = t_o < t_1 < \dots < t_N$, $t_{N-1} < T \leq t_N$, such that $t_{n+1} - t_n < \varepsilon$, $n = 0, 1, \dots, N-1$, and for each n, $y(t_n) \varepsilon \operatorname{dom} F$;

(ii) y is continuous, $y:[0, t_N] \to B$;

(iii) $\left| \frac{y(t_n) - y(t)}{t_n - t} - F(y(t_n)) \right| < \varepsilon$ for $t \varepsilon [t_{n-1}, t_n)$,

(iv) $|y(0) - x_o| < \varepsilon$.

Let Q_1 be any set such that $Q_\rho \subset Q_1 \subset Q$.

Theorem 2. Suppose that F is dissipative plus locally Lipschitzean and that F is subtangential to Q on Q_1. Assume either that hypothesis (1) holds or that F is continuous and dom F is closed. Then for $x_o \varepsilon Q_1$ there exists $T = T(x_o) > 0$ such that, given any $\varepsilon > 0$, there exists a backward ε-approximate solution y on $[0, T]$ for (E).

The time T *is independent of* ε *and* T *can be chosen continuously as a function of* x_o. *When* T *is chosen as large as possible* ($T \leq \infty$), *then* T *is lower semicontinuous.*

The backward ε-approximate solutions which we construct are piecewise linear. Next we suppose that

$$s_o = 0 < s_1 < \dots < s_i < s_{i+1} < \dots < s_{m-1} < T \leq s_m \leq t_n ,$$

$$t_o = 0 < t_1 < \dots < t_j < t_{j+1} < \dots < t_{n-1} < T \leq t_n .$$

Write $\Delta_i(s) = s_i - s_{i-1}$,

$\Delta_j(t) = t_j - t_{j-1}$.

Define $\Delta_o = \max_i \Delta_i(s)$, and suppose that $\Delta_o \leq \min_j \Delta_j(t)$.

Let $x:[0, s_m] \to B$ and $y:[0, t_n] \to B$ be piecewise linear functions satisfying

(2a) $$\left| F(x(s_i)) - \frac{x(s_i) - x(s)}{s_i - s} \right| < \eta, \text{ for } s \in [s_{i-1}, s) ,$$

(2b) $$\left| F(y(t_j)) - \frac{y(t_j) - y(t)}{t_j - t} \right| < \eta, \text{ for } t \in [t_{j-1}, t_j) ,$$

for some $\eta > 0$.

We may now state our convergence result.

Theorem 3. Suppose $F = G_D + G_L$, *where* G_D *is dissipative and* G_L *is locally Lipschitz. Then, given* $\varepsilon > 0$ *there exists* $\eta > 0$ *and* $\Delta_o > 0$ *such that if* x *and* y *are piecewise linear functions on* $[0, s_m]$ *and* $[0, t_n]$, *respectively, for which* (2a) *and* (2b) *hold, and if* Δ_o *is sufficiently small, then*

$$|x(T) - y(T)| \leq e^{(L+\eta)T}(\varepsilon + |x(0) - y(0)|)$$

for $0 < T < s < t$.

It is perhaps instructive to consider the proof of Theorem 3 in a special case. Suppose that $m=n$, and $s_i = t_i$, $i = 0, 1, 2, \ldots, m$. The following lemma may be established.

Lemma 4. $|x(s_i) - y(t_j)| \leq \varepsilon^{(L+\eta)h}\{2\eta h + |x(s_i - h) - y(t_j - h)|\}$ for $0 \leq h \leq \Delta_i(s)$, Δ_o sufficiently small, $0 \leq i \leq m$, $0 \leq j \leq n$.

If we denote $|x(s_i) - y(t_j)|$ by $d_{i,j}$, then setting $h = \Delta_i(s)$ in Lemma 4 yields

$$d_{i,i} \leq e^{(L+\eta)\Delta_i(s)}\{2\eta\, \Delta_i(s) + d_{i-1,i-1}\}.$$

Similarly, replacing i by $i-1$ yields

$$d_{i-1,i-1} \leq e^{(L+\eta)\Delta_{i-1}(s)}\{2\eta\, \Delta_{i-1}(s) + d_{i-2,i-2}\}.$$

Substituting this expression into the right hand side of the previous one yields

$$d_{i,i} \leq e^{(L+\eta)(s_i - s_{i-2})}\{2\eta\, (s_i - s_{i-2}) + d_{i-2,i-2}\}.$$

Iterating this procedure, we have in general that if $i \geq k$, then

$$d_{i,i} \leq e^{(L+\eta)(s_i - s_{i-k})}\{2\eta\, (s_i - s_{i-k}) + d_{i-k,i-k}\}.$$

If we now let $i = k = m$ we have

$$|x(s_m) - y(s_m)| \leq e^{(L+\eta)s_m}\{2\eta\, s_m + |x(0) - y(0)|\},$$

since $s_o = 0$. The result would then be established. In the general case, this simplified "proof" given above has no validity, since s_m is not necessarily a grid point for y. We know only that $t_{n-1} < s_m \leq t_n$. We may notice that y is linear on $[t_{n-1}, t_n]$, so that $y(s_m)$ may be written as the weighted average of $y(t_{n-1})$ and $y(t_n)$;

$$y(s_m) = p^m(t_n)y(t_n) + p^m(t_{n-1})y(t_{n-1})$$

where the weights are defined by

$$p^m(t_n) = \frac{s_m - t_{n-1}}{t_n - t_{n-1}}, \quad p^m(t_{n-1}) = \frac{t_n - s_m}{t_n - t_{n-1}}.$$

If we now define $p^m(t) = 0$ for $t \neq t_{n-1}, t_n$, then p^m is a finite probability destribution. It is this point of view which necessitates the probabilistic lemmas which constitute the bulk of the proof of Theorem 3.

Definition. We say $\{y_n(\cdot)\}$ is a full sequence of approximate solutions for (E) on $[0, T]$ if each $y_n(\cdot)$ is a backward $1/n$-approximate solution to (E) on $[0, T]$.

Now, suppose that for each $x_o \in Q_1$, there exists some $T_{x_o} > 0$. Let

$$S = \{(t, x): x \in Q_1, \ t \in [0, T_x)\}.$$

Definition. We say ψ is a local semi-dynamical system on Q_1 if $\psi : S \to B$ is continuous and the following hold:

(i) $\psi(\tau, \psi(s, x)) = \psi(\tau + s, x)$ for $s \in [0, T_x)$, $\tau \in [0, T_{\psi(s,x)})$, $x \in Q_1$

(ii) $\psi(0,x) = x$ for $x \in Q_1$,

(iii) each interval $[0, T_x)$ is maximal, in the sense that either $T_x = \infty$ or $|\psi(t,x)| \to \infty$ as $t \to T_x$,

(iv) the intervals $[0, T_x)$ are lower semicontinuous in x; i.e., if $x_n \to x$ then $[0, T_x) \subset \liminf [0, T_{x_n})$.

We say ψ is the backward limit approximate solution for (E) on Q_1 if it is a local semi-dynamical system and for each $x_o \varepsilon Q_1$ there is at least one full sequence of approximate solutions for (E) on $[0, T_{x_o}]$, and whenever $\{y_n(\cdot)\}$ is such a sequence, the $\lim_{n\to\infty} y_n(\tau)$ exists for each $\tau \varepsilon [0, T_{x_o})$ and equals $\psi(\tau, x_o)$. Of course, when such a ψ exists, it must be unique.

By combining our previous results, we may summarize our findings in the following theorem.

Theorem 5. Suppose that F is dissipative plus locally Lipschitz, and that F is subtangential to Q on Q_1. Assume that either hypothesis (1) holds, or that F is continuous and dom F is closed. Then there exists a backward limit approximate solution for (E).

In the case that F is continuous, Theorem 5 contains a previous result of Martin [6].

REFERENCES

[1] S. N. Chow and J. D. Schuur, An existence theorem for ordinary differential equations in Banach spaces, Bull. A.M.S., 77, No. 6 (1971), 1018-1020.

[2] M. G. Crandall, A generalization of Peano's existence theorem and flow invariance, Proc. A.M.S., 36 (1972), 151-155.

[3] M. G. Crandall and T. M. Liggett, Generation of semi-groups of nonlinear transformations on general Banach spaces, Amer. J. Math., 93 (1971), 265-298.

[4] J. Dieudonne, Deux examples singuliers d'equations differentielles, Acta Sci. Math. Szeged (Leopoldo Fejer Frederico Riesz LXX annus natis dedicatus, pars B) 12 (1950), 38-40. MR 11, 729.

[5] P. Hartman, On invariant sets and on a theorem of Wazewski, Proc. A.M.S., 32 (1972), 511-520.

[6] R. H. Martin, Generating an evolution system in a class of uniformly convex Banach spaces, J. Func. Anal. 11 (1972), 62-76.

[7] M. Nagumo, Uber die lage der integralkurven gewohnlicher differentialgleichungen, Proc. Phys. Math. Soc. Japan, 24 (1942), 551-559.

[8] T. Takahashi, Difference approximation of Cauchy problems for quasi-dissipative operators and generation of semigroups of nonlinear contractions, to appear in J. Func. Anal.

[9] J. A. Yorke, Differential inequalities and non-Lipschitz scalar functions, Math. Systems Theory, 4 (1970), 140-153.

[10] J. A. Yorke, A continuous differential equation in Hilbert space without existence, Funkcialaj Ekvac., 13 (1970), 19-21.

Boston University, Boston Massachusetts

University of Maryland, College Park, Maryland

DEGENERACY OF FUNCTIONAL DIFFERENTIAL EQUATIONS

F. Kappel

In his 1967 paper [11] on controllability of delay-differential systems L. Weiss presented the conjecture that linear difference-differential equations with constant coefficients are pointwise complete. At that time it was known that nonautonomous systems need not to be pointwise complete. If a certain matrix has full rank then contrary to the situation in the case of ordinary differential equations this only is sufficient for $\mathbb{R}^n$ null-controllability of a linear delay system. In order to make this rank condition also necessary for $\mathbb{R}^n$ null-controllability one has to assume pointwise completeness of the system. According to Weiss' suggestion many people amused themselves considering this problem. In 1971 V. M. Popov and A. M. Zverkin independently gave a negative answer presenting examples of degenerate autonomous systems [10], [13]. Popov provided necessary and sufficient conditions for pointwise completeness of autonomous Systems with one delay, whereas Zverkin obtained sufficient conditions. Another set of necessary and sufficient conditions were given by R. B. Zmood and N. H. McClamroch [12]. But these conditions did not lead to the construction of examples

---

Supported by the Deutsche Forschungsgemeinschaft, AZ. 477/436/79.

in contrast to those obtained by Popov which give a very clear picture of the algebraic structure of degenerate systems. In a series of papers [1, 2, 3] B. A. Asner and A. Halanay extending Popov's approach gave necessary and sufficient conditions for systems with several commensurable delays and clarified the algebraic structure of such degenerate systems. In case of systems with several commensurable lags also P. Charrier and Y. Haugazeau obtained necessary and sufficient conditions [4]. A. K. Choudhury considered delay-equations of neutral type [5, 6] and gave an example of a degenerate 2x2 system of neutral type with distributed lag.

In this paper we consider general autonomous linear functional differential equations and give necessary and sufficient conditions for degeneracy. In order to save space proofs are sketched only.

1. Definitions and preliminary results.

We consider the general linear system

$$\dot{x}(t) = \int_{-h}^{0} d_\theta \eta(t,\theta) x(t+\theta), \quad h > 0, \tag{1.1}$$

where the nxn matrix η is measurable in $(t,\theta) \in \mathbb{R} \times \mathbb{R}$ and is normalized, i.e. we have

$$\eta(t,\theta) = 0 \text{ for } \theta \geq 0, \ \eta(t,\theta) = \eta(t,-h) \text{ for } \theta \leq -h .$$

Furthermore η is supposed to be of bounded variation in θ and left-hand continuous in θ on $(-h, 0)$ for each t with $\operatorname{var}_{[-h,0]} \eta(t,\cdot) \leq m(t)$, $m(t)$ being locally integrable on $\mathbb{R}$.

These conditions guarantee global existence and uniqueness

of solutions to the right. Given $\sigma \in \mathbb{R}$ and $\varphi \in C([\sigma-h,\sigma], \mathbb{R}^n)$ we denote the solution of (1.1) through (σ,φ) with $x(t;\sigma,\varphi)$. We shall make use of the following representation for solutions of (1.1) given by H. T. Banks (cf. [8]):

$$x(t;\sigma,\varphi) = Y(\sigma,t)\varphi(\sigma) + \int_{\sigma-h}^{\sigma-} d_\beta\Big[\int_\sigma^t Y(\alpha,t)\eta(\alpha,\beta-\alpha)d\alpha\Big]\varphi(\beta), \tag{1.2}$$

where the nxn matrix Y is the unique solution of

$$\begin{aligned} Y(\sigma,t) &= 0 \quad \text{for } \sigma > t, \\ Y(\sigma,t) &= I - \int_\sigma^t Y(\alpha,t)\eta(\alpha,\sigma-\alpha)d\alpha \quad \text{for } \sigma \le t. \end{aligned} \tag{1.3}$$

$Y(\sigma,t)$ is absolutely continuous in t for $t \neq \sigma$ and locally of bounded variation with respect to σ.

Definition 1.1. Given $q \in \mathbb{R}^n$, $q \neq 0$, and $t_1 > \sigma$ system (1.1) is said to be degenerate with respect to q at t_1 if $q^T x(t_1;\sigma,\varphi) = 0$ for all $\varphi \in C([\sigma-h,\sigma], \mathbb{R}^n)$. System (1.1) is pointwise complete at t_1 if it is not degenerate with respect to all $q \neq 0$ at t_1.

A slight generalization of a result given by Zmood and McClamroch (Theorem 2 in [12]) is contained it

Theorem 1.1. Equation (1.1) is degenerate with respect to $q \neq 0$ at $t_1 > \sigma$ if and only if

(i) $\quad q^T Y(\sigma,t_1) = 0$

and

(ii) $\quad q^T \int_\sigma^{\beta+h} Y(\alpha,t_1)[\eta(\alpha,\beta-\alpha) - \eta(\alpha,-h)]d\alpha = 0$

for all $\beta \in [\sigma-h,\sigma)$.

The proof of this theorem is based on the observation that condition (ii) is equivalent to

$$(1.4)\qquad q^T\int_\sigma^{t_1} Y(\alpha,t_1)\eta(\alpha,\beta-\alpha)\,d\alpha = \text{const. for } \beta\in[\sigma-h,\sigma).$$

Then the result follows straightforward from the representation given in (1.2).

If equation (1.1) is specialized to

$$(1.5)\qquad \dot{x}(t) = \sum_{j=0}^{m} B_j(t)x(t-h_j),\quad 0=h_o<h_1<\dots<h_m=h,$$

$B_j(t)$ measurable with $|B_j(t)| \le m(t)$ ($m(\cdot)$ being locally integrable) on $\mathbb{R}$, then condition (ii) of Theorem 1.1 can be replaced by

$$(ii^*)\qquad q^T\sum_{j=k+1}^{m} Y(\beta+h_j,t_1)B_j(\beta+h_j) = 0 \text{ a.e. on } [\sigma-h_{k+1},\sigma-h_k],$$

$$k = 0,\dots,m-1.$$

If $m = 1$ we have equation

$$(1.6)\qquad \dot{x}(t) = A(t)x(t) + B(t)X(t-h)$$

and condition (ii*) is $q^TY(\alpha,t_1)B(\alpha) = 0$ a.e. on $[\sigma,\sigma+h]$ which can be shown to be equivalent to

$$(1.7)\qquad q^T + q^T\int_\sigma^{t_1} Y(\alpha,t_1)[A(\alpha)+B(\alpha)]\,d\alpha = 0$$

is (i) is assumed. If $B(\alpha)$ is continuous with rank $B(\alpha) = n$ on a dense subset of $[\sigma,t_1-h]$ then (ii*) implies $q^TY(\alpha,t_1) = 0$, $\alpha\in[\sigma,\sigma+h]$ which with the aid of (1.7) gives

$$q^T + q^T \int_{\sigma+h}^{t_1} Y(\alpha, t_1)[A(\alpha) + B(\alpha)]d\alpha = 0.$$

This together with $q^T Y(\sigma+h, t_1) = 0$ implies $q^T Y(\alpha, t_1)B(\alpha) = 0$ on $[\sigma+h, \sigma+2h]$. If we continue this process we arrive at $q^T Y(t_1-h, t_1) = 0$ which by regularity of $Y(\alpha, t_1)$ on $[t_1-h, t_1]$ gives $q = 0$. Therefore we have the following result given previously by Zverkin (Theorem 1 in [13]):

<u>Corollary 1.</u> Let the matrix $B(\alpha)$ be continuous on $[\sigma, t_1-h]$. If rank $B(\alpha) = n$ on a dense subset of $[\sigma, t_1-h]$ then equation (1.6) is pointwise complete at t_1.

The generalization of this result to equation (1.5) is obvious. For the rest of this paper we assume (1.1) to be autonomous, i.e. we have equation

$$\dot{x}(t) = \int_{-h}^{0} d\eta(\theta)x(t+\theta), \tag{1.8}$$

η of bounded variation on $[-h, 0]$. We always take $\sigma = 0$ and $x(t;\varphi)$ denotes the solution of (1.8) through $(0, \varphi)$, φ in $C([-h, 0], \mathbb{R}^n)$. If equation (1.8) is degenerate with respect to a vector $q \neq 0$ at $t_1 > 0$ then it is degenerate with respect to q at every $t \geq t_1$. In the following degeneracy is always ment with respect to a fixed vector $q \neq 0$. If the number t_1 cannot be replaced by a smaller number t^* such that equation (1.8) is degenerate at t^* then the intervall $[t_1, \infty)$ is called the degeneracy set of equation (1.8).

If we consider equation (1.3) in the autonomous case,

$$Y(t) = I - \int_{0}^{t} Y(t-\alpha)\eta(-\alpha)d\alpha, \quad t \geq 0 \tag{1.9}$$

(we may write $Y(t-\alpha)$ instead of $Y(\alpha-t)$), then we have (cf. [12] in case of equation (1.6))

Corollary 2. Equation (1.8) is degenerate at $t_1 > 0$ if and only if

(1.10) $$q^T Y(t) = 0 \text{ for all } t \geq t_1 .$$

Necessity of (1.10) follows immediately by Theorem 1.1, (i), since equation (1.8) is degenerate for all $t \geq t_1$. By equation (1.9) we have

$$\int_0^{t_1} Y(t_1-\alpha)\eta(\beta-\alpha)d\alpha = \int_0^{\beta} Y(t_1-\alpha)\eta(\beta-\alpha)d\alpha + I - Y(t_1-\beta)$$

for all $\beta \in [-h, 0]$. If we have in mind that $t_1-\alpha \geq t_1$ for $\alpha \in [\beta,0]$ and $t_1 - \beta \geq t_1$ for $\beta \in [-h,0]$ then by using (1.10) we obtain

$$q^T\int_0^{t_1} Y(t_1 - \alpha)\eta(\beta-\alpha)d\alpha = q^T \text{ for all } \beta \in [-h,0]$$

which is relation (1.4) equivalent to(ii).

2. The main result.

In this section we show that degeneracy of the autonomous equation (1.8) can be checked by consideration of the inverse of the matrix

(2.1) $$\Delta(\lambda) = \lambda I - \int_{-h}^{0} e^{\lambda\theta} d\eta(\theta), \ \lambda \in \mathbb{C} .$$

Since $Y(t)$ is exponentially bounded on $t \geq 0$, its Laplace-transform $\overline{Y}(\lambda)$ exists and is given by $\overline{Y}(\lambda) = \Delta^{-1}(\lambda)$.

Functions with compact support in $[0,\infty)$ and their Laplace-transform are connected in a very significant way:

Lemma 2.1 ([7], p. 225). Let $\psi(\cdot)$ be a function defined on $\mathbb{R}$, $\psi(t) = 0$ for $t < 0$, such that its Laplace-transform $f(\lambda)$ exists and let a constant $\varkappa \geq 0$ be given. Then the following statements are equivalent:

(a) $\psi(t) = 0$ a.e. on $t \geq \varkappa$.

(b) $f(\lambda)$ is an entire function and there exists a constant $\gamma > 0$ such that

$$|f(\lambda)| \leq \gamma \text{ for } \operatorname{Re} \lambda \geq 0,$$

$$|f(\lambda)| \leq \gamma e^{-\varkappa \operatorname{Re} \lambda} \text{ for } \operatorname{Re} \lambda \leq 0.$$

If $\varkappa$ is such that it cannot be replaced by a smaller number then

$$\varkappa = \limsup_{u \to \infty} \frac{1}{u} \log |f(-u)|, \quad u \in \mathbb{R}.$$

The main result of this paper is

Theorem 2.1. Equation (1.8) is degenerate (at a time $t_1 > 0$) if and only if $q^T \Delta^{-1}(\lambda)$ is an entire function. If we have degeneracy then the degeneracy set is $[t_1, \infty)$ where

$$t_1 = \limsup_{u \to \infty} \frac{1}{u} \log |q^T \Delta^{-1}(-u)|, \quad u \in \mathbb{R}.$$

The only if part of the proof is obvious by Lemma 2.1. Let $q^T \Delta^{-1}(\lambda)$ be an entire function. The coordinates of $q^T \Delta^{-1}(\lambda)$ are quotients of entire functions of exponential order ≤ 1 and not of maximum type. Therefore if $q^T \Delta^{-1}(\lambda)$ is entire then its coordinates also are of exponential order ≤ 1 and not of maximum type. Furthermore it is bounded in every strip $\alpha \leq \text{Re}\,\lambda \leq \beta$ and in $\text{Re}\,\lambda \geq 0$. The following lemma shows that conditions (b) of Lemma 2.1 hold.

Lemma 2.2 ([7], p. 227). Let $f(\lambda)$ be a holomorphic function in $\text{Re}\,\lambda < \mu$ which is bounded in every strip $\alpha \leq \text{Re}\,\lambda \leq \beta \leq \mu$. Then $\tau = \lim_{u \to \infty} \frac{1}{u} \log M(-u)$, $u \in \mathbb{R}$, exists where $M(u) = \sup_{\text{Re}\lambda = u} |f(\lambda)|$. Note that we have $\tau < \infty$ for $f(\lambda) = q^T \Delta^{-1}(\lambda)$ because this function is bounded in $\text{Re}\,\lambda \geq 0$ and not of maximum type if it is entire.

3. Differential-difference equations with commensurable lags.

In this section we consider equations of the form

$$\dot{x}(t) = \sum_{j=0}^{m} B_j x(t-jh), \quad h > 0, \tag{3.1}$$

B_j constant matrices. It should be noted that now mh corresponds to h in the previous sections. If we introduce $\mu = e^{-h\lambda}$ then we have

$$\begin{aligned} \det \Delta(\lambda) &= p_o(\lambda, \mu), \\ q^T \operatorname{adj} \Delta(\lambda) &= (p_1(\lambda, \mu), \ldots, p_n(\lambda, \mu)) \end{aligned} \tag{3.2}$$

where p_j, $j = 0, \ldots, n$, are polynomials in λ, μ. The matrix $\operatorname{adj} \Delta(\lambda)$ is uniquely determined such that $\Delta(\lambda) \operatorname{adj} \Delta(\lambda) = I \det \Delta(\lambda)$. p_o is of degree n with respect to λ and p_j, $j = 1, \ldots, n$, is of degree $\leq n-1$ with respect to λ. By (3.2) we have

$$q^T \Delta^{-1}(\lambda) = \left(\frac{p_1(\lambda,\mu)}{p_o(\lambda,\mu)}, \ldots, \frac{p_n(\lambda,\mu)}{p_o(\lambda,\mu)}\right), \quad \mu = e^{-h\lambda}.$$

We need the following

Lemma 3.1 (cf. for instance [9]). Let $r_1(\lambda,\mu)$, $r_2(\lambda,\mu)$ be two nonzero polynomials and $d(\lambda,\mu)$ their greatest common divisor. Then there exist two polynomials $R_1(\lambda,\mu)$, $R_2(\lambda,\mu)$ and a nonzero polynomial $w(\lambda)$ such that

$$r_1(\lambda,\mu)R_1(\lambda,\mu) + r_2(\lambda,\mu)R_2(\lambda,\mu) = w(\lambda)d(\lambda,\mu). \tag{3.3}$$

The following necessary condition for degeneracy of equation (3.1) was given by Zverkin [13] in the case $m = 1$. We show that this condition is a consequence of Theorem 2.1.

Theorem 3.1. If equation (3.1) is degenerate then the polynomial $p_o(\lambda,\mu)$ has a nonconstant divisor $s_o(\lambda)$ which is a polynomial in λ only.

The proof is based on

Lemma 3.2 If $p_j(\lambda,\mu)$ and $p_o(\lambda,\mu)$ are nonzero polynomials, p_o not of the form $p_o(\lambda)$, and $p_j(\lambda,\mu)/p_o(\lambda,\mu)$, $\mu = e^{-h\lambda}$, is an entire function then the polynomials $p_j(\lambda,\mu)$ and $p_o(\lambda,\mu)$ have a nonconstant greatest divisor $d(\lambda,\mu)$ which is not of the form $d(\lambda)$.

The assumptions on p_j and p_o guarantee that these polynomials have countably many common zeros (λ_ν,μ_ν), $\nu = 1, 2, \ldots$. Now assume $d(\lambda,\mu) = 1$. Then by (3.3) the nonzero polynomial $w(\lambda)$ has the zeros λ_ν which is impossible. The same reasoning shows that $d(\lambda,\mu)$ is not of the form $d(\lambda)$.

Lemma 3.2 shows that for at least one j we have

$$p_j(\lambda,\mu) = d(\lambda,\mu)s_j(\lambda,\mu),$$

$$p_o(\lambda,\mu) = d(\lambda,\mu)s_o(\lambda,\mu).$$

Note that s_o is nonconstant if p_o is not of the form $p_o(\lambda)$

because p_j has degree $\leq n-1$ with respect to λ. Since $s_j(\lambda,\mu)/s_o(\lambda,\mu)$, $\mu = e^{-h\lambda}$, is entire, every zero of s_o has to be a zero of s_j. But there can be only a finite number of zeros for s_o. Otherwise s_j and s_o would have a nontrivial common divisor and d could not be the greatest common divisor of p_j and p_j. If we take $\mu = e^{-h\lambda}$ then $d(\lambda,\mu)$ and $s_o(\lambda,\mu)$ are quasipolynomials of retarded type. But since $s_o(\lambda, \exp(-h\lambda))$ has only a finite number of zeros it has to be a polynomial in λ only.

It equation (3.1) is degenerate then its algebraic structure is well known, since there exist algebraic necessary and sufficient conditions for degeneracy (cf. for instance [10], [1], [4]). It is not the aim of this paper to give new results in that direction. But we show that the approach developed here also leads to necessary and sufficient algebraic conditions for degeneracy. These conditions are similar to those obtained in [4] which generalize conditions given in [12] for systems with one lag.

The characteristic matrix for equation (3.1) is

$$\Delta(\lambda) = \lambda I - B_o - \mu B_1 - \dots - \mu^m B_m, \; \mu = e^{-\lambda h}.$$

For γ sufficiently large we have the expansion

$$\Delta^{-1}(\lambda) = (\lambda I - B_o)^{-1} + \mu H_1 + \mu^2 H_2 + \dots$$

uniformly convergent for $\operatorname{Re}\lambda \geq \gamma$. Given a power series $K_o + \mu K_1 + \mu^2 K_2 + \dots$ we define the knxkn matrix $\tilde{K}_k$ by

$$\tilde{K}_k = \begin{pmatrix} K_o & 0 & \cdots & 0 \\ K_1 & \ddots & \ddots & \vdots \\ \vdots & \ddots & \ddots & 0 \\ K_{k-1} & \cdots & K_1 & K_o \end{pmatrix}, \quad k = 1, 2, \ldots .$$

Then corresponding to $\Delta(\lambda)\Delta^{-1}(\lambda) = E$ the relation $\tilde{\Delta}_k^{-1}(\lambda) = [\tilde{\Delta}_k(\lambda)]^{-1}$ holds. $q^T\Delta^{-1}(\lambda)$ is the Laplace-transform of $q^T Y(t)$. On the interval $[0, kh)$ $q^T Y(t)$ is obtained by inversion of $q^T[(\lambda I - B_o)^{-1} + \mu H_1 + \ldots + \mu^{k-1} H_{k-1}]$. This expression is

$$(q^T, \mu q^T, \ldots, \mu^{k-1} q^T)(\lambda I_k - A_k)^{-1} E_k$$

where I_k is the nkxnk identity matrix and the nkxnk matrix A_k and the nkxn matrix E_k are defined as

$$A_k = \begin{pmatrix} B_o & 0 & \cdots & \cdots & 0 \\ B_1 & \ddots & \ddots & & \vdots \\ \vdots & \ddots & \ddots & \ddots & \vdots \\ B_m & & \ddots & \ddots & 0 \\ 0 & \ddots & & \ddots & \\ 0 \; \cdots \; 0 & B_m & \cdots & B_1 & B_o \end{pmatrix}, \quad (\lambda I_k - A_k)^{-1} = \tilde{\Delta}_k^{-1}(\lambda),$$

$$E_k = \begin{pmatrix} I \\ 0 \\ 0 \end{pmatrix}.$$

On the intervall $[(k-1)h, kh)$ we therefore have

$$q^T Y(t) = q_k^T\left(\sum_{j=0}^{k-1} J_k^{k-1-j} e^{-jhA_k}\right) e^{A_k t} E_k = \tag{3.4}$$

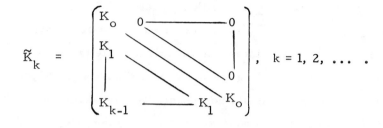

$$q_k^T(I_k - J_k e^{hA_k})^{-1} e^{(t-(k-1)h)A_k} E_k$$

where J_k is a nkxnk matrix,

$$J_k = \begin{pmatrix} 0 & \cdots & \cdots & 0 \\ I & \ddots & & \vdots \\ 0 & \ddots & \ddots & \vdots \\ \vdots & \ddots & \ddots & \vdots \\ 0 & \cdots\ 0 & I & 0 \end{pmatrix},$$

and $q_k^T = (0,\ldots,0,q^T)$. Note that J_k and A_k commute. Suppose that equation (3.1) is degenerate at t_1, i.e. $q^T Y(t) \equiv 0$ for $t \geq t_1$. We choose the integer k such that $(k-1)h \leq t_1 < kh$. Since the right-hand side of equation (3.4) defines an analytic function and has to be zero on $[t_1, hk)$, we have $t_1 = (k-1)h$ and

$$q_k^T(I_k - J_k e^{hA_k})^{-1} e^{A_k(t-(k-1)h)} E_k \equiv 0 .$$

The last condition is equivalent to

$$q_k^T(I_k - J_k e^{hA_k})^{-1} A_k^{\nu} E_k = 0, \quad \nu = 0,\ldots,nk-1. \tag{3.5}$$

On the other hand (3.5) implies $q^T Y(t) \equiv 0$ on $[(k-1)h, kh)$. If (3.5) holds then $q^T Y(t)$ on $[kh, \infty)$ is the invers transform of

$$\mu^k H_k + \mu^{k+1} H_{k+1} + \ldots, \quad \mu = e^{-\lambda h} .$$

In order to have degeneracy $q^T Y(t)$ has to be identically zero on $t \geq kh$, i.e.

$$q^T H_k = q^T H_{k+1} = \ldots\ldots = 0 . \tag{3.6}$$

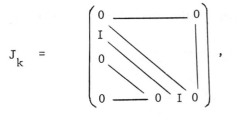

The matrices $\mathbf{H_j}$ are determined by the recursion relations

$$H_j = (H_{j-1}B_1 + \dots + H_{j-m}B_m)(\lambda I - B_o)^{-1}$$

(we define $H_\nu = 0$ for $\nu < 0$). We must have $q^T H_j = 0$ for $j = k, \dots, k+m-1$. If we define the $nk \times nm$ matrix C_k,

$$C_k = \begin{pmatrix} B_1 & \cdots & \cdots & B_m \\ \vdots & & \diagup & 0 \\ \vdots & \diagup & \diagup & \vdots \\ B_m & \diagup & & \vdots \\ 0 & & & \vdots \\ \vdots & & & \vdots \\ 0 & \cdots & \cdots & 0 \end{pmatrix},$$

then the last condition is easily seen to be equivalent to

$$q_k^T(\lambda I_k - A_k)^{-1}C_k \equiv 0. \tag{3.7}$$

This in turn is equivalent to $q^T e^{A_k t} C_k \equiv 0$ or

$$q_k^T A_k^\nu C_k = 0, \quad \nu = 0, \dots, nk-1. \tag{3.8}$$

Condition (3.7) is equivalent to

$$q_k^T S_k(\lambda) \equiv 0, \quad S_k(\lambda) = [\operatorname{adj}(\lambda I_k - A_k)]C_k. \tag{3.9}$$

We therefore have

Theorem 3.2. Equation (3.1) is degenerate with respect to $q \neq 0$ if and only if there exists an integer $k > 0$ such that conditions (3.5) and (3.8) hold. The degeneracy set is $[(k-1)h, \infty)$ where k is the smallest integer such that condition (3.9) holds.

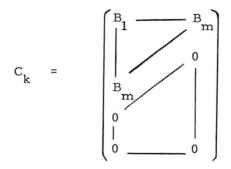

Condition (3.5) is similar to a condition given by Charrier and Haugazeau in [4] whereas condition (3.9) was given previously by Asner and Halanay ([1], Theorem 1). Without goint into details we remark that most of the results of this paper carry over to equations of neutral type.

REFERENCES

[1] Asner Jr., B. A., and A. Halanay: Algebraic theory of pointwise degenerate delay-differential systems, J. differential Eqs. 14, 293-306 (1973).

[2] -: Pointwise degenerate second order delay-differential systems, to appear.

[3] -: Delay-feedback using derivatives for minimum time linear control problems, J. Math. Analysis Appl. 48, 257-263 (1974).

[4] Charrier, P., and Y. Haugazeau: Sur la dégénérescence des équations différentielles linéarires autonomes avec retards, to appear in J. Math. Analysis Appl.

[5] Choudhury, A. K.: On the pointwise completeness of celay-differential systems of neutral type, to appear in Int. J. Control.

[6] -: Pointwise degeneracy of second order linear time-invariant delay-differential systems with distributed lag. private communication.

[7] Doetsch, G. : Handbuch der Laplace-Transformation, Vol. III, Birkhäuser 1956.

[8] Hale, J. K.: Functional Differential Equations, Springer 1971.

[9] Perron, O.: Algebra I, de Gruyter 1932.

[10] Popov, V. M.: Pointwise degeneracy of linear

time-invariant, delay-differential equations, J. differential Eqs. 11, 541-561(1972).

[11] Weiss, L.: On the controllability of delay-differential systems, SIAM J. Control 5, 575-587 (1967).

[12] Zmood, R. B., und N. H. McClamroch: On the pointwise completeness of differential-difference equations, J. differential Eqs. 12, 474-486 (1972).

[13] Zverkin, A. M.: On the pointwise completeness of systems with delay, Differencial'nye Uravnenija 9, 430-436 (1973).

University of Würzburg, D87 Würzburg, Germany

A LIAPUNOV INEQUALITY FOR NONLINEAR SYSTEMS

Leon Kotin

1. Introduction.

Liapunov's classical inequality, which applies to the following boundary value problem,

$$x'' + q(t)x = 0\,, \quad x(0) = x(a) = 0\,, \quad x \not\equiv 0\,, \tag{1}$$

is

$$\int_0^a |q(t)|\,dt > 4/a\,. \tag{2}$$

Moreover, the constant 4 cannot be increased without some $q(t)$ violating the resulting inequality; the constant 4 is thus best possible ([4, p. 246], [7]).

One can interpret (2) in various equivalent ways, to which much literature has been devoted. For instance, the negation of this inequality provides a disconjugacy criterion for the differential equation in (1). Or (2) can be used to obtain information about the distance between successive zeros of solutions of the differential equation in (1). The inequality also provides a lower bound for the eigenvalues of (1) in which q is replaced by λq. Less trivially, it can be used to determine the stability of Hill's equation.

In this paper, using an argument due to G. Birkoff (oral communication), we generalize Liapunov's inequality in two directions. We consider certain nonlinear, as well as linear, differential equations, and we minimize more general integrals. We ignore most of the aforementioned interpretations of (2); however, we obtain new criteria for the stability of Hill's equation by means of our generalization.

For instance, in the linear case we show that

$$\inf_q \int_0^a |q(t)|^\gamma dt = k_\gamma / a^{2\gamma-1} \tag{3}$$

for Eq. (1), where $k_{\gamma-1} > 0$ is independent of a and q(t). Then we show (Theorem 2, Corollary 2) that Hill's equation, with a-periodic coefficient function q(t), has only bounded solutions if q has positive mean value and $\int_0^a |q(t)|^3 dt < k_3/a^5$, where k_3 is determined.

2. The generalized inequality.

The principal result of this paper is

THEOREM 1. Let x(t) be a (nontrivial) solution of the boundary value problem

$$x'' + q(t)t^\lambda f(x, tx') + t^{-2} g(x, tx') = 0, \quad x(0) = b_0, \quad x(a) = b_1 \tag{4}$$

with $q(t) \in C^n[0, a]$ for some integer $n \geqq 0$, $t^\lambda f(x, tx') \in C^1$ and $t^{-2} g(x, tx') \in C^1$, both for $t \in [0, a]$, as functions of t, x, x'. Suppose, for some real constant α, that the function

$$t^\alpha \phi \equiv t^\alpha \phi(t^{\lambda+2} q, t^{\lambda+3} q', \ldots, t^{\lambda+n+2} q^{(n)})$$

is nonnegative and integrable as a function of t over [0, a]. Then

$$\int_0^a t^\alpha \phi \, dt \geqq k a^{\alpha+1} \tag{5}$$

where k is independent of q(t) and a, and is the best

possible constant for fixed f, g, λ, b_0 and b_1.

Proof. Define

$$I(a) \equiv \inf_q \left\{ \int_0^a t^\alpha \phi \, dt \right\}$$

over all $q \equiv q(t)$ satisfying (4), i.e., over all q for which (4) has a nontrivial solution $x(t)$. Letting $t = as$ and denoting $dq(as)/ds$ by $\dot{q}(as)$, etc., we obtain

$$I(a) = \inf_q \{a^{\alpha+1} \int_0^1 s^\alpha \phi(a^{\lambda+2} s^{\lambda+2} q(as),\ a^{\lambda+2} s^{\lambda+3} \dot{q}(as), \ldots$$

$$\ldots,\ a^{\lambda+2} s^{\lambda+n+2} q^{(n)}(as))ds\} . \qquad (6)$$

However, with the same substitution $t = as$ and $y \equiv y(s) \equiv x(as) = x(t)$, (4) becomes

$$(7) \quad y'' + a^{\lambda+2} q(as) s^\lambda f(y, sy') + s^{-2} g(s, sy') = 0,\ y(0) = b_0,\ y(1) = b_1.$$

This is of the same form as (4). Then

$$I(1) = \inf_q \{\int_0^1 s^\alpha \phi(a^{\lambda+2} s^{\lambda+2} q(as),\ a^{\lambda+2} s^{\lambda+3} \dot{q}(as), \ldots$$

$$\ldots,\ a^{\lambda+2} s^{\lambda+n+2} q^{(n)}(as))ds\} \qquad (8)$$

taken over all $q(as)$ satisfying (7). But the conditions (7) are satisfied if and only if (4) is satisfied, so the infima in (6) and (8) are taken over the same set of q's. Comparing (6) and (8), we obtain the desired result:

$$(9) \qquad I(a) = a^{\alpha+1} I(1) .$$

Denoting $I(1)$ by k gives us (5).

We remark that the boundary values in (4) can be replaced by $x' = 0$ or by $c_1 x(0) + c_2 x(a) = b_0$, $c_3 x'(0) + c_4 x'(a) = 0$.

3. Examples.

In addition to illustrating Theorem 1, the following examples provide immediate generalizations of known results, and suggest other generalizations.

Example 1. If (1) is satisfied and δ = const. $\geqq 0$, then

$$\int_0^a t^\delta |q|^\gamma \, dt \geqq ka^{\delta-2\gamma+1}$$

since we can rewrite the integrand as $t^{\delta-2\gamma}|t^2 q|^\gamma$ and apply Theorem 1. E.g., if $\gamma = 1$, then we obtain a result due to Hartman and Wintner [5] who show further that $k = 0$, 1 or $(2-\delta)^{2-\delta}/(1-\delta)^{1-\delta}$ according as $\delta > 1$, $\delta = 1$ or $\delta < 1$ respectively. On the other hand, if $\delta = 0$, then we obtain (3).

Example 2. The technique used in the proof of Theorem 1 could also be used in the case where ϕ depends on expressions of the form

$$t^{\lambda-\beta+1} \int_0^t s^\beta q(s) ds \,, \tag{10}$$

thus permitting integrals and fractional derivatives to be among the arguments of ϕ. For instance, for the boundary value problem (1), we can show that

$$\int_0^a \left(\int_0^t |q(s)|\, ds \right)^2 dt \geqq c/a$$

for some c independent of q and a. Moreover $c > 1$, for

$$\int_0^a \left(\int_0^t |q(s)|\, ds \right)^2 dt \geqq \frac{1}{a} \left(\int_0^a \int_0^t |q(s)|\, ds\, dt \right)^2$$

$$= \frac{1}{a} \left(\int_0^a (a-t)\, |q(t)|\, dt \right)^2$$

$$\geqq \frac{1}{a}$$

from Example 1 by a change in variable. More generally,

$$\inf_q \int_0^a \left(\int_0^t |q| ds \right)^{\gamma} dt = c_{\gamma} a^{1-\gamma}$$

with $c_{\gamma} \geqq 1$ for $\gamma \geqq 2$, from Hölder's inequality and the preceding result.

Example 3. Consider the boundary value problem

$$x'' + q(t)x = 0, \quad x'(0) = x(a) = 0.$$

As is Example 1, we have

$$\int_0^a |q|^{-\frac{1}{2}} dt \geqq ka^2$$

for some quitable k which cannot be improved. This is of some importance in the determination of the shape of strongest columns with given length and volume. In their investigation of this problem, I. Tadjbakhsh and J. B. Keller show [8,§7] that $k = \sqrt{3}/\pi$.

Example 4. Consider the following self-adjoint equation with homogeneous boundary values:

(11) $\quad (px')' + qx = 0, \quad p > 0, \quad x(0) = x(a) = 0.$

If we introduce the standard substitution

(12) $$s = \int_0^t du/p(u),$$

then with $y(s) \equiv x(t(s))$ we obtain

$$y'' + pqy = 0, \quad y(0) = y(s(a)) = 0.$$

From (3),

$$\int_0^{s(a)} |p(t(s))q(t(s))|^{\gamma} ds \geqq k_{\gamma} \left(\int_0^a du/p(u) \right)^{1-2\gamma}$$

whence, by a change in the variable in integration,

$$\int_0^a p^{\gamma-1} |q|^{\gamma} dt . \left(\int_0^a dt/p \right)^{2\gamma-1} \geqq k_{\gamma} .$$

In particular, if $\gamma = 1$ we conclude from Liapunov's inequality that

$$\int_0^a |q|\,dt \cdot \int_0^a dt/p > 4 .$$

Example 5. We can also apply a similar argument to obtain bounds on non-integral expressions involving the coefficients p and q in the self-adjoint case (11). Then we conclude that for a given nonnegative continuous function h(p),

$$\inf \max \{h(p)|q|^{\gamma} : t \in [0,a]\} = c/a^{2\gamma} ,$$

where the infimum is taken over all p, q for which (11) has a nontrivial solution.

The introduction of a translation in t makes it clear that a represents the distance between any two zeros of x(t). Then for $\gamma = \frac{1}{2}$ for example, we find that the distance a between consecutive zeros of any nontrivial solution of the equation in (11) satisfies

$$a \geqq c/\max(h(p)\sqrt{|q|}\,) .$$

Example 6. The same technique applies also to higher-order equations. Consider the fourth-order boundary value problem

$$x^{iv} + q(t)x = 0, \qquad x(0) = b_0, \qquad x'(0) = 0,$$
$$x(a) = b_1, \qquad x'(0) = 0 .$$

Then one can show that

$$\int_0^a |q|^{\gamma} dt \geqq k/a^{4\gamma-1}$$

where k is independent of a and is best possible.

4. Stability of Hill's equation.

We now apply the previous results to obtain an extension of Liapunov's stability criterion [3, pp. 60-61] for Hill's equation, using an argument due to G. Borg [2].

THEOREM 2. Consider the equation

(13) $$x'' + q(t)x = 0, \quad q(t+a) = q(t) \not\equiv 0,$$

where

(i) $\int_0^a q(t)dt \geqq 0$,

(ii) For some $\alpha \leqq -1$, $t^\alpha \phi$ is nonnegative, a-periodic and integrable, and

(iii) $\int_0^a t^\alpha \phi \, dt < ka^{\alpha+1}$

in the notation of Theorem 1. Then (13) is stable; i.e., all solutions are bounded on $(-\infty, +\infty)$.

Proof [G. Borg]. From Floquet's theory, there is a nontrivial solution $x(t)$ and a constant λ such that $x(t+a) = \lambda x(t)$. We know that λ is one of the two characteristic factors which are either both real or complex conjugates and whose produce is unity. Now assume that (13) is not stable. Then the λ's are real and there is a real solution $x(t)$ such that $x(t+a) = \lambda x(t) \not\equiv 0$. Thus $x(t)$ either has no zeros or an infinite number of zeros.

In the latter case, the distance between successive zeros is less than or equal to a; i.e., with t_0 and $t_1 > t_0$ consecutive zeros of x, we have $0 < t_1 - t_0 \leqq a$. Therefore, from Theorem 1 and condition (ii) above,

$$\int_0^a t^\alpha \phi \, dt \geqq \int_{t_0}^{t_1} t^\alpha \phi \, dt \geqq k(t_1 - t_0)^{\alpha+1} \geqq ka^{\alpha+1} .$$

But this contraducts (iii).

Suppose then that $x(t)$ has no zeros. Then from (13)

$$\int_0^a (x''/x)dt + \int_0^a q\,dt = 0.$$

Integrating by parts gives us

$$x'/x\Big|_0^a + \int_0^a (x'/x)^2 dt + \int_0^a q\,dt = 0.$$

Since $x(t+a) = \lambda x(t)$, the first term vanishes, whence $\int_0^a q\,dt < 0$ or $x' \equiv 0$. Either of these conclusions contradicts the hypotheses.

Consequently our assumption that (13) is not stable is untenable. This completes the proof.

Using Hölder's inequality, we can show that with $\gamma > 1$,

$$\int_0^a |q|^\gamma\,dt \geqq 4^\gamma/a^{2\gamma-1}$$

for the boundary value problem (1). This gives us the following general stability criterion.

COROLLARY 1. Consider Hill's equation (13). If there is a $\gamma > 1$ such that

$$\int_0^a |q(t)|^\gamma\,dt < 4^\gamma/a^{2\gamma-1} \quad \text{and} \quad \int_0^a q(t)dt \geqq 0,$$

then all solutions of (13) are bounded.

The constant 4^γ can, of course, be greatly increased, as shown in the following corollary, in which we borrow the result from [6] that

$$\int_0^a |q(t)|^3\,dt \geqq \Gamma^{18}(\tfrac{1}{3})/(2^4 \cdot 5\pi^6 a^5) = 656.6\ldots/a^5$$

for the boundary value problem (1).

COROLLARY 2. If

$$\int_0^a q(t)\,dt \geqq 0 \quad \text{and} \quad \int_0^a |q(t)|^3 dt < 656.6/a^5 ,$$

then all solutions of (13) are bounded.

The corresponding result in the case $\gamma = 2$ is due to Borg [1] (cf. [3], pp. 60-61).

We can apply the same argument in the case where ϕ involves integrals. Thus, using Example 2, we can show that if there is a $\gamma \geqq 2$ such that

$$\int_0^a \left(\int_0^t |q|\,ds \right)^\gamma dt < a^{1-\gamma} \quad \text{and} \quad \int_0^a q\,dt = 0,$$

then (13) is stable. (Cf. synopsis of article by T.M. Karaseva in Math. Rev., 18, p. 306).

REFERENCES

[1] G. Borg, Über die Stabilität gewisser Klassen von linearen Differentialgleichungen, Ark. Mat. Astr. Fys., 31A, no. 1 (1944), 1-39.

[2] G. Borg, On a Liapounoff criterion of stability, Amer. J. Math., 71 (1949), 67-70.

[3] L. Cesari, "Asymptotic Behavior and Stability Problems in Ordinary Differential Equations", Springer, Berlin, 1959.

[4] P. Hartman, "Ordinary Differential Equations", John Wiley & Sons, New York, 1964.

[5] P. Hartman and A. Wintner, On an oscillation criterion of Liapounoff, Amer. J. Math., 73 (1951), 885-890.

[6] L. Kotin, A generalization of Liapunov's inequality,

submitted to SIAM J. Appl. Math.

[7] Z. Nehari, Some eigenvalue estimates, J. d'Anal. Math., 7 (1959), 79-88.

[8] I. Tadjbakhsh and J. B. Keller, Strongest columns and isoperimetric inequalities for eigenvalues, J. Appl. Mech., March 1962, 159-164.

U.S. Army Electronics Command, Fort Monmouth, New Jersey

EXISTENCE AND COMPARISON RESULTS FOR DIFFERENTIAL EQUATIONS

V. Lakshmikantham

The study of the Cauchy problem for differential equations in a Banach space has taken two different directions. One direction is to find compactness type conditions that guarantee only existence of solutions and the corresponding results are extensions of the classical Peano's theorem. The other approach is to employ monotonicity (accretive or dissipative) type conditions that assure existence as well as uniqueness of solutions. In fact, this latter method shows that uniqueness conditions imply existence of solutions also [37] and therefore may be regarded as extensions of the classical Picard's theorem.

Let E be a real Banach space and let $\|\cdot\|$ denote the norm on E. We let $B(x_0,b) = [x \in E: \|x-x_0\| \leq b]$ and $R_0 = [t_0,t_0 + a] \times B(x_0,b)$. Let R^+ denote the nonnegative real line. Consider the differential equation

$$(1) \qquad x' = f(t,x), \; x(t_0) = x_0,$$

where $f \in C[R_0, E]$.

Generally, the methods of proving existence of solutions of (1) consist of three steps, namely:

(i) constructing a sequence of approximate solutions for (1);

(ii) showing the convergence of the sequence;

(iii) proving that the limit function is a solution of (1).

The steps (i) and (iii) are standard and straightforward. It is the step (ii) that deserves attention. This, in turn, leads to two possibilities:

(1) to show that the sequence of approximate solutions is a Cauchy sequence;

(2) to show the sequence of approximate solutions is relatively compact so that one can appeal to Ascoli's theorem.

These two possibilities respectively lead to monotonicity type conditions and compactness type conditions. Let us recall that if the problem (1) is in Euclidean spaces, the step (ii) does not arise.

(a) Monotonicity type conditions

A general result in this direction is the following [20, 26, 29].

Theorem 1. Assume that

(i) $f \in C[R_0, E]$ and $a > 0$, $b > 0$ and $M \geq 1$ are chosen such that $\|f(t,x)\| \leq M$ on R_0;

(ii) $g \in C[[t_0, t_0 + a] \times R^+, R]$, $g(t,0) \equiv 0$ and $u \equiv 0$ is the unique solution of

(2) $$u' = g(t, u), \quad u(t_0) = 0;$$

(iii) $V \in C[[t_0,t_0+a]\times B(x_0,b)\times B(x_0,b),R^+]$, $V(t,x,x)=0$, $V(t,x,y)>0$

if $x \neq y$,

$$|V(t,x_1,y_1) - V(t,x_2,y_2)| \leq L[\|x_1-x_2\| + \|y_1-y\|],$$

if $\{x_n\}$, $\{y_n\}$ are sequences in $B(x_0,b)$ such that

$\lim_{n\to\infty} V(t,x_n,y_n) = 0$, then $\lim_{n\to\infty}\|x_n-y_n\| = 0$ and for

$t \in [t_0,t_0+a]$, $x, y \in B(x_0,b)$,

$$D^+V(t,x,y) \equiv \limsup_{h\to 0^+} \frac{1}{h}\left[V\left(t+h,x+hf(t,x), y+f(t,y)\right) - V(t,x,y)\right]$$
$$\leq g\left(t, V(t,x,y)\right).$$

Then there exists a unique solution for the problem (1) on $[t_0,t_0+a]$.

Consider the special case $V(t,x,y) = \|x-y\|$ [25]. The condition (iii) of Theorem 1 reduces to

$$\limsup_{h\to 0} \frac{1}{h}\left[\|x-y+h[f(t,x) - f(t,y)]\| - \|x-y\|\right] \leq g(t, \|x-y\|),$$

which is clearly satisfied when one assumes Perron's type uniqueness condition

$$\|f(t,x) - f(t,y)\| \leq g(t, \|x-y\|).$$

If E is a real Hilbert space [21] and

$$\left(f(t,x) - f(t,y), x-y\right) \leq L\|x-y\|^2$$

holds, then the function $V(t,x,y) = e^{-2Lt}\|x-y\|^2$ and is an admissible candidate in Theorem 1 with $g(t,u) \equiv 0$.

Let E^* be the dual space of E and let $J:E \to 2^{E^*}$ be the the duality map defined by

$$J(x) = [x^* \in E^* : \|x^*\| = \|x\| \text{ and } x^*(x) = \|x\|^2].$$

A generalized pairing $(.,.)_- : E \times E \to R$ may be defined by means of J as

$$(x, y)_- = \inf [x^*(x) : x^* \in J(y)].$$

The pairing $(x, y)_-$ has the following properties:

(i) $(x+\alpha y, y)_- = (x, y)_- + \alpha \|y\|^2, \ \alpha \in R;$

(ii) $(x+y, z)_- \leq (x, z)_- + \|y\| \|z\|;$

(iii) if E is an inner product space and $(.,.)$ denotes the inner product, then $(x, y)_- \leq (x, y)$, equality holding everywhere if E is also complete;

(iv) if $x(t)$ is a solution of (1), then

$$D^- \|x(t)\|^2 \leq 2(f(t, x(t)), x(t))_- .$$

In place of $(x, y)_-$, one could also use $(x, y)_+ = \sup [x^*(x): x^* \in J(y)]$. The property (iv) shows that the assumption

$$(3) \qquad (f(t, x) - f(t, y), x-y))_- \leq \frac{1}{2} g(t, \|x-y\|^2)$$

implies that the choice $V(t, x, y) = \|x-y\|^2$ is admissible in Theorem 1. If on the other hand, one assumes

$$(f(t, x) - f(t, y), x-y)_- \leq g(t, \|x-y\|) \|x-y\|,$$

instead of (3), the factor $\frac{1}{2}$ occuring in (3) takes care of itself. For, letting $u = \|x-y\|$, where x and y are differentiable functions, it follows that

$$2uD^- u = D^- u^2 \leq 2(f(t, x) - f(t, y), x-y)_-$$

Consequently we have

$$\|x-y\| \, D^{-} \, \|x-y\| \leq g(t, \|x-y\|) \, \|x-y\|$$

and in the course of proof in factor $\|x-y\|$ can be canceled out. This device concerning the factor $\frac{1}{2}$ is known for a long time [10;19, p. 107]. See [11, 13, 36] for a rediscovery.

(b) Existence in closed sets.

Let $F \subset E$ be a locally closed set; that is, for each $x \in F$, there exists a $b > 0$ such that $F_0 = F \cap B(x_0, b)$ is closed in E. Suppose we consider the problem (1) with $x_0 \in F$. Assume, in place of (i) of Theorem 1, that

(i^*) $f \in C[[t_0, t_0+a] \times F, E]$ and the numbers $a, b > 0$, $M \geq 1$ are chosen such that $\|f(t,x)\| \leq M$ on $[t_0, t_0+a] \times F_0$. Suppose also that

(iv) $$\lim_{h \to 0} \frac{1}{h} d[x+hf(t,x), F] = 0, \quad (t,x) \in [t_0, t_0 + a] \times F.$$

The condition (iv) is closely related to the notion of flow invarient sets for differential equations. See [2, 3, 9, 16, 17, 31, 32]. Obviously, (iv) holds at all interior points of F. It is therefore a boundary condition. For example, (iv) holds if F is also a convex set and $f(t,x) = f_0(t,x) - x$ with $f_0(t, \partial F) \subset F$. Furthermore, (iv) is equivalent to the condition

(iv^*) if $x \in F$, $x^* \in E^* - \{0\}$ and $x^*(x) = \sup[x^*(y): y \in F]$,

then $x^*(f(t,x)) \leq 0$,

if F is convex. The quivalence of (iv) and (iv^*) follows from the duality formula

$$d(x, f) = \max [x^*(x) - \sup_F x^*(y): \|x^*\| = 1].$$

Concerning the problem (1) on the closed set F, the following statements can be made:

Proposition I. Assume

(H) (i^*) and (iv) hold. Suppose that for $t \in [t_0, t_0+a]$, $x, y \in F_0$,

$$(f(t, x) - f(t, y), x-y)_+ \leq L \|x-y\|^2. \tag{4}$$

Then F_0 convex implies the conclusion of Theorem 1.

Proposition II. If the condition (4) of (H) is strengthened to

$$(f(t, x_1) - f(s, y_1), x-y)_+ \leq L \|x-y\|^2 + \tag{5}$$

$$p(|t-s| + \|x-x_1\| + \|y-y_1\|),$$

for $t, s \in [t_0, t_0+a]$, $x_1, y_1 \in F_0$ and $x, y \in F$, where $p(u) \geq 0$ is decreasing in u and $\lim_{u\to 0} p(u) = 0$, then the convexity assumption on F_0 can be dispensed with in Proposition I.

Proposition III. If (H) holds, then the conclusion of Theorem 1 is still true.

The proofs of Propositions I and II require minor changes except in the construction of the sequence of approximation where the boundary condition (iv) has to be employed [8, 27]. The proof of Proposition III is tedious and very technical. It crucially depends on the properties of $(x, y)_+$ and specially the linearity of the comparison function $g(t,u) = Lu$. [27].

Extending the Propositions I and II in the generality of Theorem 1 does not pose special problems. However, even to replace the right hand side of (4) by a nonlinear comparison function $g(t,u)$ poses a difficult problem. One needs to develope a comparison result of the following type.

Theorem 2. Let $m(t) \geq 0$ be right continuous on $[t_0, \infty)$ with isolated discontinuities at t_k, $k = 1, 2, \ldots$ such that

$$|m(t_k) - m(t_k -)| \leq \lambda_k$$

where $\sum_{k=1}^{\infty} \lambda_k$ is convergent. Let $g \in C[R^+ \times R^+, R]$ and

$$D^+ m(t) \leq g(t, m(t)), \quad t \varepsilon [t_k, t_{k+1}) \,.$$

Then $m(t_0) \leq u_0$ implies

$$m(t) \leq r[t, t_0, u_0 + \sum_{i=1}^{\infty} \lambda_i], \quad t \geq t_0,$$

where $r(t, t_0, v_0)$ is the maximal solution of $u' = g(t,u)$, $u(t_0) = v_0$.

Having such a result at our disposal, it is possible to extend Proposition I in the spirit of Theorem 1. See [23] for details.

(c) Convergence of successive approximations

Since proving existence of solutions by means of successive approximations, rather than polygonal approximations, is constructive, we state below such a result.

Theorem 3. Assume that the hypotheses (i) and (ii) of Theorem 1 hold. Suppose further for $t \in [t_0, t_0+a]$, $x, y \in B(x_0, b)$,

$$\|f(t,x) - f(t,y)\| \leq g(t, \|x-y\|).$$

Then the successive approximations given by

$$x_{n+1}(t) = x_0 + \int_{t_0}^{t} f(s, x_n(s))ds$$

exist on $[t_0, t_0+a]$ as continuous functions and converge uniformly on this interval to the solution of (1).

Notice that we have not assumed that $g(t, u)$ is monotonic nondecreasing in u, as is usual. See for details [12, 21].

(d) Compactness type conditions.

For any bounded set A in E, let us denote by $\alpha(A)$, the measure of noncompactness of A, namely the infinum of all $d > 0$ such that there exists a finite covering of A by sets of diameter less than or equal to d. Let $\Omega = [A: A$ is a subset of $B(x_0, b)]$. A local existence theorem that employs compactness type conditions is the following. [15, 20].

Theorem 4. Assume the hypotheses (i) and (ii) of Theorem 1 hold with the additional restriction that f is uniformly continuous on R_0. Suppose that

(iii*) $V: [t_0, t_0+a] \times \Omega \to R^+$, V is continuous in t and α-continuous in A, $V(t, A) \equiv 0$ iff A is relatively compact,

$$|V(t, A) - V(t, B)| \leq L(\alpha(A-B)), \quad t \in [t_0, t_0+a], \quad A, B \in \Omega,$$

and for $t \in [t_0, t_0+a]$, $A \in \Omega$,

$$D^+V(t, A) \equiv \limsup_{h \to 0^+} \frac{1}{h} [V(t+h, A_h(f)) - V(t, A)] \leq g(t, V(t, A)),$$

where $A_h(f) = [y : y = x+hf(t, x) \, ; \, x \in A]$.

Then there exists a solution for the problem (1) on $[t_0, t_0+a]$.

Take $V(t, A) = \alpha(A)$, $A \in \Omega$. Suppose that for small $h > 0$

(6) $$\alpha(A_h(f)) \leq \alpha(A) + hg(t, \alpha(A)).$$

Then if follows that the condition (iii*) of Theorem 4 is satisfied. A stronger assumption is to assume

$$\alpha(f(t, A)) \leq g(t, \alpha(A)).$$

If $g(t, u) = ku$, $k > 0$, f is said to be α-Lipschitzian. Any Lipschitzian f is also α-Lipschitzian with $k = 0$. Thus the results in [1, 6, 7, 11, 14, 24, 33, 34, 35] are special cases of Theorem 4 if we note that instead of D^+V, one could use D^-V to obtain the same results. Also, for convenience, Perron's uniqueness criteria has been used in all places. One could replace it by Kamke's or any other uniqueness condition. For allied results in this direction see [5, 6, 24, 28].

If, as before, $R_0 = [t_0, t_0+a] \times F_0$, $F_0 = F \cap B(x_0, b)$, where $F \subset E$ is a locally closed set and $x_0 \in F$, and (iv) holds, it is known [28] that Theorem 4 is true in the special case $V(t, A) = \alpha(A)$ and $g(t, u) = ku$. Since the uniform continuity of F is not needed in some situations [14, 33] where compactness conditions are used, the question which is open is whether Theorem 4 is true, in its generality, without uniform continuity of f.

For the results concerning global existence and asymptotic equilibrium (terminal Cauchy problem) see [11, 15, 18, 23, 30].

(e) Comparison results in abstract cones.

An important technique in the theory of differential equations is concerned with estimating a function satisfying a differential inequality by means of the extremal solutions of the corresponding differential equation. As is well known [19] this comparison principle is widely employed. If we desire to develope a similar comparison principle in abstract spaces, a natural set up would be abstract cones. First of all, one has to prove the existence of extremal solutions which can then be utilized to prove comparison results. Naturally the notion of quasimonotonicity is to be introduced.

Let k be a cone in a Banach space. We consider a closed cone Q and its interior $p = Q^0$. These cones induce orderings on E. Let p^* be the set of all continuous linear functionals L on E such that $L(x) > 0$ for all $x \in p$. A function $f: E \to E$ is said to be quasimonotone if $x \leq y$ and $L(x) = L(y)$ for some $L \in p^*$, then $L(f(x)) \leq L(f(y))$. With these notions, we can state a results that gives sufficient conditions for the existence of maximal and minimal solutions for (1).

Theorem 5. Suppose that (i) and (ii) of Theorem 1 hold. Assume that the condition (6) is satisfied. Suppose also that f is uniformly continuous on R_0 and quasimonotone with respect to p for each t. Then there exist maximal and minimal solutions of (1) on $[t_0, t_0+a/2]$.

With this result at our command, it is not difficult to prove various comparison theorems. For details see [22] where as application of the abstract comparison principle to uniqueness problem is given.

REFERENCES

[1] Ambrosetti, A. Un teorema di esistenza per le equazioni differenizali negli spazi di Banach, Rend. Sem. Mat. Univ. Padova, 39 (1967), 349-360.

[2] Bony, J. M. Principe du maximum inégalité Harnack et unicité du probléme de Caushy pour les opérateures elliptiques dégénéres, Ann. Inst. Fourier, Grenoble, 19 (1969), 277-304.

[3] Brezia, H. On a characterization of flow-invariant sets, Comm. Pure. Appl. Math., 23 (1970), 261-263.

[4] Browder, F. E. Nonlinear equations of evolution, Ann. Math. 80 (1964), 485-523.

[5] Cellina, A. On the existence of solutions of ordinary differential equations in Banach spaces, Funk, Ekvac, 14 (1971), 129-136.

[6] Cellina, A. On the local existence of solutions of ordinary differential equations. Bull. Acad. Polon.

Sci., 20 (1972), 293-296.

[7] Corduneanu, C. Equazioni differenziali negli spazi di Banach teoremi di esistenza e prolungabilitz, Rend. Acc. Naz. Luicei, 23 (1957), 226-230.

[8] Crandall, M. A. Differential equations on convex sets, J. Math. Soc. Japan, 22 (1970), 443-455.

[9] Crandall, M. G. A generalization of Peano's theorem and flow-invariance, Proc. Amer. Math. Soc., 36 (1972), 151-155.

[10] Conti, R., Sansone, G. and Reissig, R. Qualitative theorie nichtlinearer differentialgleichungen, Cremonese, Rome, 1963.

[11] Deimling, K. On the existence and uniqueness for differential equations in Banach spaces, To appear.

[12] Deimling, K. On approximate solutions of differential equations in Banach spaces, To appear.

[13] Diaz, J. B. and Weinchat, R. J. On nonlinear differential equations in Hilbert spaces, Appl. Anal. 1 (1971), 31-61.

[14] Diaz, J. B. and Bownds, J. M. Euler-Cauchy polygons and the local existence of solutions to abstract ordinary differential equations, Funk. Ekvac, 15 (1972), 193-207.

[15] Eisenfeld, J. and Lakshmikantham, V. On the existence of solutions of differential equations in a Banach space, University of Texas at Arlington Technical Report. No. 8.

[16] Hartman, P. On invariant sets and on a theorem of Wazewski, Proc. Amer. Math. Soc., 32 (1972), 511-520.

[17] Ladde, G. and Lakshmikantham, V. On flow-invariant sets. Pacific J. Math, (1974).

[18] Ladas, G. and Lakshmikantham, V. Differential equations in abstract spaces, Academic Press, New York, 1972.

[19] Lakshmikantham, V. and Leela, S. Differential and integral inequalities, Vol. I and II, Academic Press, New York, 1969.

[20] Lakshmikantham, V. Stability and asymptotic behaviour of solutions of differential equations in a Banach space, Lecture notes, CIME, Italy (1974).

[21] Lakshmikantham, V. On the redundancy of monotony assumptions, University of Texas at Arlington, Tech. Report No. 12.

[22] Lakshmikantham, V., Mitchell, A. R. and Mitchell, R. W. Maximal and minimal solutions and comparison results for differential equations in abstract cones, University of Texas at Arlington, Tech. Report No. 10.

[23] Lakshmikantham, V., Mitchell, A. R. and Mitchell, R. W. Differential equations on closed sets, University of Texas at Arlington, Tech. Report No. 13.

[24] Li, T. Y. Existence of solutions for ordinary differential equations in Banach spaces, To appear.

[25] Martin, R. H. A global existence theorem for autonomous differential equations in a Banach space, Proc. Amer. Math. Soc., 26 (1970), 307-314.

[26] Martin, R. H. Lyapunov functions and autonomous differential equations in a Banach space, Math. Sys.

Theory, 7 (1973), 66-72.

[27] Martin, R. H. Differential equations on closed subsets of a Banach space, Trans. Amer. Math. Soc., 179 (1973), 399-414.

[28] Martin, R. H. Approximation and existence of solutions of ordinary differential equations in Banach spaces, Funk. Ekvac, 16 (1973), 195-211.

[29] Murakami, H. On nonlinear ordinary and evolution equations, Funk. Ekvac, 9 (1966), 151-162.

[30] Mitchell, A. R. and Mitchell, R. W. Asymptotic equilibrium of ordinary differential systems in a Banach space, University of Texas at Arlington, Tech. Report. No. 7.

[31] Nagumo, M. Über die laga der integralkurven gewöhnlicher differentialgleichungen, Proc. Phy.-Math. Soc. Japan, 24 (1942), 551-559.

[32] Redheffer, R. M. The theorems of Bony and Brezia on flow-invariant sets, Amer. Math. Monthly, 79 (1972), 740-747.

[33] Szufla, S. Some remarks on ordinary differential equations in Banach spaces, Bull. Acad. Polon. Sci., 16 (1968), 795-800.

[34] Szufla, S. On the existence of solutions of an ordinary differential equation in the case of a Banach space, Bull. Acad. Polon. Sci., 16 (1968), 311-315.

[35] Szufla, S. Measure of noncompactness and ordinary differential equations in Banach spaces, Bull. Acad. Polon. Sci., 19 (1971), 831-835.

[36] Vidossich, G. Existence, comparison and asymptotic behavior of solutions of ordinary differential equations in finite and infinite dimensional Banach spaces, To appear.

[37] Wazewski, T. Sur l'existénce et et l'unicìte des integrales des equations differentielles ordinaires au cas de l'espace de Banach, Bull. Acad. Polon. Sci., 8 (1960), 301-305.

University of Texas, Arlington, Texas

ADIABATIC INVARIANTS FOR LINEAR HAMILTONIAN SYSTEMS

Anthony Leung

n independent adiabatic invariants in involution are found for a slowly varying linear Hamiltonian system of n degrees of freedom. The system considered is $\varepsilon\dot{u} = A(t)u$ where $A(t)$ is a $2n \times 2n$ real Hamiltonian matrix with distinct, pure imaginary eigenvalues for all $t \in [-\infty, \infty]$, and $\frac{d^j A(t)}{dt^j} \in L_1(-\infty, \infty)$ for $j \geq 1$. The adiabatic invariant $I_s(u,t)$, $s = 1, \ldots, n$ are quadratic functions of u and are expressed easily in terms of the eigenvectors of $A(t)$. If $\phi(t, \varepsilon)$ is a solution of the equation and $\mathcal{I}_s(t, \varepsilon) = I_s(\phi(t, \varepsilon), \varepsilon)$ then it is shown that $\mathcal{I}_s(\infty, \varepsilon) - \mathcal{I}_s(-\infty, \varepsilon) = 0(\varepsilon^n)$ as $\varepsilon \to 0^+$ for all n.

University of Cincinnati, Cincinnati, Ohio

LIMITS OF SOLUTIONS OF INTEGRODIFFERENTIAL EQUATIONS

J. J. Levin

Consider the equation

$$(1) \quad x'(t) + \int_{-\infty}^{\infty} g(x(t-\xi))\, dA(\xi) = f(t) \quad \left(' = \frac{d}{dt}, \ -\infty < t < \infty\right),$$

where

$$(2) \begin{cases} x = (x_1 \ldots x_N) \in \mathbb{C}^N, \ g\colon \mathbb{C}^N \to \mathbb{C}^N, \ g \in C(\mathbb{C}^N), \\ A \in NBV(\mathbb{R}^1) \ \underline{\text{is an}} \ N \ \underline{\text{by}} \ N \ \underline{\text{complex valued matrix}}, \\ f\colon \mathbb{R}^1 \to \mathbb{C}^N, \ f \in L^\infty(\mathbb{R}^1), \ \lim_{t\to\infty} f(t) = f(\infty) = 0 . \end{cases}$$

Here $\mathbb{R}^1$ is the real line $(-\infty, \infty)$ and $\mathbb{C}^N$ is complex N-space. The notation $A \in NBV(\mathbb{R}^1)$ means that $A(t)$ is of bounded variation on $\mathbb{R}^1$, $A(t) = A(t-)$, and $A(-\infty) = 0$. Suppose that

$$(3) \quad x(t) \in LAC(\mathbb{R}^1) \cap L^\infty(\mathbb{R}^1) \ \underline{\text{satisfies}} \ (1) \ \underline{\text{a. e. on}} \ \mathbb{R}^1,$$

i. e., that $x(t)$ is a bounded solution of (1) which is absolutely continuous on every compact interval. In this paper we consider some questions concerning the existence or nonexistence of the limit $x(\infty) = \lim_{t\to\infty} x(t)$.

Before stating specific definitions and results we first

This research was supported by the U. S. Army Research Office.

make some general comments:

We do not assume that all solutions of (1) are bounded on $\mathbb{R}^1$; however, we only study the behavior as $t \to \infty$ of those solutions which are bounded on $\mathbb{R}^1$.

If

$$A(t) = 0 \ (-\infty < t \leq 0), \ A(t) = A_\infty \quad (0 < t < \infty),$$

where A_∞ is a prescribed matrix, then (1) becomes

$$x'(t) + g(x(t))A_\infty = f(t) \qquad (-\infty < t < \infty), \tag{4}$$

which is, of course, an ordinary differential equation. Thus our results concerning (1) are relevant to (4). Suppose, instead of (3) and (4) holding, that $x(t) \in LAC(\mathbb{R}^+) \cap L^\infty(\mathbb{R}^+)$, where $\mathbb{R}^+ = [0, \infty)$, satisfies a.e. the equation

$$x'(t) + g(x(t))A_\infty = f(t) \qquad (0 \leq t < \infty), \tag{5}$$

where g satisfies (2) and

$$f: \mathbb{R}^+ \to \mathbb{C}^N, \ f(t) \in L^\infty(\mathbb{R}^+), \ f(\infty) = 0. \tag{6}$$

It is obvious that if we set

$$\tilde{x}(t) = x(t), \ \tilde{f}(t) = f(t) \qquad (0 \leq t < \infty)$$

$$\tilde{x}(t) = x(0), \ \tilde{f}(t) = g(x(0))A_\infty \qquad (-\infty < t < 0);$$

then $\tilde{x}(t)$ satisfies a.e.

$$\tilde{x}'(t) + g(\tilde{x}(t))A_\infty = \tilde{f}(t) \qquad (-\infty < t < \infty),$$

which is of the same form as (4). Thus the study of the bounded solutions of (5) on $\mathbb{R}^+$ is subsumed under the study of the bounded solutions of (4), and thus of (1), on $\mathbb{R}^1$.

Similarly, the study of the bounded solutions of the

Volterra equation

$$(7) \qquad x'(t) + \int_0^t g(x(t-\xi))\, dA(\xi) = f(t) \qquad (0 \leq t < \infty),$$

where g satisfies (2), $f(t)$ satisfies (6), and $A(t) \in NBV(\mathbb{R}^+)$, may be subsumed under the study of the bounded solutions of (1). The details for this reduction are given in Lemma 2.2 of [8]. Theorem 5 below is an application of the present methods to equation (7).

The limit equation associated with (1) is

$$(1^*) \qquad y'(t) + \int_{-\infty}^{\infty} g(y(t-\xi))\, dA(\xi) = 0 \qquad (-\infty < t < \infty).$$

This equation plays a key role in our analysis of the bounded solutions of (1). The limit equation associated with both (4) and (5) is

$$(4^*), (5^*) \qquad y'(t) + g(y(t))A_\infty = 0 \qquad (-\infty < t < \infty),$$

which is a special case of (1^*). The limit equation associated with (7) is

$$(7^*) \qquad y'(t) + \int_{-\infty}^{\infty} g(y(t-\xi))\, d\tilde{A}(\xi) = 0 \qquad (-\infty < t < \infty),$$

where

$$\tilde{A}(t) = A(t) \quad (0 \leq t < \infty), \quad \tilde{A}(t) = 0 \quad (-\infty < t \leq 0),$$

which is of the same form as (1^*). The point being emphasized here is that even for equations defined on $\mathbb{R}^+$, the associated limit equations are defined on $\mathbb{R}^1$.

We shall investigate some possibilities for the asymptotic behavior of a bounded solution, $x(t)$, of (1) under a variety of assumed structures on the set of bounded solutions of (1^*). These assumed structures are of a geometrical nature;

illustrative examples are easily constructed for the ordinary differential equations case of (4^*). An example involving (7^*) arises in the proof of Theorem 5 below. A common theme in most of the present results is the existence or nonexistence of $x(\infty)$.

The proofs of the principal results of this paper employ Theorems 1a and 2a of Levin and Shea [8]. The latter paper was also concerned, in part, with the relationship between solutions of (1) and (1^*). Under the assumptions (2) and (3) above, Theorem 1a showed how a given solution of (1) may be represented in terms of certain solutions of (1^*), namely, in terms of the elements of $\Gamma(x(t))$ below. Theorem 1b dealt with the essentially converse problem of constructing a solution of (1) from a given family of solutions of (1^*) which satisfied certain assumptions. Theorems 1a and 2a are very general in nature. Here, as in [8], more hypothesis is required in order to obtain more detailed results.

The results of [8], insofar as they dealt with (1) and (1^*) and went beyond Theorems 1a and 2a, involved very different assumptions than the present geometrical ones. The linear case, $g(x) = x$, of (1) was treated in [8] under a variety of hypotheses on the Fourier-Stieltjes transform

$$\hat{A}(\lambda) = \int_{-\infty}^{\infty} e^{-i\lambda t}\, dA(t) \qquad (-\infty < \lambda < \infty) .$$

A nonlinear case which received considerable attention in [8] was the important one in which (1^*) is assumed to have a unique bounded solution. Since all translates of solutions of (1^*) are also solutions of (1^*), a unique bounded solution for

(1^*) must be a constant. This should be contrasted with the much richer geometrical structures considered below.

Many authors have, of course, considered the relationship between (1) and (1^*) from many different points of view. Some of these are Miller [12], who considers Volterra equations of the form (7) but with the term $x'(t)$ replaced by $x(t)$, and Hale [3] and Miller [11], who consider delay-differential equations. The case of ordinary differential equations has a large literature of its own on asymptotically autonomous equations. This topic is treated by the methods of topological dynamics by Sell [14], who gives many references to this literature. Miller and Sell [13] and Sell [15] consider, from the viewpoint of topological dynamics, equations of the form (1) and (7) except that $x'(t)$ is replaced by $x(t)$.

Under assumptions which are more or less related to those invoked in Theorem 5, equation (7) has been the subject of many investigations. Some of these are: Corduneanu [1], Halanay [2], Hannsgen [4], Levin [5], Levin and Nohel [6] (at the end of [6] there is a discussion of the applicability of the results of [1] to (7)), Levin and Shea [7] and [8], Londen [9], and MacCamy and Wong [10]. While the hypotheses of Theorem 5 concerning $g(x)$ and $a(t)$ are not the most general that have been considered, the earlier results are generally much more restrictive than Theorem 5 with respect to the assumptions made on $f(t)$. In this connection, see Londen [9].

We begin with some definitions and elementary results. It is understood that (2) and (3) hold throughout. The range of a function $z: \mathbb{R}^1 \to \mathbb{C}^N$ is denoted by:

$$Rz(t) = \{\gamma \mid \gamma = z(t_o) \text{ for some } t_o \in \mathbb{R}^1\}.$$

<u>Definitions 1.</u>

$$\Omega(x(t)) = \{\omega \mid x(t_i) \to \omega \ \underline{\text{for some}}\ t_i \to \infty\ (i \to \infty)\},$$

$$\Gamma(x(t)) = \{y(t) \mid x(t+t_i) \to y(t),\ \underline{\text{uniformly on every compact subset of}}\ \mathbb{R}^1,\ \underline{\text{for some}}\ t_i \to \infty\ (i \to \infty)\},$$

$$R\Gamma(x(t)) = \bigcup_{y(t)\in\Gamma(x(t))} Ry(t).$$

Although (1^*) plays no role in Definitions 1, it and not $x(t)$ does in the following definitions. Here $\alpha \in \mathbb{C}^N$, $\beta \in \mathbb{C}^N$, $S \subset \mathbb{C}^N$, and $C_u^1(\mathbb{R}^1)$ denotes the set of functions possessing a uniformly continuous first derivative on $\mathbb{R}^1$.

<u>Definitions 2.</u>

$$\Gamma = \{y(t) \mid y(t) \in C_u^1(\mathbb{R}^1) \cap L^\infty(\mathbb{R}^1),\ y(t) \text{ satisfies } (1^*)\},$$

$$\Gamma_S = \Gamma \cap \{y(t) \mid Ry(t) \subset S\},$$

$$\Gamma^{(\alpha,\beta)} = \Gamma \cap \{y(t) \mid y(-\infty) = \alpha,\ y(\infty) = \beta\},$$

$$\Gamma_S^{(\alpha,\beta)} = \Gamma_S \cap \Gamma^{(\alpha,\beta)}.$$

If $S \subset B$, then obviously $\Gamma_S \subset \Gamma_B$ and $\Gamma_S^{(\alpha,\beta)} \subset \Gamma_B^{(\alpha,\beta)}$.

The appearance of the letter Γ in both Definitions 1 and 2 anticipates the following lemma, where

$$B_r = \{\gamma \mid \gamma \in \mathbb{C}^N, \ |\gamma| = \sum_{i=1}^{N} |\gamma_i| \le r\}$$

for each $r \ge 0$ and

$$\|x\|_\infty = \operatorname*{ess\,sup}_{-\infty < t < \infty} |x(t)| .$$

Lemma 1. (i) $\Omega(x(t)) \subset B_{\|x\|_\infty}$ and is nonempty, compact, and connected.

(ii) $\Gamma(x(t)) \subset \Gamma_{\Omega(x(t))}$, $R\Gamma(x(t)) = \Omega(x(t))$.

Another basic definition is:

Definition 3. $c \in \mathbb{C}^N$ is called a "critical point" of (1^*) if

(8) $$g(c)A(\infty) = 0.$$

Observe that in the special case (4^*) of (1^*), this definition agrees with the usual terminology for ordinary differential equations. Concerning critical points we have the following elementary properties:

Lemma 2. (i) c is a critical point of (1^*) if and only if $\Gamma^{(c,c)} \ne \emptyset$.

(ii) If $x(\infty)$ exists, then $x(\infty)$ is a critical point of (1^*).

The first of several structures on (a subset of) Γ that we shall consider is defined as follows:

Definition 4.

If $S \subset \mathbb{C}^N$, then "Γ_S is of type 1" means that either $\Gamma_S = \emptyset$ or there exists a positive integer n such that each of the following conditions holds:

(i) (1^*) *has precisely* n *critical points* $c_i \in S\,(i=1,\ldots,n)$,

(ii) $\Gamma_S^{(c_i,c_i)} = \{c_i\}\ (i=1,\ldots,n)$,

(iii) $$\Gamma_S = \bigcup_{i=1}^{n} \Gamma_S^{(c_i,c_i)} \cup \bigcup_{(i,j)\in E} \Gamma_S^{(c_i,c_j)},$$

where, if it is not empty, $E \subset \mathbb{Z}^{(2)}$ *satisfies*

(iv) $(i,j) \in E$ *if and only if* $i \neq j$ *and* $\Gamma_S^{(c_i,c_j)} \neq \phi$,

(v) *there does not exist a sequence* $\{(i_k,j_k)\}_{k=1}^{p}$, *for any* p, *such that* $(i_k,j_k) \in E\ (k=1,\ldots,p)$ *and* $j_k = i_{k+1}\ (k=1,\ldots,p-1),\ j_p = i_1$.

Observe that:

$S \subset B$ and Γ_B of Type 1 implies that Γ_S is of Type 1.
$\Gamma_A^{(c_i,c_i)} = \{c_i\}$ does not imply that $\Gamma^{(c_i,c_i)} = \{c_i\}$.
If $(i_o,j_o) \in E$, then $(j_o,i_o) \notin E$.

Loosely speaking, Γ_S is of Type 1 means that $y(-\infty)$ and $y(\infty)$ exist for every $y(t) \in \Gamma_S$ and that there are no closed "loops", composed of elements of Γ_S, which connect a critical point of (1^*) with itself.

Theorem 1. $x(\infty)$ *exists if and only if* $\Gamma_{\Omega(x(t))}$ *is of Type 1.*

The sufficiency assertion is the nontrivial one in this theorem. For, if $x(\infty)$ exists, then $\Omega(x(t)) = \{x(\infty)\}$ and $x(\infty)$ is a critical point of (1^*). Hence $\Gamma_{\Omega(x(t))} = \{y(t) \equiv x(\infty)\}$, which is obviously of Type 1.

A setting in which Theorem 1 may be applied is the

following: Suppose it is possible to find a set $S \supset \Omega(x(t))$, perhaps $S = B_{\|x\|_\infty}$, such that Γ_S is of Type 1. Then if follows from the first observation after Definition 4 that $\Gamma_{\Omega(x(t))}$ is of Type 1; hence, by Theorem 1, $x(\infty)$ exists.

There are closed "loops" in the following structure on Γ:

Definition 5.

"Γ is of Type 2" means that either

$$\text{(i)}\quad \Gamma^{(c,c)} \neq \emptyset,\quad \Gamma^{(c,c)} \neq \{c\}$$

for some $c \in \mathbb{C}^N$, or

$$\text{(ii)}\quad \Gamma^{(c_j,c_{j+1})} \neq \emptyset \quad (j=1,\ldots,n-1),\quad \Gamma^{(c_n,c_1)} \neq \emptyset$$

for some sequence $c_j \in \mathbb{C}^N$ $(j=1,\ldots,n;\ n \geq 2)$ of distinct points.

Note that:

If Γ is of Type 2, then (i) and (ii) may hold simultaneously.

Γ may be of Type 2 and there may also exist a compact set S, which has more than one element, such that $\Gamma_S \neq \emptyset$ is of Type 1.

Theorem 2. Let Γ be of Type 2. Then there exists an $f(t) \in C_u(-\infty,\infty)$ which satisfies (2) and a solution $x(t) \in C_u^1(\mathbb{R}^1) \cap L^\infty(\mathbb{R}^1)$ of (1), with this $f(t)$, such that $x(\infty)$ does not exist. In fact, if (i) of Definition 5 holds and if

$$\tilde{y}(t) \in \Gamma^{(c,c)},\ \tilde{y}(t) \not\equiv c,$$

then $x(t)$ may be constructed so that

$$\Omega(x(t)) = \{c\} \cup R\, \tilde{y}(t).$$

If (ii) of Definition 5 holds and if

$$y^{(j)}(t) \in \Gamma^{(c_j, c_{j+1})} \; (j=1,\ldots,n-1), \; y^{(n)}(t) \in \Gamma^{(c_n, c_1)},$$

then $x(t)$ may be constructed so that

$$\Omega(x(t)) = \bigcup_{j=1}^{n} \{c_j\} \cup \bigcup_{j=1}^{n} R\, y^{(j)}(t).$$

Let $A_{ij}(t)$ be the i,j^{th} element of the matrix $A(t)$ of (1) and let

$$V(A_{ij}, (-\infty,t]) = \text{total variation of } A_{ij}(s) \text{ on } (-\infty,t]$$

$$V(t) = \sum_{i,j=1}^{N} V(A_{ij}, (-\infty,t]).$$

Corollary. In addition to the hypothesis of Theorem 2 let

(9) $g(x)$ be locally Lipschitzian

(10) $\int_{-\infty}^{0} |t| V(t)dt < \infty, \quad \int_{0}^{\infty} t[V(\infty) - V(t)]dt < \infty$

hold. Then the $f(t)$ of Theorem 2 may be constructed so that it lies in $L^1(\mathbb{R}^1)$.

If $V(t)$ is absolutely continuous, then the moment condition (10) is equivalent to

$$\int_{-\infty}^{\infty} t^2 V'(t)dt < \infty.$$

Also observe that (10) is obviously satisfied when (1) reduces to the ordinary differential equation (4).

Another hypothesis on Γ is given by:

Definition 6.

"Γ is of Type 3" means that there exists a $\varphi(t) \in \Gamma$ such that at least one of $\varphi(-\infty)$, $\varphi(\infty)$ does not exist.

Γ may, of course, be of both Types 2 and 3; simultaneously, there may exist a compact set S, which has more than one element, such that $\Gamma_S \neq \phi$ is of Type 1.

Theorem 3. Let Γ be of Type 3. Then there exists an $f(t) \in C_u(\mathbb{R}^1)$ which satisfies (2) and a solution $x(t) \in C_u^1(\mathbb{R}^1) \cap L^\infty(\mathbb{R}^1)$ of (1), with this $f(t)$, such that $x(\infty)$ does not exist.

The assertion is obviously true if $\varphi(\infty)$ does not exist; simply take $f(t) \equiv 0$ and $x(t) = \varphi(t)$. The proof when $\varphi(-\infty)$ does not exist is much harder.

Corollary. In addition to the hypothesis of Theorem 3 let (9) and (10) hold. Then the $f(t)$ of Theorem 3 can be constructed so that it lies in $L^1(\mathbb{R}^1)$.

In the next theorem a qualitative hypothesis on the structure of Γ_S, where S is compact, is shown to imply a quantitative conclusion concerning the time spent by some elements of Γ_S in going from a neighborhood of one critical point of (1^*) to a neighborhood of another one.

Theorem 4. Let $S \subset \mathbb{C}^N$ be compact and let there exist distinct $c_1, c_2 \in S$ such that

$$\Gamma_S = \Gamma_S^{(c_1,c_1)} \cup \Gamma_S^{(c_2,c_2)} \cup \Gamma_S^{(c_1,c_2)},$$

where

$$\Gamma_S^{(c_1,c_1)} = \{c_1\},\ \Gamma_S^{(c_2,c_2)} = \{c_2\},\ \Gamma_S^{(c_1,c_2)} \neq \phi .$$

For each $\varepsilon \in (0, \frac{1}{2}|c_1 - c_2|)$ and $y(t) \in \Gamma_S^{(c_1,c_2)}$ let

$$\alpha_y(\varepsilon) = \sup_t \{t \mid |y(s) - c_1| < \varepsilon \text{ on } (-\infty, t)\},$$

$$\beta_y(\varepsilon) = \inf_t \{t \mid |y(s) - c_2| < \varepsilon \text{ on } (t, \infty)\},$$

$$\rho_y(\varepsilon) = \beta_y(\varepsilon) - \alpha_y(\varepsilon) .$$

Then $\sup_{y(t) \in \Gamma_S^{(c_1,c_2)}} \rho_y(\varepsilon) < \infty$ for each such ε.

The following result is concerned with the real scalar special case of (7) given by

$$\text{(11)} \quad x'(t) + \int_0^t g(x(t-\xi))\, a(\xi)\, d\xi = f(t) \quad (0 \le t < \infty)$$

and has been commented on above; its proof employs Theorem 1 and the corollary to Theorem 2.

Theorem 5. Let

$g(x)$ be locally Lipschitzian,

$\{x \mid g(x) = 0\} \cap [-X, X]$ be finite for every $X \in \mathbb{R}^+$,

$a(t) \in C^2(0,\infty)$, $a(t) \not\equiv 0$,

$(-1)^k a^{(k)}(t) \ge 0 \quad (0 < t < \infty;\ k = 0, 1, 2)$,

$a(t),\ t\, a(t),\ t^2 a(t) \in L^1(\mathbb{R}^+)$,

$f(t) \in L^\infty(0,\infty)$, $\lim_{t \to \infty} f(t) = f(\infty) = 0$,

and let $x(t) \in LAC(\mathbb{R}^+) \cap L^\infty(\mathbb{R}^+)$ satisfy (11) a. e. on $(0,\infty)$.

Then $x(\infty) = c$, for some c such that $g(c) = 0$.

REFERENCES

[1] Corduneanu, C., Sur une équation intégrale de la théorie du réglage automatique, C. R. Acad. Sci. Paris 256 (1963), 3564-3567.

[2] Halanay, A., On the asymptotic behavior of the solutions of an integro-differential equation, J. Math. Anal. Appl. 10 (1965), 319-324.

[3] Hale, J. K., Sufficient conditions for stability and instability of autonomous functional-differential equations, J. Diff. Eqs. 1 (1965), 452-482.

[4] Hannsgen, K. B., On a nonlinear Volterra equation, Mich. Math. J. 16 (1969), 365-376.

[5] Levin, J. J., The asymptotic behavior of the solution of a Volterra equation, Proc. Amer. Math. Soc. 14 (1963), 534-541.

[6] Levin, J. J. and Nohel, J. A., Perturbations of a nonlinear Volterra equation, Mich. Math. J. 12 (1965), 431-447.

[7] Levin, J. J. and Shea, D. F., Asymptotic behavior of bounded solutions of some functional equations, Contributions to Nonlinear Functional Analysis, Academic Press, New York and London (1971), 501-522.

[8] Levin, J. J. and Shea, D. F., On the asymptotic behavior of the bounded solutions of some integral equations, I, II, III, J. Math. Anal. Appl. 37 (1972), 42-82, 288-326, 537-575.

[9] Londen, S-O., The qualitative behavior of the solutions of a nonlinear Volterra equation, Mich. Math. J. 18 (1971), 321-330.

[10] MacCamy, R. C. and Wong, J. S. W., Stability theorems for some functional equations, Trans. Amer. Math. Soc. 164 (1972), 1-37.

[11] Miller, R. K., Asymptotic behavior of nonlinear delay-differential equations, J. Diff. Eqs. 1 (1965), 293-305.

[12] Miller, R. K., Asymptotic behavior of solutions of nonlinear Volterra equations, Bull. Amer. Math. Soc. 72 (1966), 153-156.

[13] Miller, R. K. and Sell, G. R., Volterra integral equations and topological dynamics, Memoir 102 Amer. Math. Soc. (1970), 1-67.

[14] Sell, G. R., Nonautonomous differential equations and topological dynamics, I, II, Trans. Amer. Soc. 127 (1967), 241-262, 263-283.

[15] Sell, G. R., A Tauberian condition and skew flows with applications to integral equations, J. Math. Anal. Appl. 43 (1973), 388-396.

University of Wisconsin-Madison, Madison, Wisconsin

ON THE SUMMATION OF FORMAL SOLUTIONS AT AN IRREGULAR SINGULAR POINT BY FACTORIAL SERIES

Donald A. Lutz

1. Introduction.

A system of n linear differential equations of the form

$$x' = A(z)x = \left(\sum_{k=o}^{\infty} A_k z^{-k}\right)x , \tag{1.1}$$

where A_o has n distinct eigenvalues and the power series converges for $|z| > R$, has a formal, fundamental solution matrix of the form

$$F(z)z^{\Lambda'} e \quad , \tag{1.2}$$

where $F(z) = \sum_{k=o}^{\infty} F_k z^{-k}$, det $F_o \neq 0$, and Λ' and Λ are constant diagonal matrices. It is well known (see, for example, [13] pp. 60-61) that there exists an actual fundamental solution matrix of the form $\hat{F}(z)z^D \exp[\Lambda z]$ such that $\hat{F}(z)$ has the formal series $F(z)$ as its asymptotic expansion as $z \to \infty$ in a half-plane. It has been shown by Horn [3], [4] (in this case that A_o has n distinct eigenvalues) and also by Trjitzinsky [9] and Turrittin [10] (who treated somewhat more general situations), that the formal series $F(z)$ can be replaced by a convergent generalized factorial series so as to obtain a convergent

Supported in part by the National Science Foundation under Grant GP-28149.

expression for $\hat{F}(z)$. Such factorial series have, as their natural domain of convergence, half-planes. This problem is closely related to the Borel summability of the formal series $F(z)$ to the function $\hat{F}(z)$ and is usually approached via Laplace transformations.

For the convenience of the reader, we state this result on convergence of generalized factorial series here as

Theorem A. (See Wasow [13], p. 338.)

Let λ_j, $1 \leq j \leq n$, denote the (distinct) eigenvalues of A_o *and let α be any angle for which $\alpha \neq -\arg(\lambda_\nu - \lambda_\mu)$, $1 \leq \nu \neq \mu \leq n$. Then there exist positive numbers $\omega_o \geq 1$ and κ such that for $\omega \geq \omega_o$ the differential system* (1.1) *possesses in the half-plane*

$$\text{(1.3)} \qquad H: \quad \operatorname{Re}(ze^{-i\alpha}) > \kappa$$

a fundamental solution $x(z)$ of the form

$$\text{(1.4)} \qquad x(z) = \mathfrak{F}(z) z^{\Lambda'} \exp[\Lambda z] ,$$

where

$$\text{(1.5)} \qquad \mathfrak{F}(z) = \mathfrak{F}_o + \sum_{r=o}^{\infty} \frac{\mathfrak{F}r}{\left(\frac{ze^{-i\alpha}}{\omega}\right)\left[\left(\frac{ze^{-i\alpha}}{\omega}\right)+1\right]\cdots\left[\left(\frac{ze^{-i\alpha}}{\omega}\right)+r\right]}$$

and $\mathfrak{F}_o$ is non-singular. The series $\mathfrak{F}(z)$ is called a generalized factorial series and it converges in the half-plane H.

One natural requirement on κ is that $\kappa > R$, but aside from this, the theorem does not yield convenient and useful estimates for the values of both ω_o and κ. Such information would predict which generalized factorial series we should employ and the location of its abscissa of convergence (i.e., the boundary of its half-plane of convergence). In the proof of Theorem A, via a Laplace integral representation, an associated system of integral equations, and an estimate on the growth of the determining function as $t \to \infty$ along the real axis, the assumption that $A(z)$ is holomorphic at ∞, as Wasow remarks ([13], pp. 330-340), is not completely utilized. Wasow does give an estimate on ω_o in terms of the requirement that the point set

$$\omega_o(\lambda_\nu - \lambda_\mu)e^{i\alpha}, \quad 1 \leq \nu \neq \mu \leq n,$$

must be outside a region Γ, the image of the disk $|1 - s| < 1$ under the mapping $t = \log 1/s$, where the branch of the logarithm is selected which is real for positive s. A corresponding lower bound for κ, however, is given in terms of quantities which appear to be generally inaccessible and therefore Wasow questions the usefulness of the estimate ([13]; p. 340).

In this note, a proof for Theorem A is given which yields some explicit estimates for the quantities ω_o and κ. The proof begins with a reduction of the original system (1.1) to a Birkhoff canonical form. The asymptotic behavior of the formal series in a formal solution for the reduced system is then given via an associated system of difference equations. Using a result of Sibuya, the formal series is shown to asymptotically represent an analytic function in a sector larger than a half-plane. Then a result of Watson is applied to conclude that the formal series may be summed as a convergent generalized factorial series. Finally, the formal series of the original system is shown to be summed as a convergent generalized factorial series.

Another question which arises from Theorem A concerns the necessity of the hypothesis that A_o has all distinct eigenvalues. Of course, in this case formal solutions generally look different from (1.2), but they still asymptotically represent actual solutions in some sector of sufficiently small opening. It is unknown, however, when A_o has some equal eigenvalues, whether there exists a suitably modified type of generalized factorial series which would converge in the same sector. Trjitzinsky [9] and Turrittin [10] have treated some cases of equal eigenvalues with positive results. (See Turrittin [10], pp. 27-66 for a discussion of the known results as well as the simplest undecided cases.) When $A(z)$ has a pole of order q at ∞ and A_o has all distinct eigenvalues, then a generalized factorial series still converges (with z replaced by z^{q+1}) in an appropriate sector which is sufficiently far removed from the origin. (See Wasow [13], pp. 340-344.)

2. Reduction to a Birkhoff "canonical" form.

We begin by stating a result due to Turrittin on the simplification of the original system (1.1) (in our case $q = 0$).

<u>Theorem B.</u> (See Turrittin [11], p. 492.)

If the eigenvalues of A_o are distinct, there exists a transformation $x = T(z)y$, where $T(z)$ is meromorphic in a neighborhood of ∞ with $\det T(z) \not\equiv 0$ in this neighborhood, which reduces the differential system (1.1) to the "canonical" form

$$(2.1) \qquad y' = B(z)y = (B_0 + B_1 z^{-1})y .$$

Such transformations $T(z)$ are called meromorphic and they can be factored as the product

$$(2.2) \qquad T(z) = P(z)z^D Q(z) ,$$

where $P(z)$, $P^{-1}(z)$, $Q(z)$, and $Q^{-1}(z)$ are all holomorphic at ∞ and $D = \operatorname{diag}\{d_1, \ldots, d_n\}$, where the d_i are integers and can be ordered in increasing order. For our application of Turrittin's result, we need an estimate on the span of D, i.e., $d_n - d_1$, of some such transformation $T(z)$. There are many meromorphic matrices $T(z)$ which transform (1.1) into a system of the form (2.1), and it is generally not possible to obtain a universal bound on the span of D for all such matrices $T(z)$. However, using a modification in the proof of Turrittin's result, it can be shown that there exists a $T(z)$ with the factorization (2.2) and $d_n - d_1 \leq n - 1$, provided the system (1.1) does not split into systems of smaller dimensions. We say the system (1.1) splits into smaller dimensional systems if there exists a transformation $x = T(z)w$ which takes (1.1) into a system of the form

$$(2.3) \qquad w' = \begin{bmatrix} \tilde{A}_{11}(z) & 0 \\ \tilde{A}_{21}(z) & \tilde{A}_{22}(z) \end{bmatrix} w ,$$

with square diagonal blocks of positive dimension. For such systems, the problem of summing the formal series as convergent generalized factorial series is equivalent to summing the series for the systems corresponding to the diagonal blocks and those in the off-diagonal blocks can be obtained by quadrature: the estimate for the location of the abscissa of convergence will depend additively upon the dimension n of the system, hence for systems which split, the same estimate may be used.

We now sketch a proof of this modification* in Turrittin's proof. From one form of Birkhoff's factorization lemma (see [6], Section 2) it follows that the system (1.1) is equivalent by means of $x = P(z)w$ to $w' = C(z)w$, where $P(z)$ and $P^{-1}(z)$ are holomorphic at ∞ and $C(z)$ has the form

$$C(z) = (c_{ij}(z)) = \left(\sum_{k=0}^{1+k_j-k_i} c_{ij}^{(k)} z^{-k}\right),\ 1 \leq i, j \leq n, \tag{2.4}$$

and the k_i are integers such that $k_1 \geq k_2 \geq \cdots \geq k_n$. In those entries of $C(z)$ above the main diagonal for which $k_j - k_i < -1$, the summand is taken to be zero. If $k_1 - k_n \geq n$, then a combinational argument shows that in $C(z)$ there is a block of zero entries in the upper right hand corner which partitions $C(z)$ to have at least two square diagonal blocks. Hence in this case the system $w' = C(z)w$ would have the form (2.3) which means that the original system splits. Therefore, if no splitting occurs, $k_1 - k_n \leq n - 1$. In the sub-diagonal entries of $C(z)$, there may be terms of the form z^{-k} with $2 \leq k \leq n$. The object of the next transformation is to annihilate these terms. Since A_o has n distinct eigenvalues, and $C_o = (c_{ij}^{(o)})$ is similar to A_o,

*Done jointly with W. B. Jurkat.

there exists a constant lower triangular matrix M with ones on the diagonal such that $M^{-1}C_o M$ is upper triangular. A shearing transformation of the form $\mathrm{diag}\{z, z, \ldots, z, 1, \ldots, 1\}$ (with span one) can then be utilized to produce a new system with coefficient in the same form as (2.4) and new integers $k_1^*, \ldots, k_n^*$ such that $k_1^* - k_n^*$ is decreased by one. In at most $n - 1$ such steps this procedure yields systems of the form (2.1) and the resulting transformation has a span of at most $n - 1$.

If the eigenvalues of A_o are denoted by λ_j, $1 \leq j \leq n$ and if $\Lambda = \Lambda_A = \mathrm{diag}\{\lambda_1, \ldots, \lambda_n\}$, then it is easy to see by comparing the exponential terms of the formal solutions corresponding to (1.1) and (2.1) that the eigenvalues of B_o must coincide with those of A_o. Hence, without loss in generality we may assume that $A_o = \Lambda = B_o$, for this may be accomplished by constant transformations which do not alter the span of $T(z)$ and leave the system (2.1) with a coefficient matrix of the same type.

The system (1.1) thus has a formal fundamental solution matrix of the form

$$F_A(z) z^{\Lambda'_A} \exp[\Lambda z], \tag{2.5}$$

where $F_A(z) = I + \sum_1^\infty F_k[A] z^{-k}$ and $\Lambda'_A = \mathrm{diag}\ A_1$, and the system (2.1) has a formal fundamental solution matrix of the same form as (2.5) with A replaced by B.

The asymptotic behavior of the coefficients $F_k[B]$ will be determined (in the next section) via an associated system of difference equations and depends upon the matrices A

and Λ'_B. In order to turn this over to obtain the asymptotic behavior of $F_k[A]$, we need to know how Λ'_A and Λ'_B are related.

Returning to the factorization (2.2), define $\hat{A}(z) = P^{-1}(z)A(z)P(z) - P^{-1}(z)P'(z)$ and $\hat{B}(z) = Q(z)B(z)Q^{-1}(z) + Q'(z)Q^{-1}(z)$. Since $\hat{x}' = \hat{A}(z)\hat{x}$ and $\hat{y}' = \hat{B}(z)\hat{y}$ are related by $\hat{x} = z^D\hat{y}$, then

$$\hat{B}(z) = z^{-D}\hat{A}(z)z^D - Dz^{-1}, \tag{2.6}$$

and considering the diagonal terms of (2.6) we have

$$\operatorname{diag} \hat{A}_o = \operatorname{diag} \hat{B}_o \quad \text{and} \quad \operatorname{diag} \hat{B}_1 = \operatorname{diag} \hat{A}_1 - D.$$

If D is blocked along constant stretches of the d_i, then since both $\hat{A}(z)$ and $\hat{B}(z)$ are holomorphic at ∞, the matrix $\hat{A}_o$ must be blocked lower triangular and $\hat{B}_o$ is blocked upper triangular. Hence blocked triangular matrices with identity blocks along the main diagonal may be used to block diagonalize $\hat{A}_o$ and $\hat{B}_o$. Within the diagonal blocks, $A(z)$ and $B(z)$ coincide aside from a scalar multiple of the identity divided by z. These blocked triangular matrices leave the diagonal blocks of $\hat{A}_1$ and $\hat{B}_1$ unchanged, so if we let $\Lambda'_{\hat{A}}$ and $\Lambda'_{\hat{B}}$ denote respectively the diagonals of the coefficients in z^{-1} when $\hat{A}(z)$ and $\hat{B}(z)$ have been transformed so that their leading coefficients are both equal to Λ, then $\Lambda'_A = \Lambda'_{\hat{A}}$ and $\Lambda'_B = \Lambda'_{\hat{B}}$. Also, from the above discussion it follows that $\Lambda'_{\hat{B}} = \Lambda'_{\hat{A}} - D$, hence we have obtained

$$\Lambda'_A = \Lambda'_B + D. \tag{2.7}$$

If $F_B(z)z^{\Lambda'_B}\exp[\Lambda z]$ denotes a formal fundamental solution matrix for $y' = B(z)y$, then a formal fundamental solution matrix for $\hat{y}' = \hat{B}(z)\hat{y}$ is $Q(z)F_B(z)z^{\Lambda'_B}\exp[\Lambda z]$. Likewise, a formal fundamental solution matrix for $\hat{x}' = \hat{A}(z)\hat{x}$ is $P^{-1}(z)F_A(z)z^{\Lambda'_A}\exp[\Lambda z]$. Hence there exists a constant, non-singular matrix C such that

$$P^{-1}(z)F_A(z)z^{\Lambda'_A}\exp[\Lambda z]C = z^D Q(z)F_B(z)z^{\Lambda'_B}\exp[\Lambda z]. \tag{2.8}$$

Since Λ is diagonal with distinct eigenvalues, C must be diagonal and from (2.8) we obtain

$$P^{-1}(z)F_A(z)C = z^D Q(z)z^{\Lambda'_B - \Lambda'_A} \tag{2.9}$$

and view if (2.7) this becomes

$$F_A(z)C = P(z)z^D Q(z)F_B(z)z^{-D}. \tag{2.10}$$

In the next section, the asymptotic behavior of the coefficients in the formal series $F_B(z)$ is determined and from (2.10) this information can be utilized to obtain the asymptotic behavior of the coefficients of $F_A(z)$.

3. An associated system of difference equations.

As we remarked in the previous section, the differential system $y' = (\Lambda + B_1 z^{-1})y$ has a formal fundamental solution matrix of the form $F_B(z)z^{\Lambda'_B}\exp[\Lambda z]$, where $\Lambda'_B = \text{diag } B_1$ and

$$F_B(z) = I + \sum_{\nu=1}^{\infty} F_\nu[B]z^{-\nu},$$

The matrices $F_\nu[B]$ are determined recursively from the equations

$$(3.1)\quad -\nu F_\nu[B] = \Lambda F_{\nu+1}[B] - F_{\nu+1}[B]\Lambda + B_1 F_\nu[B] - F_\nu[B]\Lambda'_B ,$$

for $\nu \geq 0$. If $f^j(\nu)$ denotes the (n-1) dimensional vector consisting of the (n-1) components of the j-th column of $F_\nu[B]$ with the diagonal entry removed, then from the equations (3.1) we have by direct verification

Lemma 3.1. The vector $f^j(\nu)$ satisfies the (n-1) dimensional system of linear difference equations

$$(3.2)\qquad f^j(\nu+1) = (\nu C_o^j + C_1^j + \frac{1}{\nu} C_2^j) f^j(\nu) ,$$

where $C_o = \text{diag}\{(\lambda_j - \lambda_1)^{-1}, \ldots, (\lambda_j - \lambda_n)^{-1}\}$,

$$\text{diag } C_1 = \text{diag}\left\{\frac{\mu'_j - \mu'_1}{\lambda_1 - \lambda_j}, \ldots, \frac{\mu'_j - \mu'_n}{\lambda_n - \lambda_j}\right\},$$

and μ'_j, $1 \leq j \leq n$, denote the entries of Λ'_B.

Note that our assumption that the λ_j are distinct implies that the eigenvalues of C_o are distinct and, furthermore, none of them are zero. The diagonal entries of $F_\nu[B]$ are determined from (3.1) to be

$$(3.3)\qquad -\nu(F_\nu[B])_{jj} = \sum_{\substack{m=1\\ m\neq j}}^{n} b_{jm}(F_\nu[B])_{mj} .$$

Hence, if we obtain the asymptotic behavior of the off-diagonal elements of the j-th column of $F_\nu[B]$, then the asymptotic behavior of $(F_\nu[B])_{jj}$ is determined from (3.3) as a linear combination of the off-diagonal entries divided by ν. Thus any asymptotic estimates for the growth of off-diagonal entries of the j-th column of $F_B(z)$ will also hold for the diagonal entry, which grows at a slower rate as $\nu \to \infty$.

We proceed now to consider the asymptotic behavior of the vectors $f^j(\nu)$ as $\nu \to \infty$. The system of difference equations (3.2) has an irregular singular point at ∞. The system (3.2) has a formal fundamental solution matrix of the form

$$x(\nu) = \Gamma(\nu)S(\nu)\nu^{\tilde{D}}(C_o)^{\nu} \quad , \tag{3.4}$$

where $\Gamma(\nu)$ denotes the Gamma function, $\tilde{D} = \mathrm{diag}\{\mu_j' - \mu_1', \ldots, \mu_j' - \mu_n'\}$, and $S(\nu)$ is a formal power series in ν^{-1} with non-singular leading coefficient. This representation can be obtained by formally diagonalizing the system (3.2) and then solving the resulting scalar equations. (See Harris [2] for a discussion and references.) Since we only are concerned with a solution at the positive integers, for our purposes it is sufficient to know that in a small sector containing the positive real axis, there exists an actual solution of (3.2) having the formal solution (3.4) as its asymptotic expansion. This follows from an asymptotic existence theorem for non-linear difference systems. (Again see Harris [2], pp. 14-16.) Therefore, there exists a column vector c such that

$$f^j(\nu) \approx \Gamma(\nu)S(\nu)\nu^{\tilde{D}}C_o^{\nu}\, c$$

as $\nu \to \infty$ (ν a positive integer). Hence, each component of $f^j(\nu)$ will have an asymptotic expansion of the form

$$\Gamma(\nu) \sum_{\substack{i=1 \\ i \neq j}}^{n} s_i(\nu)\nu^{\mu_j' - \mu_i'}(\lambda_j - \lambda_i)^{-\nu} \quad \text{as} \quad \nu \to \infty \quad , \tag{3.5}$$

where $s_i(\nu)$ denote asymptotic expansions in ν^{-1}.

From (3.5) we see that the coefficients of the formal series $F_B(z) = I + \sum_1^n F_\nu[B]z^{-\nu}$ satisfy

$$(3.6) \qquad (F_\nu[B])_{ij} = O(\Gamma(\nu)\rho^\nu) \text{ as } \nu \to \infty,$$

where ρ is a positive real number such that

$$(3.7) \qquad \rho > \rho_o = \max_{j\neq i} \left| \frac{1}{\lambda_j - \lambda_i} \right| .$$

If we let $d = \max_{i\neq j} \mathrm{Re}(\mu'_j - \mu'_i)$, we have a more precise estimate

$$(3.8) \qquad (F_\nu[B])_{ij} = O(\Gamma(\nu)\nu^d \rho_o^\nu) \text{ as } \nu \to \infty .$$

We now turn to equation (2.10), which relates $F_A(z)$ and $F_B(z)$. Since $Q(z)$ and $Q^{-1}(z)$ are holomorphic at ∞, it can be shown that the coefficients of the formal series $Q(z)F_B(z)$ satisfy the same relations (3.6) and (3.8). The (i,j) component $z^D[Q(z)F_B(z)]z^{-D}$ can be expanded in the formal power series

$$\sum_{k=o}^{\infty} f_{ij}(k) z^{d_i - d_j - k} = \sum_{m=d_i-d_j}^{\infty} f_{ij}^{(d_i-d_j+m)} z^{-m} ,$$

where $(Q(z)F_B(z))_{ij} = \sum_o^\infty f_{ij}^{(k)} z^{-k}$ and the coefficients are zero if $d_i - d_j < 0$. From (3.8) we have

$$f_{ij}^{(d_i-d_j+m)} = O(\Gamma(d_i-d_j+m)(d_i-d_j+m)^d \rho_o^{d_i-d_j+m})$$

$$= O(\Gamma(m) m^{d+d_i-d_j} \rho_o^m) \text{ as } m \to \infty .$$

If $\max_{i \neq j}(d_i - d_j) = K$, then the coefficients of the formal series $z^D [Q(z) F_B(z)] z^{-D}$ and hence the coefficients of the formal power series $\sum_0^\infty F_\nu [A] z^{-\nu}$ satisfy

$$(3.9) \qquad (F_\nu [A])_{ij} = O(\Gamma(\nu) \nu^{d+K\nu} \rho_o^\nu) \text{ as } \nu \to \infty .$$

Recall that μ_j', $1 \leq j \leq n$, are the entries of Λ_B' and from (2.7) we have $\Lambda_A' = \Lambda_B' + D$. Hence $\mu_j' - \mu_i' = \lambda_j' - \lambda_i' + d_j - d_i$. If $x' = A(z)x$ does not split into lower dimensional systems, then $K \leq n-1$ and we have

$$d = \max_{i \neq j} \mathrm{Re}(\mu_j' - \mu_i') \leq \max_{i \neq j} \mathrm{Re}(\lambda_j' - \lambda_i') + n - 1 .$$

Therefore, in this case

$$(3.10) \qquad (F_\nu [A])_{ij} = O(\Gamma(\nu) \nu^{d_o} \rho_o^\nu) \text{ as } \nu \to \infty ,$$

where

$$(3.11) \qquad d_o = \max_{i \neq j} \mathrm{Re}(\lambda_j' - \lambda_i') + 2(n-1) .$$

Since ρ in (3.6) can be taken to be any fixed positive number satisfying (3.7), then we also have

$$(3.12) \qquad (F_\nu [A])_{ij} = O(\Gamma(\nu) \rho^\nu) \text{ as } \nu \to \infty ,$$

where ρ satisfies (3.7). We summarize these results in

Theorem 1.

Let $P^{-1} A_o P = \Lambda = \mathrm{diag}\{\lambda_1, \ldots, \lambda_n\}$, where the λ_i

are distinct and $\operatorname{diag}\{P^{-1}A_1P\} = \Lambda'_A = \operatorname{diag}\{\lambda_1', \ldots, \lambda_n'\}$. Then the differential system $x' = A(z)x$ has a formal fundamental solution matrix of the form $F_A(z)z^{\Lambda'_A}\exp[\Lambda z]$, where $F_A(z) = \sum_0^\infty F_\nu[A]z^{-\nu}$, $F_0[A]$ is non-singular, and

$$F_\nu[A] = O(\Gamma(\nu)\rho^\nu) \text{ as } \nu \to \infty,$$

where ρ is any positive number satisfying (3.7). More precisely,

$$F_\nu[A] = O(\Gamma(\nu)\nu^d\rho_0^\nu) \text{ as } \nu \to \infty,$$

where d is a non-negative real number. In case $x' = A(z)x$ does not split into lower dimensional systems, then d may be taken to be

$$d \geq d_0 = \max_{i \neq j} \operatorname{Re}(\lambda_i' - \lambda_j') + 2(n-1).$$

4. A proof of Theorem A.

As an application of the results of the previous section, a result of Sibuya, and a theorem of G. N. Watson, we will now obtain a proof of Theorem A which yields an estimate on ω_0 and $\varkappa$. We first state the result of Watson as

Theorem C. (Watson, [12], p. 45).

Let $f(z)$ be a function which is analytic for $\operatorname{Re} z > 0$ and let $f(z)$ be analytic in the sector

$$\text{(4.1)} \qquad S: \ |z| \geq \gamma \text{ and } |\arg z| < \tfrac{1}{2}\pi + \beta,$$

where $0 < \beta \leq \frac{\pi}{2}$. Assume that $f(z)$ possesses the asymptotic expansion

$$(4.2) \qquad f(z) = a_o + a_1 z^{-1} + \cdots + a_k z^{-k} + R_k$$

as $z \to \infty$, $z \in S$, where $|a_\nu| < A\rho^\nu \nu!$ and $|R_\nu z^{\nu+1}| < B\sigma^\nu \nu!$, A, B, ρ, and σ denoting constants independent of ν. Let $0 < p < 1 + \exp[-\pi \cot \beta]$ be selected and let M_o denote the largest positive root of

$$(4.3) \qquad \exp\left[\frac{-2\cos\beta}{\rho M_o}\right] - 2\cos\left(\frac{\sin\beta}{\rho M_o}\right)\exp\left[\frac{-\cos\beta}{\rho M_o}\right] + 1 - p^2 = 0.$$

Then for all $0 < M < M_o$, $f(z)$ can be expanded into a generalized factorial series of the form

$$(4.4) \qquad f(z) = b_o + \sum_{k=1}^{\infty} \frac{b_k}{(Mz+1)(Mz+2)\cdots(Mz+k)}$$

which converges when $\operatorname{Re} z > 0$.

After an appropriate rotation, this result will be applied to the function $\hat{F}_B(z)$, which is defined to be the function whose asymptotic expansion is the formal series $F_B(z)$ corresponding to the reduced system $y' = B(z)y = (B_o + B_1 z^{-1})y$. The standard asymptotic existence theorems for this situation yield such a function which has $F_B(z)$ as its asymptotic expansion in a half-plane. Since $B(z)$ has singularities only at 0 and at ∞, then this half-plane can be taken with boundary through the origin. But in order to be able to apply Watson's result, we need to find a positive number β such that the asymptotic expansion is valid in a sector of the form S, which is larger than a half-plane by the amount 2β.

To obtain such a sector, we apply a procedure due to

Sibuya. The author thanks Prof. Sibuya for describing to him how the general procedure specializes to the situation treated here. See P. F. Hsieh and Y. Sibuya [5], p. 100, for a similar application.

Let $\omega_{jk} = \arg(\lambda_j - \lambda_k)$, $j \neq k$, with some fixed choice for the argument, not necessarily between 0 and 2π, and define

$$(4.5) \qquad \omega_+ = \max_{j,k} \omega_{jk} \quad \text{and} \quad \omega_- = \min_{j,k} \omega_{jk} .$$

If we consider the sector

$$(4.6) \qquad S^*: \{z: \frac{3\pi}{2} - \omega_- < \arg z < \frac{3\pi}{2} - \omega_+\} ,$$

then according to Sibuya, there exists for γ sufficiently a large function $\hat{F}_B(z)$ analytic for $|z| > \gamma$ and $z \in S^*$ such that $\hat{F}_B(z)$ has $F_B(z)$ as its asymptotic expansion as $z \to \infty$, $z \in S^*$. The opening of the sector S^* is $3\pi - (\omega_+ - \omega_-)$, hence it satisfies the requirements of the sector S given by (4.1) (under a suitable rotation) with any number β satisfying

$$(4.7) \qquad 0 < \beta < \pi - (\omega_+ - \omega_-)/2 .$$

Let α denote the argument of the bisector of the sector S^* defined by (4.6). Under the rotation $z^* = ze^{-i\alpha}$, the sector S^* is changed so that it is bisected by the positive real axis. The result of Theorem C now yields M_o, defined to be the largest positive root of (4.3) such that (componentwise) the function $\hat{F}_B(z)$ can be expressed as a generalized factorial series of the form (4.4) (with z replaced by $ze^{-i\alpha}$), which converges for $\mathrm{Re}(ze^{-i\alpha}) > 0$.

In what follows, we denote z^* be z and make the inverse rotation later.

In order to obtain a convergent factorial series for the formal series $F_A(z)$ of the original differential system, recall the equation (2.10) relating $F_A(z)$ and $F_B(z)$. Since $Q(z)$ is holomorphic at ∞, it can be represented as a convergent factorial series (see argument below for $P(z)$). But in the proof of our modification of Turrittin's result, $Q(z)$ actually turns out to be a polynomial in z^{-1}, hence it has a factorial series which converges for $\operatorname{Re} z > 0$. Such a factorial series can be expressed as a generalized factorial series of the form (4.4) which still converges for $\operatorname{Re} z > 0$ ([8], p. 358). Since the product of two factorial series, each which converges for $\operatorname{Re} z > \lambda, \lambda'$ respectively, converges when $\operatorname{Re} z > \max\{0, \lambda, \lambda'\}$ (see [7], p. 254), then $Q(z)F_B(z)$ has a generalized factorial series which converges for $\operatorname{Re} z > 0$. The entries of the matrix $z^D Q(z) F_B(z) z^{-D}$ consist of terms of the form $z^d f(z)$, where d is an integer and $f(z)$ has a convergent generalized factorial series expansion for $\operatorname{Re} z > 0$. If $d < 0$, then, as before, the product consists of a polynomial in z^{-1} and a convergent generalized factorial series, and such a product converges for $\operatorname{Re} z > 0$. However, if $d > 0$, then we may have to move to the right in order to obtain convergence. To see what happens, let $(Mz + 1)$ (4.4) be replaced by ζ, which we now write as an "ordinary" factorial series

$$g(\zeta) = \sum_{\nu=0}^{\infty} \frac{b_\nu \nu!}{\zeta(\zeta+1)\cdots(\zeta+\nu)} . \tag{4.8}$$

Assume that the series (4.8) has 0 as its abscissa of convergence. Then it converges for $\zeta=1$, which implies that its general term (with $\zeta=1$) approaches zero, i.e.,

$$\frac{b_\nu \nu!}{(\nu+1)!} \to 0 , \quad \text{hence} \quad b_\nu = 0(\nu) .$$

The function $\zeta g(\zeta)$ has the factorial series expansion (see [1], p. 218)

$$b_0 + \sum_0^\infty \frac{[(\nu+1)b_{\nu+1} - \nu b_\nu]\nu!}{\zeta(\zeta+1)\cdots(\zeta+\nu)} , \tag{4.9}$$

which may be viewed as the difference of two factorial series. Since each one of them has the form (4.8) with b_ν replaced by νb_ν and $(\nu+1)b_{\nu+1}$, respectively, then the coefficients satisfy $b_\nu = 0(\nu^2)$ as $\nu \to \infty$. According to Landau's Theorem (see [14], pp. 97-99), the factorial series (4.9) converges for $\operatorname{Re} z > 3$. This argument may be repeated inductively to conclude that the function $\zeta^d g(\zeta)$, d is a positive integer, converges for $\operatorname{Re}\zeta > 3d$. Hence the generalized factorial series $z^d g(Mz+1)$ converges for $\operatorname{Re} z > 3d/M$. Of course, it is assumed that the first d coefficients of $g(\zeta)$ are zero, but this must happen automatically since according to (2.10), the products we consider all lead to formal power series in z^{-1}.

Finally, the formal series $F_A(z)$ is given by

$$F_A(z) = P(z)[z^D Q(z) F_B(z) z^{-D}]C^{-1}$$

and the matrix $P(z)$ consists of functions holomorphic in a neighborhood of ∞. If $A(z)$ is holomorphic for $|z| > R$, then according to Birkhoff's factorization lemma, $P(z)$ is likewise holomorphic (componentwise) in $|z| > R$, and such a function may be expanded into a generalized factorial series expansion of the form which converges for $\operatorname{Re} z > R$. This follows since a function $p(z) = \sum_{0}^{\infty} p_k z^{-k}$ which is holomorphic at ∞ is a generating function, i.e.,

$$p(z) = p_o + \int_o^\infty e^{-zt} \varphi(t)dt .$$

Then $\varphi(t)$, its determining function, is given by

$$\varphi(t) = \sum_{k=1}^{\infty} \frac{p_k}{(k-1)!} t^{k-1}$$

and turns out to be an entire function or order one and type R. Hence([1], p. 221) $p(z)$ may be expressed as a factorial series which converges for $\operatorname{Re} z > R$. Therefore $F_A(z)$ can be expressed as a generalized factorial series of the form (4.4) which converges when $\operatorname{Re}(ze^{-i\alpha}) > \mu = \max\{R, 3d/M\}$. We summarize these results as

Theorem 2.

Let $A(z) = A_o + \sum_{1}^{\infty} A_\nu z^{-\nu}$ converge for $|z| > R$ and assume A_o has all distinct eigenvalues, which are denoted by λ_j, $1 \leq \lambda \leq n$. Let $\omega_{jk} = \arg(\lambda_j - \lambda_k)$, $j \neq k$, for some choice of the argument, and define ω_+ and ω_- by (4.5). Select β satisfying to (4.7), p satisfying $0 < p < 1 + \exp[-\pi \cot \beta]$, and ρ satisfying (3.7). Let α denote the argument of the bisector of the sector given by (4.6), and *select* M_o *to be the largest positive root of* (4.3). *Then for all* $0 < M < M_o$, *the differential system* $x' = A(z)x$ *has a fundamental solution matrix of the form*

$$\hat{F}(z) z^{\Lambda'} \exp[\Lambda z],$$

where $\hat{F}(z) = \mathfrak{F}_0 + \sum_{\nu=1}^{\infty} \frac{\mathfrak{F}_\nu}{(Mze^{-i\alpha}+1)\cdots(Mze^{-i\alpha}+\nu)}$

and this generalized factorial series converges for $\operatorname{Re}(ze^{-i\alpha}) > \mu = \max\{R, 3d/M\}$. *If* $x' = A(z)x$ *does not split into lower dimensional systems, then the integer* d *does not exceed* $n - 1$.

REFERENCES

[1] G. Doetsch, *Handbuch der Laplace - Transformation, Band II,* Birkhäuser Verlag, Basel, 1955.

[2] W. A. Harris, Jr., *Analytic theory of difference equations,* Lecture Notes in Mathematics No. 183, Springer - Verlag, 1971, pp. 46-58.

[3] J. Horn, *Fakultätenreihen in der Theorie der linearen Differentialgleichungen,* Math. Ann. **71**(1912) pp. 510-532.

[4] J. Horn, *Integration linear Differentialgleichungen durch Laplaceshe Integrale und Fakultätenreihen,* Jahresber. Deut. Math. Ver. **24**(1915) pp. 309-325, **25**(1915) pp. 74-83.

[5] P. F. Hsieh and Y. Sibuya, *Note on regular*

perturbations of linear ordinary differential equations at irregular singular points, Funkcialaj Ekvacioj, 8(1966) pp. 99-108.

[6] W. Jurkat, D. Lutz, and A. Peyerimhoff, Birkhoff invariants and effective calculations for meromorphic linear differential equations, I, J. Math. Anal. Appl. (to appear).

[7] N. Nielsen, Handbuch der Theorie der Gammafunktion, Teubner Verlag, Leipzig, 1906.

[8] N. E. Nörlund, Sur les séries de facultes, Acta Math. 37(1914) pp. 327-387.

[9] W. J. Trjitzinsky, Laplace integrals and factorial series in the theory of linear differential and difference equations, Trans. Amer. Math. Soc., 37(1935) pp. 80-146.

[10] H. L. Turrittin, Convergent solutions of ordinary linear homogeneous differential equations in the neighborhood of an irregular singular point, Acta Math., 93(1955) pp. 27-66.

[11] H. L. Turrittin, Reduction of ordinary differential equations to the Birkhoff canonical form, Trans. Amer. Math. Soc., 107(1963) pp. 495-507.

[12] G. N. Watson, The transformation of an asymptotic series into a convergent series of inverse factorials, Cir. Mat. Palermo, Rend., 34(1912) pp. 41-88.

[13] W. Wasow, Asymptotic Expansions for Ordinary Differential Equations, J. Wiley, New York, 1965.

[14] D. V. Widder, The Laplace Transform, Princeton, Univ. Press, Princeton, N. J., 1946.

University of Wisconsin, Milwaukee, Wisconsin

INVARIANT SETS FOR EVOLUTION SYSTEMS*

Robert H. Martin, Jr

1. Introduction.

In recent years many results have been obtained concerning the determination and characterization of invariant sets for ordinary differential equations, both in finite dimensional spaces and in general Banach spaces. A sample of these results and techniques may be found in Bony [1]; Brezis [2]; Crandall [4], [5]; Hartman [7]; Martin [9], [10], [12]; Redheffer [15]; Redheffer and Walter [16], [17]; Volkmann [19]; Walter [21]; and Yorke [22], [23]. These rechniques may also be applied to differential inequalities as well (see Martin [13]; Redheffer and Walter [16]; and Vokmann [18], [19], [20]). The underlying (and crucial) condition either directly or indirectly associated with each of these techniques goes back to Nagumo [14].

In [3, Chapter IV, § 4], Brezis establishes invariance criteria for semigroups of nonlinear operators in a Hilbert space. It is the purpose of this paper to obtain invariance criteria for nonlinear evolution systems in general Banach spaces. Also, we indicate an application of our results to

*This work was supported by a NATO Fellowship in Science and the U.S. Army Research Office, Durham, North Carolina.

systems of nonlinear parabolic equations.

§ 2. Preliminaries.

Suppose that E is a real Banach space with norm $|\cdot|$ and that E^* is the dual space of E with the norm on E^* also denoted $|\cdot|$. Now let D be a closed subset of E and let ω be a real number. A family $U = \{U(t, s): 0 \leq s \leq t\}$ of mappings from D into D is said to be an evolution system of type ω if the following conditions are fulfilled:

(U1) $U(t, t)x = x$ for all $x \in D$ and $t \geq 0$;

(U2) $U(t, s)U(s, r)x = U(t, r)x$ for all $x \in D$ and $0 \leq r \leq s \leq t$;

(U3) for each $x \in D$ the mapping $(t, s) \rightarrow U(t, s)x$ is continuous from $\Delta = \{(t, s) : 0 \leq s \leq t\}$ into E; and

(U4) $|U(t, s)x - U(t, s)y| \leq |x - y|e^{\omega(t-s)}$ for all $x, y \in D$ and $0 \leq s \leq t$.

We say that U is a linear evolution system of type ω if $D = E$ and $U(t, s)$ is a bounded linear operator from E into E for each $0 \leq s \leq t$. If the family U satisfies (U1) - (U4) and $t \geq 0$, define the operator $G(t)$ by

$$D(G(t)) = \{x \in D: \lim_{h \rightarrow 0+} [U(t+h), t)x - x]/h \text{ exists}\}$$

exists

$$G(t)x = \lim_{h \rightarrow 0+} [U(t+h, t)x - x]/h \text{ for all } x \in D(G(t))$$

If $D(G(t))$ is nonempty for each $t \geq 0$, we say that G is the <u>infinitesimal generator of U</u>. Roughly speaking, if $s \geq 0$

and $x \in D(G(s))$, the function $t \to U(t,s)x$, $t \geq s$, may be regarded as a generalized solution to the initial value problem

$$u'(t) = G(t)u(t), \quad u(s) = x, \quad t \geq s. \tag{2.1}$$

If A is a subset of $E \times E$ we set $Ax = \{y \in E : (x,y) \in A\}$, $D(A) = \{x \in E : Ax \text{ is nonempty}\}$, and $R(A) = \bigcup_{x \in D(A)} Ax$. Moreover, if A and B are subsets of $E \times E$ and $\lambda \in \mathbb{R}$, then $\lambda A = \{(x, \lambda y) : (x,y) \in A\}$, $A^{-1} = \{(y,x) : (x,y) \in A\}$. and $A + B = \{(x, y+z) : (x,y) \in A \text{ and } (x,z) \in B\}$. When convenient we identify functions from E into E with their graphs. In particular, the identity map I on E is written as $I = \{(x,x) : x \in E\}$. The class $\mathcal{A}(\omega)$ denotes the class of all subsets A of $E \times E$ with the property that if $\lambda > 0$ and $\lambda\omega < 1$, then

$$|x_1 - x_2 - \lambda(y_1 - y_2)| \geq (1 - \lambda\omega)|x_1 - x_2| \quad \text{for all } (x_1, y_1), (x_2, y_2) \in A.$$

An important and very general example of "generating" an evolution system is given by Crandall and Pazy [6]. Suppose that $\{A(t) : t \geq 0\}$ is a family of subsets of $E \times E$ with the following properties:

(A1) $A(t) \in \mathcal{A}(\omega)$ for $t \geq 0$;

(A2) $D(A(t)) = D(A(0))$ for $t \geq 0$ and $\overline{D(A(0))} = D$;

(A3) there is a $\lambda_o > 0$ such that $\lambda_o \omega < 1$ and $R(I - \lambda A(t)) \supset D$ for $\lambda \in (0, \lambda_o]$ and $t \geq 0$; and

(A4) there is an increasing function $L : [0,\infty) \to [0,\infty)$ such that

$$|(I - \lambda A(t))^{-1}x - (I - \lambda A(s))^{-1}x| \leq \lambda |t-s| L(|x|)(1 + \|A(s)x\|)$$

for all $\lambda \in (0, \lambda_o]$, $t, s \geq 0$, and $x \in D$.

The definition of $\|A(s)x\|$ in (A4) is as follows: $\|A(s)x\| = \lim_{\lambda \to 0+} |A_\lambda(s)x|$ where $A_\lambda(s)x = [(I - \lambda A(s))^{-1}x - x]/\lambda$ (see [6, Definition 1.1]). If follows from Crandall and Pazy [6, Theorem 2.1] that if $\{A(t) : t \geq 0\}$ satisfies (A1) - (A4) then

$$(2.2) \qquad U(t, s)x = \lim_{n \to \infty} \prod_{i=1}^{n} [I-n^{-1}(t-s)A(s+in^{-1}(t-s))]^{-1}x$$

exists for $0 \leq s \leq t$ and $x \varepsilon D$, and U is an evolution system of type ω.

If $\{A(t) : t \geq 0\}$ satisfies (A1) - (A4) and E and E^* are uniformly convex, then $A(t)x$ is a closed convex subset of E for all $t \geq 0$ and $x \varepsilon D(A(t))$. Moreover, if $A(t)^o x$ is the element of minimal norm in $A(t)x$ and U is defined by (2.2), then $U(t, s) : D(A(0)) \to D(A(0))$ for all $0 \leq s \leq t$ and the infinitesimal generator G of U satisfies $G(t)x = A(t)^o x$ for all $t \geq 0$ and $x \varepsilon D(G(t)) = D(A(t))$. In particular,

$$d^+U(t, s)x/dt = A(t)^o U(t, s)x \quad \text{for } 0 \leq s \leq t \text{ and } x \varepsilon D(A(s))$$

(see Crandall and Pazy [6, Remark 3.4] and Martin [11]).

If C is a subset of E and $x \varepsilon E$, define $d(x;C) = \inf\{|x-y| : y \varepsilon C\}$. Note that if C is convex and $x, z \varepsilon E$, then $h \to d(x+hz;C)$ is convex from $\mathbb{R}$ into $\mathbb{R}$. Consequently,

$$(2.3) \qquad \lim_{h \to 0+} [d(x+hz;C) - d(x;C)]/h$$

exists whenever C is convex. Whenever C is a closed convex subset of E and x is in E define

$$(2.4) \quad J(x;C) = \{\alpha \varepsilon E^* : |\alpha| = 1 \text{ and } \alpha(x-y) \geq d(x;C) \text{ for } y \varepsilon C\}.$$

Note that $J(x;C)$ may be empty if $x \in C$. However, if $x \notin C$ then $J(x;C)$ is always nonempty (see Kothe [8, (6), p. 345]). Note also that if $C = \{\theta\}$, then $|x|J(x;C)$ is the duality mapping from E into the set of subsets of E^*. We have the following fundamental connection between (2.3) and (2.4):

Proposition 1. Suppose that C is a closed convex subset of E and that x and z are in E. Also, let $J(x;C)$ be defined by (2.4). Then

$$\lim_{h \to 0+} [d(x+hz;C)-d(x;C)]/h = \begin{cases} \sup\{\alpha(z): \alpha \in J(x;C)\} & \text{if } x \notin C \\ \max\{0, \sup\{\alpha(z) : \alpha \in J(x;C)\}\} & \text{if } x \in C \end{cases}$$

where $\sup\{\alpha(z) : \alpha \in J(x, C)\} = 0$ if $J(x;C)$ is empty.

Proof. Set $\Gamma = \lim_{h \to 0+} [d(x+hz;C) - d(x;C)]/h$ and $\Gamma^* = \sup\{\alpha(z) : \alpha \in J(x:C)\}$. Then

$$d(x+hz;C) = d(x;C) + h\Gamma + o(h),$$

where $o(h)/h \to 0$ as $h \to 0+$. Thus for each $h > 0$ there is a $y_h \in C$ such that

$$|x + hz - y_h| = d(x;C) + h\Gamma + o(h).$$

If $J(x;C) \neq \emptyset$ and $\alpha \in J(x;C)$ then

$$\alpha(x + hz - y_h) \leq d(x;C) + h\Gamma + o(h),$$

and since $\alpha(x - y_h) \geq d(x;C)$ it follows that

$$h\alpha(z) \leq d(x;C) - \alpha(x - y_h) + h\Gamma + o(h) \leq h\Gamma + o(h).$$

Thus $\alpha(z) \leq \Gamma + o(h)/h$ and letting $h \to 0+$ shows that $\alpha(z) \leq \Gamma$ for all $\alpha \in J(x;C)$. If $J(x;C) = \emptyset$ then $d(x;C) = 0$ and it is clear that $\Gamma \geq 0$. Consequently, $\Gamma^* \leq \Gamma$. Now we consider two cases. First suppose that there is a $\delta > 0$ with $x + hz \notin C$ for all $h \in (0, \delta]$. Then for each $h \in (0,\delta]$ there is an $\alpha_h \in J(x + hz; C)$. Therefore, for each $y \in C$

$$\alpha_h(x+hz-y) \geq d(x+hz;C) = d(x;C) + h\Gamma + o(h),$$

and it follows that

$$h\alpha_h(z) \geq d(x;C) - \alpha_h(x-y) + h\Gamma + o(h)$$

$$\geq d(x;C) - |x-y| + h\Gamma + o(h)$$

for all $y \in C$. Taking the infimum over $y \in C$ we obtain $\alpha_h(z) \geq \Gamma + o(h)/h$ for all $h \in (0,\delta]$. By weak star compactness we may assume that there is an $\alpha \in E^*$ and a sequence $\{h_n\}_1^\infty$ in $(0,\delta]$ such that $h_n \to 0$ as $n \to \infty$ and $\alpha_{h_n}(w) \to \alpha(w)$ for all $w \in E$. Then $|\alpha| \leq 1$ and

$$\alpha(z) = \lim_{n \to \infty} \alpha_{h_n}(z) \geq \lim_{n \to \infty} \Gamma + o(h_n)/h_n = \Gamma.$$

Moreover, if $y \in C$ then

$$\alpha(x-y) = \lim_{n \to \infty} \alpha_{h_n}(x+h_n z-y) \geq \lim_{n \to \infty} d(x;C) + h_n\Gamma + o(h_n)$$

$$= d(x;C).$$

Thus if $d(x;C) > 0$ then $|\alpha| = 1$, $\alpha \in J(x;C)$ and $\Gamma^* \geq \alpha(z) \geq \Gamma$. If $d(x;C) = 0$ and $\Gamma = 0$ then clearly $\max\{0,\Gamma^*\} \geq \Gamma$. If $d(x;C) = 0$ and $\Gamma > 0$ then $\alpha(z) \geq \Gamma > 0$ so $\alpha \neq \theta$. Thus

$\alpha/|\alpha| \in J(x;C)$ and $\Gamma^* \geq \alpha(z)/|\alpha| \geq \Gamma$ since $\alpha(z) \geq \Gamma$ and $|\alpha| \leq 1$. This establishes the proposition in the case when $x + hz \notin C$ for all sufficiently small $h > 0$. If there is a sequence $\{h_n\}_1^\infty$ in $(0, \infty)$ such that $\lim_{n \to \infty} h_n = 0$ and $x + h_n z \in C$, then clearly $x \in C$ and $\Gamma = 0$. Since $\Gamma^* \leq \Gamma$ it is immediate that $\max\{0, \Gamma^*\} = \Gamma$ and the proof of Proposition 1 is complete.

In certain cases Proposition 1 may be phrased in terms of the duality mapping on E. For each $x \in E$ let $Fx = \{\alpha \in E^* : \alpha(x) = |x|^2 = |\alpha|^2\}$. Note that if $x \neq \theta$ then $\alpha \in J(x; \{\theta\})$ only in case $|x|\alpha \in Fx$. For each pair $(x, y) \in E \times E$ define

$$[x, y] = \sup\{\alpha(y) : \alpha \in Fx\}. \tag{2.5}$$

The duality mapping F is connected to $J(\cdot;C)$ in the following manner:

Proposition 2. Suppose that C is a closed convex subset of E, $x \in E - C$, and there is a $y \in C$ such that $d(x;C) = |x-y|$. Then

$$|x - y| J(x;C) \subset F(x - y),$$

and hence

$$|x-y| \lim_{h \to 0+} [d(x + hz;C) - d(x;C)]/h \leq [x-y, z]$$

for all $z \in E$. In particular, if $F(x-y)$ consists of exactly one member (e.g., if E^* is strictly convex) then

$d(x;C)\,J(x;C) = F(x - y)$ and

$$|x-y| \lim_{h \to 0+} [d(x + hz;C) - d(x;C)]/h = [x - y, z]$$

for all $z \in E$.

Proof. Let $\alpha \in J(x;C)$ and let $\beta(w) = |x-y|\alpha(w)$ for all $w \in E$. Then $\beta \in E^*$ and $|\beta| = |x-y| = d(x;C)$. Moreover,

$$\beta(x-y) = |x-y|\alpha(x-y) \geq |x-y|d(x;C) = |x-y|^2,$$

so $\beta(x-y) = |x-y|^2 = |\beta|^2$ and hence $\beta \in F(x-y)$. Consequently $|x-y|J(x;C) \subset F(x-y)$ and the remaining assertions follow easily.

§ 3. Invariance Critera.

In this section we prove our main results. Throughout we assume that D is a closed subset of E, ω is a real number, and $U = \{U(t, s) : 0 \leq s \leq t\}$ is an evolution system of type ω which satisfies properties (U1) - (U4) in § 2. Also, we assume that C is a closed subset of E and $C \subset D$. Our aim is to develop criteria which assure that C is invariant for U -- that is, $U(t,s) : C \to C$ for all $0 \leq s \leq t$. Our fundamental result is the following:

Theorem 1. Suppose that C is a closed subset of D. Then the following statements are equivalent:

(i) $U(t, s) : C \to C$ for all $0 \leq s \leq t$.

(ii) $d(U(t, s)x; C) \leq d(x; C)e^{\omega(t-s)}$ for $x \in D$ and $0 \leq s \leq t$.

(iii) $\liminf_{h \to 0+} [d(U(t+h, t)x; C) - d(x; C)]/h \leq \omega d(x; C)$ for all $x \in D$.

(iv) $\liminf_{h \to 0+} d(U(t+h, t)x; C)/h = 0$ for all $x \in C$.

For the proof of Theorem 1 we first show that (i) - (iii) are equivalent, since this is straightforward. Suppose first that (i) holds, $x \in D$ and $0 \leq s \leq t$. By (U4), for each $y \in C$ we have that

$$d(U(t, s)x; C) \leq |U(t, s)x - U(t, s)y| \leq |x-y|e^{\omega(t-s)},$$

since $U(t, s)y \in C$. Since this holds for all $y \in C$ we see that (i) implies (ii). If (ii) holds and $x \in D$ and $t \geq 0$, then

$$d(U(t+h, t)x; C) - d(x; C) \leq (e^{\omega h}-1)\, d(x; C)$$

for all $h > 0$, and (iii) follows by dividing each side of this inequality by h and letting $h \to 0^+$. Now suppose that (iii) holds and that $x \in D$ and $s \geq 0$. For each $t \geq s$ define $p(t) = d(U(t, s)x; C)$. If $h > 0$ and $t \geq s$, $[p(t+h) - p(t)]/h = [d(U(t+h, t)U(t, s)x; C) - d(U(t, s)x; C)]/h$. Thus by (iii) we see that if $D_+p(t)$ is the lower right Dini derivative of p at t, then

$$D_+p(t) \leq \omega d(U(t, s)x; C) = \omega p(t) \text{ for } t \geq s.$$

Since p is continuous we obtain that $p(t) \leq p(s)e^{\omega(t-s)}$ for all $t \geq s$, and hence (iii) implies (ii). The fact that (ii) implies (i) is immediate and we have shown that statements (i) - (iii) are equivalent. The fact that (iii) implies (iv) is trivial, so to complete the proof of Theorem 1 we show that (iv) implies (i). To establish this we first prove the following lemma.

Lemma 1. Suppose that (iv) in Theorem 1 holds, that $x \in C$, that $s \geq 0$, and that $T, \varepsilon > 0$. Then there is an increasing sequence $\{t_n\}_0^\infty$ in $[s, s+T]$ such that $t_o = s$ and $\lim_{n \to \infty} t_n = s + T$, and a function $\beta: [s, s+T] \to D$ with the following properties:

(a) $\beta(s) = x$, $\beta(t_n) \in C$ and $\beta(t) = U(t, t_n)\beta(t_n)$ for $t \in [t_n, t_{n+1})$;

(b) $d(\beta(t); C) \leq \varepsilon$ for $t \in [s, s+T]$; and

(c) $|\beta(t_i) - U(t_i, t_j)\beta(t_j)| \leq \varepsilon(t_i - t_j)\exp(|\omega|(t_i - t_j))$ for $0 \leq j \leq i$.

Proof. The construction of $\{t_n\}_0^\infty$ and β is by induction. Define $t_o = s$ and $\beta(t_o) = x$, and suppose that t_n is defined and that β is constructed on $[s, t_n]$ and satisfies (a) - (c) with T replaced by $t_n - s$. If $t_n = s + T$ set $t_{n+1} = s + T$ and if $t_n < s + T$ let δ_n be the largest number in $(0, s + T - t_n]$ such that (a)′ $d(U(t_n + h, t_n)\beta(t_n);\ C) \leq \varepsilon$ for $h \varepsilon [0, \delta_n]$ and

(b)′ $d(U(t_n + \delta_n, t_n)\beta(t_n);\ C) \leq \delta_n \varepsilon/2$.

Note that $\delta_n > 0$ since $\beta(t_n) \varepsilon C$ and (iv) holds. Set $t_{n+1} = t_n + \delta_n$ and by (b)′ choose $\beta(t_{n+1}) \varepsilon C$ such that $|U(t_{n+1}, t_n)\ \beta(t_n) - \beta(t_{n+1})| \leq \varepsilon(t_{n+1} - t_n)$. Also set $\beta(t) = U(t, t_n)\beta(t_n)$ for $t \varepsilon [t_n, t_{n+1})$. Clearly (a) and (b) hold on $[s, t_{n+1}]$. Moreover, by the induction hypothesis, if $j \leq n$ and $i = n+1$ then

$$|\beta(t_i) - U(t_1, t_j)\beta(t_j)| \leq |\beta(t_{n+1}) - U(t_{n+1}, t_n)\beta(t_n)|$$

$$+ |U(t_{n+1}, t_n)\beta(t_n) - U(t_{n+1}, t_n)U(t_n, t_j)\beta(t_j)|$$

$$\leq \varepsilon(t_{n+1} - t_n) + |\beta(t_n) - U(t_n, t_n)\beta(t_j)| \exp(\omega(t_n - t_j))$$

$$\leq \varepsilon(t_{n+1} - t_n) + \varepsilon(t_n - t_j)\exp(|\omega|(t_{n+1} - t_j)).$$

Thus we see that (c) holds on $[0, t_{n+1}]$. Thus we can construct the sequence $\{t_n\}_o^\infty$ and β on $[s, r]$ where $r = \lim_{n\to\infty} t_n$. The claim is that $r = s + T$, for suppose $r < s + T$ and set $w = \lim_{n\to\infty} \beta(t_n)$. To see that w exists, let $\varepsilon' > 0$ and by (c) choose $j \geq 0$ such that $|\beta(t_i) - U(t_i, t_j)\beta(t_j)| \leq \varepsilon'/3$ for $i \geq j$. Now let $i, k \geq j$ be sufficiently large so that

$$|U(t_i, t_j)\beta(t_j) - U(t_k, t_j)\beta(t_j)| \leq \varepsilon'/3.$$

It then follows that $|\beta(t_i) - \beta(t_j)| \leq \varepsilon'$ and hence that w exists. Also, $w \in C$ since $\beta(t_n) \in C$ and C is closed. By the continuity of U we see that there is a $\gamma \in (0, s + T - r)$ such that

$$d(U(t_n + h, t_n)\beta(t_n); C) \leq \varepsilon/2 \text{ if } h \in [0, r + \gamma - t_n]$$

and n is sufficiently large. Since δ_n was chosen as large as possible in (a)′ and (b)′ we see that

$$d(U(r+\eta, t_n)\beta(t_n); C) > (r+\eta - t_n)\varepsilon/2$$

for $\eta \in (0, \gamma]$ and large n(or we could have chosen $\delta_n = r + \eta - t_n$). Letting $n \Rightarrow \infty$ we see that this implies that $d(U(r+\eta, r)w; C) \geq \eta\varepsilon/2$ for all $\eta \in (0, \gamma]$. However, this is a contradiction to (iv) since $w \in C$. Thus $\lim_{n \to \infty} t_n = s + t$ and defining $\beta(s+T) = \lim_{n \to \infty} \beta(t_n)$ completes the proof of Lemma 2.

Completion of the proof of Theorem 1. As we have indicated, it suffices to show that (iv) implies (i). So let $x \in C$ and let $0 \leq s < t$. Now let $\varepsilon > 0$ and set $T = t - s$ in Lemma 1. Using the fact that $\beta(t_i) \in C$ we have by setting $j = 0$ in (c) of Lemma 1 that

$$d(U(t, s)x; C) = \lim_{i \to \infty} d(U(t_i, s)x; C)$$

$$\leq \limsup_{i \to \infty} |U(t_i, t_o)x - \beta(t_i)|$$

$$\leq \limsup_{i \to \infty} \varepsilon (t_i - t_o) \exp(|\omega|(t_i - t_o))$$

$$= \varepsilon (t - s) \exp(|\omega|(t-s)).$$

Since this holds for all $\varepsilon > 0$ we see that $U(t, s)x \varepsilon C$ and the proof of Theorem 1 is complete.

Remark 1. The proof techniques in Theorem 1 show easily that one may replace $\liminf_{h \to 0+}$ by $\limsup_{h \to 0+}$ in parts (iii) and (iv).

Remark 2. The only tedious part of the proof of Theorem 1 was showing that (iv) implied any of the conditions (i) - (iii). This part may be simplified in certain circumstances. For suppose that for each $x \varepsilon D$ there is a $y \varepsilon C$ such that $d(x;C) = |x-y|$ (e.g., if C is weakly compact). Now assume that (iv) holds and that $x \varepsilon D$ and $y \varepsilon C$ is such that $d(x;C) = |x-y|$. Then, for each $t \geq 0$ and $h > 0$,

$$d(U(t+h, t)x;C) \leq |U(t+h, t)x - U(t+h, t)y| + d(U(t+h, t)y;C)$$
$$\leq e^{\omega h} d(x;C) + d(U(t+h, t)y;C).$$

The fact that (iv) implies (iii) follows easily from this inequality.

Note that (iv) of Theorem 1 gives invariance criteria in terms of the behavior of U on the boundary of C. One may also establish criteria in terms of the behavior of U in the exterior of C.

Proposition 3. Suppose that $D_o \subset D$, $\bar{D}_o \supset C$ and that $U(t, s): D_o \to D_o$ for $0 \leq s \leq t$. If

(3.1) $$\liminf_{h \to 0+} [d(U(t+h,t)x; C) - d(x;C)]/h \leq \omega d(x;C)$$

for all $x \in D_o - C$ and $t \geq 0$, then $U(t, s): C \to C$ for all $0 \leq s \leq t$.

Proof. Let $x \in D_o$, $s \geq 0$, and define $p(t) = d(U(t, s)x; C)$ for $t \geq s$. If $U(t, s)x \notin C$ then $U(t, s)x \in D_o - C$ and it follows from (3.1) that

$$D_+p(t) = \liminf_{h \to 0+} [d(U(t+h, t)U(t, s)x; C) - d(U(t,s)x;C]/h$$

$$\leq \omega d(U(t, s)x; C) = \omega p(t).$$

From this it follows easily that

(3.2) $$d(U(t, s)x; C) \leq d(x; C)e^{\omega(t-s)} \quad \text{for } t \geq s.$$

For suppose that $d(U(T, s)x; C) > d(x; C)e^{\omega(T-s)}$ for some $T > s$. By continuity and the fact that $U(T, s)x \notin C$, it follows that there is an $r \in [s, T)$ such that $U(T, s)x \in D_o - C$ for $t \in (r, T]$. Also, if $r > s$ then necessarily $U(r, s)x \in C$. Thus for all $t \in (r, T)$ $D_+p(t) \leq \omega p(t)$ and it follows that

$$d(U(T, s)x; C) \leq d(U(r,s)x;C)e^{\omega(T-r)} .$$

If $r = s$ we obtain a contradiction since $U(r, s)x = x$ and if $r > s$ we also obtain a contradiction since $d(U(r, s)x; C) = 0$. Thus (3.2) holds for all $x \in D_o$ and since $\bar{D}_o \supset C$ we see that (3.2) holds for all $x \in C$. Proposition 3 is immediate from this.

From Proposition 3 we may establish invariance criteria in terms of the infinitesimal generator of U.

Theorem 2. Suppose that U has an infinitesimal generator $\{G(t) : t \geq 0\}$ with $D(G(t)) = D_o$ for all $t \geq 0$. Suppose further that $\bar{D}_o \supset C$ and $U(t,s) : D_o \rightarrow D_o$ for $0 \leq s \leq t$. Then these are equivalent:

(i) $U(t, s) : C \Rightarrow C$ for all $0 \leq s \leq t$.

(ii) $\liminf_{h \Rightarrow 0+} [d(x + hG(t)x; C) - d(x; C)]/h \leq \omega d(x; C)$ for all $x \in D_o - C$.

Proof. Note that if $x \in D_o$, $t \geq 0$, and $h > 0$,

$$|x + hG(t)x - U(t+h, t)x| = h\,|G(t)x - [U(t+h, t)x - x]/h|,$$

and hence

$$d(x + hG(t)x; C) = d(U(t+h, t)x; C) + o(h)$$

where $\lim_{n \to 0+} o(h)/h = 0$. Thus (ii) is equivalent to (3.1) in Proposition 3. The conclusions of Theorem 2 follow easily from this observation and Proposition 3.

As one consequence of the above theorem we obtain the following:

Corollary 1. Suppose that C is convex, E and E^* are uniformly convex, $\{A(t) : t \geq 0\}$ is a family of subsets of E x E satisfying (A1) - (A4) in § 2, and U is defined by (2.2). Then these are equivalent:

(i) $U(t, s) : C \to C$ for all $0 \leq s \leq t$.

(ii) $\liminf_{h \to 0+} [d(x + hA(t)^o x; C) - d(x; C)]/h \leq \omega d(x; C)$ for all $t \geq 0$ and $x \in D(A(t)) - C$.

(iii) $[x - Px, A(t)^o x] \leq \omega |x - Px|^2$ for all $t \geq 0$ and $x \in D(A(t))$, where Px is the projection of x onto C and $[\cdot, \cdot]$ is defined by (2.5).

Proof. Since $\{A(t)^o : t \geq 0\}$ is the infinitesimal generator of U, the equivalence of (i) and (ii) follows from Theorem 2, and the equivalence of (ii) and (iii) follows from Proposition 2.

Remark 3. Corollary 1 extends a criterion of Brezis [3, Proposition 4.5, p.131], where E is assumed to be a Hilbert space and $\{A(t) : t \geq 0\}$ independent of t.

In certain cases when C is convex and U is defined by (2.2), one may obtain further results. In each of the two propositions below, it is assumed that $\{A(t) : t \geq 0\}$ is a family of subsets of E x E which satisfy (A1) - (A4) and that the evolution system U is defined by (2.2). We also assume that J is defined by (2.4).

Proposition 4. Suppose that C is convex and that the interior C^o of C is nonempty. Suppose also that for each

$t \geq 0$, $x \in D(A(t)) \cap \partial C$ and $y \in A(t)x$ there is an $\alpha \in J(x;C)$ such that $\alpha(y) \leq 0$. Then $(I - \lambda A(t))^{-1} : C \to C$ for each $\lambda \in (0, \lambda_o]$ and $U(t, s) : C \to C$ for each $0 \leq s \leq t$.

Remark 4. Note that $J(x; C)$ is nonempty for each $x \in \partial C$ since C^o is nonempty. Note also that, by Proposition 1, the suppositions of Proposition 4 are fulfilled if

$\liminf_{h \to 0+} d(x + hy; C)/h = 0$ for all $x \in D(A(t)) \cap \partial C$ and all $y \in A(t)x$.

Proof of Proposition 4. Let $z \in C^o$ and let $t \geq 0$. Set $x_h = (I - hA(t))^{-1} z$ for $h \in [0, \lambda_o]$ and let γ be the largest number in $[0, \lambda_o]$ such that $x_h \in C$ for $h \in [0, \gamma]$. Since $h \to x_h$ is continuous and $x_o = z \in C^o$, we see that $\gamma > 0$ and $x_\gamma \in D(A(t)) \cap \partial C$ if $\gamma < \lambda_o$. Assume that $\gamma < \lambda_o$ and let $y \in A(t)x_\gamma$ be such that $x_\gamma - \gamma y = z$. By the hypothesis, let $\alpha \in J(x_\gamma; C)$ be such that $\alpha(y) \leq 0$. Since $z \in C^o$ it follows that $\alpha(x_\gamma - z) > 0$. Thus

$$\alpha(x_\gamma) > \alpha(z) = \alpha(x_\gamma - \gamma y) = \alpha(x_\gamma) - \gamma\alpha(y) \geq \alpha(x_\gamma),$$

which is impossible. Thus $\gamma = \lambda_o$ and it follows that $(I - \lambda A(t))^{-1} : C^o \to C$ for all $\lambda \in (0, \lambda_o]$. The continuity of $(I - \lambda A(t))^{-1}$ along with the fact that C^o is dense in C shows that $(I - \lambda A(t))^{-1} : C \to C$ for all $\lambda \in (0, \lambda_o]$ and $t \geq 0$. The final assertion is immediate from this fact and (2.2).

Proposition 5. Suppose that C is convex and for each

$t \geq 0$, $x \in D(A(t)) - C$ and $y \in A(t)x$ there is an $\alpha \in J(x;C)$ such that $\alpha(y) \leq \omega d(x;C)$. Then $(I - \lambda A(t))^{-1}: C \to C$ for each $\lambda \in (0, \lambda_o]$ and $U(t, s): C \to C$ for each $0 \leq s \leq t$.

Proof. Suppose, for contradicition, that for some $z \in C$ and $h \in (0, \lambda_o]$, $x_h = (I - hA(t))^{-1} z \notin C$, and let $y \in Ax_h$ be such that $x_h - hy = z$. If $\alpha \in J(x;C)$ and $\alpha(y) \leq \omega d(x;C)$, it follows that

$$\alpha(x_h) \geq \alpha(z) + d(x_h;C) = \alpha(x_h - hy) + d(x_h; C)$$

$$= \alpha(x_h) - h\alpha(y) + d(x_h;C) \geq \alpha(x_h) + (1-h\omega)d(x_h;C).$$

But this is impossible since $1 - h\omega > 0$ and $d(x_h;C) > 0$. Thus $(I - hA(t))^{-1} : C \to C$ and the assertions of Proposition 5 follow.

§4. Examples.

Here we indicate a situation where these results may be applied. Let N be a positive integer and let $\Omega \subset \mathbb{R}^N$ be a bounded region with smooth boundary $\partial\Omega$. Also, let M be a positive integer and let $|\cdot|$ denote a norm on the space $\mathbb{R}^M$. Suppose that $1 < p < \infty$ and denote by X the space $L^p(\Omega, \mathbb{R}^M)$ of all measurable functions $u = (u_i)_1^M$ of Ω into $\mathbb{R}^M$ with

$$|u| = [\int_\Omega |u(x)|^p dx]^{1/p} < \infty .$$

Now define the operator L on X by

$$D(L) = \{ u \in X : u_i \in H^{2,p}(\Omega) \text{ for } i = 1, \ldots, M\} \text{ and}$$

(4.1)

$$Lu = (\Delta u_i)_1^M \text{ for } u \in D(L),$$

where $H^{2,p}(\Omega)$ is the space of real valued $L^p(\Omega)$ functions having 2 derivatives in $L^p(\Omega)$ and Δ is the Laplacian operator on Ω. Also, for each $\psi = (\psi_i)_1^M \in D(L)$ define the operator L_ψ on X by

$$D(L_\psi) = \{u \in D(L) : u_i - \psi_i \in H_o^{1,p}(\Omega) \text{ for } i = 1, \ldots, M\} \text{ and}$$

(4.2)

$$L_\psi u = Lu \text{ for } u \in D(L_\psi),$$

where $H_o^{1,p}(\Omega)$ is the space of real valued $L^p(\Omega)$ functions having one derivative in $L^p(\Omega)$ and vanishing on $\partial\Omega$. It follows that for each $\psi \in D(L)$ L_ψ is the generator of a nonlinear semigroup $T_\psi = \{T_\psi(t) : t \geq 0\}$ of type 0 on $\overline{D(L_\psi)} = X$(i.e. If $U(t, s) = T(t-s)$ for $0 \leq s \leq t$, then U is an evolution system of type 0). Moreover, $R(I - hL_\psi) = X$ for $h > 0$ and

$$(4.3) \quad T_\psi(t)z = \lim_{n \to \infty} (I - n^{-1}tL_\psi)^{-1}z \text{ for } t \geq 0 \text{ and } z \in X.$$

Note also that if $o(x) = \theta$ for all $x \in \Omega$ then T_o is a semigroup of linear operators on X and for each $\psi \in D(L)$

(4.4) $\quad T_\psi(t)w = T_o(t)(w-\psi) + \int_o^t T_o(\tau)L\psi d\tau + \psi$ for $t \geq 0$

and $w \in X$.

Throughout this section we suppose that Λ is a closed, convex subset of $\mathbb{R}^M$ and defined

(4.5) $\quad K(\Lambda) = \{u \in X: u(x) \in \Lambda$ for almost all $x \in \Omega\}$.

Before proceding further we first establish the following lemma on the behavior of the semigroup T_ψ.

<u>Lemma 2.</u> Suppose that $\phi, \psi \in D(L)$ and $\phi(x) - \psi(x) \in \Lambda$ for $x \in \partial\Omega$. If $z, w \in X$ and $z - w \in K(\Lambda)$, then $T_\phi(t)z - T_\psi(t)w \in K(\Lambda)$ for all $t \geq 0$.

<u>Proof.</u> By (4.3) it surfices to show that if $z - w \in K(\Lambda)$ and $h > 0$, then $(I - hL_\phi)^{-1}z - (I - hL_\psi)^{-1} w \in K(\Lambda)$. So let $z_1 = (I - hL_\phi)^{-1}z$ and $w_1 = (I - hL_\psi)^{-1}w$, and let Z be a class of affine mappings from $\mathbb{R}^M$ to $\mathbb{R}$ such that $\Lambda = \{\xi \in \mathbb{R}^M: \alpha(\xi) \leq 0$ for all $\alpha \in Z\}$. It then follows that

$$\alpha(z_1 - w_1) - h\Delta\,\alpha(z_1 - w_1) = \alpha(z-w)$$

for all $\alpha \in Z$. Since $\alpha(z(x) - w(x)) \leq 0$ for $x \in \Omega$ and $\alpha(z_1(x) - w_1(x)) = \alpha(\phi(x) - \psi(x))$ for $x \in \partial\Omega$, one obtains from the maximum principle that $\alpha(z_1(x) - w_1(x)) \leq 0$ for all $x \in \Omega$ and $\alpha \in Z$. Thus $z_1 - w_1 \in K(\Lambda)$ and the assertion follows.

Now we consider a special class of nonlinear operators on X. Let ω be a real number and denote by Y_ω the class of all functions $f: [0, \infty) \times \bar{\Omega} \times \mathbb{R}^M \to \mathbb{R}^M$ with the following properties:

(P1) f is continuous and there is a number $L > 0$ such that $|f(t,x,\xi) - f(s,x,\xi)| \le L|t-s|(1+|\xi|)$ for all $t, s \in [0, \infty)$, $x \in \bar{\Omega}$ and $\xi \in \mathbb{R}^M$;

(P2) there are continuous functions $a, b: [0, \infty) \to [0, \infty)$ such that $|f(t, x, \xi)| \le a(t) |\xi| + b(t)$ for all $t \in [0, \infty)$, $x \in \bar{\Omega}$ and $\xi \in \mathbb{R}^M$; and

(P3) $|\xi - \eta - h[f(t, x, \xi) - f(t, x, \eta)]| \ge (1-h\omega) |\xi - \eta|$ for all $h > 0$, $t \in [0, \infty)$, $x \in \bar{\Omega}$ and $\xi, \eta \in \mathbb{R}^M$.

For each $f \in Y_\omega$ define the family $B_f = \{B_f(t) : t \ge 0\}$ of operators on X by

(4.6) $[B_f(t)u](x) = f(t,x,u(x))$ for $t \in [0, \infty)$, $x \in \Omega$ and $u \in X$.

It follows that B_f is continuous from $[0, \infty) \times X$ into X and

$|u-v-h[B_f(t)u - B_f(t)v]| \ge (1-h\omega) |u-v|$ for $h > 0$, $t \in [0, \infty)$ and $u, v \in X$.

Moreover, if $A(t)u = L_\psi u + B_f(t)u$ for all $u \in D(A(t)) = D(L_\psi)$ and $t \ge 0$, then $\{A(t): t \ge 0\}$ satisfies (A1) - (A4) in

§2. Also, if $U_{\psi,f}$ is defined by (2.2) then

(4.7 $$U_{\psi,f}(t,s)z = T_{\psi}(t-s)z + \int_s^t T_o(t-\tau)B_f(\tau)U_{f,\psi}(\tau,s)z\,d\tau$$

for all $0 \leq s \leq t$ and $z \in X$. Furthermore,

$\{L_{\psi} + B_f(t) : t \geq 0\}$ is the infinitesimal generator of $U_{f,\psi}$.

For $\eta \in \mathbb{R}^M$ write $d_1(\eta:\Lambda) = \inf\{|\eta - \xi| : \xi \in \Lambda\}$ and if

$f, g \in Y_{\omega}$ we write $f \sim_{\Lambda} g$ if

(4.8) $$\lim_{h \to 0+} d_1(\xi - \eta + h[f(t,x,\xi) - g(t,x,\eta)];\Lambda)/h = 0$$

for all $t \in [0,\infty)$, $x \in \bar{\Omega}$ and $\xi, \eta \in \mathbb{R}^M$ with $\xi - \eta \in \Lambda$.

For some applications of the condition $f \sim_{\Lambda} g$ to differential inequalities, see Martin [13]. Our fundamental result is given by the following

<u>Proposition 6.</u> Suppose that $f, g \in Y_{\omega}$, $f \sim_{\Lambda} g$, and that $\phi, \psi \in D(L)$ with $\phi(x) - \psi(x) \in \Lambda$ for $x \in \partial\Omega$. Then

$$U_{f,\phi}(t,s)z - U_{g,\psi}(t,s)w \in K(\Lambda)$$

for all $0 \leq s \leq t$ and all $z, w \in X$ with $z - w \in K(\Lambda)$.

Before indicating the proof of Proposition 6 we first consider a few special cases. If $\phi \in D(L)$, $\phi(x) \in \Lambda$ for $x \in \partial\Omega$, and

$$\lim_{h \to 0+} d_1(\xi + hf(t,x,\xi);\Lambda)/h = 0 \text{ for } (t,x,\xi) \in [0,\infty) \times \bar{\Omega} \times \mathbb{R}^M,$$

then $U_{f,\phi}(t,s) : K(\Lambda) \to K(\Lambda)$ for all $0 \leq s \leq t$. This follows directly from Proposition 6 by taking $\psi = 0$, $g = \theta$ and $w = \theta$. If $f \in Y_\omega$ and there is a number $r > 0$ such that

$$\lim_{h \to 0+} [\,|\xi - \eta + h[f(t,x,\xi) - f(t,x,\eta)]| - |\xi - \eta|\,]/h \leq 0$$

for all $t \geq 0$, $x \in \bar{\Omega}$ and $\xi, \eta \in \mathbb{R}^M$ with $|\xi - \eta| = r$, then

$$|U_{f,\phi}(t,s)z(x) - U_{f,\psi}(t,s)w(w(x)| \leq r \text{ for } 0 \leq s \leq t$$
$$\text{and } x \in \Omega,$$

whenever $|z(x) - w(x)| \leq r$ for $x \in \Omega$ and $|\phi(x) - \psi(x)| \leq r$ for $x \in \partial\Omega$. This follows from Proposition 6 by taking $f = g$ and $\Lambda = \{\xi \in \mathbb{R}^M : |\xi| \leq r\}$.

One of the most important cases is when Λ is a cone (i.e., $\xi, \eta \in \Lambda$ and $t \geq 0$ implies $\xi + \eta, t\xi \in \Lambda$). So assume Λ is a cone and write $\xi \geq \eta$ (or $\eta \leq \xi$) whenever $\xi - \eta \in \Lambda$. Then $K(\Lambda)$ is a cone in X, so write $z \geq w$ (or $w \leq z$) whenever $z - w \in K(\Lambda)$. Thus if $f \in Y_\omega$, $f \sim_\Lambda f$, and $\phi, \psi \in D(L)$ with $\phi(x) \geq \psi(x)$ for $x \in \partial\Omega$, then $U_{f,\phi}(t,s)z \geq U_{f,\psi}(t,s)w$ whenever $0 \leq s \leq t$ and $z \geq w$. Note that if $\Lambda = \{\xi \in \mathbb{R}^M : \xi_i \geq 0$ for $i = 1, \ldots, M\}$ and $f = (f_i)_1^M \in Y_\omega$, then $f \sim_\Lambda f$ only in case $f_i(t,x,\xi) \geq f_i(t,x,\eta)$ whenever $\xi - \eta \in \Lambda$ and $\xi_i = \eta_i$. In particular, if $M = 1$ and $\Lambda = [0, \infty)$ then $f \sim_\Lambda f$ for all $f \in Y_\omega$. Note also that if $M > 1$ and

$\Lambda = \{ \xi \varepsilon \mathbb{R}^M : \xi_1 \geq 0 \}$, then $f^{\sim}_{\Lambda} f$ only in case $f_1(t, x, \xi) \geq f_1(t, x, \eta)$ whenever $\xi_1 = \eta_1$.

Now we give a brief indication of the proof of Proposition 6. Set $E = X \times X$ with $|(u, v)| = |u| + |v|$ for all $(u, v) \varepsilon E$ and set $C = \{(u, v) \varepsilon E: u - v \varepsilon K(\Lambda)\}$. Now define the evolution system V on E by

$$V(t, s)(u, v) = (U_{f,\phi}(t, s)u, U_{g,\psi}(t, s)v) \tag{4.9}$$

for $0 \leq s \leq t$ and $(u, v) \varepsilon E$. Note that the assertion of Proposition 6 is equivalent to showing that $V(t, s) : C \to C$ for $0 \leq s \leq t$. Applying (iv) of Theorem 1 it suffices to show that

$$\lim_{h \to 0+} d^* (V(t+h, t)(z,w); C)/h = 0 \tag{4.1}$$

$$\text{for } t \geq 0 \text{ and } (z, w) \varepsilon C,$$

where $d^*(u, C) = \inf \{ |u-v| : v \varepsilon C\}$ for all $u \varepsilon E$. However, by (4.9) and (4.7) one sees that (4.10) is equivalent to

$$\lim_{h \to 0+} d^*((T_\phi(h)z, T_\psi(h)w) + h(B_f(t)T_\phi(h)z, B_g(r)T_\psi(h)w); C)/h = 0. \tag{4.11}$$

To see that this is so note that by continuity one obtains

$$T_\phi(h)z + \int_t^{t+h} T_o(t+h-\tau)B_f(\tau)U_{f,\phi}(\tau, s)z d\tau = T_\phi(h)z + hB_f(t)z + o(h)$$

$$= T_\phi(h)z + hB_f(t)\, T_\phi(h)z + o(h),$$

where $o(h)/h \to 0$ as $h \to 0+$. To establish (4.11) it is sufficient to show that

$$\lim_{h \to 0+} d_1^*((\xi,\eta) + h(f(t,x,\xi), g(t,x,\eta)); \Lambda^*)/h = 0 \tag{4.12}$$

for all $t \geq 0$, $x \in \bar{\Omega}$, and $(\xi,\eta) \in \Lambda^*$ where $\Lambda^* = \{(\xi,\eta) \in \mathbb{R}^M \times \mathbb{R}^M : \xi - \eta \in \Lambda\}$ and $d^*((\xi,\eta);\Lambda^*) = \inf\{|\xi-\xi^1| + |\eta-\eta^1| : (\xi^1,\eta^1) \in \Lambda^*\}$. (See Martin [12, Lemma 2 and Proposition 5]). However, noting that $d^*((\xi,\eta);\Lambda^*) = d(\xi-\eta;\Lambda)$ for all $\xi,\eta \in \mathbb{R}^M$, one sees that (4.12) is immediate from the definition of $f \sim_\Lambda g$. This completes the proof indication of Proposition 6.

REFERENCES

[1] J. M. Bony, Principe du maximum, inegalite de Harnack et unicite du problemes de Cauchy pour les operateurs elliptique degeneres, Ann. Inst. Fourier Grenoble 19(1969), 277-304.

[2] H. Brezis, On a characterization of flow-invariant sets, Comm. Pure App. Math. 23 (1970), 261-263.

[3] H. Brezis, "Operateurs Maximaux Monotones", North-Holland, Amsterdam, 1973.

[4] M. G. Crandall, Differential equations on convex sets, J. Math. Soc. Japan, 22(1970), 443-455.

[5] M. G. Crandall, A generalization of Peano's Theorem and flow invariance, Proc. Amer. Math. Soc. 36 (1972), 151-155.

[6] M. G. Crandall and A. Pazy, Nonlinear evolution equations in Banach spaces, Israel J. Math. 11(1972), 57-94.

[7] P. Hartman, On invariant sets and a theorem of Wazewski, Proc. Amer. Math. Soc. 32 (1972), 511-520.

[8] G. Kothe, "Topological Vector Spaces", Vol. 1, Springer-Verlag, New York, 1969.

[9] R. H. Martin, Jr., Differential Equations on closed subsets of a Banach space, Trans. Amer. Math. Soc. 179 (1973), 399-414.

[10] R. H. Martin, Jr., Approximation and existence of solutions to ordinary differential equations in Banach spaces, Funk. Ekv. 16 (1973), 195-213.

[11] R. H. Martin, Jr., Generating an evolution system in a class of uniformly convex Banach spaces, J. Funct. Anal. 11 (19720~ 62-76.

[12] R. H. Martin, Jr., Invariant sets for perturbed semigroups of linear operators, Annali Mat. Pura Appl. (to appear).

[13] R. H. Martin, Jr., Remarks on differential inequalities in Banach spaces, (to appear).

[14] M. Nagumo, Uber die Lage der Integralkurven gewohnlicher Differential-gleichungen, Proc. Phys. Math. Soc. Japan. 24 (1942), 551-559.

[15] R. M. Redheffer, The theorems of Bony and Brezis on flow invariant sets, Amer. Math. Monthly 79 (1972), 740-747.

[16] R. M. Redheffer and W. Walter, Flow-invariant sets and differential inequalities in normed space, J. Appl. Anal. (to appear).

[17] R. M. Redheffer and W. Walter, A differential inequality for the distance function in normed linear spaces, (to appear).

[18] P. Volkmann, Gewohnliche Differentialungleichungen mit quasimonoton wachsenden Funktionen in topologischen Vektorraumen, Math. Zeitschr. 127, 157-164, (1972).

[19] P. Volkmann Uber die Invarianz konvexer Mengen und Differentialungleichungen in normierten Raum, Math. Annalen 203 (1973), 201-210.

[20] P. Volkmann, Gewohnliche Differentialungleichungen mit quasimonoton wachsenden Funktionen in Banachraumen, Proceedings of the Conference on Differential Equations, Dundee, Scotland, 1974 (to appear).

[21] W. Walter, Ordinary differential inequalities in ordered Banach spaces, J. Diff. Eqns. 9(1971), 253-261.

[22] J. A. Yorke, Invariance for ordinary differential equations, Math. Systems Theory 1 (1967), 353-372.

[23] J. A. Yorke, Differential inequalities and non-Lipschitz scalar functions, Math. Systems Theory 4 (1970), 140-153.

North Carolina State University, Raleigh, North Carolina

RECENT RESULTS ON PERIODIC SOLUTIONS OF DIFFERENTIAL EQUATIONS

by
Jean Mawhin

1. Introduction.

The use of functional analytic methods in the study of periodic solutions or of boundary value problems for ordinary or functional differential equations has been extensively developed those last years and the interested reader will find useful to consult the books of Hale [14], Reissig, Sansone and Conti [32], Rouche and Mawhin [36], as well as the survey papers by Krasnosel′skii [20], Antosiewicz [1] and Cesari [5]. In particular, degree arguments have been systematically used in [36] to give simple and unified proofs of a number of existence theorems, e.g. Krasnosel′skii guiding functions method, Hartman-Knobloch type results and more specific criteria for Duffing and Lienard systems.

The aim of this paper is to continue this program and give a survey of recent theorems which, even if their original proof is quite different, fall into the scope of coincidence degree theory [24], and specially of the corresponding continuation theorem. As for the usual Leray-Schauder continuation theorem, the more difficult point in applying the result is to obtain a priori estimates for the possible solutions. It is this fact which will be mainly emphasized here and various ways, depending upon the order of the equation, will be sketched, the remaining of the proof being; generally omitted.

The following notations will be used throughout the paper. If A is any set, ∂A will denote its boundary and $\bar{A}$ its closure. If E is a normed space, the norm of the element x will be denoted by $|x|$, which will be the Euclidian norm if $E = R^n$. Also, in R^n, $x.y$ will denote the inner product of the elements x and y. Lastly, P_T^k will be the (Banach) space of continuous and T-periodic mappings $x : R \to R^n$ having continuous and T-periodic derivatives up to the order k, with the norm

$$|x| = \max_{j=0,\ldots,k} \sup_{t \in R} |x^{(j)}(t)| .$$

2. First order vector differential equations.

The following well known result of the theory of convex sets in R^n will be useful in sections 2 and 3 of this paper.

Proposition 2.1. If $G \subset R^n$ is an open, bounded and convex set containing the origin, then, for each $x \in \partial G$, there exists a (not necessarily unique) $n(x) \in R^n \setminus \{0\}$ such that:

(i) $n(x).x \neq 0$, $x \in \partial G$,

(ii) $\bar{G} \subseteq \{y \mid (\mathrm{sign}\ n(x).x)\, n(x).(y-x) \leq 0\}$.

Such a $n(x)$ will be called a normal at x to ∂G (an outer normal (resp. inner normal) if $n(x).x > 0$ (resp. < 0).

Let $f : R \times R^n \to R^n$, $(t,x) \mapsto f(t,x)$ be T-periodic with respect to t and continuous and let us consider the differential equation

$$(2.1) \qquad x' = f(t,x).$$

The following result, which is proved in [28] (resp. in [27]) for the ordinary (resp. the functional) case, extends and unifies previous results of Knobloch [16], Krasnosel'skii [19], Gustafson and Schmitt [12] given below. Although they have often a corresponding formulation for functional differential equations, the theorem will always be formulated here, for the sake of simplicity, in the case of ordinary differential equations.

Theorem 2.1. Let $G \subset R^n$ be an open bounded convex set containing the origin and suppose that for each $x \in \partial G$ we can find a normal $n(x)$ to ∂G such that, for every $t \in R$,

$$n(x).\, f(t, x) \geq 0. \tag{2.2}$$

Then, if the Brouwer degree

$$d[g, G, 0]$$

with

$$g : R^n \to R^n, \quad a \Rightarrow T^{-1} \int_0^T f(t, a)\,dt,$$

is different from zero, equation (2.1) has at least one T-periodic solution x such that, for each $t \in R$, $x(t) \in \bar{G}$.

The case where (2.2) holds with a non-strict inequality can be obtained, by a limit process, from the case where the inequality in (2.2) is strict. Then, if we introduce the open bounded set of P_T^0,

$$\Omega = \{ x \in P_T^0 \mid x(t) \in G, \ t \in R \}, \tag{2.3}$$

one has

$$\partial\Omega \subset \{x \in P_T^0 \mid x(t) \in \bar{G},\ t \in R \text{ and } x(t_0) \in \partial G$$
$$\text{for some } t_0\}$$

and, if $\lambda \in]0,1]$, each possible T-periodic solution of

$$x' = \lambda f(t,x)$$

does not belong to $\partial\Omega$. Because, if it is the case, one deduces easily from condition (ii) in Proposition 2.1 that

$$n(x(t_0)).x'(t_0) = 0,$$

which, after replacing $x'(t_0)$ by $\lambda f(t_0, x(t_0))$, contradicts (2.2).

Corollary 2.1. (Gustafson and Schmitt [12]). The conclusions of Theorem 2.1 hold if only (2.2) is required with $n(x)$ either an outer or an inner normal for each $x \in \partial G$.

In this case the homotopy $\pm(1-\lambda)a+\lambda g(a)$, $\lambda \in [0,1]$, with + or - according to n is an outer or an inner normal, does not vanish on ∂G and hence $d[g, G, 0] = (\pm 1)^n$.

Corollary 2.2. (Krasnosel'skii [19]). Suppose that the open bounded convex set $G \subset R^n$ can be defined by a finite number of inequalities $\Phi_i(x) \leq 0$, $i = 1, \ldots, r$, where the C^1-functions Φ_i are such that Φ_i and grad Φ_i cannot vanish simultaneously on ∂G. Then, if

$$\text{grad } \Phi_i(x).f(t,x) \leq 0 \ (\text{or} \geq 0),\ t \in R,$$

for each i for which $\Phi_i(x) = 0$ and each $x \in \partial G$, then the conclusion of Theorem 2.1 holds.

Corollary 2.3. (Knobloch [16]). Suppose that there exist C^1 - T-periodic functions α_i, β_i such that $\alpha_i \leq \beta_i$ and

$$[\alpha_i'(t) - f_i(t, x_1, \ldots, \alpha_i(t), \ldots, x_n)],$$

$$[\beta'_i(t) - f_i(t, x_1, \ldots, \beta_i(t), \ldots, x_n)] \leq 0$$

when $\alpha_j(t) \leq x_j \leq \beta_j(t)$, $t \in R$, $j = 1, \ldots, n$, $j \neq i$, $i = 1, \ldots, n$. Then equation (2.1) has at least one T-periodic solution x such that

$$\alpha_i(t) \leq x_i(t) \leq \beta_i(t), \quad t \in R, \quad i = 1, \ldots, n.$$

The proof of Corollary 2.3 consists in defining $\tilde{f}: R \times R^n \to R^n$ by

$$\tilde{f}\,(t, x) = f(x, X_1(t, x_1), \ldots, X_n(t, x_n))$$

where

$$X(t, u) = \begin{cases} \alpha_i(t) & \text{if } u < \alpha_i(t), \\ u & \text{if } \alpha_i(t) \leq u \leq \beta_i(t), \\ \beta_i(t) & \text{if } u > \beta_i(t), \qquad i = 1, \ldots, n. \end{cases}$$

and showing, using Theorem 2.1 and elementary arguments, that equation

$$x' = \tilde{f}(t, x)$$

has a T-periodic solution x such that $\alpha_i(t) \leq x_i(t) \leq \beta_i(t)$, $t \in R$, $i = 1, \ldots, n$, which is then necessarily a T-periodic solution of (2.1).

To end this section, let us also note that, by coincidence degree techniques, J. Cronin [6] has got results on the existence of at least two T-periodic solutions for vector first order functional differential equations and that the author and Muñoz [29] has studied the problem of estimating the number and the stability of T-periodic solutions of ordinary differential equations.

3. Second order vector differential equations.

We shall be interested in this section in the T-periodic solutions of the vector second order differential equations

$$x'' = f(t, x, x') \tag{3.1}$$

where $f: R \times R^n \times R^n \to R^n$ is T-periodic with respect to t and continuous. For the sake of simplicity, we shall limit ourself to the special case

$$x'' = f(t, x), \tag{3.2}$$

the assertions and proofs being easily extended to (3.1) when f satisfies with respect to x′ a growth condition of Nagumo type as developed for example by Hartman [15] and Schmitt [41].

The following theorem was first proved by Bebernes and Schmitt [4] by the use of Poincare's translation operator, and then by Bebernes [2] using Leray-Schauder degree. Extensions to functional differential equations are due to Gustafson and Schmitt [11]. For the details of the proof considered here, see [27, 28], and for a special case with $n = 2$, see Knobloch [18].

Theorem 3.1. Let $G \subset R^n$ be an open bounded convex set containing the origin and suppose that, for every $x \in \partial G$, there exists an outer normal $n(x)$ to ∂G such that

$$n(x) \cdot f(t, x) \geq 0 \tag{3.3}$$

for every $t \in R$. Then, equation (3.2) has at least one T-periodic solution x such that $x(t) \in \bar{G}$, $t \in R$.

As in Theorem 2.1, one reduces easily to the case of a strict inequality in (3.3). Then, the idea of the obtention of an a priori estimate is to introduce the open bounded set of P_T^0 defined in (2.3) and to show that, if $\lambda \in]0,1]$ and if x is a T-periodic solution of

$$x'' = \lambda f(t, x),$$

then $x \notin \partial\Omega$. Because, if it is the case, one has, with t_0 such that $x(t_0) \in \partial G$,

$$n(x(t_0)) \cdot x'(t_0) = 0$$

and hence, by Taylor formula, for each $h \in R$,

$$(x(t_0+h)-x(t_0)) \cdot n(x(t_0)) = \lambda\, n(x(t_0)) \cdot \int_0^1 f(t_0+sh, x(t_0+sh))(h^2/2)\, ds$$
$$> 0$$

by (3.3) for sufficiently small $|h|$, a contradicition with condition (ii) of Proposition 2.1.

The following Corollary, due to Knobloch [17], has been proved subsequently by Schmitt [37] and Muldowney and Willett [31] and Gudkov and Lepin [10]. Extensions to the case of functional differential equations are due to

Muldowney and Willett [30] and Schmitt [40].

Corollary 3.1. *Suppose that* $n = 1$ *and that there exists* C^2- *T-periodic real functions* α, β *such that* $\alpha \leq \beta$ *and*

$$\alpha''(t) \geq f(t, \alpha(t)), \quad \beta''(t) \leq f(t, \beta(t)), \quad t \in R. \tag{3.4}$$

Then, equation (3.2) *has at least one T-periodic solutions* x *such that* $\alpha \leq x \leq \beta$.

A proof consists in defining $\tilde{f} : R \times R \to R$ by

$$\tilde{f}(t, x) = \begin{cases} f(t, \alpha(t)) + x - \alpha(t) & \text{if } x < \alpha(t), \\ f(t, x) \text{ if } \alpha(t) \leq x \leq \beta(t), \\ f(t, \beta(t)) + x - \beta(t) \text{ if } x > \beta(t), \end{cases}$$

and in showing, using Theorem 3.1 with $G =]-R, R[$, R sufficiently great, that equation

$$x'' = \tilde{f}(t, x)$$

has at least one T-periodic solution. This solution is then shown to be necessarily situated between α and β, and is therefore a T-periodic solution of (3.2).

When $\alpha(t) = -R$, $\beta(t) = R$, $t \in R$, conditions (3.4) reduce to

$$xf(t, x) \geq 0, \quad |x| = R, \quad t \in R.$$

The case where this inequality is reversed is more delicate and, using coincidence degree theory, Gaines [9] has recently obtained results in this line. The a priori bounds are obtained under assumption that

$$xf(t, x) < 0$$

for all t and $|x| \geq R$ and by requiring that f satisfies a one-sided growth condition. A very simple example of Gaines growth condition is that there exists $R_1 > 0$ such that, for all t and $|x| \geq R_1$,

$$xf(t,x) > -|x|(A + C|x|)$$

with A, C positive constants and with $C^{1/2} < \pi/(2T)$.

4. Vector differential equations of order greater than two.

The techniques esquissed in sections 2 and 3 for getting a priori estimates have evident common points but seem difficult to be extended to differential equations of higher order.

In a recent paper, Bebernes, Gaines and Schmitt [3] have used coincidence degree theory to study the T-periodic solutions of the scalar third and fourth order equations

$$x''' = f(t, x, x', x''),$$

$$x^{(iv)} = f(t, x, x', x'', x''').$$

To obtaine the a priori bounds, they require growth conditions on f with respect to the variables corresponding to the derivatives of x and conditions of non-vanishing of f for large values of $|x|$. They are then able to use sub- and superfunction techniques of Jackson and Schrader (from which comes the restriction to third and fourth order equations). We refer to the original paper for explicit statement of the theorems.

Let us consider now the differential equation

$$x^{(m)} + \sum_{j=1}^{m} A_j x^{(m-j)} = f(t, x, x', \ldots, x^{(m-1)}) \tag{4.1}$$

where f is T-periodic with respect to t, continuous and such that

$$|u|^{-1} |f(t, u)| \to 0 \quad \text{if} \quad |u| \to \infty ,$$

uniformly in t, and where the matrices A_j are such that the linear part of (4.1) has no nontrivial T-periodic solution. It is easy to prove, using Schauder or Granas theorem for example, that (4.1) has at least one T-periodic solution. A simple case where the condition upon the linear part is not satisfied is to assume that $A_m = 0$ and that the $A_1, \ldots, A_{m-1}$ are such that the linear part of (4.1) only admits the constant T-periodic solutions. In this case, the following theorem has been proved in [25] by the author (also for the case of functional differential equations) by coincidence degree techniques.

Theorem 4.1. *Suppose that the following supplementary assumptions hold for* (4.1)

(a) *There exists* $R > 0$ *such that*

$$T^{-1} \int_0^T f(t, x(t), x'(t), \ldots, x^{(m-1)}(t))dt \neq 0$$

for each $x \in P_T^m$ *such that* $|x(t)| \geq R$ *for each* $t \in R$.

(b) *The Brouwer degree* $d[g, B(O, R), 0]$, *with*

$$g : R^n \to R^n, \quad a \to T^{-1} \int_0^T f(t, a, 0, \ldots, 0)dt,$$

is different from zero.

Then, equation (4.1) has at least one T-periodic solution.

As shown in [25], Theorem 4.1 contains as special cases results of Lazer for second order equations, of Ezeilo [7], Sedsiwy [42] and Villari [47] for third order scalar equat ions, and of Reissig [33, 34, 35] and Sedsiwy [43, 44] for higher-order scalar or vector equations.

Another existence theorem, recently obtained by the author [26] using Leray-Schauder and coincidence degree arugments, concerns differential equations of the form

$$\sum_{\substack{j=0 \\ j \neq \ell-1}}^{m} A_j x^{(j)}(t) = \text{grad } V(x^{(\ell-1)}(t)) + e(t, x(t), x'(t), \ldots, x^{(\ell-1)}(t)) \tag{4.2}$$

with matrices A_j symmetric, $V : R^n \to R$ of class C^1 and e T-periodic in t, continuous and bounded. When $m = 2$, (4.2) specializes to a Duffing vector equation when $\ell = 1$ and a Lienard vector equation when $\ell = 2$.

Theorem 4.2. Suppose that the following conditions hold.

(a) there exist integers k, p satisfying the relations

$$\ell \leq k \leq m, \; k \equiv \ell \pmod 2), \; 0 \leq p \leq k-1$$

such that :

(i) A_k is positive or negative definite ;

(ii) $(-1)^{(j-\ell)/2} A_j$ is semi-definite of the same sign that $(-1)^{(k-\ell)/2} A_k$, $p \leq j \leq m$, $j \equiv \ell \pmod 2)$;

(iii) if $p \geq 1$,

$$\sum_{\substack{j=0 \\ j=\ell (\mathrm{mod}\ 2)}}^{p-1} |A_j| \omega^{j-k} < \sum_{\substack{j=k \\ j=\ell (\mathrm{mod}\ 2)}}^{m} \alpha j \omega^{j-k} \quad (\omega = 2\pi / T)$$

where the α_j is the smallest of the absolute values of the eigenvalues of A_j.

(b) when $\ell \geq 2$, A_m and A_0 are nonsingular.

(c) when $\ell = 1$, A_m is nonsingular, there exists $R > 0$ such that, for each $x \in P_T^m$ satisfying $|x(t)| \geq R$, $t \in R$,

$$T^{-1} \int_0^T (\mathrm{grad}\ V(x(t)) + e(t, x(t)))\, dt \neq 0,$$

and the Brouwer degree $d[w, B(O, R), 0]$ is nonzero, with

$$w : R^n \to R^n,\ a \to R^{-1} \int_0^T (\mathrm{grad}\ V(a) + e(t, a))dt.$$

Then (4.2) has at least one T-periodic solution.

For Theorem 4.2, the a priori bounds are obtained via L_2-norm estimates for the possible T-periodic solutions of the equations obtained from (4.2) by multiplying the right-hand member by λ, with $\lambda \in]0, 1]$. To get those L_2-norm estimates, one multiplies scalarly the equation by a convenient derivative of x, integrates over one period T and one makes use of the periodicity and of various classical inequalities for integrals.

Theorem 4.2 contains as very special cases earlier results of Faure [8], Villari [46] and the author [23] for second and third order equations. It is to be noted that, when $\ell \geq 2$, no condition is required for the nonlinear term grad $V(x^{(\ell-1)})$.

5. Some perturbation type results.

Coincidence degree approach can also be used to extend and simplify considerably the treatment of some old and recent perturbations type results for periodic solutions of differential equations.

As an example, let us consider the vector differential equation

$$x^{(m)} = X_0(t,x) + X_1(t,x,\varepsilon) \tag{5.1}$$

where $\varepsilon \in R$ is a parameter, the X_i are T-periodic with respect to t and continuous and such that, $p > 1$ and $q \geq 0$ being integers,

$$X_0(t,x) = X_0^{(p)}(t,x) + X_0^{(p+1)}(t,x) + \cdots + X_0^{(p+q)}(t,x) + R(t,x), X_1(t,x,0) = 0,$$

where the $X_0^{(j)}$ are homogeneous of order j in x and $R(t,x) = o(|x|^{p+q})$ uniformly in t. The following theorem is an easy consequence of results on the computation of the coincidence index due to the author and Laloux ([21, 22]) and of basic properties of coincidence degree. It generalizes a theorem of Halanay [13] who requires $m = 1$, $q = 0$ and X_0 and X_1 analytic.

Theorem 5.1. *Suppose satisfied the following conditions.*

(a) *for every* $i \neq q$ *and every* $a \in R^n$, $\int_0^T x_0^{(p+i)}(t, a)dt = 0$

(b) *for every* $a \in \partial B(0,1)$, $\int_0^T x_0^{(p+q)}(t, a)dt \neq 0$.

(c) $p - 1 > q$.

(d) *The Brouwer index* $i[x_0, 0]$ *is non zero, with*

$$x_0 : a \to T^{-1} \int_0^T x_0^{(p+q)}(t, a)dt.$$

Then, for $|\varepsilon|$ *sufficiently small, equation* (5.1) *has at least one* T-*periodic solution.*

For example, the scalar equation

$$x'' = \sum_{j=p}^{p+q} a_j(t)x^j + \varepsilon f(t)$$

where the a_j and f are T-periodic and continuous, $\int_0^T a_i(t)dt = 0$, $i = p, \ldots, p+q-1$, $\int_0^T a_{p+q}(t)dt \neq 0$, $p + q$ is odd and $p > q + 1$, has at least one T-periodic solution for sufficiently small $|\varepsilon|$.

Related perturbational results for functional differential equations have been given by Strygin [45] and used by Gustafson and Schmitt [11].

To end this section, we shall also note that recent theorems of Schmitt [38, 39] on periodic solutions of small period can also be proved very easily by coincidence degree techniques, without the assumption of prolongability of the solutions.

REFERENCES

[1] H. A. Antosiewicz, A general approach to linear problems for nonlinear ordinary differential equations, in "Ordinary differential equations", L. Weiss ed., Academic Press, New York, 1971.

[2] J. W. Bebernes, A simple alternative problem for finding periodic solutions of second order ordinary differential systems, Proc. Amer. Math. Soc. 42 (1974), 121-127.

[3] J. W. Bebernes, R. Gaines and K. Schmitt, Existence of periodic solutions for third and fourth order ordinary differential equations via coincidence degree, Ann. Soc. Sci. Bruxelles 88 (1974) 25-36.

[4] J. W. Bebernes and K. Schmitt, An existence theorem for periodic boundary value problems for systems of second order differential equations, Arch. Math. (Brno), to appear.

[5] L. Cesari, Nonlinear analysis, in "Nonlinear Mechnaics", CIME, 1972, Cremonese, Roma, 1973.

[6] J. Cronin, Periodic solutions of nonautonomous equations, Boll. Un. Mat. Ital. (4) 6 (1972) 45-54.

[7] J. O. C. Ezeilo, On the existence of periodic solutions of a certain third-order differential equation, Proc. Cambridge Phil. Soc. 56(1960) 381-389.

[8] R. Faure, Solutions periodiques d'equations differentielles et methode de Leray-Schauder (cas des vibrations forcees), Ann. Inst. Fourier

(Grenoble) 14 (1964) 195-204.

[9] R. E. Gaines, Existence of periodic solutions to second order nonlinear ordinary differential equations, J. Differential Equations, to appear.

[10] V. V. Gudkov and A. Ja. Lepin, On necessary and sufficient conditions for the solvability of certain boundary-value problems for a second-order ordinary differential equation, Soviet Math. Dokl. 14 (1973 800-803.

[11] G. B. Gustafson and K. Schmitt, Periodic solutions of hereditary differential systems, J. Differential Equations 13 (1973) 567-587.

[12] G. B. Gustafson and K. Schmitt, A note on periodic solutions for delay-differential systems, Proc. Amer. Math. Soc. 42 (1974) 161-166.

[13] A. Halanay, Solutions periodiques des systemes non-lineaires a petit parametre, Acc. Naz. Lincei, Rend. Cl. Sci. fis. mat. Natur. (8) 22 (1957) 30-32.

[14] J. K. Hale, "Ordinary differential equations", Wiley-Interscience, New York, 1969.

[15] P. Hartman, "Ordinary differential equations", Wiley, New York, 1964.

[16] H. W. Knobloch, An existence theorem for periodic solutions of nonlinear ordinary differential equations, Michigan Math. J. 9(1962) 303-309.

[17] H. W. Knobloch, Eine neue Methode zur Approximation periodischer Losungen nicht-linearer Differentialgleichungen zweiter Ordnung, Math. Z. 82(1963)

177-197.

[18] H. W. Knobloch, On the existence of periodic solutions for second order vector differential equations, J. Differential Equations 9 (1971) 67-85.

[19] M. A. Krasnosel'skii, "Translation along trajectories of differential equations", Amer. Math. Soc., Providence, RI, 1968.

[20] M. A. Krasnosel'skii, The theory of periodic solutions of non-autonomous differential equations, Russian Math. Surveys 21 (1966) 53-74.

[21] B. Laloux, On the computation of the coincidence index in the critical case, Ann. Soc. Sci. Bruxelles 88 (1974) 176-182.

[22] B. Laloux and J. Mawhin, Coincidence index and multiplicity, to appear.

[23] J. Mawhin, Degre topologique et solutions periodiques des systemes differentiels non lineaires, Bull. Soc. R. Sci. Liege 38 (1969) 308-398.

[24] J. Mawhin, Equivalence theorems for nonlinear operator equations and coincidence degree theory for some mappings in locally convex topological vector spaces, J. Differential Equations 12 (1972) 610-636.

[25] J. Mawhin, Periodic solutions of some vector retarded functional differential equations, J. Math. Anal. Appl. 45 (1974) 588-603.

[26] J. Mawhin, L_2-estimates and periodic solutions of some nonlinear differential equations, Bol. Un. Mat. Ital., to appear.

[27] J. Mawhin, Nonlinear perturbations of Fredholm mappings in normed spaces and applications to differential equations, Univ. de Brasilia, Trabalho de Matem. n° 61, May 1974.

[28] J. Mawhin, Nonlinear functional analysis and periodic solutions of ordinary differential equations, Summer School "Diford 74", Stara Lesna, Czechoslovaquia, to appear.

[29] J. Mawhin and C. Muñoz, Application du degre topologique a l'estimation du nombre des solutions periodiques d'equations differentielles, Ann. Mat. Pura Appl. (4) 96 (1973) 1-19.

[30] J. S. Muldowney and D. Willett, An intermediate value property for operators with applications to integral anddifferential equations, to appear.

[31] J. S. Muldowney and D. Willett, An elementary proof of the existence of solutions to second order nonlinear boundary value problems, to appear.

[32] R. Reissig, G. Sansone and R. Conti, "Nonlinear differential equations of higher order", Noordhoff, Leyden, 1974.

[33] R. Reissig, On the existence of periodic solutions of a certain non-autonomous differential equations, Ann. Mat. Pura Appl. (4) 85 (1970) 235-240.

[34] R. Reissig, Note on a certain non-autonomous differential equations, Atti Accad. Naz. Lincei Rend. Cl. Sci. fis. mat. natur. (8) 48 (1970) 246-248.

[35] R. Reissig, Periodic solutions of a nonlinear n-th order vector differential equation, Ann. Mat. Pura Appl. (4) 87 (1970) 111-124.

[36] N. Rouche and J. Mawhin, "Equations differentielles ordinaires", 2 volumes, Masson, Paris, 1973.

[37] K. Schmitt, Periodic solutions of nonlinear second order differential equations, Math. Z. 98 (1967) 200-207.

[38] K. Schmitt, A note on periodic solutions of second order ordinary differential equations, SIAM J. Appl. Math. 21 (1971) 491-494.

[39] K. Schmitt, Periodic solutions of small period of systems of n-th order differential equations, Proc. Amer. Math. Soc. 36(1972) 459-463.

[40] K. Schmitt, Intermediate value theorems for non-linear periodic functional differential equations, in "Equations differentielles et fonctionnelles non linearires", P. Janssens, J. Mawhin and N. Rouche ed., Hermann, Paris, 1973.

[41] K. Schmitt, Randwertaufgaben fur gewohnliche Differentialgleichungen, in "Proc. Steiermark. Marthematisch Symposium", Graz, Austria, 1973.

[42] S. Sedsiwy, On periodic solutions of a certain third-order nonlinear differential equation, Ann. Polon. Math. 17 (1965) 147-154.

[43] S. Sedsiwy, Asymptotic properties of solutions of nonlinear differential equations of the higher-order, Zeszyty Nauk Uniw Jagiell. n° 131 (1966) 69-80.

[44] S. Sedsiwy, Asymptotic properites of solutions of a certain n-th order vector differential equations, Atti Accad. Naz. Lincei, Rend. Cl. Sci. Fis. Mat. Natur. (8) 47 (1969) 472-475.

[45] V. V. Strygin, A theorem concerning the existence of periodic solutions of systems of differential equations with delayed arguments, Math. Notes 8 (1970) 600-602.

[46] G. Villari, Contributi allo studio dell'esistenza di soluzioni periodiche per i sistemi di equazioni differenziali ordinarie, Ann. Mat. Pura Appl. (4) 69 (1965) 171-190.

[47] G. Villari, Soluzioni periodiche di una classe di equazioni differenziali, Ann. Mat. Pura Appl. (4) 73 (1966) 103-110.

Universite de Louvain, Institut Mathematique
B-1348 Louvain-la-Neuve Belgium

FACTORIZATION OF FREDHOLM OPERATORS ON ANALYTIC FUNCTIONS

Alex McNabb

and

Alan Schumitzky

1. Introduction. Let Ω denote a simply connected domain in the complex plane and $A(\Omega^n)$ the space of functions which are defined and analytic on $\Omega^n = \Omega \times \cdots \times \Omega$ (n times). We are concerned with the theory of Volterra factorization of class of Fredholm operators on $A(\Omega)$ [and also $A(\Omega^n)$], which is analogous to factorization theories on the real line developed by the authors in [1, 2], and earlier by Gohberg and Krein [cf. 3].

This paper presents a summary of our main results and ideas; complete details will be published elsewhere.

The most basic and obvious extension of classical Fredholm theory to the complex analytic setting is consideration of equations of the form:

This research was supported in part by the National Science Foundation under Grant No. GP-20130.

$$(1) \qquad u(t) = g(t) + \int_a^b k(t, s)\, u(s)\, ds\,,$$

where the kernel $k \in A(\Omega^2)$, the points $a, b \in \Omega$ and the forcing function $g \in A(\Omega)$. Such kernels, as in (1), we call holomorphic and these form an essential ingredient of the theory developed herein. However this class of problems by itself is somewhat restricted and does not include common kernels arising from the theory of ordinary differential equations. For example, the boundary value problem

$$\frac{dy}{dt}(t) = A(t)\, y(t), \quad \lambda\, y(a) + \mu\, y(b) = \xi$$

where ξ, $y(t)$ are n-vectors, $A(t)$, λ, μ are $n \times n$ matrices, $\lambda + \mu$ = Identity and the components of y and A are elements of $A(\Omega)$, can be transformed to an equivalent Fredholm equation,

$$y(t) = \xi + \int_a^t \lambda\, A(s)\, y(s)\, ds + \int_b^t \mu\, A(s)\, y(s)\, ds,$$

incompatible with the form (1) since $\lambda + \mu \neq 0$. This suggests a richer extension would be obtained by considering the theory of equations of the form

$$y(t) = g(t) + \int_a^t k_1(t, s)\, y(s)\, ds + \int_b^t k_2(t, s)\, y(s)\, ds$$

where k_1 and $k_2 \in A(\Omega^2)$, $g \in A(\Omega)$, or more generally equations of the form

$$(2)\qquad y(t) = g(t) + \sum_{i=1}^{n} \int_{a_i}^{t} k_i(t, s)\, y(s)\, ds\ ,$$

$$k_i \varepsilon\ A(\Omega^2),\ i = 1, 2, \ldots, n.$$

A special case of such an equation would arise as the multi-point analogue of the example above where the boundary condition is replaced by

$$\sum_{i=1}^{n} \lambda_i\, y(a_i) = \xi\ ,$$

where λ_i are such that $\sum_{i=1}^{n} \lambda_i$ = Identity. The equivalent Fredholm equation is

$$y(t) = \xi + \sum_{i=1}^{n} \int_{a_i}^{t} \lambda_i\, A(s)\, y(s)\, ds.$$

The fundamental elements of the theory developed here are integral operators of the form $K_a \equiv [k(t, s)]_a$, $k \varepsilon\ A(\Omega^2)$, $a \varepsilon\ \Omega$, and multiplication operators of the form $M = [m(t)]$, $m \varepsilon\ A(\Omega)$, which map $A(\Omega)$ into itself in the following way: for any $u \varepsilon\ A(\Omega)$,

$$K_a u = v, \text{ where } v(t) = \int_a^t k(t, s)\, u(s)\, ds\ ,\quad t \in \Omega$$

$$Mu = w, \text{ where } w(t) = m(t)u(t),\quad t \in \Omega.$$

If we write I for the identity operator on $A(\Omega)$ the typical Fredholm problem (2) can be written as $(I-K)u = g$, where

(3) $$K = \sum_{i=1}^{n} K_{a_i} = \sum_{i=1}^{n} [k_i(t, s)]_{a_i} .$$

In the case $n = 1$, I-K is a Volterra operator and it is shown that I-K is always invertible with an inverse of the same form; when $n = 2$ and $k_1(t, s) + k_2(t, s) \equiv 0$, I-K is a holomorphic Fredholm operator as in (1). In general K is called holomorphic if and only if

(4) $$\sum_{i=1}^{n} k_i(t, s) \equiv 0.$$

2. Algebraic results and holomorphic factorization. It is shown that the class of integral operators $\mathcal{J}$ and the multiplication operators $\mathcal{M}$ form a direct-sum subring of the ring of continuous linear operators on $A(\Omega)$. [$A(\Omega)$ is endowed with the usual topology of uniform convergence on compact subsets.] Multiplication in this ring is that induced by ordinary operator multiplication and it is shown that Fredholm operators of the form M+K multiply according to the formulae,

$$(M_1 + K_1)(M_2 + K_2) = M_3 + K_3$$

where $M_i = [m_i(t)]$, $K_i = \sum_{j=1}^{n} [k_j^i(t, s)]_{a_i}$, $i = (1, 2, 3)$ and

(5) $$m_3(t) = m_1(t)\, m_2(t),\ k_\ell^3(t, s) = k_\ell^1(t, s)\, m_2(s) + m_1(t)\, k_\ell^2(t, s) + \omega_\ell(t, s),$$

with

(6) $$\omega_\ell(t,s) = \sum_{j=1}^{n} \left[\int_{a_j}^{t} k_j^1(t,\theta)\, k_\ell^2(\theta,s)\, d\theta - \int_{a_\ell}^{s} k_\ell^1(t,\theta) k_j^2(\theta,s) d\theta\right].$$

For a given set of n points $\alpha = (a_1, a_2, \ldots, a_n)$, $a_i \varepsilon \Omega$ we denote by $\mathcal{J}(\alpha)$ or $\mathcal{J}(\alpha,\Omega)$ the ring of operators K of the form (3), by $\mathcal{F}(\alpha)$ those of the form M+K, $M \varepsilon \mathcal{M}$, $K \varepsilon \mathcal{J}(\alpha)$ and by $\mathcal{H}(\alpha)$, the subring of holomorphic operators $K \varepsilon \mathcal{J}(\alpha)$ which satisfy equation (4). $\mathcal{H}(\alpha)$ is shown to be a two-sided ideal in $F(\alpha)$. Moreover, if $K \varepsilon \mathcal{J}(\alpha)$ then I-K may be uniquely factored in the form

(7) $$I - K = (I - V)\,(I - H)$$

where $V \varepsilon \mathcal{J}(a_1)$ and $H \varepsilon \mathcal{H}(\alpha)$. Since $\mathcal{J}$-V is a Volterra operator, hence invertible, the invertibility of I-K is equivalent to the invertibility of I-H. Thus the factorization (7) has important practical significance in solving Fredholm equations of type (2) and is currently be used for numerical purposes.

3. Factorization and generalized Wiener-Hopf equations.

Our primary interest is focused on the problem of factoring I-K, $K \varepsilon \mathcal{J}(\alpha)$ in the form

(8) $$I - K = (I - V_1)\,(I-V_2)\cdots(I-V_n),\ V_i \varepsilon I(a_i)\,.$$

Natural norms are defined on the elements $\mathcal{F}(\alpha)$ and it is shown that if K has sufficiently small norm then factorization of I-K in the form (8) is possible. Since the right hand side of expression (8) is invertible such a form cannot be expected to hold in general. However what does hold in general is a certain "Wiener-Hopf" equation which is equivalent to (8) whenever (8) is valid. (For results connecting factorization and Wiener-Hopf equations, see [4, 5] and most recently [6].)

An example follows which illustrates this point in the simplest case:

Consider equation (1) for the case $a = 0$ and b in Ω, $k(t, s) \equiv 1$ in Ω^2, so that in the notation of (3)

$$K = [1]_0 - [1]_b.$$

A factorization of the form (8) is possible and in fact

$$\begin{aligned} (I-K) &= (I - V_1)(I - V_2) \\ &= [I - [\tfrac{1}{1-s}]_0][I + [\tfrac{1}{1-t}]_b] \end{aligned} \tag{9}$$

provided Ω does not contain the point. 1. Equation (9) can be interpreted as a factorization in terms of meromorphic functions in Ω^2. To obtain the classical Wiener-Hopf equation equivalent to (9), we write

$$(I-K)(I-V_2)^{-1} = (I-V_1),$$

that is

$$(I - [1]_0 + [1]_b)[I - [\tfrac{1}{1-s}]_b] = [I - [\tfrac{1}{1-s}]_0].$$

The generalized equation is then gotten by multiplying both sides on the right by $[(1-t)] \in M$. We thus arrive at the factored form

$$(I - [1]_0 + [1]_1)\,([1-t] - [1]_b) = ([1-t] - [1]_0)$$

valid in the whole region Ω of analyticity of K. The function 1-t may be characterised as a certain "determinant" of I-K and is denoted by $d_K(t)$. Note that the zeros of d_K are the points of singularity of $(I-K)^{-1}$.

For a general operator I-K, the function d_K is uniquely defined and shown to have the multiplicative and zero properties expected of a determinant of a Fredholm operator (cf. [7]). Our main result is to prove that such an operator admits, in the manner of the above example, a generalized Wiener-Hopf factorization.

To this end, we define a projection π mapping $\mathcal{J}(a_1, \ldots, a_n)$ onto $\mathcal{J}(a_1, \ldots, a_{n-1})$ as follows :

$$\pi\left(\sum_{i=1}^{n} [k_i]_{a_i}\right) = \sum_{i=1}^{n-1} [k_i]_{a_i} + [k_n]_{a_{n-1}} .$$

We note that this projection induces a factorization structure on $\mathcal{J}(\alpha)$ as described in [1].

Theorem. If $K \in \mathcal{J}(a_1, \ldots, a_n)$ then there exist an element $W \in \mathcal{J}(a_n)$ such that

$$(10) \qquad (I-K)\,(M-W) = (I-\pi K)\,(M-\pi W),$$

where $M = [d_K(t)] \in \mathfrak{m}$.

Since $I - \pi K \in \mathscr{I}(a_1, \ldots, a_{n-1})$, the Theorem can be used recursively to factor $I-\pi K$ in a similar way and so on ; thus building a chain of Wiener-Hopf factorizations as the analogne of (8).

4. Sketch of proof. An essential ingredient of the derivation of this general result is an imbedding process in which we are led to consider families of operators K in which the a_i's are parameters. Since kernels derives as products using the formula (5-6) are analytic functions of a_i we must consider the functions k_i as functions of n+2 variables in $A(\Omega^{n+2})$ and $[k_i]_{a_i}$ as operators on $A(\Omega^{n+1})$. The ring product formulae (5,6) leads naturally to a kernel Algebra $\mathscr{I}_n(\Omega)$ of elements $(k_1, k_2, \ldots, k_n)$, which we write alternatively as $\sum_{i=1}^{n} [k_i]_i$ where $k_i \equiv k_i(t, s, a_1, \ldots, a_n) \in A(\Omega^{n+2})$. This is isomorphic to the subring of $\mathscr{I}_n(\Omega_1)$, $\Omega_1 \subset \Omega$ which is obtained by considering elements of $\mathscr{I}_n(\Omega)$ restricted to Ω_1^{n+2} as elements of $\mathscr{I}_n(\Omega_1)$. If Ω_1 is small enough, then any given $K \in \mathscr{I}_n(\Omega)$ will have sufficiently small norm, as an element of $\mathscr{I}_n(\Omega_1)$, to ensure factorizability of I-K in the form (8) in $\mathscr{I}_n(\Omega_1)$. Next we define the "determinant" of a Volterra operator $(I-[v(t, s, \alpha)]_i)$ as $\exp\left[-\int_{a_i}^{t} v(\theta, \theta, \alpha)d\theta\right]$ and then show that it is consistent to define the determinant of I-K in $\mathscr{I}_n(\Omega_1)$ as the product of its Volterra factors.

This determinant function in Ω_1^{n+1} has an analytic extension into Ω^{n+1} obtained by considering a class of degenerate kernels $\mathcal{D}_n(\Omega)$ which are dense in $\mathcal{K}_n(\Omega)$, and given an explicit representation for their determinants. The operator $K \in \mathcal{K}_n(\Omega)$ is degenerate if for some integer p,

$$\text{(11)} \qquad k_i(t, s, \alpha) = \sum_{j=1}^{p} a_j^i(t, a)\, b_j^i(s, a) \quad (t, s, \alpha) \in \Omega^{n+2}$$

where a_j^i, $b_j^i \in A(\Omega^{n+1})$ for $i = 1, 2, \ldots, n$; $j = 1, 2, \ldots, p$. More generally, if $K \in \mathcal{J}_n(\Omega)$ and condition (11) is satisfied, then K is said to be separable. Once the determinant is globally defined, the basic foctorization theorem (10) is proved for separable kernels. The results are extended to general K as follows: The region Ω can be represented as the union of an infinite nested family of compact simply connected open sets Ω_r, such that $\Omega_0 \subset \Omega_1 \subset \ldots \subset \Omega$, $\Omega = \cup \Omega_r$. For any given r, $H \in \mathcal{K}_n(\Omega)$ has a representation

$$H = D_r + N_r$$

where $D_r \in \mathcal{D}_n(\Omega)$ and N_r as an element in $\mathcal{K}_n(\Omega_r)$ has sufficiently small norm. We use this decomposition together with the results on factoring degenerate operators and operators of small norm to obtain a factorization in $\mathcal{J}_n(\Omega_r)$. The factorizations formulae (10) are analytic in Ω_r^{n+2} and unique in Ω_0^{n+2} if Ω_0 is chosen small enough and therefore define a factorization in $\mathcal{J}_n(\Omega)$. A factorization for general K follows from the result described by equation (7).

In conclusion it should be noted that all results and proofs are readily translated to the matrix case where the elements of $A(\Omega^{n+1})$ are vectors and those of $A(\Omega^{n+2})$ are matrices.

REFERENCES

[1] A. McNabb and A. Schumitzky, Factorization of operators - I: Algebraic theory and examples, J. Functional Analysis 9(1972), 262-295.

[2] A. McNabb and A. Schumitzky, Factorization and imbedding for general linear boundary value problems, J. Differential Equations 14(1973), 518-546.

[3] I. C. Gohberg and M. G. Krein, Theory and Applications of Volterra Operators in Hilbert Space, Translations of Math. Monographs, vol. 24, Amer. Math. Soc. 1970.

[4] A. Schumitzky, On the equivalence between matrix Riccati equations and Fredholm resolvents, J. Comput. System Sci 2(1968), 76-87.

[5] T. Kailath, Fredholm resolvents, Wiener-Hopf equations, and Ricatti differential equations, IEEE Trans. Infor. Theory, IT-15 (1969), 665-672.

[6] L. H. Brandenburg, Covariance factorization: Some unified results encompassing both stationary and nonstationary processes, IEEE Trans. Infor. Theory. (to appear)

[7] A. Grothendieck, La theorie de la Fredholm, Bull. Soc. Math. France 84(1956), 319-384.

The Australian National University, Wellington, New Zealand
University of Southern California, Los Angeles, California

SOME FUNDAMENTAL THEORY OF VOLTERRA INTEGRAL EQUATIONS

R. K. Miller

1. Introduction.

The purpose here is to discuss recent results of several authors on existence, uniqueness, continuity, continuability and (in nonunique cases) properties of the solution of Volterra integral equations

$$\text{(VI)} \quad x(t) = f(t) + \int_0^t g(t, s, x(s))\, ds$$

and of integrodifferential equations. The next section contains a discussion of equations in real or complex finite dimensional space. The last section contains some results for linear integrodifferential equations in a Banach space.

2. Finite dimensional spaces.

I first wish to discuss some interesting results of N. Kikuchi and S. Nakagiri [1, 2], W. G. Kelly [3], Z. Artstein [4] and R. C. Grimmer and W. N. Hunsaker [unpublished].

Kikuchi and Nakagiri have greatly generalized the results in [5, 6] for integral equations of the form

$$(2.1) \quad x(t) = f(t) + \int_0^t a(t, s) g(s, x(s))\, ds .$$

This work was supported by the National Science Foundation under Grant No. GP-31184X.

They use the following notation and hypotheses. Let R^+ be the interval $0 \leq t < \infty$ and let W be an open set in R^n = real n-dimensional Euclidean space with norm $|x|$. Let $L_p(J, K)$ denote L_p functions $f: J \to K$ with the usual norm $\|f\|_p$. Let $B_p(J) = L_p(J, R^n)$, let $B_p^*(J)$ be its adjoint space and let $B_p^*(J)^n = B_p^*(J) \oplus \cdots \oplus B_p^*(J)$ (n factors). Assume the following

(A) $f : R^+ \to W$ is continuous

(B) Let p satisfy $1 \leq p < \infty$ and let $g(t, x)$ be a function on $R^+ \times W$ to R^n such that (1) for each fixed x in W, g is measurable in t, (2) for each fixed t in R^+, g is continuous in x, and (3) for each compact set $K \subset W$ and each compact interval $J \subset R^+$ there exists $m(t)$ in $L_p(J, R^1)$ such that $|g(t, x)| \leq m(t)$ for all (t, x) in $J \times K$.

(C_p^*) Let $a(t, s)$ map $R^+ \times R^+$ into M^n, the space of linear operators on R^n, such that (1) for any compact interval $J \subset R^+$ and any t in R^+ the map $S: B_p(J) \to R^n$ defined by

$$S : x(\cdot) \to \int_J a(t, s)x(s)ds$$

is a bounded linear map and (2) the map $I \to B_p^*(J)^n$ defined by $t \to a(t, \cdot)$ is continuous in the weak $*$-topology on $B_p^*(J)^n$. The next three results are reported in [1] and [2]. The authors give two different proofs of Theorem 1 below. The second of their proofs is an interesting and unusual argument involving compactness in $B_1(J)$.

<u>Theorem 1.</u> <u>There exists</u> $\beta > 0$ <u>and a continuous function</u> $x(t)$ <u>such that</u> $x(t)$ <u>solves</u> (2.1) <u>on</u> $0 \leq t \leq \beta$.

By a noncontinuable $x(t)$ of (2.1), or of any other Volterra equation, we mean a solution $x(t)$ defined on $[0,\alpha)$ where either $\alpha = +\infty$ or $x(t)$ tends to the boundary ∂W of W on a sequence $t_n \nearrow \alpha$.

Theorem 2. Any solution of (2.1) can be extended to a noncontinuable solution of (2.1).

Let $\Sigma[0, b] = \{x(\cdot) \in C[0, b] : x(t)$ solves (2.1) on $0 \leq t \leq b\}$, i.e. $\Sigma[0, b]$ is the cone of solutions of (2.1) up to time b. Let $\Sigma(c) = \{y \in R^n : \exists\, x \in \Sigma[0, c]$ with $x(c) = y\}$, i.e. $\Sigma(c)$ is the cross section at c of the cone of solutions.

Theorem 3. If all solutions of (2.1) exist on $[0, b]$, then for any c in the interval $0 < c \leq b$ the cross section $\Sigma(c)$ is compact and connected.

Artstein [4] studied existence, continuity and continuation for a general Volterra problem of the form (VI). He assumes:

(G1) $g : R^+ \times R^+ \times R^n \to R^n$ is measurable in s (t and x fixed) and continuous in x (t and s fixed) while $g(t, s, x) = 0$ if $s > t$.

(G2) For any $b > 0$ and any $N > 0$ there exists a function $m(t, s)$, integrable in $s \in [0, t]$ such that $|g(t, s, x)| \leq m(t, s)$ when $|x| \leq N$ and $0 \leq s \leq t \leq b$.

(G3) For any $b > 0$ and any $N > 0$, if $t \geq 0$, then as $\tau \to t$.

$$\sup \{ | \int_0^b [g(t, s, \varphi(s)) - g(\tau, s, \varphi(s))] ds | : \|\varphi\| \leq N, \varphi \in C[0,b]\} \to 0.$$

Here $\|\varphi\|$ denotes the uniform norm over $0 \leq t \leq b$.

Several remarks are in order. First note that (VI)

with (G1), (G2) and (G3) true includes and generalizes the case of (2.1) with (B_p) and (C_p^*) true. Artstein's hypotheses are certainly more general than those used by the author in [7, Chapter II] or by Kelly [3]. Artstein points out that the proofs of Theorem 1.1 and 2.2 in [7, Chapter II] remain valid under his weaker hypotheses. Thus the following result is true.

Theorem 4. *If* $f \in C(R^+)$ *and if* (G1-G3) *are true, then there exists* $b > 0$ *such that* (VI) *has a continuous solution* $x(t)$ *on* $0 \leq t \leq b$. *Moreover any solution of* (VI) *can be extended to a noncontinuable solution.*

Remarkably Theorem 4 contains Theorems 1 and 2 above (they were originally proved by quite different means). It is also worth noting that Theorem 3.1 of [3] remains true under hypotheses (G1-G3).

Theorem 5. *If* $f \in C(R^+)$, *if* (G1-G3) *are true, and if all solutions of* (VI) *exist on* $[0, b]$, *then the cone* $\Sigma[0, b]$ *is compact and connected as a subset of* $C[0, b]$ *with the uniform norm.*

This result is somewhat stronger than Theorem 3 above. In particular it will follow that Theorem 3 can be extended to (VI) with (G1-G3) true. We note that Kelly's proof in [3] works for Theorem 5 with only one change. His argument for uniform boundedness of $\{x_k(\cdot)\}$ at the top of page 189 is not valid. However this set of functions is equicontinuous and satisfies $x_k(0) = f(0)$ for all k. Thus the uniform boundedness remains true.

Continuity type results for (VI), without uniqueness, are proved in [7, Theorem II.4.2] and in [3, Theorem 2.1]. Artstein has generalized these results in a very satisfying way. His result can be used to reduce previous results for (VI) and also for ordinary differential equations to easy exercises. By $U(t,\tau,b,N)$ we will denote any nonnegative function such that limit $U(t,\tau,b,N) = 0$ as $t \to \tau$. For a fixed U let $\mathcal{G}$ be the set of all functions g which satisfy (G1), (G2) and the following uniform version of (G3):

(G3′) for any $\varphi \in C[0,b]$ such that $\|\varphi\| \leq N$ and for any t and τ such that $0 \leq t, \tau \leq b$ one has

$$\left|\int_0^b [g(t,s,\varphi(s)) - g(\tau,s,\varphi(s))]\,ds\right| \leq U(t,\tau,b,N).$$

Definition. A topology $\mathcal{J}$ of $\mathcal{G}$ is called jointly continuous if for any fixed $t > 0$ the map

$$(g,\varphi) \to \int_0^t g(t,s,\varphi(s))\,ds$$

from $\mathcal{G} \times C[0,t]$ to R^n is continuous in the product topology of $\mathcal{J}$ on $\mathcal{G}$ and the uniform norm on $C[0,t]$.

For (VI) and also a sequence of equations

$$(\mathrm{VI}_k) \qquad x(t) = f_k(t) + \int_0^t g_k(t,s,x(s)ds,$$

Artstein proves the following.

Theorem 6. Given a $\mathcal{G}$ and a topology $\mathcal{J}$, $\mathcal{J}$ will be jointly continuous if and only if the following continuity condition is true:

(C) Suppose $\{g_k\}$ and $\{f_k\}$ are generalized sequences such that $g_k \to g$ in $\mathcal{J}$ and $f_k(t) \to f(t)$ uniformly for t on compact sets. Let $x_k(t)$ be a noncontinuable solution of

(VI_k) with domain $[0,\alpha_K)$. Then there exists a noncontinuable solution $x(t)$ of (VI) defined on an interval $[0,\alpha)$, there exists a countable set of generalized subsequences $\{x_{hk}\}$ such that $\{x_k\} \supset \{x_{1k}\} \supset \{x_{2k}\} \supset \cdots$ and there exists $\beta_n \nearrow \alpha$ such that $x_{nk}(t) \to x(t)$ as $k \to \infty$ uniformly on $[0,\beta_n]$. In particular if $0 < d < \alpha$, then $d < \alpha_k$ for k sufficiently large.

As pointed out above, if (G1-G3) are true and if f is continuous, then any solution of (VI) can be extended to a noncontinuable solution. If $x(t)$ is noncontinuable but only defined on a finite interval $[0,\alpha)$, then necessarily there must be a sequence $t_n \nearrow \alpha$ on which $|x(t_n)| \to \infty$. For a time it was an open problem whether or not $|x(t)| \to \infty$ as $t \to \alpha$. In an appendix of [4] Artstein constructs a rather complicated example for which $0 < \alpha < \infty$, $\limsup x(t) = +\infty$ as $t \to \alpha$ and the $\liminf x(t) = 0$. In a private communication R. C. Grimmer had already pointed out to me the following easier counterexample. It is known (see [14]) that there exists a positive continuous function $\gamma(t)$ and initial conditions y_0 and y_0' such that the problem

$$y'' + \gamma(t)\, y^3 = 0, \quad y(0) = y_0, \; y'(0) = y_0'$$

has finite escape time α. We need only integrate twice to obtain

$$y(t) = (y_0 + y_0' t) - \int_0^t (t-s)\gamma(s) y^3(s)\, ds .$$

W. N. Hunsaker and R. C. Grimmer have privately communicated a different type of continuity result. Under hypotheses (H1-H4) in [7, Chapter II], if all solutions of (VI) exists on $0 \leq t \leq b$, then they claim that the cross section $\Sigma(c)$ is a continuous function of c in $[0, b]$ to sets with the Hausdorff topology. It would be interesting to know if their result is true under the weaker assumptions (G1-G3).

3. Banach space valued problems.

A recent paper by T. Y. Li [8] contains a readable account of existence theory for continuous Volterra equations in a Banach space. Our interest here will be to report some results for noncontinuous equations. In particular we shall consider linear Volterra integrodifferential equations with bounded operator coefficients:

$$(3.1) \quad x'(t) = Ax(t) + \int_0^t B(t-s)x(s)\,ds + f(t), \quad x(0) = x_0.$$

Friedman and Shinbrot [9] and Hannsgen [10, 11, 12] have obtained existence and uniqueness Theorems for (3.1). Hannsgen's results depend on special properties of Hilbert space and require very special classes of forcing functions, i. e. $f(t) = \alpha+\beta t$ a linear function. The results in [9] apply to a general class of "smooth" initial conditions (x_0, f). Theorem 7 below is typical of the type of results which can be proved by the methods in [9].

Let $A{:}D(A) \to X$ be a linear map with domain $D(A)$ dense in the B-space X. Assume

(A1) A is the infinitesimal generator of a C_0 - semigroup.

(A2) $B(t) = b(t)A$ with $b(t) \in C^1(R^1) \cap L^1(R^+)$ real valued.

(A3) $f(t)$ is strongly continuously differentiable on R^+.

From (A2) we see that (3.1) can be written as

$$(3.2) \quad x'(t) = Ax(t) + \int_0^t b(t-s)\,Ax(s)\,ds + f(t), \; x(0) = x_0.$$

Theorem 7. If (A1-A3) are true and if $x_0 \in D(A)$, then (3.2) has a unique solution $x(t)$ defined on R^+.

Roughly speaking the theorem says that on a dense set of "smooth" initial conditions (x_0, f) equation (3.2) has a unique solution. We shall see below that such a dense set of smooth initial conditions exists under very weak hypotheses. However the particular dense set prescribed in Theorem 7 is "large enough" so that the solution $x(t, x_0, f)$ of (3.2) is continuous in $(t, x_0, f) \in R^+ \times X \times C(R^+)$. This continuity allows us to define generalized solutions of (3.2) for all other values of (x_0, f) in $X \times C(R^+)$. The details of this theory are in [13]. Assumption (A2) can be generalized for (3.1) as follows. Assume that $A: D(A) \to X$ is linear with dense domain, $B(t)$ is a linear map defined on $D(A)$ (or on a larger set) for each fixed $t \geq 0$, and $B(t)x_0$ is measurable in $t \geq 0$ for any fixed x_0 in $D(A)$. Moreover assume $B(t)$ is subordinate to A in the sense that there exists a real valued function $\beta(t) \in L^1(R^+)$ such that

$$\|B(t)x\| \leq \beta(t)\ [\|x\| + \|Ax\|]\quad (t \geq 0,\ x \in D(A)).$$

A solution of (3.1) on an interval $[0, T]$ will be a strongly differentiable function $x: [0, T] \to D(A)$ such that $x(t)$, $x'(t)$ and $Ax(t)$ are continuous, $x(0) = x_0$ and (3.1) is true in $[0, T]$.

Define

$$B^*(s)x = \int_0^\infty \exp(-st)B(t)x\,dt$$

for $\operatorname{Re} s \geq 0$ and x in $D(A)$ and define

$$\rho(s) = (sI - A - B^*(s))^{-1}$$

whenever $\operatorname{Re} s \geq 0$ and the inverse exists as a bounded linear map on X to X. Let $BU(R^+) = \{f: R^+ \to X: f$ is bounded and uniformly continuous$\}$.

Theorem 8. If A is a closed map, $\rho(s)$ exists for all s in a half plane $\operatorname{Re} s \geq s_0 > 0$ and $\|\rho(s)\| = O(1+|s|^k)$ for some $k \geq 0$ as $|s| \to \infty$ in $\operatorname{Re} s \geq s_0$, then there exists a dense set $Y \subset X \times BU(R^+)$ such that (3.1) has a strong solution on R^+ when $(x_0, f) \in Y$.

For example suppose $b(t)$ is a scaler function whose Laplace transform $b^*(s)$ exists for $\operatorname{Re} s \geq b$ where $b \geq 0$. Suppose the resolvent $R(s, A) = (sI-A)^{-1}$ exists when $\operatorname{Re} s \geq b$ and satisfies $\|R(s, A)\| \leq M(1+|s|)^k$ for some $M > 0$ and $k \geq 0$. Then $\rho(s) = (1+b^*(s))^{-1}R(s(1+b^*(s))^{-1}, A)$ exists and for some $b_1 \geq b$ and $M_1 > 0$ one has

$$\|\rho(s)\| \leq |1+b^*(s)|^{-1}M(1+|s||1+b^*(s)|^{-1})^k$$

$$\leq M_1 (1+|s|)^k \qquad (\operatorname{Re} s \geq b_1).$$

Thus Theorem 8 will apply for $B(t) = b(t)A$.

If the hypotheses of Theorem 8 are true and if for any $T > 0$ and any $f \in C[0, T]$ such that $Af \in C[0,T]$ one has

$$(3.3) \qquad \int_0^T B(t)f(s)ds = B(t)\int_0^T f(s)ds.$$

then solutions of (3.1) will be unique whenever they exist. In particular if $B(t) = b(t)A$, $b(t)$ is scaler and A is closed, then (3.3) is always true. Therefore we see that existence and uniqueness of solutions of (3.1) for a dense set of initial conditions can be achieved under relatively weak hypotheses on A and $B(t)$. Now we shall consider a special dense set D_c defined as follows:

$D_c = \{(x, f): x \in D(A),\ f \in BU(R^+),\ f'$ exists a.e. and $f' + Bx \in BU(R^+)\}$. If A is closed, then it can be shown that D_c is dense in $X \times BU(R^+)$.

Theorem 9. *Suppose* A *is closed,* $\rho(s)$ *exists for at least one* s *in the half plane* $\operatorname{Re} s > 0$ *and for each pair* (x_0, f) *in* D_c, (3.1) *has a unique solution* $x(t, x_0, f)$ *on* $0 \leq t < \infty$. *Then given* $T > 0$ *there exists* $K = K(T) > 0$ *such that*

$$\|x(t, x_0, f)\| \leq K[\|x_0\| + |||f|||]$$

on $0 \leq t \leq T$ *and* $(x_0, f) \in D_c$. *Here* $|||f||| = \sup \{\|f(t)\|: 0 \leq t \leq \infty\}$.

If A and $B(t) = b(t)A$ satisfy (A1-A3) then they satisfy the general hypotheses on A and $B(t)$ listed above, A will be closed and the operator $\rho(s)$ will exist when $\operatorname{Re} s$ is sufficiently large with $\rho(s) = (1+b^*(s))^{-1} R(s(1+b^*(s))^{-1}, A)$. Since $B(t)x = b(t)Ax$ is continuous in $t \geq 0$ if $x \in D(A)$, then for (x_0, f) in D_c, f' must be continuous ($f' + Bx_0$ is in $BU(R^+)$). Therefore Theorem 7 implies that the solution of (3.2) exists and is unique for any pair (x_0, f) in D_c. In particular Theorem 9 can be applied to prove continuity of solutions $x(t, x_0, f)$ on $R^+ \times X \times BU(R^+)$.

REFERENCES

[1] Norio Kikuchi and Shin'ichi Nakagiri, An existence theorem of solutions of non-linear integral equations, Fund. Ekvac. 15 (1972), 131-138.

[2] _____________, Kneser's property of solutions of non-linear integral equations, Funk, Ekvac., 17 (1974), 57-66.

[3] Walter G. Kelly, A Kneser theorem for Volterra integral equations, Proc. Amer. Math. Soc. 40 (1973), 183-190.

[4] Zvi Artstein, Continuous dependence of solutions of Volterra integral equations, SIAM Math. Anal., to appear.

[5] R. K. Miller and G. R. Sell, Existence, uniqueness and continuity of solutions of integral equations, Ann. Mat. Pura ed Appl. IV, LXXX, 135-152.

[6] _____________, Volterra Integral Equations and topological dynamics, Memoirs Amer. Math. Soc., Number 102, 1970, Amer. Math. Soc., Providence, R.I.

[7] R. K. Miller, Nonlinear Volterra Integral Equations, W. A. Benjamin, Inc., Menlo Park, California, 1971.

[8] Tien-Yien Li, Existence of solutions for ordinary differential equations in Banach space, technical report, Institute for Fluid Dynamics and Applied Mathematics, University of Maryland.

[9] A. Friedman and Marvin Shinbrot, Volterra integral equations in Banach space, Trans, Amer. Math. Soc. 126 (1967), 131-179.

[10] K. B. Hannsgen, A Volterra equation with parameter, SIAM J. Math. Anal. 4 (1973), 22-30.

[11] _____________, A Volterra equation in Hilbert space, SIAM J. Math. Anal. to appear.

[12] _____________, A linear Volterra equation in Hilbert space, SIAM J. Math. Anal. to appear.

[13] R. K. Miller, Volterra integral equations in a Banach space, submitted.

[14] C. V. Coffman and D. F. Ulbrich, On the continuability of solutions of a certain nonlinear differential equation, monatsheft für Math. 71 (1967), 385-392.

[15] S. Nakagiri and H. Murakami, Kneser's property of solution families of nonlinear Volterra Integral Equations, Proc. Japan, Acad. 50 (1974), 296-300.

[16] T. L. Herdman, Existence and continuation properties of solutions of a nonlinear Volterra integral equation, Ph. D. Thesis, University of Oklahoma, Norman, May 1974.

Iowa State University, Ames Iowa

ON THE GLOBAL BEHAVIOR OF A NONLINEAR VOLTERRA EQUATION

J. A. Nohel and D. F. Shea

1. Introduction.

In this lecture we discuss by frequence domain methods the global existence, boundedness and the behavior as $t \to \infty$ of solutions of the scalar nonlinear Volterra equation

$$(*) \qquad x'(t) = -\int_0^t g(x(\tau))a(t-\tau)d\tau + f(t) \qquad (x(0) = x_0),$$

and of some other closely related equations on the interval $0 \leq t < \infty$. Frequency domain methods were developed by V. M. Popov [15], [16] and C. Corduneanu [2] for problems in feedback control theory involving ordinary differential equations. Equations (*) and variants of it arise in such applied areas as control theory, nuclear dynamics, population genetics, and viscoelasticity. The results below extend as well as unify research of several authors concerning stability properties of (*). Proofs will appear in [14].

2. Statements of results and discussion.

Concerning the given real functions $a(t)$, $f(t)$, $g(x)$ we assume

Partially supported by ARO Grant DAH C04-74-G0012.
Partially supported by NSF Grant GP021340.

$$(2.1)\quad \begin{cases} a(t)e^{-\sigma t} \in L^1(0,\infty) \text{ for all } \sigma > 0 \\ f(t) \in L^1(0,\infty), \quad g(x) \in \mathcal{C}(-\infty,\infty) . \end{cases}$$

Let $\Pi = \{s \in \mathbb{C} : \text{Re } s > 0\}$ and define

$$G(x) = \int_0^x g(\xi)\,d\xi, \quad \hat{a}(s) = \int_0^\infty e^{-st} a(t)dt \ (s \in \Pi),$$

$$U(i\tau) = \liminf_{\substack{s \to i\tau \\ s \in \Pi}} \text{Re } \hat{a}(s) \qquad (-\infty \le \tau \le \infty);$$

by $U(i\tau)$, $\tau = +\infty$, we mean $\liminf_{\sigma \to 0^+,\ \eta \to +\infty} \text{Re } \hat{a}(\sigma + i\eta)$ and similarly at $\tau = -\infty$. Also define

$$U_0(i\tau) = \liminf_{\sigma \to 0^+} \text{Re } \hat{a}(\sigma + i\tau) .$$

Theorem 1. (i) If, in addition to (2.1), the conditions

$$(2.2)\qquad U(i\tau) \ge 0 \qquad (-\infty \le \tau \le \infty)$$

(2.3) $|g(x)| \le M[1 + |G(x)|]$ for some $M > 0$ and

$$\liminf_{x \to \pm\infty} G(x) > -\infty$$

are satisfied, then (*) has at least one solution $x(t)$ on $0 \le t < \infty$, and for any solution $x(t)$ the function $g(x(t))$ is bounded; if also $\limsup_{x \to \pm\infty} G(x) = +\infty$, then all solutions of (*) are bounded.

(ii) Let $x(t)$ be a bounded solution of (*) and, in addition to (2.2) let

$$(2.4)\quad \begin{cases} U_0(i\tau) \ge \dfrac{\eta}{1+\tau^2} \text{ for some } \eta > 0 \text{ a.e. for } -\infty < \tau < \infty \\ a(t) \in BV[1,\infty) . \end{cases}$$

Then

$$\lim_{t\to\infty} g(x(t)) = 0, \tag{2.5}$$

$$\lim_{t\to\infty} (x'(t) - f(t)) = 0. \tag{2.6}$$

If also $g(x)$ vanishes on no interval, or if $ta(t) \in L^1(0,\infty)$, then

$$\tilde{x} = \lim_{t\to\infty} x(t) \tag{2.7}$$

exists and $g(\tilde{x}) = 0$.

We remark that the BV condition on $a(t)$ in (2.4) can be relaxed to $a(t) = b(t) + c(t)$, where $b(t) \in L^1(0,\infty)$ and $c(t) \in BV[0,\infty)$.

We also remark that our method can be used to discuss the initial value problem for Volterra equations with infinite time lags:

$$\begin{cases} x'(t) = -\int_{-\infty}^{t} g(x(\tau))a(t-\tau)d\tau + f(t) & (0 < t < \infty) \\ x(t) = \phi(t) & (-\infty < t \leq 0), \end{cases}$$

where the given initial function ϕ is such that $g(\phi(t)) \in L^\infty(0,\infty)$, and where $ta(t) \in L^1(0,\infty)$; the latter insures that

$$\int_{-\infty}^{0} g(\phi(\tau))a(t-\tau)d\tau \in L^1(0,\infty).$$

It is further assumed that a, f, g satisfy the hypotheses of Theorem 1. This extends a result of Hale [4].

Before discussing the relationship of Theorem 1 to other results in the literature, it will be useful to give a very brief sketch of the proof.

Define

$$Q_a[v;T] = \int_0^T v(t) \int_0^t v(\tau)a(t-\tau)d\tau\, dt \quad (T > 0, v \in \mathcal{C}[0,T]). \tag{2.8}$$

We first show that (2.2) implies $Q_a[v;T] \geq 0$. Define

$$v_T(t) = \begin{cases} v(t) & (0 \leq t \leq T) \\ 0 & (t < 0,\ t > T) \end{cases}, \quad a(t) = a(-t) \quad (t < 0),$$

$$a_\sigma(t) = a(t)e^{-\sigma|t|} \quad (\sigma > 0), \quad v_T * a(t) = \int_{-\infty}^{\infty} v_T(\tau)a(t-\tau)d\tau .$$

Then the following calculation is straightforward:

$$Q_a[v;T] = \frac{1}{2}\int_0^T v(t) \int_0^T v(\tau)a(t-\tau)d\tau \, dt$$

$$= \frac{1}{2}\int_{-\infty}^{\infty} v_T(t)[v_T * a](t)dt$$

$$= \lim_{\sigma \to 0+} \frac{1}{2}\int_{-\infty}^{\infty} v_T(t)[v_T * a_\sigma](t)dt$$

where the last equality requires Lebesgue's theorem. Since $a_\sigma \in L^1(-\infty,\infty)$ for $\sigma > 0$ and v_T has compact support, Parseval's theorem gives

$$Q_a[v;T] = \frac{1}{4\pi} \lim_{\sigma \to 0+} \int_{-\infty}^{\infty} |\tilde{v}_T(y)|^2 \tilde{a}_\sigma(y)dy ,$$

where $\sim$ represents the usual Fourier transform. Moreover, letting $\wedge$ denote the Laplace transform,

$$U(\sigma + i\tau) = \operatorname{Re} \hat{a}(\sigma + i\tau) = \int_0^{\infty} a(t)e^{-\sigma t} \cos t\tau \, dt$$

$$= \frac{1}{2}\int_{-\infty}^{\infty} a(t)e^{-\sigma|t|} e^{-it\tau} dt = \frac{1}{2}\tilde{a}_\sigma(\tau) ,$$

and therefore,

(2.9) $$Q_a[v;T] = \frac{1}{2\pi} \lim_{\sigma \to 0+} \int_{-\infty}^{\infty} |\tilde{v}_T(y)|^2 U(\sigma + iy)dy .$$

Assuming $a(t) \not\equiv 0$, (2.2), the fact that $U(\sigma + i\tau)$ is harmonic for $\sigma > 0$ and the maximum principle imply $U(\sigma + i\tau) > 0$ $(\sigma > 0)$. Thus Fatou's lemma applied to (2.9) gives

$$Q_a[v;T] \geq \frac{1}{2\pi} \int_{-\infty}^{\infty} |\tilde{v}_T(y)|^2 U_0(iy)dy \, . \tag{2.10}$$

But $U_0(iy) \geq U(iy)$ and thus using (2.2)

$$Q_a[v;T] \geq 0 \, . \tag{2.11}$$

(i) Now multiply (*) by $g(x(t))$ and integrate the resulting equation from 0 to T. Letting $v(t) = g(x(t))$ for any solution $x(t)$ existing on $[0, T]$ one obtains

$$G(x(T)) + Q_a[v;T] = G(x_0) + \int_0^T v(t)f(t)dt \, . \tag{2.12}$$

In view of (2.11) there results the equality

$$G(x(T)) \leq G(x_0) + \int_0^T v(t)f(t)dt \; ;$$

Theorem 1(i) is then established using (2.3) and the Gronwall inequality, followed by a continuation argument. Variants of this technique have been used by Halanay [3], Londen [10], [11], MacCamy and Wong [13].

(ii) The following argument is based on an idea of Halanay [3], (and see also [13]), but it requires generalizations at several steps. For the bounded solution $x(t)$ of (*) we use inequality (2.10) and the first assumption in (2.4), where $v(t) = g(x(t))$, to obtain

$$Q_a[v;T] \geq \frac{1}{2\pi} \int_{-\infty}^{\infty} |\tilde{v}_T(y)|^2 \frac{\eta}{1 + y^2} dy = Q_b[v;T] \tag{2.13}$$

since, if one lets $b(t) = \eta e^{-|t|}$, then $\hat{b}(s) = \frac{\eta}{1 + s}$,

$\operatorname{Re}\hat{b}(iy) = \dfrac{\eta}{1+y^2}$. Therefore, from (2.12), (2.13)

$$0 \le Q_b[v;T] \le Q_a[v;T] = G(x_0) - G(x(T)) + \int_0^T v(t)f(t)dt.$$

Since $v(t) = g(x(t)) \in L^\infty(0,\infty)$ and $f(t) \in L^1[0,\infty)$, there exists an a priori constant K such that

(2.14) $$0 \le Q_b[v;T] \le K \qquad [0 \le T < \infty).$$

It is also useful to note that

$$Q_b[v;T] = \int_0^T v(t)\Gamma(t)dt, \text{ where } \Gamma(t) = \eta\int_0^t v(\tau)e^{-(t-\tau)}d\tau$$

and that $\Gamma(t)$ satisfies the ordinary differential equation

(2.15) $$\Gamma'(t) + \Gamma(t) = \eta v(t), \ \Gamma(0) = 0,$$

and also the equation

(2.16) $$\frac{\Gamma^2(t)}{2} + \int_0^t \Gamma^2(s)ds = Q_b[v;t];$$

(2.16) is obtained by multiplying (2.15) by $\Gamma(t)$ and integrating.

Conclusion (2.5) of Theorem 1(ii) is established from (2.14) and (2.16), which yield $\Gamma \in L^2(0,\infty) \cap L^\infty(0,\infty)$, $\Gamma(t) \to 0$ as $t \to \infty$. One next proves that $g(x(t))$ is uniformly continuous on $[0,\infty)$; this is immediate if $a(t) \in L^1(0,\infty)$. For the general case we proceed by showing that under out hypotheses

(2.17) $$\int_0^t g(x(\tau))a(t-\tau)d\tau = 0(1) \qquad (t \to \infty).$$

Thus $|x'(t)| \le |f(t)| + 0(1)$ and so $x(t)$ is uniformly continuous. Now (2.5) follows from tauberian arguments applied to

$$\int_0^t g(x(\tau))e^{-(t-\tau)}d\tau = \eta^{-1}\Gamma(t) \to 0 \qquad (t \to \infty).$$

Once (2.5) is known, (2.17) can be sharpened to

$$(2.18) \qquad \int_0^t g(x(\tau))a(t-\tau)d\tau = o(1) \qquad (t \to \infty)$$

which is equivalent to (2.6).

The proof of (2.7) requires the following technical result.

Lemma. Let $a(t) \in L^1(0,\infty)$, $ta(t) \in L^1(0,\infty)$, let $g(x) \in \mathcal{C}(-\infty,\infty)$, $f(t) \in L^1(0,\infty)$. If $x(t) \in L^\infty(0,\infty)$ is a solution of (*) which satisfies (2.5), then (2.7) holds and $g(\tilde{x}) = 0$.

It is of interest to note that if a solution $x(t)$ of (*) satisfies (2.5) and if the hypothesis $f(t) \in L^1(0,\infty)$ is changed to $\lim_{t\to\infty} f(t) = 0$, then (2.7) need not hold. Indeed, let $g(x) \equiv 0$ for $|x| \le 1$ and otherwise $g(x)$ is arbitrary subject to the assumptions of Theorem 1. Let $a(t)$ have compact support in $[0,T]$, $T > 0$, and let $x(t) = \sin\sqrt{t}$ $(t \ge t_0)$. Define $f(t)$ by equation (*), so that

$$f(t) = \begin{cases} x'(t) + \int_0^T g(x(t-\xi))a(\xi)d\xi & \text{if } t \ge T \\ x'(t) = \dfrac{\cos\sqrt{t}}{2\sqrt{t}} & \text{if } t \ge T + t_0 . \end{cases}$$

Clearly $f(t) \to 0$ as $t \to \infty$, $\lim_{t\to\infty} g(x(t)) = 0$, but $x(t)$ fails to converge.

Although it is not obvious at a glance, Theorem 1 contains as a special case-many earlier result concerning the asymptotic behavior of solutions of (*). Of primary interest in this discussion are assumptions (2.2) and (2.4) concerning the kernel $a(t)$. It is easy to check that $a(t) = t^{-\alpha}e^{-\beta t}\cos\gamma t$ $(0 \le \alpha < 1,\ \beta \ge 0,\ \gamma$ real), or a linear combination of such functions with positive coefficients satisfies (2.2), and if $\alpha > 0$ or $\beta > 0$ such $a(t)$ satisfy (2.4). [Such kernels arise naturally in the theory of automatic controls.] In order to discuss other kernels which satisfy our assumptions we define, similar to Halanay [3], MacCamy and Wong [13], and Wong [19], the following notion of positivity (see (2.8) above for the definition of Q_a).

Definition 1. A real function $a(t) \in L^1_{loc}(0,\infty)$ is of positive type if $Q_a[v;T] \geq 0$ for every $v \in \mathcal{C}[0,T]$ and for every $0 < T < \infty$.

Since this definition is rarely easy to check directly the following characterization is important.

Theorem 2. The following statements are equivalent.

(i) $a(t)$ is of positive type

(ii) $\operatorname{Re} \hat{a}(s) \geq 0 \qquad (s \in \Pi)$

(iii) $U(i\tau) \geq 0 \qquad (-\infty \leq \tau \leq +\infty)$,

where $U(i\tau) = +\infty$ is permitted in (iii).

The implication (iii) ⇒ (i) is sketched following (2.8) above. That (ii) implies (iii) is obvious; for (i) implies (ii) see [14]. Positive definite functions in the sense of Bochner [17] are functions of positive type; another such class are the functions $a(t)$ on $(0,\infty)$ which are locally integrable, positive, nonincreasing and convex (see Hannsgen [5, Lemma 5]); in particular if $a(t) \in L^1_{loc}(0,\infty)$ and $(-1)^k a^{(k)}(t) \geq 0$ $(0 < t < \infty; k = 0,1,2)$, $a(t)$ is of positive type (see Levin [7], Levin & Nohel [9], Londen [10]). As we have seen above it is (i) which is useful in proving boundedness; however (iii) is a much more tractable condition to verify.

We note is passing the representation formula for functions of positive type (see Cooper [1])

$$a(t) = \lim_{T\to\infty} \int_0^T (1 - \frac{\xi}{T})\cos \xi t \, d\mu(\xi) \tag{2.19}$$

which holds a. e. on $(0, \infty)$ for some nondecreasing function $\mu(t)$, $\mu(0) = 0$, $\mu(\tau) = \frac{1}{2}[\mu(\tau + 0) - \mu(\tau - 0)]$ for $\tau > 0$, with

$$\int_0^\infty \frac{d\mu(\tau)}{1+\tau^2} < \infty \quad \text{or even} \quad \lim_{\tau\to\infty} \frac{\mu(\tau)}{\tau} = 0 .$$

One observes that if μ is extended as an odd function to $(-\infty, \infty)$ then (2.19) is equivalent to

$$a(t) = \frac{1}{2}\int_{-\infty}^{\infty} e^{-i\xi t} d\mu(\xi) \qquad (\mathcal{C},1) \text{ a. e. },$$

so that Bochner's theorem for positive definite functions [17] still holds for functiosn of positive type, but only in a summability sense. Moreover, (2.19) also shows that a function of positive type is positive definite in the sense of Bochner if and only if (2.19) holds for a bounded function $\mu(t)$.

For the discussion of asymptotic behavior Halanay [3] introduced the concept of strong positivity (see (2.13) above).

Definition 2. A real function $a(t)$ is strongly positive if there exists a constant $\eta > 0$ such that $b(t) = a(t) - \eta e^{-t}$ is of positive type.

The following characterization of strong positivity is useful for the application of this concept and is a consequence of Theorem 2.

Corollary 2.1. The kernel $a(t)$ is strongly positive if and

<u>and only if</u> $a(t)$ <u>is of positive type and there exists</u> $\eta > 0$ <u>such that</u> $U_0(i\tau) \geq \frac{\eta}{1+\tau^2}$ a. e. <u>in the interval</u> $-\infty < \tau < \infty$.

This corollary together with Theorems 1 and 2 shows that Theorem 1 contains as a special case the result of Halanay [3] (where also $f(t) \equiv 0$ is taken), and the results for scalar equations studied by MacCamy and Wong [13, Theorems 5.2, 5.3].

Another useful application of Theorem 2 and of Corollary 2.1 is

<u>Corollary 2.2</u>. <u>Let</u> $a(t) \in L^1_{loc}(0,\infty)$ <u>be ≥ 0, not identically constant, nonincreasing, convex and such that</u> $da'(t)$ <u>is not a purely singular measure</u>. <u>Then</u> $a(t)$ <u>is strongly positive</u>.

Corollary 2.2 shows that Theorem 1 contains as a special case the boundedness and asymptotic behavior results for equation (*) established in Levin [7], Levin and Nohel [9], Hannsgen [6], Londen [10]. We point out that in [9] other types of perturbations $\notin L^1(0,\infty)$ are considered which are not readily discussed by the method of Theorem 1; likewise, in [6] Hannsgen discusses the behavior of solutions in the case that the measure $da'(t)$ has no absolutely continuous component.

We also mention that Corollary 2.2 is the correct version of Theorem 4.3 of MacCamy and Wong [13]; they assert that $a(t) \in L^1_{loc}(0,\infty)$ is strongly positive if $a(t) \in \mathcal{C}'(0,\infty)$, $a'(t)$ is nondecreasing and $a(t) \not\equiv$ constant. A counterexample to this assertion is constructed in [14]; there we take μ to be a continuous purely singular measure of total mass 1 concentrated on a (compact) Kroenecker set [17, pp. 97-100] P and define

$$\varphi(t) = \int_t^\infty d\mu(\xi), \quad a(t) = \int_t^\infty \varphi(\xi)d\xi \ ;$$

then a(t) satisfies the hypotheses of Theorem 4.3 of [13]. But because of the arithmetic properties of Kroenecker sets [17, p. 113, Theorem 5.5.2 (b)] one has

$$\overline{\lim_{\tau \to \infty}} \sup \int_0^\infty \cos \tau t \, da'(t) = 1$$

and therefore also

$$\lim_{\tau \to \infty} \inf \tau^2 \operatorname{Re} \hat{a}(i\tau) = \lim_{\tau \to \infty} \inf \int_0^\infty (1 - \cos \tau t) da'(t) = 0 ,$$

which contradicts the estimate (4.18) in [13].

A striking improvement of the growth condition (2.4) has recently been discovered by Olof Staffans [18], a graduate student at Wisconsin, who shows that the hypothesis of strong positivity in Theorem 1 (ii) can be relaxed to what he calls strict positivity: There exists a function $b(t) \in L^1(0, \infty)$ with $\operatorname{Re} \hat{b}(i\tau) > 0$ for $-\infty < \tau < \infty$ such that $a - b$ is of positive type. His application of this condition to (*) and related equations is still in progress at present he requires the assumption (see [18, Theorem 2]) that the bounded solution x(t) be uniformly continuous on $[0, \infty)$. However, it is clear that combining his result with our Theorem 1 (ii) one obtains the following improvement.

Let x(t) be a bounded solution of (*) and, in addition to (2.2), let

$$\text{(2.20)} \qquad \begin{cases} U(i\tau) > 0 & (-\infty < \tau < \infty) \\ a(t) \in BV[1,\infty) . \end{cases}$$

<u>Then the conclusion of Theorem</u> 1 (ii) <u>holds.</u>

The proof uses the auxiliary result that the first condition in (2.20) implies the existence of a function $b(t) \in L^1(0,\infty)$ such that $\mathrm{Re}\, \hat{b}(i\tau) > 0$ $(-\infty < \tau < \infty)$. The second condition in (2.20) is used to establish the uniform continuity of $g(x(t))$ as in the proof of Theorem 1 (ii).

The method of Theorem 1 enables us to discuss the asymptotic behavior of solutions on $0 \le t < \infty$ of the Volterra-Stieltjes equation

$$(**) \qquad x'(t) = -\int_0^t g(x(t-\tau))dA(\tau) + f(t) \qquad (x(0) = x_0),$$

in which $g(x)$ and $f(t)$ are as in (*) and

$$A(t) \in BV[0,\infty), \; A(0) = 0, \; A(t) = A(t^-) \quad (0 \le t < \infty).$$

Note that if one takes

$$A(t) = \theta I(t) + \int_0^t a(\sigma)d\sigma \qquad (\theta > 0, \; 0 < t < \infty),$$

where $a(t)$ is as in (*) and $I(t)$ is the unit step function, equation (**) becomes

$$x'(t) = -\theta g(x(t)) - \int_0^t g(x(t-\tau))a(\tau)d\tau + f(t)$$

which is of particular interest in viscoelasticity [12], [13]. For equation (**) a result similar to Theorem 1 holds, but with the boundedness condition (2.2) replaced by

$$\int_0^\infty \cos ty \, dA(t) \geq 0 \qquad (0 \leq y \leq \infty)$$

and with the stability condition (2.4) replaced by

$$\int_0^\infty \cos ty \, dA(t) \geq \frac{\eta}{1 + y^2} \qquad (\eta > 0, \, 0 \leq y < \infty) .$$

For a different result concerning (**) see Londen [11].

Finally, let us mention that this last result can be used to discuss solutions of the Volterra equation

$$x(t) = - \int_0^t [\alpha(t - \tau) + \rho] g(x(\tau)) d\tau + F(t)$$

which is of importance in control theory and has been studied from a somewhat different point of view by Corduneanu [2]; for details see [14].

REFERENCES

[1] Cooper, J. L. B., Positive definite functions of a real variable, Proc. London Math. Soc (3) 10 (1960), 53-66.

[2] Corduneanu, C., Integral Equations and Stability of Feedback Systems, Academic Press, New York and London, 1973.

[3] Halanay, A., On the asymptotic behavior of the solutions of an integro-differential equation, J. Math. Anal. Appl. 19 (1965), 319-324.

[4] Hale, J. K., Dynamical systems and stability, J. Math. Anal. Appl. 26 (1969), 39-59.

[5] Hannsgen, K. B., Indiredt Abelian theorems and a linear Volterra equations, Trans. Amer. Math. Soc.

142 (1969), 539-555.

[6] Hannsgen, K. B., On a nonlinear Volterra equation, Mich. Math. J. 16 (1969), 365-376.

[7] Levin, J. J., The asymptotic behavior of the solution of a Volterra equation, Proc. AMS 14 (1963), 534-540.

[8] Levin, J. J. and Nohel, J. A., Note on a nonlinear Volterra equation, Proc. AMS 14 (1963), 924-929.

[9] Levin, J. J. and Nohel, J. A., Perturbations of a nonlinear Volterra equation, Mich. Math. J. 12 (1965), 431-447.

[10] Londen, S. O., The qualitative behavior of the solutions of a nonlinear Volterra equation, Mich. Math. J. 18 (1971), 321-330.

[11] Londen, S. O., On the variation of the solutions of a nonlinear integral equation (to appear).

[12] MacCamy, R.C., Stability theorems for a class of functional differential equations (to appear).

[13] MacCamy, R. C. and Wong, J. S. W., Stability theorems for some functional equations, Trans. AMS

[14] Nohel, J. A. and Shea, D. F., Frequency Domain Methods for Volterra Equations, Advances in Math. (to appear).

[15] Popov, V. M., Criterii de stabilitate pentru sistemele neliniare de reglare automata bazate pe utilizarea transformatei Laplace, Stud. Cerc. Energetica IX (1959), 119-135.

[16] _______________, Absolute stability of nonlinear systems of automatic control, Autom. Remote Control 22 (1961), 857-875.

______________, Hiperstabilitatea Sistemelov Automate, Editura Academiei R. S. R., Bucaresti, 1966.

[17] Rudin, W., Fourier Analysis on Groups, Interscience Publishers, New York, 1967.

[18] Staffans, O., Nonlinear Volterra integral equations with positive definite kernels, Proc. AMS (to appear).

[19] Wong, J. S. W., Positive definite functions and Volterra integral equations, Bull. Amer. Math. Soc. 80 (1974), 679-682.

University of Wisconsin, Madison, Wisconsin

THE SINGULAR PERTURBATION APPROACH TO SINGULAR ARCS

R. E. O'Malley, Jr.

1. Introduction.

We shall use singular perturbation methods to obtain the asymptotic solution of the optimal control problem

$$\dot{x} = Ax + Bu \tag{1}$$

with the prescribed initial n vector $x(0)$ and the scalar cost functional

$$J(\epsilon) = \frac{1}{2}\int_0^1 [x^T Qx + \epsilon^2 u^T Ru]dt \tag{2}$$

to be minimized by selection of the r-dimensional control u. Here, ϵ is a small, positive parameter; A, Q and R are smooth functions of the independent variable t; Q and R are symmetric matrices which are positive semi-definite and positive definite, respectively, and the super-script T denotes transposition.

The limiting problem with $\epsilon = 0$ is a singular problem of optimal control (cf. Jacobson (1971), Bryson and Ho (1969), and Anderson and Moore (1971)). The problem for each fixed $\epsilon > 0$ is one of the most familiar control problems and has a unique readily obtained solution (cf. Kalman (1960)

Work supported by the Office of Naval Research under Contract Number N00014- 67 A-0209-0022.

and, e. g., Anderson and Moore). It has therefore been natural for control theorists to study the singular problem as the limit of the problem with $\epsilon > 0$ (cf. Jacobson, Gershwin, and Lele (1970) and Jacobson and Speyer (1971)). Care is required for such discussions since, as we shall find, a singular perturbation and nonuniform convergence at endpoints are involved. Closely related problems occur in other control contexts; for example, in analyzing the limiting possibilities for regulators and filters (cf. Kwakernaak and Sivan (1972), Friedland (1971), and Hutton (1973)). Lions (1973) refers to analogous problems for distributed parameter systems as problems of "cheap control" since the cost of the control u in (2) is cheap relative to that of the state x (for nontrivial matrices Q). We recall that in all these problems a reduction in dimensionality occurs (this aspect of the singular arc problem is emphasized by Johnson (1965) and Ho (1972)). It is felt that singular perturbation theory will be useful in illuminating the structure of solutions to many other control problems involving reduction in dimensionality.

For each $\epsilon > 0$, the calculus of variations implies that the optimal control satisfies

$$u = -\frac{1}{\epsilon^2} R^{-1} B^T \tilde{p} \tag{3}$$

where the n-dimensional costate $\tilde{p}$ satisfies

$$\dot{\tilde{p}} = -Qx - A^T\tilde{p}, \quad \tilde{p}(1) = 0 \tag{4}$$

(cf., e.g., Anderson and Moore). Eliminating u in (1) through (3), then, results in the singularly perturbed two-point problem

$$(5)\qquad \begin{cases} \epsilon^2 \dot{x} = \epsilon^2 Ax - BR^{-1}B^T\tilde{p}, & x(0) \text{ prescribed} \\ \dot{\tilde{p}} = -Qx - A^T\tilde{p}, & \tilde{p}(1) = 0. \end{cases}$$

(Alternatively, one could let $\tilde{p} = kx$ where k satisfies a terminal value problem for a singularly perturbed matrix Riccati equation.) A previous singular perturbation solution to the problem (1)-(2) was given by O'Malley and Jameson (to appear) for the case where A, B, Q and R are time-invariant. They first transformed B to the form $B = \begin{bmatrix} 0 \\ I \end{bmatrix}$, presuming the $n \times r$ matrix B had rank $r \leq n$. The present approach follows another transformation technique, as used by Moylan and Moore (1971) and Friedland for the singular arc problem. It will be thoroughly developed elsewhere by Jameson and O'Malley. Related discussion of other singular perturbation problems in control is contained in Yackel and Kokotović (1973), Wilde and Kokotović (1973), O'Malley and Kung (1974, 1975), and Jameson and O'Malley (to appear).

2. The transformed problem.

We introduce the new variables

$$(6)\qquad u_1 = \int_0^t u(s)ds$$

and

$$(7)\qquad x_1 = x - Bu_1 .$$

From (1) and (2), we have the transformed problem

$$(8)\qquad \dot{x}_1 = Ax_1 + B_1u_1, \quad x_1(0) = x(0)$$

with the cost functional

$$(9) \qquad J(\epsilon) = \frac{1}{2}\int_0^1 [x_1^T Q x_1 + 2u_1^T B^T Q x_1 + u_1^T B^T Q B u_1 + \epsilon^2 u^T R u]dt$$

for

$$B_1 \equiv AB - \dot{B}.$$

Unlike the original problem, the transformed problem (8)-(9) will not be singular when $\epsilon = 0$ provided

$$(10) \qquad B^T Q B \text{ is positive definite.}$$

We will assume that (10) holds throughout this paper. The condition has often been used (e.g., in Ho (1971)). It allows the simple problem where B and Q are invertible, as well as many more frequent situations where B is of rank $r < n$.

We now reformulate the problem as a regulator problem of state dimension $n + r$. Thus,

$$(11) \qquad \begin{pmatrix} \dot{x}_1 \\ \dot{u}_1 \end{pmatrix} = \mathcal{A}\begin{pmatrix} x_1 \\ u_1 \end{pmatrix} + \mathcal{B}u, \quad \begin{pmatrix} x_1(0) \\ u_1(0) \end{pmatrix} = \begin{pmatrix} x(0) \\ 0 \end{pmatrix}$$

with

$$(12) \qquad J(\epsilon) = \frac{1}{2}\int_0^1 [\begin{pmatrix} x_1 \\ u_1 \end{pmatrix}^T \mathcal{Q}\begin{pmatrix} x_1 \\ u_1 \end{pmatrix} + \epsilon^2 u^T R u]dt .$$

Here,

$$\mathcal{A} = \begin{pmatrix} A & B_1 \\ 0 & 0 \end{pmatrix}, \quad \mathcal{B} = \begin{pmatrix} 0 \\ I \end{pmatrix},$$

and

$$\mathcal{Q} = \begin{pmatrix} Q & QB \\ B^T Q & B^T Q B \end{pmatrix}.$$

Moreover, $\mathcal{Q}$ is positive semi-definite since

$$Q - QB(B^TQB)^{-1}B^TQ = E^TQE$$

for

$$E = I - B(B^TQB)^{-1}B^TQ .$$

Introducing a corresponding costate

(13) $$p(t) = \begin{pmatrix} p_1 \\ p_2 \end{pmatrix}$$

satisfying

(14) $$\dot{p} = -2\begin{pmatrix} x_1 \\ u_1 \end{pmatrix} - \mathcal{A}^T p , \quad p(1) = 0 ,$$

the optimal control will be given by

(15) $$u = -\frac{1}{\epsilon^2}R^{-1}\mathcal{B}^T p = -\frac{1}{\epsilon^2}R^{-1}p_2 .$$

Along the optimal path, then, we have the singularly perturbed two-point problem

(16) $$\begin{cases} \dot{x}_1 = Ax_1 + B_1u_1 , & x_1(0) = x(0) \\ \epsilon^2\dot{u}_1 = -R^{-1}p_2 , & u_1(0) = 0 \\ \dot{p}_1 = -Qx_1 - QBu_1 - A^Tp_1 , & p_1(1) = 0 \\ \dot{p}_2 = -B^TQx_1 - B^TQBu_1 - B_1^Tp_1 , & p_2(1) = 0 . \end{cases}$$

Using standard singular perturbation methods (cf. Wasow (1965) or O'Malley (1974)), we will obtain the asymptotic solution to (16). It will prove simpler to work with this (temporarily) higher order problem than with the more immediate problem (5).

3. The asymptotic solution of the transformed problem.

The asymptotic solution of (16) will consist of an outer expansion (which will represent the solution asymptotically within $0 < t < 1$) plus boundary layer corrections at both endpoints. Experience (cf. O'Malley and Kung (1974)) leads us to seek a solution of the form

$$(17)\quad \begin{cases} x_1(t,\epsilon) = X_1(t,\epsilon) + \epsilon m_1(\tau,\epsilon) + \epsilon^2 n_1(\sigma,\epsilon) \\ u_1(t,\epsilon) = U_1(t,\epsilon) + v_1(\tau,\epsilon) + \epsilon w_1(\sigma,\epsilon) \\ p_1(t,\epsilon) = P_1(t,\epsilon) + \epsilon q_1(\tau,\epsilon) + \epsilon^2 s_1(\sigma,\epsilon) \\ p_2(t,\epsilon) = \epsilon^2 P_2(t,\epsilon) + \epsilon q_2(\tau,\epsilon) + \epsilon^2 s_2(\sigma,\epsilon) \end{cases}$$

where the outer solution, $(X_1, U_1, P_1, \epsilon^2 P_2)$, the left boundary layer correction, $(\epsilon m_1, v_1, \epsilon q_1, \epsilon q_2)$, and the right boundary layer correction $(\epsilon^2 n_1, \epsilon w_1, \epsilon^2 s_1, \epsilon^2 s_2)$, all have asymptotic power series expansions as $\epsilon \to 0$ and the boundary layer corrections tend to zero as the appropriate boundary layer variable

$$\tau = t/\epsilon \quad \text{or} \quad \sigma = (1-t)/\epsilon$$

tends to infinity.

(i) The outer expansion. Suppose

$$(18)\qquad (X_1(t,\epsilon),\ U_1(t,\epsilon),\ P_1(t,\epsilon),\ P_2(t,\epsilon)) \sim \sum_{j=0}^{\infty} (X_{1j}, U_{1j}, P_{1j}, P_{2j})\epsilon^j .$$

Since the expansion must satisfy the system (16) within (0,1), we will obtain the term of (18) by successively equating coefficients in (16). In particular, when $\epsilon = 0$

(19)
$$\begin{cases} \dot{X}_{10} = AX_{10} + B_1U_{10}, & X_{10}(0) = x(0) \\ \dot{U}_{10} = -R^{-1}P_{20} & \\ \dot{P}_{10} = -QX_{10} - QBU_{10} - A^TP_{10}, & P_{10}(1) = 0 \\ 0 = -B^TQX_{10} - (B^TQB)U_{10} - B_1^TP_{10} & . \end{cases}$$

Thus

(20)
$$\begin{cases} U_{10} = -(B^TQB)^{-1}(B^TQX_{10} + B_1^TP_{10}) \\ P_{20} = -R\dot{U}_{10} \end{cases}$$

and there remains the two-point problem

(21)
$$\begin{cases} \dot{X}_{10} = (A - B_1(B^TQB)^{-1}B^TQ)X_{10} \\ \qquad - B_1(B^TQB)^{-1}B_1^TP_{10}, & X_{10}(0) = x(0) \\ \dot{P}_{10} = -(Q - QB(B^TQB)^{-1}B^TQ)X_{10} \\ \qquad - (A^T - QB(B^TQB)^{-1}B_1^T)P_{10}, & P_{10}(1) = 0. \end{cases}$$

The solution to (21) can be found by setting

(22)
$$P_{10} = K_{10}X_{10}$$

where the symmetric, positive definite matrix K_{10} satisfies the Riccati equation

(23)
$$\begin{aligned} &\dot{K}_{10} + K_{10}(A - B_1(B^TQB)^{-1}B^TQ) \\ &+ (A - B_1(B^TQB)^{-1}B^TQ)^TK_{10} \\ &- K_{10}B_1(B^TQB)^{-1}B_1^TK_{10} + (Q - QB(B^TQB)^{-1}B^TQ) = 0, \\ &\qquad K_{10}(1) = 0. \end{aligned}$$

Existence of a unique solution K_{10} is guaranteed since E^TQE is positive semi-definite (cf. Anderson and Moore). Thus, there remains only the linear initial value problem

$$(24) \qquad \dot{X}_{10} = [A - B_1(B^TQB)^{-1}(B^TQ - B_1^TK_{10})]X_{10}, \quad X_{10}(0) = x(0)$$

for the limiting outer trajectory. The first terms of the expansion (18) are thereby completely and uniquely determined since $B^TQB > 0$.

It is easy to show that any solution of (23) satisfies

$$(25) \qquad K_{10}B = 0,$$

the familiar singular arc condition (cf. Bryson and Ho) for the transformed problem. It implies that the limiting solution within (0, 1) is determined by a dynamical system of order $n - r$ ((23) and (25)) provided $n \geq r$ (cf. Ho, Jameson and O'Malley and, especially, Kwatny (1974)).

Higher order terms of the expansion (18) will likewise satisfy linear problems of the form

$$(26) \qquad \begin{cases} \dot{X}_{1j} = AX_{1j} + B_1U_{1j}, & X_{1j}(0) = -m_{1,j-1}(0) \\ \dot{U}_{1j} = -R^{-1}P_{2j} & \\ \dot{P}_{1j} = -QX_{1j} - QBU_{1j} - A^TP_{1j}, & P_{1j}(1) = -s_{1,j-2}(0) \\ \dot{P}_{2,j-2} = -B^TQX_{1j} - B^TQBU_{1j} - B_1^TP_{1j}. & \end{cases}$$

Since (19) has a unique solution, the Fredholm alternative implies that (26) can also be uniquely solved for successive indices j provided the boundary values

$$m_{1,j-1}(0) \text{ and } s_{1,j-2}(0)$$

of preceding terms of the boundary layer corrections are known.

(ii) The initial boundary layer correction. By linearity, the initial boundary layer correction must satisfy the system (16). Thus, we have

$$
(27)\quad \begin{cases}
\dfrac{dm_1}{d\tau} = \epsilon A m_1 + B_1 v_1 \\[2ex]
\dfrac{dv_1}{d\tau} = -R^{-1} q_2 , \qquad v_1(0,\epsilon) = -U_1(0,\epsilon) \\[2ex]
\dfrac{dq_1}{d\tau} = -\epsilon Q m_1 - QBv_1 - \epsilon A^T q_1 \\[2ex]
\dfrac{dq_2}{d\tau} = -\epsilon B^T Q m_1 - B^T QBv_1 - \epsilon B_1^T q_1
\end{cases}
$$

Moreover, these terms should decay to zero as $\tau \to \infty$. When $\epsilon = 0$, then,

$$
(28)\quad \begin{cases}
\dfrac{dm_{10}}{d\tau} = B_1(0) v_{10} \\[2ex]
\dfrac{dv_{10}}{d\tau} = -R^{-1}(0) q_{20} , \qquad v_{10}(0) = -U_{10}(0) \\[2ex]
\dfrac{dq_{10}}{d\tau} = -Q(0)B(0)v_{10} \\[2ex]
\dfrac{dq_{20}}{d\tau} = -B^T(0)Q(0)B(0)v_{10}
\end{cases}
$$

for $0 \leq \tau < \infty$. Thus,

$$\frac{d^2 v_{10}}{d\tau^2} = B^T(0)Q(0)B(0)R^{-1}(0)v_{10}$$

and we have

$$v_{10}(\tau) = -R^{\frac{1}{2}}(0)e^{-C_o^T} R^{-\frac{1}{2}}(0)U_{10}(0) \tag{29}$$

where C_0 is the positive-definite matrix

$$C_0 = \sqrt{R^{-1/2}(0)B^T(0)Q(0)B(0)R^{-1/2}(0)}\,.$$

We obtain the other leading terms uniquely by integration, e.g.,

$$m_{10}(\tau) = B_1(0)R^{1/2}(0)C_0^{-1}R^{-1/2}(0)v_{10}(\tau)\,.$$

Higher order terms are easily obtained as decaying solutions of the linear problem

$$\left\{\begin{aligned}
\frac{dm_{1j}}{d\tau} &= B_1(0)v_{1j} + \alpha_j \\
\frac{dv_{1j}}{d\tau} &= -R^{-1}(0)q_{2j} + \beta_j, \quad v_{1j}(0) = -U_{1j}(0) \\
\frac{dq_{1j}}{d\tau} &= -Q(0)B(0)v_{1j} + \gamma_j \\
\frac{dq_{2j}}{d\tau} &= -B^T(0)Q(0)B(0)v_{1j} + \delta_j
\end{aligned}\right. \tag{30}$$

where α_j, β_j, γ_j, and δ_j are successively known exponentially decaying functions. Note that the unique solution to (30) implies the initial value $m_{1j}(0)$ needed to specify the $(j+1)$st terms of the outer expansion (18).

(iii) The terminal boundary layer correction. Proceeding as in (ii), we find that the leading terms of the terminal correction must be decaying solutions of the linear system

$$(31)\quad \begin{cases} \dfrac{dn_{10}}{d\sigma} = -B_1(1)w_{10} \\ \dfrac{dw_{10}}{d\sigma} = R^{-1}(1)s_{20} \\ \dfrac{ds_{10}}{d\sigma} = Q(1)B(1)w_{10} \\ \dfrac{ds_{20}}{d\sigma} = B^T(1)Q(1)B(1)w_{10}\,, \qquad s_{20}(0) = -P_{20}(1)\,. \end{cases}$$

Thus,

$$(32)\qquad s_{20} = -R^{1/2}(1)e^{-C_1\sigma}R^{-\frac{1}{2}}(1)\,P_{20}(1)$$

where

$$C_1 = \sqrt{R^{-1/2}(1)B^T(1)Q(1)B(1)R^{-1/2}(1)} > 0\,.$$

The decaying vectors n_{10}, w_{10} and s_{10} uniquely follow by integration, as do higher order terms.

(iv) The optimal control and optimal cost. The control relation (15) and the representation (17) imply that the optimal control u for both the original and transformed problems has the asymptotic form

$$(33)\qquad \begin{aligned} u(t,\epsilon) &= -R^{-1}(t)P_2(t,\epsilon) \\ &\quad -\tfrac{1}{\epsilon}R^{-1}(\epsilon\tau)q_2(\tau,\epsilon) - R^{-1}(1-\epsilon\sigma)s_2(\sigma,\epsilon)\,. \end{aligned}$$

Thus, the optimal control within $(0,1)$ tends to $U_0(t) \equiv -R^{-1}(t)P_{20}(t)$, which by (19), (20) and (22) is easily represented in the feedback form

$$(34) \qquad U_0(t) = \gamma(t)X_{10}(t)$$

for the appropriate matrix γ. In the immediate vicinity of $t = 0$, however, the control is generally dominated by the initial impulse

$$(35) \qquad -\frac{1}{\epsilon}R^{-1}(0)q_{20}(\tau) = -\frac{1}{\epsilon}C_0^2R^{3/2}(0)C_0^{-1}e^{-C_0t/\epsilon}U_{10}(0)$$

which behaves like a multiple of a delta function in the limit $\epsilon \to 0$.

Finally, the asymptotic representations (17) and (33) and the cost functional (12) imply that the optimal cost $J^*(\epsilon)$ for both problems has an asymptotic power series in ϵ, i.e.,

$$(36) \qquad J^*(\epsilon) \sim \sum_{\ell=0}^{\infty} J_\ell^*\epsilon^\ell .$$

This follows since the integrand of (12) has a generalized asymptotic expansion consisting of an outer expansion and boundary layer corrections at both endpoints. In particular, we find

$$(37) \qquad J_0^* = \frac{1}{2}\int_0^1 (X_{10}^TQX_{10} + 2X_{10}^TQBU_{10} + U_{10}^TB^TQBU_{10})dt$$

$$= x^T(0)K_{10}(0)x(0)$$

(cf. Moylan and Moore). Thus, the leading term of the optimal cost is unaffected by the boundary layer corrections (nonuniform convergence), even though the optimal control is initially unbounded as $\epsilon \to 0$.

The asymptotic expansions for the transformed states x_1 and u_1 imply a corresponding asymptotic expansion for the optimal trajectory $x(t,\epsilon)$ of (1), (2). It has the form

$$x(t,\epsilon) = X(t,\epsilon) + m(\tau,\epsilon) + \epsilon n(\sigma,\epsilon) . \tag{38}$$

It converges nonuniformly at $t = 0$ and has the limiting solution

$$\begin{aligned} X_0(t) &= X_{10}(t) + BU_{10}(t) \\ &= (I - B(B^TQB)^{-1}(B^TQ + B_1^TK_{10}))X_{10} . \end{aligned} \tag{39}$$

4. Summary and Comments.

We have shown the following:

THEOREM: *Consider the state regulator problem*

$$\dot{x} = Ax + Bu, \quad x(0) \text{ prescribed}$$

where

$$J(\epsilon) = \frac{1}{2}\int_0^1 [x^TQx + \epsilon^2 u^TRu]dt$$

is to be minimized and

$$B^TQB$$

is positive definite. The optimal control, the corresponding trajectories, and the optimal cost take the asymptotic form

$$\begin{cases} u(t,\epsilon) = U(t,\epsilon) + \frac{1}{\epsilon}v(\frac{t}{\epsilon},\epsilon) + w(\frac{1-t}{\epsilon},\epsilon) \\ x(t,\epsilon) = X(t,\epsilon) + m(\frac{t}{\epsilon},\epsilon) + \epsilon n(\frac{1-t}{\epsilon},\epsilon) \\ J^*(\epsilon) \end{cases}$$

for each $\epsilon > 0$ *sufficiently small. These expressions all have asymptotic series expansions in* ϵ *and the functions* *of $\tau = t/\epsilon$ and $\sigma = (1-t)/\epsilon$ tend to zero as τ or σ, respectively, tend to infinity.*

Comments. 1. If we limit the differentiability of coefficients in (1)-(2), these expansions will have to be terminated after an appropriate finite number of terms.

2. If B^{-1} exists, we will have $U = w = X = n = J^*(0) = 0$.

3. The corresponding fixed endpoint problem could also be examined. It would generally feature impulsive controls at both endpoints. Likewise, the corresponding dual stochastic problems might be analogously studied (cf. Rauch (1973)).

4. The transformation approach has enabled us to completely determine the terms of the relevant asymptotic expansions. Direct substitution of these expansions into the original problem would generally seem quite intractable however.

5. A hierarchy of more restricted problems could be studied (cf. Moylan and Moore). One will find, for example, that the optimal control will behave initially like a combination of delta function and its derivative when $B^TQB = 0$ and $B_1^TQB_1$ is positive definite.

6. The limiting outer solution can be readily identified with the solution when $\epsilon = 0$ (cf. O'Malley and Jameson).

REFERENCES

[1] B. D. O. Anderson and J. B. Moore, *Linear Optimal Control*, Prentice-Hall, Englewood Cliffs, 1971.

[2] A. E. Bryson, Jr. and Y. C. Ho, Applied Optimal Control, Blaisdell, Waltham, 1969.

[3] B. Friedland, "Limiting forms of optimum stochastic linear regulators," J. Dynamic Systems, Measurement, and Control, Trans, ASME, Series G, 93 (1971), 134-141.

[4] Y. C. Ho, "Linear stochastic singular control problems," J. Optimization Theory Appl. 9 (1972), 24-31.

[5] M. F. Hutton, "Solutions of the singular stochastic regulator problem," J. Dynamic Systems, Measurement, and Control, Trans, ASME, Series G, 95 (1973), 414-417.

[6] D. H. Jacobson, "Totally singular quadratic minimization problems," IEEE Trans. Automatic Control 16 (1970), 651-658.

[7] D. H. Jacobson, S. B. Gershiwn, and M. M. Lele, "Computation of optimal singular controls," ibid.,15 (1970), 67-73.

[8] D. H. Jacobson and J. L. Speyer, "Necessary and sufficient conditions for optimality for singular control problems: A limit approach," J. Math. Anal. Appl., 34 (1971), 239-266.

[9] A. Jameson and R. E. O'Malley, Jr., "Cheap control of the time-invariant regulator," Applied Math. and Optimization to appear.

[10] C. Johnson, "Singular solutions in problems of optimal control," Advances in Control Systems, 2(1965), 209-267.

[11] R. E. Kalman, "Contributions to the theory of optimal control," Bull. Soc. Mat. Mexicana, 5 (1960), 102-119.

[12] H. Kwakernaak and R. Sivan, "The maximally

achieveable accuracy of linear optimal regulators and linear optimal filters," IEEE Trans. Automatic Control, 17 (1972), 79-86.

[13] H. G. Kwatny, "Minimal order observers and certain singular problems of optimal estimation and control," ibid., 19 (1974), 274-276.

[14] J. L. Lions, Perturbations Singulières dans les Problèmes aux Limites et en Contrôle Optimal, Lecture Notes in Mathematics 323, Springer-Verlag, Berlin, 1973.

[15] P. J. Moylan and J. B. Moore, "Generalizations of singular optimal control theory," Automatica, 7 (1971), 591-598.

[16] R. E. O'Malley, Jr., Introduction to Singular Perturbations, Academic Press, New York, 1974.

[17] R. E. O' Malley, Jr. and A. Jameson, "Singular perturbations, and singular arcs I," IEEE Trans. Automatic Control, 20(1975) to appear.

[18] ~~R. E.~~ O'Malley, Jr. and C. F. Kung, "The singularly perturbed linear state regulator problem II," SIAM J. Control, 13 (1974),

[19] R. E. O'Malley, Jr. and C. F. Kung, "The matrix Riccati approach to a singularly perturbed regulator problem," J. Differential Equations, 16 (1974)

[20] H. E. Rauch, "Application of singular perturbation to optimal estimation," Proceedings, Eleventh Annual Allerton Conference on Circut and System Theory, October 1973, 718-728.

[21] W. Wasow, Asymptotic Expansions for Ordinary

Differential Equations, Interscience, New York, 1965.

[22] R. R. Wilde and P. V. Kokotović, "Optimal open and closed loop control of singularly perturbed linear systems," IEEE Trans. Automatic Control, 18 (1973), 616-626.

[23] R. A. Yackel and P. V. Kokotović, "A boundary layer method for the matrix Riccati equation," ibid., 18 (1973), 17-24.

University of Arizona, Tucson, Arizona

EXISTENCE THEORY IN OPTIMAL CONTROL PROBLEMS - THE UNDERLYING IDEAS

Czesław Olech

Introduction.

The existence theory in mathematical optimal control theory is quite an extensive subject. In this paper we discuss some aspects of this theory, namely that part of it which is an extension of the so called direct method in calculus of variations initiated by Hilbert and developed by Tonelli, McShane and others.

The first existence theorems were given for linear time optimal control problem by LaSalle and others. For the non-linear time optimal control problem the first existence theorem was given by Filippov.

In the last ten years existence theory has been developed substantially; in particular, the Tonelli - McShane theory has been extended mainly by Cesari, to cover various optimal control problems.

In this paper we do intend neither to give a survey of the literature concerning the subject nor to enumerate and compare numerous particular existence theorems proved recently. Thus also the list of references is far from being complete.

The aim of the paper is to explain the general structure of a class of existence theorems modeled on the classical Weierstrass theorem concerning the existence of a minimum of a real valued function. In such theorems two conditions are essential: the so called lower closure property of lower semi-continuity and compactness of the minimizing sequence. The first of the two properties is discussed in more detail and in particular an effort is made to give a full characterization of this property. In section 1 we give a control-free formulation of a Lagrange type optimal problem to which the further discussion is confined. Section 2 contains a general remark concerning the structure of existence theorems for such problems. In section 3 we give a full characterization of lower closure property in a special case. In particular a necessary and sufficient condition for lower semi-continuity of some integral functionals is stated in the case when the functionals are independent of the state variable. In section 4 lower closure property is discussed in the case when the problem depends on the state variable. There we give a lemma which we believe is instrumental in establishing lower closure property.

Finally in the last section we discuss the role of the so called growth condition.

The paper does not contain proofs. Some of the results are new and will be published in detail elsewhere.

An optimal problem.

In this section we give the formulation of an optimal problem which will serve us as a model for further discussion. First let us fix some notations and basic assumptions.

Let Q be a map from $J \times R^n$ into subsets of R^{n+1}, J is an interval say [0, 1]. Below the following will always be assumed about Q.

<u>Assumption H.</u> If $(q, q_o) \in Q(t, x)$, q_o - the last component of a point from $Q(t, x)$, and $r > q_o$ then $(q, r) \in Q(t, x)$ also. The graph of $Q(t, x)$ is a $\mathcal{L} \times \mathcal{B}$ - measurable subset of $J \times R^n \times R^{n+1}$ where $\mathcal{L}$ is the σ - field of Lebesgue measurable subsets of J and $\mathcal{B}$ the Borel σ - field.

Let Ω be a set, called admissible, of pairs (x,u), both being integrable functions from J into R^n. The problem we wish to discuss is the following:

(P) minimize $\{\int_J v(t)dt \mid (u(t), v(t)) \in Q(t, x(t)), (x, u) \in \Omega\}$.

The Lagrange problem of calculus of variations is a special case of (P). Indeed suppose we wish to minimize the functional

$$I(x, u) = \int_J f_o(t, x(t), u(t))dt \tag{1}$$

on $\Omega = \{(x, u) \mid x \in X, u = \dot{x}\}$, where X is a fixed subset of the space AC of absolutely continuous functions. Then this problem reduces to (P) if we define Q as the epigraph of $f(t, x, \cdot)$; that is

$$Q(t, x) = \{(q, q_o) \mid q_o \geq f_o(t, x, q)\}. \tag{2}$$

Similarly the optimal control problem which consists of minimizing (1) on Ω given by the following conditions

$$x \in X, \ x(t) \in A(t), \ t \in J,$$
$$(3) \qquad u(t) \in U(t, x(t))$$
$$\dot{x}(t) = f(t, x(t), u(t))$$

can be reduced to P by putting

$$(4) \quad Q(t,x) = \{(q, q_o) \mid q_o \geq f_o(t,x,u), q = f(t,x,u), u \in U(t,x)\}.$$

In fact under quite mild regularity conditions for f_o, f and the set valued maps U and A, the above problems can be shown to be equivalent. In showing this a theorem concerning the existence of measurable selectors to set valued functions (cf. for example [8]) is instrumental.

Problem P as formulated is quite general and covers most interesting cases. It is a natural control-free formulation of optimal control problems and in this or similar form has been studied by many authors including Filippov, Roxin, Ważewski for the case of bounded control domains and for unbounded case by Cesari and others. The Lagrange problem with the integrand function assuming also $+\infty$ values has been investigated by Rockafellar and we refere to his paper [14] for the relation of it with other optimization problems. For simplicity's sake we restrict ourselves to Lagrange type optimal problems and on a fixed interval. However we also believe that this generality is quite enough to cover essential difficulties which one has to face in existence theory of optimal solution as well as to explain the main ideas and tools being used in this theory. We shall restrict also the discussion to an interval with Lebesgue measure on it, though this is not much of a restriction because most of the results to be discussed here hold as well for J being an arbitrary topological Hausdorff space, compact and metrizable with a positive, non-atomic regular Borel measure on it.

Existence theorem.

The literature concerning existence of optimal solutions in both control and classical variational problems is quite extensive. We do not intend to give a complete review of this literature, which would be quite a difficult task, also. Instead we wish to emphasize the usual structure of such theorems and also to present some more recent results, mostly concerning lower semi-continuity or lower closure property, which lead to existence theorems.

There are two main conditions which are instrumental in existence theorems we have in mind:

C_1. The set

$$D = \{(x,u,a) \mid (u(t), v(t)) \in Q(t,x(t)) \text{ a.e. in } J,\ a = \int_J v(t)dt\}$$

is closed.

C_2. There is an a such that $\Omega_a = \{(x,u) \in \Omega \mid$ there is v such that $(u(t), v(t)) \in Q(t,x(t))$ a.e. in J and $\int_J v(t)dt \leq a\}$ is not empty and precompact.

Existence theorem. If Ω is closed and both C_1 and C_2 hold then there is an optimal solution to problem (P).

This theorem as stated is of course quite trivial. It becomes less trivial if we try to formulate conditions for Q and Ω which would imply C_1 and C_2. We did not specify the meaning of closedness or compactness in the above statement and this naturally has to be done if we want to get a meaningful statement. However we wish to emphazise that no particular topology on mode of convergence is suggested by the problem itself, that we may choose any topology we like provided that we are able to prove in a particular case of interest that Ω is closed and C_1 and C_2 hold.

In choosing topologies in x - space and u-space we are faced with a conflict between condition C_1 and C_2. The weaker the topology we choose the more probable it is that condition C_2 would hold, while for condition C_1 the situation is opposite.

Let us make one more remark. There are two types of specifications of the above theorem. One is when condition C_2 holds for reasons independent of the structure of Q. For example if we know beforehand the topology in which Ω is compact, then the only assumptions we need to impose on Q should be such that C_1 holds with respect to this particular topology. The other type is such that properties of Q we seek should imply both C_1 and C_2 and this is the case when Ω is quite large.

Condition C_1 is an equivalent formulation of the so called lower closure property/l.c.p. - for short/of orientor field $(u(t), v(t)) \in Q(t, x(t))$ introduced by L. Cesari; that is the property that if $x_\alpha \to x_o$, $u_\alpha \to u_o$, $\int_J v_\alpha(t)dt \to a$, and $(u_\alpha(t), v_\alpha(t)) \in Q(t, x_\alpha(t))$ a.e. in J, then there is $v_o \in L_1$ such that $(u_o(t), v_o(t)) \in Q(t, x_o(t))$ and $\int_J v_o(t)dt \leq a$. If

Q is given by (2) then C_1 is nothing else than lower semi-continuity of I given by (1). Condition C_2 is to guarantee that the called minimizing sequence is precompact. In the next two sections we shall discuss the l. c. p., first for the case when Q does not depend on x and then for general case.

Lower closure property: Q independent of x.

In this section we shall consider set valued functions Q from J only into subsets of R^{n+1} satisfying assumption H and we give a characterization of l. c. p.. This characterization will be given with respect to the weak topology $w(L_1, S)$, where $S = L_\infty$ or C (the space of continuous functions) or C^∞. In the statements which follow if S is not specified then it means that the statement holds for each of the above three cases.

As in C_1, let

$$D = \{(u,a) \,|\, u \in L_1(J,R^n),\ a = \int_J v(t)dt \text{ and } (u(t), v(t)) \in Q(t)\}.$$

Notice that if for some u_o, $(\{u_o\} \times R) \subset D$ and D is $w(L_1, S)$ closed then the same holds for each u such that $(\{u\} \times R) \cap D$ is not empty.

Indeed, let $(u(t), v(t)) \in Q(t)$ and be integrable. Then there is a sequence of sets A_n whose measure tends to zero and a sequence $\tilde{v}_n \in L_1$ such that $(u_o(t), \tilde{v}_n(t)) \in Q(t)$ and $\int_{A_n} \tilde{v}_n(t)\,dt \to -\infty$. Putting $u_n(t) = u_o(t)$ if $t \in A_n$ and $u(t)$ otherwise and similarly $v_n(t) = \tilde{v}_n(t)$ if $t \in A_n$ and $v(t)$ if $t \in J \setminus A_n$ we see that $u_n \to u$ strongly in L_1 while

$\int_J v_n(t)\,dt \to -\infty$.

In what follows, we say that D is bounded from below if for each u the $\inf\{a \mid (u,a) \in D\}$ is finite.

Theorem 1. Assume condition H, then the following conditions are equivalent:

(i) D is bounded from below and closed in $L_1 \times R$ with $w(L_1, S)$ topology on L_1.

(ii) D is bounded from below, closed and convex in $L_1 \times R$ with $w(L_1, S)$ topology in L_1.

(iii) The set $K = \{(u,v) \mid u \in L_1(J,R^n),\ v \in L_1(J,R)$ and $(u(t), v(t)) \in Q(t)\}$ is convex and closed in $w(L_1, S)$ topology and there is $\varphi \in S$ and $\psi \in L_1$ such that

(5) $-v(t) + \langle u(t), \varphi(t)\rangle \le \psi(t)$ a.e. in J for each $(u,v) \in K$.

(iv) Q(t) is closed and convex a.e. in J and admits the representation.

(6) $Q(t) = \bigcap_{\varphi \in \Phi} \{(q, q_o) \mid -q_o + \langle q, \varphi(t)\rangle \le \psi_\varphi(t)\}$ a.e. in J,

where $\Phi \subset S$ is not empty and at most denumerable and ψ_φ is integrable for each $\varphi \in \Phi$.

(v) For each generalized sequence $\{(u_\alpha, v_\alpha)\} \subset K$ such that $u_\alpha \to u_o - w(L_1, S)$, $\int_J v_\alpha(t)\,dt \to a$ there is $v_o \in L_1(J,R)$ such that $(u_o, v_o) \in K$ and $\int_J v_o(t)\,dt \le a$.

(vi) There is a map f(t,u) from $J \times R^n$ into $R \cup \{+\infty\}$ whose graph is $\mathcal{L} \times \mathcal{B}$ measurable, convex and l.s.c. in u such that Q(t) is equal to the epigraph of $f(t,\cdot)$ a.e. in J, the functional $I(u) = \int_J f(t, u(t))\,dt$ is well defined from $L_1(J,R^n)$ into $R \cup \{+\infty\}$ and I(u) is l.s.c. in u with respect to $w(L_1, S)$ topology.

Moreover the equivalence still holds if the $w(L_1,S)$ topology is replaced by $w(L_1,S)$ sequencial convergence in (i), (ii), (iii), (v) or (vi).

The proof of this theorem is based on the characterization of the $w(L_1,S)$ closure of a set $K \subset L_1(J,R^n)$ with the property that if $w_1, w_2 \in K$ and $A \subset J$ is measurable then

(7) $$w_1\chi_A + w_2\chi_{J\setminus A} \in K,$$

where χ_A stands for the characterixtic function of A. This characterization is given by the following

<u>Theorem 2.</u> Let $K \subset L_1(J,R^n)$ satisfy (7) and let $\bar{K}$ be the closure of K in the $w(L_1,S)$ topology.

Then there is a measurable set valued function $Q(t)$ and a denumerable set $\Phi \subset S$ such that

(8) $$K = \{w \mid w \in L_1,\ w(t) \in Q(t) \text{ a.e. in } J\}$$

and

(9) $$Q(t) = \bigcap_{\varphi \in \Phi} \{q \mid \langle q, \varphi(t)\rangle \leq \psi_\varphi(t)\},$$

where $\psi_\varphi(t)$ = essential supremum of $\langle w(t), \varphi(t)\rangle$, over all $w \in K$.

Theorem 2, for the case $S = C$, has been proved by the author in [11]; for $S = C^\infty$ the proof is quite analogous and for $S = L_\infty$ it is enough to prove the existence of Q such that (8) holds, since (9) is true for any measurable Q if Φ is a subset of L_∞.

From Theorem 1 it follows that if the orientor field $w(t) = (u(t), v(t)) \in Q(t)$ has l. c. p. then $Q(t)$ is the epigraph of function $f(t,\cdot)$, where f maps $J \times R^n$ into $R \cup \{+\infty\}$ and l.c.p. is equivalent to l.s.c. of the integral functional

$$I(u) = \int_J f(t,u(t))\,dt$$

provided the latter is finite for at least one $u \in L_1$. We shall state now a necessary and sufficient condition for I to be l.s.c. in $w(L_1, L_\infty)$ and $w(L_1, C)$, respectively in terms of the conjugate function to f; that is the function

$$f^*(t,p) = \sup_u \left(-f(t,u) + \langle u,p\rangle\right). \tag{10}$$

This function can assume $+\infty$ values and it is known to be convex and l.s.c. in p. If f is measurable in t then so is f^*. Denote by

$$A(t) = \{p \mid f(t,p) < +\infty\}. \tag{11}$$

The set $A(t)$ is called the effective domain of $f^*(t,\cdot)$ and is convex for each t.

Corollary 1. Assume that $I(u)$ is finite for some $u \in L_1$ and that f is $L \times B$ measurable. Then I is l.s.c. with respect to weak topology on L_1 if and only if f is convex and l.s.c. in u and there is $\varphi \in L_\infty$ such that $f^*(t,\varphi(t))$ is integrable. The latter is equivalent to weak sequencial l.s.c. of I.

Indeed, in this case only convexity and closedness of $Q(t)$ is needed for weak closedness of K and integrability of $f^*(t,\varphi(t))$ for some $\varphi \in L_\infty$ is to guarantee that I is well defined (or never equal $-\infty$). Corollary 1 for f continuous in u and finite everywhere **was** been obtained by Poljak [13]. A similar characterization of l.s.c. in weak topology of L_p with $p > 1$ is given there.

If one considers the $w(L_1, C)$ or $w(L_1, C^\infty)$ topology then convexity itself of Q is not enough for K to be closed as the following example shows.

Let $n = 1$ and put $Q(t) = [1, +\infty)$ if t belongs to a Cantor set of positive measure contained in J and $(-\infty, +\infty)$ otherwise. Then the closure of the corresponding set K with respect to $w(L_1, C)$ or $w(L_1, C^\infty)$ as well consists of all integrable functions, hence is different from K (cf. [g]).

<u>Corollary 2.</u> Let f be as in corollary 1. Then I is l. s. c. in the $w(L_1, C)$(or $w(L_1, C)$) topology if and only if there is a set $N \subset J$ of measure zero such that for each $t \in J \setminus N$ and each a from intrinsic interior of A(t) - the effective domain of f^* there is $\varphi \in C$ (or $\varphi \in C^\infty$) such that $\varphi(t) = a$ and $f^*(t, \varphi(t))$ is integrable.

This corollary for the case of $w(L_1, C)$ is given in author's paper [12].

In contrast to the $w(L_1, L_\infty)$ - closedness of K, where pointwise properties of Q are sufficient, that is convexity and closedness of values and measurability in t, in the case of the $w(L_1, C)$ or $w(L_1, C^\infty)$ topology the behavior of Q(t) in t has to be more regular in the sense that the effective domain A(t) cannot be only measurable but up to a set of measure zero it is, as follows from corollary 2, l. s. c. in t. To be more precise $A(t) = \tilde{A}(t)$ a. e. in J, where $\tilde{A}(t)$ is l. s. c., that is for each open G the set $A - G = \{t \mid A(t) \cap G \neq \phi\}$ is open.

On the other hand if Q(t) is upper semi-continuous in the sense that the graph of it is closed the effective domain

$A(t)$ is l. s. c. as the following proposition shows.

Proposition 1. If $Q(t)$ is convex, the graph of Q is closed or equivalently if $f(t,u)$ is l. s. c. in both variables and convex in u and there is measurable and bounded w such that $w(t) \in Q(t)$ a. e. in J then $A(t)$ is l. s. c.. This proposition, Corollary 2 and some properties of conjugate function lead to the following result.

Corollary 3. If f is l. s. c. in both variables, convex in u, and there is a bounded function u such that $f(t,u(t))$ is bounded from above, then I is l. s. c. with respect to $w(L_1,C)$ topology.

A stronger u. s. c. property introduced in this context by Cesari is the so called condition (Q):

$$\text{(Q)} \qquad Q(t_o) = \bigcap_{\varepsilon>0} \text{cl co} \bigcup_{|t-t_o|<\varepsilon} Q(t),$$

where cl co stands for closed convex hull.

If property (Q) is assumed for the epigraph $Q(t)$ of $f(t,\cdot)$ then a stronger l. s. c. property of I can be established, namely in this case the following can be proved [g].

Corollary 4. If Q satisfies property (Q) then I is l. s. c. with respect to the $w(L_1,C^\infty)$ topology, thus also with respect to the $w(L_1,C)$ and $w(L_1,L_\infty)$ topologies.

In general closedness of the graph does not imply property (Q) unless the maximal linear manifold $E(t)$ contained in $Q(t)$ is such that $E(t_o) \supset E(t)$ for t from some neighborhood of t_o. The relation between those two notions of u. s. c. in the case when $Q(t)$ does not contain a line for each t is described by the following.

Proposition 2. If $Q(t)$ is convex, and does not contain a line for each t, the graph of Q is closed and there is a bounded selector of Q, then (Q) holds for each $t_o \in J$.

The above and Corollary 4 implies.

Corollary 5. If $Q(t)$ is convex and closed, and does not contain a line for each t and there is a bounded selector of Q or equivalently $f(t,u)$ is l. s. c. in both variables, convex in u, $f(t,u(t))$ is bounded from above for some $u(t)$ bounded and for each $t_o \in J$ there is $\varepsilon > 0$ and constant $a \in R$, $b \in R^n$ such that

$$f(t_o, u) \geq a + \langle b, u \rangle + \varepsilon |u| \tag{12}$$

then I is l. s. c. in the $w(L_1, C^\infty)$ topology.

Inequality (12) is related to the semi-normality condition of McShane - Tonelli and property (Q) is a geometric version of this condition. For the complete discussion of the relation between those two conditions we refer the reader to the paper of Cesari [4].

Lower closure property: Q dependent on x.

Let now $Q(t, x)$ be the set valued map and again we assume assumption H from section 1.

We are interested in l. c. p. of the 'entor field

$$w(t) = (u(t), v(t)) \in Q(t,x(t)) \text{ a. e. in } J, \tag{13}$$

where w and x are integrable.

With each function x we shall associate the subset $K(x)$ of L_1 given by

(14) $K(x) = \{w \mid w \text{ integrable and satisfying (13)}\}$.

The orientor field (13) has l. c. p. if for each (generalized) sequences $\{x_\alpha\}$ and $\{w_\alpha\} = \{(u_\alpha, v_\alpha)\}$ such that $x_\alpha \to x_o$, $u_\alpha \to u_o$, $\int_J v_\alpha(t)\,dt \to a$ and $w_\alpha \in K(x_\alpha)$, there is $w_o = (u_o, v_o) \in K(x_o)$ such that $\int_J v_o(t)\,dt \leq a$.

As before, the convergence in u-space is in the sense of the weak topology $w(L_1, S)$ with $S = L_\infty$ or C or C^∞. In each of those three cases if $v_\alpha \to v_o$ $w(L_1,S)$ then $\int_J v_\alpha(t)\,dt \to \int_J v_o(t)\,dt$. The l. c. p. of (13) is related to the upper semi-continuity of the map K in the sense that the graph of K is closed. The author believes that the two properties are actually equivalent in some cases.

A necessary condition for l. c. p. of (13) is that for each fixed x, $Q(t) = Q(t,x(t))$ has l. c. p. with respect to the $w(L_1,S)$ topology. In particular, as follows from Theorem 1, l. c. p. of (13) implies that for each fixed function x_o, $K(x_o)$ is $w(L_1,S)$ - closed, convex and it is the epigraph of a function $f(t,x,\cdot)$, where the latter is $\mathcal{L} \times \mathcal{B}$ measurable, l. s. c. and convex with respect to the last variable. Moreover

(15) $$Q(t,x_o(t)) = \bigcap_{\varphi \in \Phi} \{(q,q_o) \mid -q_o + \langle q, \varphi(t)\rangle \leq \psi_\varphi(t) \text{ a. e. in } J\},$$

where $\Phi \subset S$ is not empty and at most denumerable and ψ_φ is integrable for each $\varphi \in \Phi$.

We have actually

(16) $$\psi_\varphi(t) = f^*(t, x_o(t), \varphi(t)),$$

where

$$f^*(t,x,p) = \sup_u(-f(t,x,u) + \langle u,p\rangle) = \sup\{-q_o + \langle q,p\rangle \mid (q,q_o) \in Q(t,x)\}.$$

This suggests that to obtain l. c. p. of (13) a useful and natural tool should be the following.

Lemma. Let $Q(t,x)$ be closed and convex, and satisfy assumption H. Let $x_\alpha \to x_o$ in some fixed sense, $u_\alpha \to u_o$ in the $w(L_1,S)$ topology $\int_J v_\alpha(t)dt \to a$, where $(u_\alpha, v_\alpha) \in K(x_\alpha)$ and index α is from a directed set / possibly not denumerable /.

Assume that (15) holds and for each $\varphi \in \Phi$ from (15) there is $\varphi_\alpha \in L_1$ such that

(17) $\sup(-q_o + \langle q, \varphi_\alpha(t)\rangle) = f^*(t,x_\alpha(t), \varphi_\alpha(t)) \leq \psi_\alpha(t)$ a. e. in J,

where the sup is taken for $(q,q_o) \in Q(t,x_\alpha(t))$, and that

$$\langle \varphi_\alpha(t), u(t)\rangle \to \langle \varphi(t), u_o(t)\rangle \text{ in the } w(L_1,C) \text{ sense,}$$

while

(18) $\psi_\alpha(t) \leq \psi(t)$, $\psi \in L_1$ and $\limsup \psi_\alpha(t) \leq \psi_\varphi(t)$ a. e. in J.

Suppose further that v_α converges to a measure in m in the weak* topology $w(C^*,C)$ and let m_a, m_s be the absolutely continuous part and singular part of m, respectively. Then

(19) $m_s \geq 0$ and $-dm_a/dt(t) + \langle u_o(t), \varphi(t)\rangle \leq \psi_\varphi(t)$ a. e. in J.

In fact if the lemma holds then from (15) and (19), putting $v_o = dm_a/dt$, it follows that $(u_o(t), v_o(t)) \in Q(t,x_o(t))$ a. e. in J and $\int_J v_o(t)\,dt = a - m_s(J) \leq a$, which gives l. c. p. for (13).

In what follows, we state some lower closure theorems which can be obtained by application of this lemma. Before however we want to make the following remark. In previous sections when Q was independent of x there was no difference between topological convergence and sequential with respect to $w(L_1,S)$. This is not the case if Q depends on x. Indeed, let $f(t,x,u) \equiv x\,u$ where both x and u are reals. In this case if $x_n, x_o \in C$, $u_n, u_o \in L_1$, $x_n \to x_o$ uniformly and $u_n \to u_o$ weakly in L_1 $(w(L_1,L_\infty))$ then $x_n u_n \to x_o u_o$ weakly in L_1, which is not the case if we consider generalized sequences. In fact there is a sequence $\{x_n\} \subset C$, $x_n(t) \to 0$ uniformly and such that for any real function $v \in L_1$ (also negative) and any $w(L_1,L_\infty)$ neighborhood N of (u_o,v), for each $n \geq n_o(N)$ there is $u_u^N \in L_1$ such that $(u_u^N, x_n u_n^N) \in N$. Note that in this case $f^*(x,p) = 0$ if $p = -x$ and $+\infty$ otherwise, thus the lemma can be applied if x_n and u_n are denumerable sequences with $\varphi_n = -x_n$ but not in the case of topological $w(L_1,L_\infty)$ convergence.

In obtaining sufficient conditions for l. c. p. a useful assumption for $Q(t,x)$ is property (Q). The first lower closure result will be under this assumption assumed with respect to both variables. That is we assume

$$Q(t,x) = \bigcap_{\varepsilon>0} \operatorname{cl}\operatorname{co} \bigcup_{|y-x|<\varepsilon,\ |t-\tau|<\varepsilon} Q(\tau,y) \quad \text{for each } t,x. \tag{20}$$

Condition (20) implies that $Q(t,x(t))$ has property (Q) if $x(t)$ is continuous and that if $p \in$ intr. int. $A(t_o,x_o)$, where

$$A(t,x) = \{p \mid f^*(t,x,p) < +\infty\} \tag{21}$$

then for each $\varepsilon > 0$ there is a neighborhood N of (t_o,x_o) such that

(22) $f^*(t,x,p) \leq f^*(t_o,x_o,p) + \varepsilon$ if $(t,x) \in N$.

This implies that for any continuous $\varphi(t) \in$ intr. int. $A(t,x_o(t))$ the function $f^*(t,x_o(t),\varphi(t))$ is upper semi-continuous and finite hence integrable and so is $f^*(t,x(t),\varphi(t))$ if $|x(t) - x_o(t)| \leq \varepsilon$ and ε is small enough. Moreover because of (22) they are uniformly bounded by an integrable function. This allows us to use the lemma with $\varphi_\alpha = \varphi$ for each α, to prove the following lower closure theorem.

Theorem 3. If $Q(t,x)$ has property (Q) with respect to both variables and $A(t,x)$ is not empty for each (t,x) then (13) has l. c. p. with respect to uniform convergence in x-space and weak $w(L_1,C)$ topology in u-space.

In the theorem above the property (Q) is used through properties it implies with respect to the effective domain $A(t,x)$ of $f^*(t,x,\cdot)$. One of them being lower semi-continuity and the other: that if $\varphi(t) \in$ intr. int. $A(t,x(t))$ then $f^*(t,x(t),\varphi(t))$ is integrable. In fact a more general theorem could be stated. Let $\tilde{A}(t_o,x_o) = \{p \mid$ there is $\varepsilon > 0$ and $\psi \in L_1$ such that $f^*(t,x,p) < \psi(t)$ for $|t-t_o| < \varepsilon$ and $|x-x_o| < \varepsilon\}$. This set is convex and depends in a lower semi-continuous manner on t and x.

Theorem 3. If Q satisfies assumption H and property (Q) with respect to x for fixed t and $\text{cl}\,\tilde{A}(t_o,x_o) = \text{cl}\,A(t,x)$ for each x and a.e. in J then the conclusion of theorem 3 holds.

Of course in both theorem 3 and 3 the $w(L_1,C)$ topology can be replaced by $w(L_1,L_\infty)$ or weak sequencial convergence in

L_1 for function u. In theorem 3 we relaxed condition Q in the sense that we assumed it only with respect to x but instead because of the $w(L_1,C)$ topology in the conclusion we have to make an assumption about A(t,x). None of this was needed if we require the l.c.p. with respect to $w(L_1,L_\infty)$. Theorems of that kind can be found in papers by Cesari [6] and also Berkovitz [1]. However in this case much less is needed as the following theorem shows.

<u>Theorem 4.</u> Let Q satisfy assumption H, Q(t,x) is convex and closed, for each fixed t and graph $Q(t,\cdot)$ is closed and assume there is $M, M_1 > 0$ and $\psi \in L_1$ such that for each integrable x there is $p \in L_\infty$ such that

$$f^*(t,x(t),p(t)) \leq \psi(t) + M_1|x(t)|, |p(t)| \leq M \text{ a.e. in } J.$$

Then (13) has l.c.p. with respect to strong convergence in L_1 for x and weak sequencial for u.

If in the above assumption $M_1 = 0$, then the conclusion holds with pointwise convergence for x.

This theorem is essentially due to Berkovitz [2] and Cesari [5]. If contains also a lower semi-continuity theorem by Poljak [13] for functionals of the form (1) with f finite and satisfying the Carathéodory assumption.

If Q(t,x) in theorem 4 is the epigraph of $f(t,x,\cdot)$ then assumption for Q mean that f is measurable in t for fixed x,u, lower semi-continuity with respect to u,x for fixed t, convex in u, and that the inequality

$$f(t,x,u) \geq \psi_1(t) - M|u| - M_1|x|$$

holds, where $\psi_1 \in L_1$ and M, M_1 are nonnegative constants.

Growth condition in existence theory

Roughly speaking a scalar function $\Phi(u)$ satisfies growth condition if it grows quicker that any linear function; that is $\Phi(u)/|u| \to +\infty$ as $|u| \to +\infty$.

This condition when assumed for f, in (1), with respect to u and uniformly in t and x, that is $f(t,x,u) \geq \Phi(u)$ with satisfying the above condition, implies the existence of absolute minimum of the functional (1) if x is from a compact set X and $u \in L_1$, arbitrary. This condition has been introduced and used in calculus of variations by Nagumo, McShane, Tonelli.

Here we shall give a geometric version of the growth condition concerning the set valued function Q.

Let us consider first the case when Q depends only on t. Again values of Q are in R^{n+1} with the last component in R^{n+1} being distinguished by assumption H concerning Q.

We say that such Q satisfies the growth condition if for each $p \in R^n$ there is an integrable function ψ such that

(23) $-q_o + \langle p,q \rangle \leq \psi(t)$ a.e. in J and for each $(q,q_o) \in Q(t)$.

If Q(t) is the epigraph of a function $f(t,\cdot)$ then the growth condition can be expressed equivalently in terms of conjugate function f^*, namely (23) is equivalent to

(24) $f^*(t,p(t))$ is integrable for each $p \in L_\infty$.

In particular, if (24) holds then the effective domain A(t) of $f^*(t,\cdot)$ is equal almost everywhere to the open half-space $\{(q,q_o) \mid q_o < 0\}$ or equivalently that a.e. in J

$Q(t)$ contains only rays which are translations of positive half axis $\{(q,q_o) \mid q = 0,\ q_o \geq 0\}$. However the latter conditions are not equivalent to (24), because they do not imply integrability of $f^*(t,p(t))$ but only that the latter function is finite a.e. in J.

Growth condition (24) implies compactness asked for in condition C_2 of the existence theorem. Indeed, the following implications of (24) make it useful in existence theory compared with a lemma in [10].

Proposition 3. If $Q(t)$ is the epigraph of $f(t,\cdot)$ and the conjugate function f^* to f has property (24) then: (a) for each integrable selector (u,v) of Q the inequality $\int_J v(t)\,dt \leq M$ implies $\int_J |u(t)|\,dt \leq N$, where M and N are constants and N depends only on M; (b) if $\{(u_\alpha,v_\alpha)\}$ is a (generalized) sequence of selectors of Q such that $\int_J v_\alpha(t)\,dt \leq M$ then there is a subsequence, still denoted by $\{(u_\alpha,v_\alpha)\}$ converging in the weak* topology of C^* and the limit is of the form (u_o, m) where $u_o \in L_1$ and m is a scalar measure whose singular part m_s is nonnegative; (c) if $Q(t)$ is additionally assumed to be convex and closed for each t then (u_o,v_o), where $v_o(t) = dm_a/dt\,(t)$, with m_a being absolutely a continuous part of m, is a selector of Q.

Proposition 3 implies existence of absolute minimum of the functional $I(u) = \int_J f(t,u(t))\,dt$ or of the minimum of I for $u \in \Omega \subset L_1$ provided Ω is $w(L_1,S)$ closed and f is l.s.c. and convex in u, measurable in t and (24) holds.

In the case Q depends on x an analogue of (24) is the

following inequality

(25) $\quad - q_o + \langle p,q \rangle \leq \psi(t,p,r)$, for each $(q,q_o) \in Q(t,x)$ and $|x| \leq r$, where ψ is assumed to be integrable in t for each fixed p and r fixed. If $Q(t,x)$ is the epigraph of $f(t,x,\cdot)$, inequality (25) can be expressed equivalently in terms of the conjugate function as follows

(26) $$f^*(t,x,p) \leq \psi(t,p,r).$$

Growth condition (25) or (26) plus closedness and compactness of $Q(t,x)$ implies l.c.p. for (13) with respect to pointwise convergence in x on bounded subsets of x-space and $w(L_1,S)$ topology in u-space. It implies also compactness condition C_2. To be more precise, it implies the precompactness of the set $\{u \mid$ there is x such that $(x,u) \in \Omega_a\}$ in the $w(L_1,S)$ topology provided that the set $\{x \mid$ there is u such that $(x,u) \in \Omega_a\}$ is bounded. However in most special cases of problem (P), in particular in the special cases mentioned in section 1, x and u are connected and $\dot{x} = u$. Thus compactness in u - component implies in general compactness in x - component of Ω_a.

To be more specific we state an existence theorem in the case $\Omega = \{(x,u) \mid x \in X \subset AC,\ u = \dot{x}\}$.

Theorem 5. Let $Q(t,x)$ be closed and convex, satisfy assumption H and growth condition (25). Let $X \subset AC$ be bounded and closed in uniform topology, and $\Omega = \{(x,u) \mid x \in X,\ u = \dot{x}\}$. Under those assumptions there exists an optimal solution of problem (P).

The boundedness condition on X may be relaxed if the

growth condition is uniform with respect to x, that is if ψ in (25) or (26) does not depend on r, or if one imposes some extra conditions on ψ, for example that

(27) $$\psi(t,p,r) \leq \psi_1(t,p) + |x| (\alpha(t) + \beta(t)|p|)$$

where α, β and $\psi_1(t,p(t))$ for each $p \in L_\infty$ are integrable. In those cases instead of closedness and boundness of X it is enough to assume in theorem 5 that the set $\{x(t_o) \mid x \in X\} \subset R^n$ is compact for some $t_o \in J$ fixed.

Theorems of that kind, the reader can find, for example in [3], [7], [10], [14].

The latter two papers contain also a relaxation of theorem 5, when growth condition (25) with ψ satisfying (27) is assumed, while in the paper [7] a more general existence theorem is given in the sense that the growth condition is not assumed everywhere but the set where (25) does not hold, has certain properties.

Theorem 5 is an example of existence theorems of the second kind / cf. section 2 / when conditions on Q imply both C_1 and C_2.

REFERENCES

[1] Berkovitz, L. D., Existence and lower closure theorems for abstract control problems, SIAM J. on Control 12(1974), pp. 27-42.

[2] ___________, Lower Semicontinuity of Integral Functionals., to appear in Trans. A. M. S.

[3] Cesari, L., Existence theorems for weak and usual

optimal solutions in Lagrange problems with unilateral constraints. I and II, Trans. Am. Math. Soc. 124, (1966), 369-412, 413-430.

[4] __________, Seminormality and upper semicontinuity in optimal control, J. Optim. Th. Appl. 6(1970), 114-137.

[5] __________, Lower semicontinuity and lower closure theorems without semicormality conditions, Annal. Mat. Pura ed. Appl. 98(1974), pp. 381-397.

[6] __________, Closure theorems for orientor fields and weak convergence, Arch. Ratl. Mech. Anal. to appear.

[7] Cesari, L., La Palm, J. R. and Sanchez, D. A., An existence theorem for Lagrange problems with unbounded controls and a slender set of exceptional points, SIAM J. Control, 9(1971), pp. 590-605.

[8] Kuratowski, K. and Ryll-Nardzewski, C., A general theorem on selectors, Bull. Acad. Polon. Sci., Ser. scien., math., astr. et phys., 6(1965), 397-403.

[9] Lasota, A. and Olech, C., On Cesari's semicontinuity condition for set valued mappings, Bull. Acad. Polon. Scien., Ser. Scienc. math., astr. et phys., 9(1968), 711-716.

[10] Olech, C., Existence theorems for optimal problems with vector-valued cost function, Trans. Am. Math. Soc., 136(1969), 159-179.

[11] __________, The characterization of the weak closure of certain sets of integrable functions SIAM J. Control 12(1974).

[12] __________, A necessary and sufficient condition for

lower semicontinuity of certain integral functionals, to appear.

[13] Poljak, B. T., Semicontinuity of integral functionals and existence theorems for extremal problems, Math. Sbor., 78 1969 65-84 / in Russian /

[14] Rockafellar, R. T., Existence theorems for general control problems of Bolza and Lagrange, to appear in Advances in Math.

Polish Academy Science, 00-950 Warsaw, Poland

RECENT ERROR ANALYSIS OF ASYMPTOTIC SOLUTIONS OF LINEAR DIFFERENTIAL EQUATIONS

F. W. J. Olver

1. Introduction.

A comprehensive account, up to 1973, of error analyses of asymptotic solutions of linear ordinary differential equations is included in the author's recent text [5]. The purpose of the present paper is to describe briefly more recent results that have been obtained by the author. These results all apply to second-order equations. They concern two coalescing simple turning points (Section 2), a single fractional turning point (Section 3), and connection formulas for multiple turning points (Section 4).

2. Two turning points.

Consider the equation

$$d^2w/dx^2 = \{u^2 f(a,x) + g(a,x)\}w \tag{2.1}$$

in which u is a large positive parameter, x is a real variable ranging over an open, possibly infinite, interval (x_1,x_2), a is a bounded real parameter, and the functions $f(a,x)$ and $g(a,x)$ are free from singularity within (x_1,x_2). We further assume that $f(a,x)$ has exactly two zeros z_1 and z_2, say, within (x_1,x_2), which depend continuously on a

and are simple and distinct, except that as a tends to a critical value a_o, z_1 and z_2 coalesce into a double zero z_o. The zeros of $f(a,x)$ are the so-called turning points of equation (2.1).

An example of the situation just described is provided by the equation

$$\frac{d^2L}{dx^2} = \left\{ u^2 \frac{x^2-a^2}{(1-x^2)^2} - \frac{3+x^2}{4(1-x^2)^2} \right\} L \tag{2.2}$$

satisfied by $L = (1-x^2)^{1/2} P_n^m(x)$ or $(1-x^2)^{1/2} Q_n^m(x)$, where $P_n^m(x)$ and $Q_n^m(x)$ denote the associated Legendre functions and

$$u = n + \tfrac{1}{2}, \qquad a^2 = 1 - (m^2/u^2).$$

For large m and n, there are turning points at $x = \pm a$, which coalesce into a single double turning point when $a = 0$, that is, when $m = n + \frac{1}{2}$.

Returning to the general case, we note that when $a \neq a_o$ the x-interval can be subdivided at a conveniently chosen point d lying between z_1 and z_2, for example, $d = \frac{1}{2}(z_1+z_2)$. Then for each of the partly closed intervals $(x_1, d]$ and $[d, x_2)$ we may use well-known methods to construct asymptotic solutions of (2.1), in terms of Airy functions, which are uniform in x. As $a \to a_o$, however, these Airy-function approximations break down owing to nonuniformity with respect to a.

To derive asymptotic solutions that are uniform with respect to a, as well as x, we first apply a Liouville transformation to convert (2.1) into one of the forms

$$d^2W/d\zeta^2 = \{\pm u^2(\zeta^2-\alpha^2) + \psi(\alpha,\zeta)\}W. \tag{2.3}$$

Here α is a new parameter that is a continuous function of a and vanishes as $a \to a_0$; ζ is a new variable which depends continuously on x and a, with x_1 corresponding to $\zeta = -\infty$, and x_2 to $\zeta = +\infty$; and W is a new dependent variable closely related to w. The function $\psi(\alpha,\zeta)$ is expressible in terms of f(a,x) and g(a,x), and is continuous in the (α,ζ)-plane, including the transformed critical point $\alpha = \zeta = 0$.

Approximate solutions of (2.3) are found by neglecting the term $\psi(\alpha,\zeta)$. The differential equation is then solvable exactly in terms of parabolic cylinder functions. For example, with the upper choice of the ambiguous sign, suitable solutions are $U(-\frac{1}{2}u\alpha^2, \zeta\sqrt{2u})$ and $\overline{U}(-\frac{1}{2}u\alpha^2, \zeta\sqrt{2u})$, where U(b,x) and $\overline{U}$(b,x) are the standard solutions of the equation

$$d^2w/dx^2 = (\tfrac{1}{4}x^2+b)w$$

defined and discussed by Miller [3]. Similarly, when the lower sign is taken in (2.3) we employ as approximants Miller's modified parabolic cylinder functions $W(\frac{1}{2}u\alpha^2, \zeta\sqrt{2u})$ and $W(\frac{1}{2}u\alpha^2, -\zeta\sqrt{2u})$, also given in the same reference.*

Typical of the bounds that have been obtained for the errors in the approximate solutions is the following result.

*Miller's solutions have not yet passed firmly into the literature, which is surprising since they are much more convenient when the variables are real than the commonly used and older standardization of Whittaker [10].

We consider the equation

$$(2.4) \qquad d^2W/d\zeta^2 = \{u^2(\zeta^2+\alpha^2) + \psi(\alpha,\zeta)\}W.$$

This does not belong to either of the types (2.3) owing to changes of signs. The turning points in fact are now situated at $\zeta = \pm i\alpha$, and only become real when $\alpha = 0$. Nevertheless equation (2.4) has the same essential features as (2.3), and is selected because the results are easier to describe. Denote

$$(2.5) \quad F(u,\alpha,\zeta) = \Gamma(\tfrac{1}{2}+\tfrac{1}{2}u\alpha^2)\int U(\tfrac{1}{2}u\alpha^2,\zeta\sqrt{2u})U(\tfrac{1}{2}u\alpha^2,-\zeta\sqrt{2u})\psi(\alpha,\zeta)d\zeta.$$

Then equation (2.4) possesses a solution**

$$W(u,\alpha,\zeta) = U(\tfrac{1}{2}u\alpha^2, \zeta\sqrt{2u}) + \varepsilon(u,\alpha,\zeta),$$

where

$$(2.6) \quad |\varepsilon(u,\alpha,\zeta)| \le U(\tfrac{1}{2}u\alpha^2,\zeta\sqrt{2u})[\exp\{\tfrac{1}{2}(\pi u)^{-1/2}\mathcal{V}_{\zeta,\infty}(F)\}-1].$$

Here $\mathcal{V}_{\zeta,\infty}(F)$ denotes the total variation of the function F over the interval (ζ,∞); thus

$$\mathcal{V}_{\zeta,\infty}(F) = \Gamma(\tfrac{1}{2}+\tfrac{1}{2}u\alpha^2)\int_{\zeta}^{\infty}|U(\tfrac{1}{2}u\alpha^2,t\sqrt{2u})U(\tfrac{1}{2}u\alpha^2,-t\sqrt{2u})\psi(\alpha,t)|dt.$$

To derive an asymptotic estimate for $\varepsilon(u,\alpha,\zeta)$ from this result when u is large, we assume that $|\psi(\alpha,\zeta)/\zeta|$ is integrable at $\zeta = \pm\infty$, uniformly with respect to α. Then by use of Stirling's formula and the uniform asymptotic expansions of parabolic cylinder functions of large order given in [4], it is provable that $\mathcal{V}_{-\infty,\infty}(F)$, that is, the total variation of $F(u,\alpha,\zeta)$ over $(-\infty,\infty)$, is $O(u^{-1/2}\ln u)$, uniformly with respect to α. Since $\mathcal{V}_{\zeta,\infty}(F) \le \mathcal{V}_{-\infty,\infty}(F)$ it

**A second solution can be obtained by replacing ζ by $-\zeta$ throughout.

immediately follows by substitution in (2.6) that the ratio of the error term $\varepsilon(u, \alpha, \zeta)$ to the approximation $U(\frac{1}{2}u\alpha^2, \zeta\sqrt{2u})$ is $O(u^{-1}\ell n\, u)$ as $u \to \infty$, uniformly with respect to $\zeta \in (-\infty, \infty)$ and α lying in a closed interval that contains $\alpha = 0$. This is the desired result.

The conditions that have been stated for $\psi(\alpha, \zeta)$ are not the most general. Indeed, a major advantage in having explicit bounds for the error term of the type given above (and also in the following sections) is that we can decide a posteriori what conditions on $\psi(\alpha, \zeta)$ suffice to ensure that the ratio of $\varepsilon(u, \alpha, \zeta)$ to $U(\frac{1}{2}u\alpha^2, \zeta\sqrt{2u})$ is uniformly $O(u^{-1}\ell n\, u)$, or even just $o(1)$. For example, it was tacitly assumed in the foregoing that $\psi(\alpha, \zeta)$ is independent of u. But as far as the inequality (2.6) is concerned, this is irrelevant. The approximation $U(\frac{1}{2}u\alpha^2, \zeta\sqrt{2u})$ to the solution $W(u, \alpha, \zeta)$ remains meaningful when $\psi(\alpha, \zeta)$ depends on u, provided that $\mathcal{V}_{-\infty, \infty}(F)$ is $o(u^{1/2})$ as $u \to \infty$, uniformly with respect to α.

For further details, including an historical survey of earlier work on the problem, see [6], and for an application to associated Legendre functions of large degree and order, via (2.2), see [7]. Other promising applications of the new theory include approximations for Whittaker functions with both parameters large, and wave-scattering problems involving elliptic cylinders or prolate spheroids in regions of resonance.

3. Fractional turning points.

If we set $a = a_0$, or equivalently, $\alpha = 0$ in the results of the previous section, we obtain asymptotic solutions, complete with error bounds, of a second-order differential equation with a double turning point. Moreover, in these circumstances the parabolic cylinder functions are expressible in terms of Bessel functions of order $\pm\frac{1}{4}$. The methods that were used to establish the bound (2.6) are capable of extension to differential equations having turning points of multiplicity higher than two, or even turning points of "fractional multiplicity". Thus we consider now equations of the form

$$d^2w/dx^2 = \{u^2 f(x) + g(x)\}w \tag{3.1}$$

in which u is again a large positive parameter, and x is a real variable ranging over a partly closed, possibly semi-infinite, interval $[z_o, x_2)$. In this interval the function $f(x)$ is assumed to be nonvanishing and free from singularity, except possibly at z_o, and $g(x)$ is free from singularity. Furthermore, at z_o the quotient $f(x)/(x-z_o)^\lambda$ is nonvanishing and finite, λ being an unrestricted real constant. The point z_o may then be said to be a <u>turning point of equation</u> (3.1) <u>of order</u> λ. When $\lambda \leq -2$ the problem of constructing asymptotic solutions, complete with error bounds, is solvable in terms of elementary functions; see [5], Chapters 6 and 10. Consequently in what follows we shall restrict $\lambda > -2$.

By use of an appropriate Liouville transformation, equation (3.1) transforms into

$$d^2W/d\zeta^2 = \{\pm\tfrac{1}{4}(\lambda+2)^2 u^2 \zeta^\lambda + \phi(\zeta)\}W, \tag{3.2}$$

in which the new variable ζ ranges over the interval $[0,\infty)$ and $\phi(\zeta)$ is continuous. The ambiguous sign here depends on the sign of $f(x)$ in (z_o, x_2). Approximate solutions of (3.2) are found by neglecting the term $\phi(\zeta)$. The differential equation is then solvable exactly in terms of Bessel functions of order $1/(\lambda+2)$.

To illustrate the error bounds that have been found for the approximate solutions, we fix the ambiguous sign in (3.2) to be positive, write $\mu = \lambda+2$ (> 0) for brevity, and define

$$F(u,\zeta) = (2/\mu)\int I_{1/\mu}(u\zeta^{\mu/2})K_{1/\mu}(u\zeta^{\mu/2})\zeta\phi(\zeta)d\zeta;$$

compare (2.5). Here $I_{1/\mu}$ and $K_{1/\mu}$ denote the modified Bessel functions of order $1/\mu$ in the usual notation. Then in the interval $(0,\infty)$ equation (3.2) possesses solutions

$$W_1(u,\zeta) = \zeta^{1/2}\{I_{1/\mu}(u\zeta^{\mu/2}) + \varepsilon_1(u,\zeta)\},$$

$$W_2(u,\zeta) = \zeta^{1/2}\{K_{1/\mu}(u\zeta^{\mu/2}) + \varepsilon_2(u,\zeta)\},$$

where

$$|\varepsilon_1(u,\zeta)| \leq I_{1/\mu}(u\zeta^{\mu/2})[\exp\{\mathcal{V}_{o,\zeta}(F)\} - 1],$$

$$|\varepsilon_2(u,\zeta)| \leq K_{1/\mu}(u\zeta^{\mu/2})[\exp\{\mathcal{V}_{\zeta,\infty}(F)\} - 1].$$

As in Section 2, uniform asymptotic properties of the error terms immediately derive from these error bounds whenever the total variation of the function $F(u,\zeta)$ over the interval (o,∞) is $o(1)$ as $u \to \infty$.

Similar results are available for the case in which the lower sign is used in (3.2): in place of the modified Bessel functions $I_{1/\mu}$ and $K_{1/\mu}$, we employ as approximants the Bessel functions $J_{1/\mu}$ and $Y_{1/\mu}$. Similar results can also be found for complex variables. Furthermore, among the various generalizations that can be made in a fairly straightforward manner, a particularly important one is to permit the function $g(x)$ in the original equation (3.1) to have a singularity at the fractional turning point z_o, provided that $(x-z_o)^2 g(x)$ is finite there.

Full details of these results, including a discussion of earlier work by Langer [2] and Riekstinss [9], will appear in [8].

4. Multiple turning points.

Another outgrowth of the theory described in Section 2 has been the development of an effective and mathematically rigorous theory of connection formulas for second-order differential equations having multiple turning points. Most previous investigations of this problem have been of a heuristic nature; see for example [1], Chapter V. Multiple turning points are of course included in the theory outlined in Section 3, but an important difference when $\lambda = m$, a nonnegative integer, is that all solutions of the differential equation are nonsingular at $x = z_o$. In contrast, for other values of λ solutions that are real on the real axis on one side of z_o generally become complex when continued to the other side.

We therefore suppose that the given multiple turning point z_o is interior to an open, possibly infinite, interval

(x_1, x_2), within which the functions $f(x)$ and $g(x)$ appearing in the differential equation (3.1) are free from singularity; moreover $f(x)/(x-z_o)^m$ is finite and nonvanishing. By analysis similar to that used for deriving the results of Section 3, asymptotic solutions can be constructed in terms of Bessel functions, or modified Bessel functions, of order $1/(m+2)$ which are uniform with respect to $x \in (x_1, x_2)$. Connection formulas across the turning point are then found by the well-known procedure of replacing the Bessel functions by their asymptotic approximations for large argument. This yields the desired approximations to the connection formulas, complete with error bounds. Several cases arise, depending on the sign of $f(x)/(x-z_o)^m$ and also whether the order m of the turning point is odd or even. This work is being readied for publication.

REFERENCES

[1] J. Heading, An introduction to phase-integral methods. Wiley, New York, 1962.

[2] R. E. Langer, On the asymptotic solutions of ordinary differential equations, with reference to the Stokes' phenomenon about a singular point. Trans. Amer. Math. Soc. 37 (1935), pp. 397-416.

[3] J. C. P. Miller, Tables of Weber parabolic cylinder functions. H.M. Stationery Office, London, 1955.

[4] F. W. J. Olver, Uniform asymptotic expansions for Weber parabolic cylinder functions of large orders. J. Res. Nat. Bur. Standards Sec. B. 63 (1959),

pp. 131-169.

[5] ____________, Asymptotics and special functions. Academic Press, New York, 1974.

[6] ____________, Second-order linear differential equations with two turning points. Philos. Trans. Roy. Soc. London [In press.].

[7] __________, Legendre functions with both parameters large. Philos. Trans. Roy. Soc. London [In press].

[8] ___________, Second-order differential equations with fractional transition points. [In preparation.]

[9] E. Riekstinss, On the method of tabulated functions (Russian). Latvijas Valsts Univ. Zinatn. Raksti 20, No. 3 (1958), pp. 65-86.

[10] E. T. Whittake, On the functions associated with the parabolic cylinder in harmonic analysis, Proc. London Math. Soc. 35(1903), pp. 417-427.

University of Maryland, College Par, Maryland and
U.S. National Bureau of Standards

ACCESSIBLE SETS IN CONTROL THEORY

A. Pliś

Introduction.

Consider a given control system on $R^n \times R$. Under weak assumptions, for each point $(x, t) \in R^n \times R$, there exists a trajectory passing through the point. Let S be a subset of $R^n \times \{0\}$. The union E of all trajectories $x = x(t)$ satisfying the condition $x(0) \in S$ is called an emission zone of set S in respect to the control system. The intersection of set E with the hyperplane $t = s$ is called an accessible set from S in time s. We give a sufficient condition for uniform convexity of accessible sets.

Notations and Definitions.

We consider control systems in the form (introduced by T. Ważewski [3]).

$$x' \in F(t, x), \tag{1}$$

where $x = (x_1, \ldots, x_n)$ and $F(t, x)$ is a set valued function. Sets $F(t, x)$ are supposed to be non-empty, convex and compact. A function $x = x(t)$, defined on an open interval I is called a trajectory of (1) if it is absolutely continuous on any compact subinterval of I and satisfies condition $x'(t) \in F(t, x(t))$ for almost every $t \in I$.

Let e be a positive constant. We call A an e-regular set if for each pair of points $a, b \in A$ and any number τ, $0 < \tau < 1$ the ball

(2) $$\{x: |x-(\tau a+(1-\tau)b)| \leq e\tau(1-\tau)|a-b|^2\}$$

is contained in A.

Remark 1. It is easy to see that for A compact it is enough to assume that $(2) \subset A$ only for a, b from a dense subset of A.

Remark 2. A uniformly convex set is e-regular for a certain e and a e-regular set is uniformly convex (the more strictly convex). A set consisting of a single point is e-regular for any positive e.

Let A be a strictly convex and compact set and w non-vanishing vector. Denote by p(w, A) such a point of A that $p(w, A)w = \max pw$ for $p \in A$, where pw is a scalar product. We denote

$$s(A, B) = \max|p(w, A)-p(w, B)| \quad \text{for } w \neq 0 .$$

The author has introduced function s(A, B) in [2] for formulation of a theorem on the local uniqueness of optimal trajectories. An example in [1] shows that in this theorem s(A, B) can not be replaces by Hausdorff's distance f(A, B). We have $s(A, B) \geq r(A, B)$.

We say that a function F(t, x) is of class L if there exists a positive constant C that for any points $y, z, \in R^n$ any t and any number τ, $0 \leq \tau \leq 1$, we have

(3) $$r(F(t,\tau y+(1-\tau)z), \tau F(t,y)+(1-\tau)F(t,z)) \leq C\tau(1-\tau)|y-z|^2 .$$

Theorem. Let F(t,x) be continuous on R^{n+1} (with

Hausdorff's topology in the space of subsets of R^n), of class L and satisfy the Lipschitz condition

$$(4) \qquad s(F(t,x),\ F(t,y)) \leq K|x-y|$$

where K is a positive constant. Let the sets F(t,x) be e-regular with the same e and the set $A \subset R^n \times \{0\}$ be e-regular (if considered as a subset of R^n) and compact.

Under these assumptions the sets accessible from A for sufficiently small positive time s are uniform
In the proof we use the following

Lemma. Let sets A, B be e-regular $a \in A$, $b \in B$, $|a-b| > 2s(a, B)$, τ a constant $0 < \tau < 1$ then the ball

$$x: \ |x-(\tau a+(1-\tau)b)| \leq M\tau(1-\tau)|a-b|^2$$

is contained in the set $\tau A+(1-\tau)B$, where M is a positive constant (which may depend on e).

Proof of the Theorem. In virtue of (4) each trajectory passing through A can be approximated by trajectories of class C^1 passing through A. We shall see that the accessible set for time s is q-regular with $q = \tilde{e} - Ns$, where N is sufficiently large constant and $s < e/N$.

Consider two trajectories y(t), z(t) (passing through A) of class C^1 on the interval $0 \leq t \leq s$ and a number τ, $0 < \tau < 1$. We shall prove that the set

$$(5) \qquad |x-v(t)| \leq (\tilde{e}-Nt)\tau(1-\tau)|y(t)-z(t)|^2, \quad 0 < t \leq s$$

where $v(t) = \tau y(t) + (1-\tau)z(t)$, is contained is the emission zone of A. Without loss of generality we can assume that $y(t) \neq z(t)$ for $t > 0$.

We have $y'(t) \in F(t,y(t))$, $z'(t) \in F(t,z(t))$

$$s(F(t,y(t)),\ F(t,z(t))) \leq K|y(t)-z(t)|.$$

The set (5) is of the form

$$(6) \qquad |x-v(t)| \leq h(t) \qquad 0 < t < s.$$

Denote $w = (x-v(t))|x-v(t)|^{-1}$. The vectors $(1, v'(t)+gw)$ attached to a boundary point x of (6) possess transversal position in respect to the boundary, pointing to the exterior of (6) for $wg > h'(t)$. The set $G(t,x)$ consisting of these vectors for $wg \geq h'(t) + \epsilon(t)$, where $\epsilon(t) > 0$ is a sufficiently small function, is closed and convex. The function $G(t,x)$ is continuous on the boundary of (6) for $t > 0$. We claim that for sufficiently large N (and small $\epsilon(t)$) the intersection $F(t,x) \cap G(t,x)$ is non-empty. If so, $F(t,x)$ being continuous $G(t,x)$ can be extended as a continuous function to the layer $0 < t < s$, x arbitrary in such a way that $F(t,x) \cap G(t,x)$ is non-empty on the layer. For the new control system $F(t,x) \cap G(t,x)$ semicontinuous on the layer an existence theorem can be applied. The trajectory of the new system starting from any point of (6) remains to the left in (6) in virtue of the inequality $wg > h'(t)$ and we get that (6) is contained in the emission zone.

The non-emptiness of $F \cap G$ follows for $|y'(t)-z'(t)| \geq P|y(t)-z(t)|$, where P is sufficiently large constant, from the Lemma. For $|y'(t)-z'(t)| \leq P|y(t)-z(t)|$ it can be obtained by putting N sufficiently large. It completes the proof.

REFERENCES

[1] A. Pliś, Generalized ordinary differential equations and control theory. Mathematica Balkanica, 3 (1973), 111-114.

[2] A. Pliś, Uniqueness of optimal trajectories for non-linear control systems, to appear in Annales Polonici Mathematici.

[3] T. Wazewski, On an optimal control problem. Proc. of the Conference on Differential Equations, held in Prague September 1962.

Centro de Investigación del IPN, México
On leave from Institute of Mathematics,
Polish Academy of Sciences.

A CLASS OF OPTIMAL CONTROL PROBLEMS INVOLVING HIGHER ORDER ISOTONIC APPROXIMATION

William T. Reid

1. Introduction.

In an earlier paper [6] the author has considered the problem of minimizing a suitable type of integral functional $J[y]$ in the class of real, monotone non-increasing functions $y(t)$, $t \in [a,b]$, which are Lipschitzian, and the relation of this problem to the minimization of $J[y]$ in the larger class of real, monotone non-increasing functions on $[a,b]$. In the present paper, certain of the earlier results are extended to a corresponding problem of minimizing an integral functional

$$(1.1) \qquad J[y] = \int_a^b F(t,y(t), y'(t), \dots, y^{[\ell]}(t))dt,$$

where $0 \leq \ell < m$, in the class of functions $y(t)$ satisfying on a given compact interval $[a,b]$ on the real line the differential equation

$$(1.2) \qquad L[y](t) \equiv a_m(t)y^{[m]}(t) + \dots + a_o(t)y(t) = \gamma(t), \quad m \geq 1,$$

where γ is a real, monotone non-increasing function on $[a,b]$. For the case of γ Lipschitzian it is shown that the extremizing function satisfies in the sense of Filippov [2] a

This research was supported by the National Science Foundation under Grant GP-36120.

differential system involving discontinuous functions. Also, special attention is devoted to the problem in which $J[y]$ is the integral of $\frac{1}{2}[y - h(t)]^2$, where h is a bounded Lebesgue measurable function on $[a,b]$.

For a comprehensive treatment of isotonic regression, and the relation of the results of [6] to this subject, the reader is referred to [1].

Matrix notation is used extensively; in particular, matrices of one column are called vectors. All matrices are supposed to be real-valued. The symbol R^q is used to denote the linear vector space of q-dimensional real vectors $w = (w_i)$, $(i = 1, \ldots, q)$, with $|w|$ signifying the Euclidean norm $(w_1^2 + \cdots + w_q^2)^{1/2}$. The transpose of a matrix M is denoted by M^*; the symbol 0 is used indiscriminately for the zero matrix of any dimensions. A matrix function is called continuous, integrable, etc., when each element of the matrix possesses this property. If $M = [M_{ik}]$ and $N = [N_{jk}]$ are $m \times r$ and $n \times r$ matrices, then $(M; N)$ denotes the $(m+n) \times r$ matrix $S = [S_{\alpha k}]$ with $S_{ik} = M_{ik}$, $(i = 1, \ldots, m;\ k = 1, \ldots, r)$ and $S_{m+j, k} = N_{j, k}$, $(j = 1, \ldots, n; k = 1, \ldots, r)$. If $w(t)$ is a continuous vector function on $[a,b]$, then $\|w\|_{[a,b]}$ denotes the supremum norm Max $\{|w(t)| \mid a \leq t \leq b\}$. If a matrix function $M(t)$ is a.c., (absolutely continuous), on $[a,b]$ then $M'(t)$ signifies the matrix of derivatives at values where these derivatives exist and the zero matrix elsewhere. Correspondingly, if $M(t)$ is (Lebesgue) integrable on $[a,b]$ then $\int_a^b M(t)dt$ denotes the matrix of integrals of respective elements. If $M(t)$ and $N(t)$ are equal a.e. (almost everywhere) on their domain of existence, we write simply $M(t) = N(t)$.

The symbols $\underline{C}[a,b]$, $\underline{C}^{[q]}[a,b]$, $\underline{L}[a,b]$ and $\underline{L}^{\infty}[a,b]$ denote the classes of real-valued functions on $[a,b]$ which are respectively continuous, continuous and possessing continuous derivatives of the first q orders, (Lebesgue) integrable, (Lebesgue) measurable and essentially bounded. If $y \in \underline{L}^{\infty}[a,b]$, then $\|y\|_{\infty}$ denotes the essential supremum of $|y(t)|$ on $[a,b]$.

2. Basic hypotheses.

Throughout the paper it is assumed that the following conditions hold.

$\underline{H}_1$) The coefficient functions $a_o(t), \ldots, a_m(t)$ of $L[y]$ are real Lebesgue measurable, with $a_m(t) \neq 0$ for all $t \in [a,b]$, while $1/a_m$ and a_β/a_m, $(\beta = 0, 1, \ldots, m-1)$, belong to $\underline{L}^{\infty}[a,b]$.

$\underline{H}_2$) The real integrand function $F(t, r) = F(t, r_o, \ldots, r_\ell)$ is such that:

(i) for fixed $r = (r_\nu) \in R^{\ell+1}$, $F(\cdot, r)$ is measurable on $[a,b]$;

(ii) for fixed $t \in [a,b]$, $F(t, \cdot)$ is continuous on $R^{\ell+1}$;

(iii) for $P > 0$ there exists a non-negative integrable function $\kappa_p(t)$ on $[a,b]$ such that

$$|F(t,r)| \leq \kappa_p(t), \text{ for } t \in [a,b], \ |r| \leq P; \tag{2.1}$$

(iv) $\lim_{|r| \to \infty} F(t,r) = +\infty$, uniformly for $t \in [a,b]$; that is, for $0 < K < +\infty$ there exists a value $L = L(K)$ such that

(a) $L(K) \geq K$; (b) $F(t,r) > K$ for $t \in [a,b]$ if $|r| > L(K)$.

In particular, $\underline{H}_2$-iv) implies the existence of a value $L_o > 0$ such that $F(t,r) > 0$ if $|r| > L_o$, and consequently

$$(2.2) \qquad F(t,r) \geq -\kappa_{L_o}(t), \text{ for } (t,r) \in [a,b] \times R^{\ell+1}.$$

In the following, $\underline{M}$ will denote the class of real-valued functions $y \in \underline{C}^{[m-1]}$ with $y^{[m-1]}$ a.c. on this interval, and satisfying (1.2) with γ a real-valued monotone non-increasing function on (a,b) which is integrable on $[a,b]$; that is, if $f(t) = \gamma(t)$ for $t \in (a,b)$, $f(a) = +\infty$, $f(b) = -\infty$, then f is Lebesgue integrable on $[a,b]$. The subclass of functions $y \in \underline{M}$ for which the corresponding γ is bounded is denoted by $\underline{MB}$, and if $|\gamma(t)| \leq k$ on $[a,b]$ we write $y \in \underline{MB}_k$. Also, for $k > 0$ the subclass of functions $y \in \underline{M}$ satisfying (1.2) with a γ for which $0 \leq \gamma(t_1) - \gamma(t_2) \leq k[t_2 - t_1]$ for $a \leq t_1 \leq t_2 \leq b$ will be denoted by $\underline{M}'_k$. To indicate that $y \in \underline{M}$ with corresponding γ, we write $y \in \underline{M}:\gamma$, with analogous meanings for $y \in \underline{MB}:\gamma$, $y \in \underline{MB}_k:\gamma$, and $y \in \underline{M}'_k:\gamma$. Since $0 \leq \ell \leq m-1$, if $y \in \underline{M}$ then $y, y', \ldots, y^{[\ell]}$ are continuous, so that $J[y]$ exists; moreover, in view of (2.2) we have

$$(2.3) \qquad J[y \geq -\int_a^b \kappa_{L_o}(t)dt.$$

For $y \in \underline{M}$ we shall denote by ϕ_y the $(\ell+1)$-dimensional vector function $\phi_y(t) = (y^{[\nu]}(t))$, $(\nu = 0,1,\ldots,\ell)$. Clearly ϕ_y is an a.c. vector function with derivative function

$\phi'_y(t) = (y^{[\nu+1]}(t))$, $(\nu = 0,1,\ldots,\ell)$; in particular, if $\ell = m-1$ then a.e. on $[a,b]$ we have $y^{[\ell+1]}(t) = y^{[m]}(t) = [1/a_m(t)]\gamma(t) - \Sigma_{\beta=0}^{m-1} [a_\beta(t)/a_m(t)]y^{[\beta]}(t)$.

If $y \in \underline{M}'_k : \gamma$, then on $[a,b]$ the function γ is Lipschitzian, and hence absolutely continuous. Therefore, it follows readily that $y \in \underline{M}'_k : \gamma$ if and only if $(\eta;\eta_{m+1}) = (y;\gamma)$ is a solution in the Carathéodory sense, (see, for example, [7; Ch. II]), of the differential system

$$L[\eta](t) - \eta_{m+1}(t) = 0, \quad \eta'_{m+1}(t) = -u(t),$$

with $u(t)$ a measurable function satisfying $0 \leq u(t) \leq k$, for $t \in [a,b]$.

3. Properties of $\underline{M}'_k$.

Prefatory to the proof of existence theorems for $J[y]$ on $\underline{M}'_k$ and $\underline{MB}_k$, certain basic properties of the class $\underline{M}'_k$ will be established.

LEMMA 3.1. *There exist distinct values t_α, $(\alpha = 1,\ldots,m)$, on $[a,b]$ such that the multipoint boundary problem*

$$L[y] = 0, \quad y(t_\alpha) = 0, \quad (\alpha = 1,\ldots,m), \tag{3.1}$$

has only the trivial solution $y(t) \equiv 0$.

Indeed, if the result of the lemma were not true, then for each δ satisfying $0 < \delta < (b-a)/(m-1)$ there would exist a non-trivial solution $y(t:\delta)$ of (3.1) with $t_\alpha = a + [\alpha-1]\delta$, $(\alpha = 1,\ldots,m)$, and chosen normed so that $\int_a^b y^2(t:\delta)dt = 1$. An elementary limit argument would then yield the existence of a solution y of $L[y] = 0$ such that $y^{[\alpha]}(a) = 0$, $(\alpha = 0, 1, \ldots, m-1)$, and the contradictory condition $\int_a^b y^2(t)dt = 1$.

From the fact that (3.1) has only the solution $y(t) \equiv 0$ it then follows that for each set of m values k_α, $(\alpha = 1, \ldots, m)$, there exists a unique solution of $L[y] = 0$ satisfying $y(t_\alpha) = k_\alpha$, and consequently for each $\beta = 1, \ldots, m$ there is a unique solution $y = u_\beta(t)$ of the system

(3.2) $$L[y] = 0, \quad y(t_\alpha) = \delta_{\alpha\beta}, \quad (\alpha, \beta = 1, \ldots, m).$$

In particular, $u_1(t), \ldots, u_m(t)$ is a basis for the linear vector space of solutions of $L[y] = 0$, and the solution of this equation satisfying $y(t_\alpha) = k_\alpha$, $(\alpha = 1, \ldots, m)$, is $y(t) = \Sigma_{\beta=1}^m k_\beta u_\beta(t)$.

The non-homogeneous system

(3.3) $$L[y] = \gamma(t)$$

is equivalent to the first order vector differential equation

(3.3′) $$\eta' + A(t)\eta = \psi(t)$$

in $\eta(t) = (y^{[\alpha-1]}(t))$, $(\alpha = 1, \ldots, m)$, where the $m \times m$ matrix function $A(t) = [A_{ij}(t)]$ is such that $A_{i,\, i+1}(t) = -1$, $(i = 1, \ldots, m-1)$, $A_{mj} = a_{j-1}(t)/a_m(t)$, $(j = 1, \ldots, m)$, and $\psi(t)$ is the m-dimensional vector function with $\psi_i(t) = 0$, $(i = 1, \ldots, m-1)$, and $\psi_m(t) = \gamma(t)/a_m(t)$. The corresponding homogeneous adjoint vector differential equation $-\zeta'(t) + A^*(t)\zeta(t) = 0$ in component form is

$$-\zeta_1' + [a_o(t)/a_m(t)]\zeta_m = 0,$$

(3.4)

$$-\zeta_{i+1}' - \zeta_i + [a_i(t)/a_m(t)]\zeta_m = 0, \quad (i = 1, \ldots, m-1),$$

and $(\zeta_\alpha(t))$, $(\alpha = 1, \ldots, m)$, is a solution of (3.4) if and only if $\zeta_i(t) = a_i(t)\hat{\zeta}(t) - \zeta_{i+1}'(t)$, $(i = 1, \ldots, m-1)$, where $\hat{\zeta} = \zeta_m/a_m$ is a solution on $L^*[\hat{\zeta}] = 0$, and $L^*[\hat{\zeta}] = a_o\hat{\zeta} - [a_1\hat{\zeta} - \{a_2\hat{\zeta} - \cdots\}']'$ is the formal adjoint of $L[y]$.

Now $U(t) = [u_\beta^{[\alpha-1]}(t)]$, $(\alpha, \beta = 1, \ldots, m)$, is a fundamental matrix solution of the homogeneous equation $y' + A(t)y = 0$. Let $V(t) \equiv [V_{\alpha\beta}(t)]$ be the $m \times m$ matrix function whose column vectors are independent solutions of the adjoint equation $-\zeta' + A^*(t)\zeta = 0$, and the constant $m \times m$ matrix function $V^*(t)U(t)$ is the identity matrix. From the variation of parameters formula it follows that $\eta(t)$ is a solution of (3.3') if and only if $\eta(t)$ is of the form

$$\eta(t) = U(t)\,[c + \int_a^t V^*(s)\psi(s)ds],$$

where c is a constant vector. As $u_\beta(t_\alpha) = \delta_{\alpha\beta}$, it then follows that $y(t)$ is a solution of (3.3) if and only if

(3.5) $$y(t) = \Sigma_{\beta=1}^m u_\beta(t)[y(t_\beta) + \int_{t_\beta}^t [V_{m\beta}(s)/a_m(s)]\gamma(s)ds],$$

and in this case for $\alpha = 1, \ldots, m$ we have

$$(3.5')\ y^{[\alpha-1]}(t) = \Sigma_{\beta-1}^{m} u_{\beta}^{[\alpha-1]}(t)[y(t_{\beta}) + \int_{t_{\beta}}^{t} [V_{m\beta}(s)/a_m(s)]\gamma(s)ds].$$

If $y(t)$ and $y^{\#}(t)$ denote the $(m+1)$-dimensional vector functions with $y_{\alpha}(t) = y_{\alpha}^{\#}(t) = y^{[\alpha-1]}(t)$, $(\alpha = 1, \ldots, m)$, and $y_{m+1}(t) = L[y](t)$, $y_{m+1}^{\#}(t) = y^{[m]}(t)$, then in view of hypothesis $\underline{H}_1$) there exists a constant $\varkappa > 0$ such that $|y^{\#}(t)| \leq \varkappa\ |y(t)|$ for $t \in [a, b]$. Consequently in view of (3.5') we have the following lemma.

LEMMA 3.2. *Suppose that* $\gamma \in \underline{L}[a, b]$ *and* y *is a solution of* (3.3). *Then there exist positive constants* $d_{o\alpha}, d_{1\alpha}$ *such that*

$$(3.6)\quad |y^{[\alpha-1]}(t)| \leq d_{o\alpha} \|y\|_{[a,b]} + d_{1\alpha} \int_a^b |\gamma(t)|dt,\ (\alpha=1, \ldots, m).$$

If $\gamma \in \underline{L}^{\infty}[a, b]$, *then there exist positive constants* d_o, d_1 *such that* $|y^{\#}(t)| \leq d_o \|y\|_{[a,b]} + d_1 \|\gamma\|_{\infty}$, *for* $t \in [a, b]$, *and consequently there exist positive constants* $d_1, d_o \geq 1/(b-a)$ *such that*

$$(3.6')\quad |\phi_y'(t)| \leq d_o' \|y\|_{[a,b]} + d_1 k, \text{ for } t \in [a, b], \text{ if } y \in \underline{MB}_k.$$

With the aid of this lemma, we may establish the following result.

THEOREM 3.1. If $Q > \int_a^b F(t, 0)dt$, *then the class* $\Gamma(\underline{MB}_k : Q) = \{y \mid y \in \underline{MB}_k,\ J[y] < Q\}$ *is non-empty, and if* $K > 0$ *is such that*

(3.7) $$K^2 - (2kd'_o + d_1 k)(Q + \int_a^b \kappa_{L_o}(t)dt) > 0,$$

then for $y \in \Gamma(\underline{MB}_k : Q)$ we have $\|\phi_y\|_{[a,b]} < 2L(K)$, where $L(K)$ is defined in $\underline{H}_2$-iv).

Since $y(t) \equiv 0$ belongs to $\underline{M}_k$ for arbitrary $k > 0$, clearly $\Gamma(\underline{MB}_k : Q)$ is non-empty when $Q > \int_a^b F(t, 0)dt$.

Now suppose that $K > 0$ satisfies (3.7), and that there exists a $y \in \Gamma(\underline{MB}_k : Q)$ with $\|\phi_y\|_{[a,b]} \geq 2L$, where for brevity we write merely L for $L(K)$. If $t_o \in [a, b]$ is such that $|\phi_y(t_o)| = \|\phi_y\|_{[a,b]}$, then in view of (3.6'), and the fact that $-|\phi'_y| \leq |\phi_y|' \leq |\phi'_y|$ a.e. on $[a, b]$, it follows that if $t \in [a, b]$ and $|t - t_o| \leq 1/d'_o$, then

$$|\phi_y(t)| \geq |\phi_y(t_o)| - (d'_o |\phi_y(t_o)| + d_1 k)|t-t_o|$$

$$> L, \text{ if } |t - t_o| < L/(2Ld'_o + d_1 k).$$

As $L/(2Ld'_o + d_1 k) < 1/(2d'_o) \leq (b-a)/2$, there exists a subinterval I of $[a, b]$ with t_o as an end-point, and of length $L/(2Ld'_o + d_1 k)$, such that $|\phi_y(t)| \geq L$ for $t \in I$. Hence $F(t, y(t)) \geq K$ for $t \in I$, and consequently

$$Q > J[y] \geq (KL)/(2L\alpha_o' + \alpha_1 k) - \int_a^b \varkappa_{L_o}(t)dt,$$

$$\geq K^2/(2K\alpha_o' + \alpha_1 k) - \int_a^b \varkappa_{L_o}(t)dt,$$

which implies for K an inequality contrary to (3.7).

For the consideration of the minimization of $J[y]$ in the class $\underline{M}_k'$, use is made of the following result analogous to that of Lemma 3.1.

LEMMA 3.3. *If t_α, $(\alpha = 1, \ldots, m)$, are values on $[a, b]$ satisfying the conditions of Lemma 3.1, then there exists a $t_{m+1} \in [a, b]$ distinct from $t_1, \ldots, t_m$ and such that the multipoint boundary problem*

$$\text{(3.8)} \quad \begin{aligned} &\text{a)} \quad L[y] - y_{m+1} = 0, \; y_{m+1}' = 0, \\ &\text{b)} \quad y(t_\sigma) = 0, \; (\sigma = 1, \ldots, m+1), \end{aligned}$$

has only the trivial solution $y(t) \equiv 0$, $y_{m+1} \equiv 0$ on $[a, b]$, and consequently there exists a set of linearly independent solutions $(y; y_{m+1}) = (\eta_\sigma; \rho_\sigma)$, $(\sigma = 1, \ldots, m+1)$ of (3.8a) satisfying

$$\text{(3.9)} \quad \eta_\sigma(t_\tau) = \delta_{\sigma\tau}, \quad (\sigma, \tau = 1, \ldots, m+1).$$

The general solution of (3.8a) is $y = \Sigma_{\sigma=1}^{m+1} c_\sigma u_\sigma(t)$, $y_{m+1} = c_{m+1}$, where $y(t) = u_\alpha(t)$, $(\alpha = 1, \ldots, m)$, is the solution of (3.2), and $y = u_{m+1}(t)$ is a solution of $L[y] = 1$. Then $y_{m+1}(t) = u_{m+1}(t) - \Sigma_{\alpha=1}^{m} u_{m+1}(t_\alpha)u_\alpha(t)$ is the unique solution of $L[y] = 1$ satisfying $y_{m+1}(t_\beta) = 0$, $(\beta = 1, \ldots, m)$. In particular, $y_{m+1}(t) \not\equiv 0$ on $[a, b]$, and there exists a value $t_{m+1} \neq t_\alpha$, $(\alpha = 1, \ldots, m)$, such that $c = y_{m+1}(t_{m+1}) \neq 0$. Consequently, (3.9) is satisfied by $(\eta_{m+1}(t); \rho_{m+1}(t)) = ([1/c]y_{m+1}(t); 1/c)$ and

$$(\eta_\alpha(t); \rho_\alpha(t)) = (u_\alpha(t) - u_\alpha(t_{m+1})\eta_{m+1}(t); -u_\alpha(t_{m+1})\rho_{m+1}(t)),$$

$$(\alpha = 1, \ldots, m).$$

For a given function g integrable on $[a, b]$ the non-homogeneous differential system

$$(3.10) \qquad L[y] - y_{m+1} = 0, \; y'_{m+1} = g,$$

may be written in terms of the (m+1)-dimensional vector function $y(t) = (y^{[\alpha-1]}(t); y_{m+1}(t))$, $(\alpha = 1, \ldots, m)$, as the vector equation

$$(3.11) \qquad y'(t) + B(t)y(t) = f(t),$$

where $B(t) = [B_{\sigma\tau}(t)]$ is the $(m+1) \times (m+1)$ matrix function with $B_{\alpha,\alpha+1}(t) = -1$, $\alpha = 1, \ldots, m-1$; $B_{m\beta}(t) = a_{\beta-1}(t)/a_m(t)$, $(\beta = 1, \ldots, m)$, $B_{m,m+1}(t) = -1/a_m(t)$, and $B_{\alpha\beta}(t) \equiv 0$

otherwise, while $f(t) = (f_\sigma(t))$ is the $(m+1)$-dimensional vector function with $f_\alpha(t) \equiv 0$, $(\alpha = 1, \ldots, m)$, and $f_{m+1}(t) = g(t)$. The $(m+1) \times (m+1)$ matrix function $Y(t) = [\eta_\tau^{[\alpha-1]}(t); \rho_\tau]$, $(\alpha = 1, \ldots, m;\ \tau = 1, \ldots, m+1)$, is a fundamental matrix solution of the homogeneous vector differential equation $y'(t) + B(t)y(t) = 0$. Also, the corresponding homogeneous adjoint vector differential equation $-\zeta'(t) + B^*(t)\zeta(t) = 0$ in component form is

$$-\zeta_1' + [a_0(t)/a_m(t)]\zeta_m = 0,$$

$$(3.12) \qquad -\zeta_{\alpha+1}' - \zeta_\alpha + [a_\alpha(t)/a_m(t)]\zeta_m = 0, \quad (\alpha = 1, \ldots, m-1),$$

$$-\zeta_{m+1}' - [1/a_m(t)]\zeta_m = 0,$$

and $(\zeta_\sigma(t))$, $(\sigma = 1, \ldots, m+1)$, is a solution of (3.12) if and if $\hat{\zeta} = \zeta_m/a_m$ satisfies the equation $\zeta_\alpha = a_\alpha\hat{\zeta} - \zeta_{\alpha+1}'$, $(\alpha = 1, \ldots, m-1)$, and

$$(3.13) \qquad L^*[\hat{\zeta}] = 0, \quad \zeta_{m+1}(t) = \zeta_{m+1}(a) - \int_a^t \hat{\zeta}(s)ds.$$

Now let $Z(t)$ be the $(m+1) \times (m+1)$ matrix function whose column vectors are solutions of (3.12), and such that the constant matrix $Z^*(t)Y(t)$ is the $(m+1)\times(m+1)$ identity matrix. Proceeding in a fashion analogous to that used above for the equation (3.3), we now obtain from the variation of parameter formula and Lemma 3.3 that for a given g integrable on $[a, b]$ the $(m+1)$-dimensional vector function $(y(t); y_{m+1}(t))$ is a solution of (3.10) if and only if

$$y(t) = \Sigma_{\tau=1}^{m+1} \eta_\tau(t)[y(t_\tau) + \int_{t_\tau}^{t} Z_{m+1,\tau}(s)g(s)ds],$$

(3.14)

$$L[y](t) \equiv y_{m+1}(t) = \Sigma_{\tau=1}^{m+1} \rho_\tau [y(t_\tau) + \int_{t_\tau}^{t} Z_{m+1,\tau}(s)g(s)ds],$$

and for such a solution

$$y^{[j]}(t) = \Sigma_{\tau=1}^{m+1} \eta_\tau^{[j]}(t)[y(t_\tau) + \int_{t_\tau}^{t} Z_{m+1,\tau}(s)g(s)ds], \tag{3.15}$$

$$(j = 0, 1, \ldots, m).$$

Also, corresponding to Lemma 3.2 and Theorem 3.1 we now have the following results.

LEMMA 3.4. _If_ $g \in L[a, b]$ _and_ $(y; y_{m+1})$ _is a solution of_ (3.10), _then there exist positive constant_ c_{oj}, c_{1j} _such_

$$|y^{[j]}(t)| \leq c_{oj} \|y\|_{[a, b]} + c_{1j} \int_a^b |g(t)| dt, \quad (j = 0, 1, \ldots, m).$$

_If_ $g \in L^\infty[a, b]$, _then there exist positive constants_ c_o, c_1 _such that_ $|y^{\#}(t)| \leq c_o \|y\|_{[a, b]} + c_1 \|g\|_\infty$ _for_ $t \in [a, b]$, _and consequently there exist positive constants_ $c_1, c'_o \geq 1/(b-a)$ _such that_ $|\phi'_y(t)| \leq c'_o \|y\|_{[a, b]} + c_1 k$, _for_ $t \in [a, b]$ if $y \in M'_k$.

THEOREM 3.2. *If* $Q > \int_a^b F(t, 0)dt$ *then the class* $\Gamma(\underline{M}'_k; Q) = \{y \mid y \in \underline{M}'_k, J[y] < Q\}$ *is non-empty, and if* $K > 0$ *is such that*

$$K^2 - (2Kc'_o + c_1 k)(Q + \int_a^b \varkappa_{L_o}(t)dt) > 0,$$

then for $y \in \Gamma(\underline{M}'_k; Q)$ *we have* $\|\phi_y\|_{[a,b]} < 2L(K)$, *where* $L(K)$ is defined in $\underline{H}_2$-iv).

4. *Existence theorems.* For $k > 0$, let λ_k denote the infimum of $J[y]$ on $\underline{M}'_k$, and μ_k the infimum of $J[y]$ on $\underline{MB}_k$; moreover, denote by λ the infimum of $J[y]$ on $\underline{M}$. In view of (2.3), we have that $\lambda \geq -\int_a^b \varkappa_{L_o}(t)dt$. If $k_2 > k_1 > 0$ then clearly $\lambda_{k_1} \geq \lambda_{k_2} \geq \lambda$ and $\mu_{k_1} \geq \mu_{k_2} \geq \lambda$. Also, if $y \in \underline{M}'_k : \gamma$ and $k' = |\gamma(b)| + k(b-a)$, then $|\gamma(t)| \leq k'$ for $t \in [a, b]$, so that $y \in \underline{MB}_{k'} : \gamma$, and hence $\lambda_k \geq \mu_{k'} \geq \lambda$. Moreover, if $\lambda_\infty = \lim_{k \to \infty} \lambda_k$ and $\mu_\infty = \lim_{k \to \infty} \mu_k$, then clearly $\lambda_\infty \geq \mu_\infty \geq \lambda$.

THEOREM 4.1. *For* $k > 0$ *there exists a* $\check{y}_k \in \underline{MB}_k$ *such that* $J[\check{y}_k] = \mu_k$, *and a* $\hat{y}_k \in \underline{M}'_k$ *such that* $J[\hat{y}_k] = \lambda_k$.

Suppose that $\{y_n\}$, $(n = 1, 2, \ldots,)$, is a minimizing sequence for $J[y]$ on $\underline{MB}_k$. For a given $Q > \int_a^b F(t, 0)dt$ we may suppose that $J[y_n] < Q$, $n = 1, 2, \ldots$, and by Theorem 3.1 the sequence $\{|\phi_{y_n}(t)|\}$ is uniformly bounded on $[a, b]$. If t_α, $(\alpha = 1, \ldots, m)$, are as in Lemma 3.1, then for each α the sequence $\{y_n(t_\alpha)\}$ is bounded. Consequently there exists a subsequence, which will still be denoted by $\{y_n\}$, such that each sequence $\{y_n(t_\alpha)\}$, $n = 1, 2, \ldots$, is convergent; for brevity, let $k_\alpha = \lim_{n \to \infty} y_n(t_\alpha)$. Now if $y_n \epsilon \underline{MB}_k : \gamma_n$, then $\{\gamma_n\}$ is a sequence of monotone functions satisfying $|\gamma_n(t)| \le k$ on $[a, b]$, and by the Helly-Bray theorem, [3; Theorem 33, p. 291], there exists a subsequence $\{\gamma_{n_j}(t)\}$ converging to a limit function $\check{\gamma}(t)$ on $[a, b]$. Clearly $\check{\gamma}$ is non-increasing and $|\check{\gamma}(t)| \le k$, so that by the bounded convergence theorem of Lebesgue the sequence $\{\int_a^b |\gamma_{n_j}(t) - \check{\gamma}(t)|dt\}$ converges to zero. If $\check{y}_k$ is defined as

$$\check{y}_k(t) = \Sigma_{\beta=1}^m u_\beta(t)[k_\beta + \int_{t_\beta}^t [V_{m\beta}(s)/a_m(s)]\check{\gamma}(s)ds], \quad t \epsilon [a, b],$$

then from the formulas (3.5), (3.5′) for $y = y_{n_j}$, $\gamma = \gamma_{n_j}$, it follows that $\check{y} \epsilon \underline{MB}_k : \check{\gamma}$, the sequence $\{\phi_{y_{n_j}}(t)\}$ converges to $\phi_{\check{y}_k}(t)$ uniformly on $[a, b]$, and consequently $\mu_k = \lim J[y_{n_j}] = J[\check{y}_k]$.

The proof of the existence of a $\hat{y}_k \in \underline{M}'_k$ such that $J[\hat{y}_k] = \lambda_k$ follows the same pattern. If $\{y_n\}$ is a minimizing sequence for $J[y]$ on $\underline{M}'_k$, and $y_n \in \underline{M}'_k: \gamma_n$, then the Arzelá-Ascoli theorem, [3; Theorem 28, p. 122], assures the existence of a subsequence $\{\gamma_{n_j}\}$ which converges uniformly on $[a, b]$ to a monotone non-increasing function γ which is Lipschitzian with constant k.

THEOREM 4.2. $\lim_{k \to \infty} \lambda_k = \lambda$, and $\lim_{k \to \infty} \mu_k = \lambda$.

As noted above, we have $\lambda_\infty \geq \mu_\infty \geq \lambda$. Also, in order to establish the result of the theorem it clearly suffices to show that if $y \in \underline{M}$, and $\varepsilon > 0$, then there exists a $k > 0$ and a $y_\varepsilon \in \underline{M}'_k$ such that $|J[y_\varepsilon] - J[y]| < \varepsilon$. In turn, this result is a ready consequence of hypotheses $\underline{H}_1$), $\underline{H}_2$), and the following auxiliary results:

(i) <u>if</u> $y \in \underline{M}$ <u>and</u> $\delta > 0$, <u>then there exists</u> <u>a</u> $y_\delta \in \underline{MB}$ <u>such that</u> $\|\phi_y - \phi_{y_\delta}\|_{[a, b]} < \delta$;

(ii) If $y \in \underline{MB}$ and $\delta > 0$, then there exists a $k > 0$ and a $y \in \underline{M}'_k$ such that $\|\phi_y - \phi_{y_\delta}\|_{[a, b]} < \delta$.

For $y \in \underline{M}: \gamma$ and N a positive integer, let γ_N be the truncated function satisfying $\gamma_N(t) = \gamma(t)$ if $|\gamma(t)| \leq N$, $\gamma_N(t) = N$ if $\gamma(t) > N$, and $\gamma_N(t) = -N$ if $\gamma(t) < -N$. Since $\gamma \in \underline{L}[a, b]$, it follows that $\int_a^b |\gamma(s) - \gamma_N(s)|\, ds \to 0$ as $N \to \infty$. Consequently, if $t_1, \ldots, t_m$ are as in Lemma 3.1, and y_N is defined by

$$y_N(t) = \Sigma_{\beta=1}^{m} u_\beta(t)\{y(t_\beta)+\int_{t_\beta}^{t} [V_{m\beta}(s)/a_m(s)]\gamma_N(s)ds\},$$

we have that $y_N \in \underline{MB}_N: \gamma_N$, with $y_N(t_\beta) = y(t_\beta)$, ($\beta = 1, \ldots, m$). Moreover, from relation (3.5′) for y and y_N it follows that there exists a constant C such that for $\alpha = 1, \ldots, m$ we have

$$|y^{[\alpha-1]}(t) - y_N^{[\alpha-1]}(t)| \leq C \int_a^b |\gamma(s)-\gamma_N(s)|ds, \quad t \in [a, b].$$

Consequently, $\|\phi_y - \phi_{y_N}\|_{[a,b]} \leq (h+1)C \int_a^b |\gamma(s) - \gamma_N(s)| ds$,

and $\|\phi_y - \phi_{y_N}\|_{[a,b]} \to 0$ as $N \to \infty$.

In order to establish (ii), for $y \in \underline{MB}_k: \gamma$, let the definition of γ be extended to $(-\infty, \infty)$ by setting $\gamma(t) = \gamma(a)$ for $t \leq a$, $\gamma(t) = \gamma(b)$ for $t \geq b$, and for $\sigma > 0$ let $\gamma_\sigma(t) = (1/2\sigma) \int_{-\sigma}^{\sigma} \gamma(t+s)ds$. Then $\gamma_\sigma(t)$ is monotone non-increasing, $|\gamma_\sigma(t)| \leq k$ on $[a, b]$, and $\int_a^b |\gamma(s)-\gamma_\sigma(s)| ds \to 0$ $\sigma \to 0$. Consequently, upon proceeding as in the proof of (i) we obtain a function y_σ such that $y_\sigma \in \underline{MB}_k: \gamma_\sigma$, with $y_\sigma(t_\beta) = y(t_\beta)$, ($\beta = 1, \ldots, m$), and a constant C

such that $|y^{[\alpha-1]}(t) - y_\sigma^{[\alpha-1]}(t)| \leq C \int_a^b |\gamma(s) - \gamma_\sigma(s)| ds$,

for $\alpha = 1, \ldots, m$, so that $\|\phi_y - \phi_{y_\sigma}\|_{[a,b]} \to 0$ as $\sigma \to 0$.

Moreover, a. e. on $[a, b]$ we have that $\gamma'_\sigma(t) =$ $[\gamma(t+\sigma) - \gamma(t-\sigma)]/(2\sigma)$ so that $y_\sigma \in M'_{k'} : \gamma_\sigma$ with $k' = k/\sigma$.

5. Characterization of a minimizing function for $J[y]$ on M'_k. In addition to hypotheses H_1) and H_2) we now assume the following condition.

H_3) On $[a, b] \times R^{\ell+1}$ the function F has partial derivatives $F_{r_\nu}(t, r)$, $(\nu = 0, 1, \ldots, \ell)$, and each of the functions F_{r_ν} satisfies the conditions of (i), (ii), (iii) of H_2).

THEOREM 5.1. If F satisfies hypotheses H_1), H_2), H_3), and $\hat{y} \in M'_k : \hat{\gamma}$ is a minimizing function for $J[y]$ on M'_k, then there exist functions ζ_σ, $(\sigma = 1, \ldots, m+1)$, which satisfy in the sense of Filippov [2; Sect. 1] with $y = \hat{y}$, $y_{m+1} = \hat{\gamma}$ the differential system

$$
\begin{aligned}
&L[y] - y_{m+1} = 0, \\
&y'_{m+1} = -G_k(t, \zeta_{m+1}), \\
(5.1)\qquad &-\zeta'_1 + [a_0(t)/a_m(t)]\zeta_m = \chi_1, \\
&-\zeta'_{\alpha+1} - \zeta_\alpha + [a_\alpha(t)/a_m(t)]\zeta_m = \chi_{\alpha+1}, \quad (\alpha = 1, \ldots, m-1), \\
&-\zeta'_{m+1} - [1/a_m(t)]\zeta_m = 0,
\end{aligned}
$$

(5.2) $$\zeta_\sigma(a) = 0 = \zeta_\sigma(b), \quad (\sigma = 1, \ldots, m+1),$$

where

(5.3) $$G_k(t, \zeta_{m+1}) = 0 \text{ if } \zeta_{m+1} > 0,$$

$$G_k(t, \zeta_{m+1}) = k \text{ if } \zeta_{m+1} < 0,$$

and

(5.4) $$\begin{aligned} \chi_{\nu+1}(t) &= -F_{r_\nu}(t, y(t)), \quad \nu = 0, 1, \ldots, \ell, \\ &= 0, \quad \ell + 1 \le \nu \le m-1. \end{aligned}$$

Let $\zeta = (\zeta_\sigma)$ denote the solution of the system of differential equations in (5.1) which satisfies the initial condition $\zeta_\sigma(b) = 0$, $(\sigma = 1, \ldots, m+1)$. If $\hat{y} \in \underline{M}'_k : \hat{\gamma}$ is a minimizing function for $J[y]$ on $\underline{M}'_k$, and $y \in \underline{M}'_k : \gamma$, let $\hat{u} = -\hat{\gamma}'$ and $u = -\gamma'$. An integration be parts, and use of the basic identity between a vector linear differential operator and its adjoint then yields the relation

$$J[\hat{y}] - J[y] = \int_a^b \Gamma_F(t, \hat{y}(t), y(t))dt$$

$$+ \int_a^b \zeta_{m+1}(t)[u(t) - \hat{u}(t)]\, dt$$

$$+ \Sigma_{\sigma=1}^{m+1} \zeta_\sigma(a)[\hat{y}_\sigma(a) - y_\sigma(a)],$$

where in the last sum $\hat{y}_i(a) = \hat{y}^{[i-1]}(a)$, $y_i(a) = y^{[i-1]}(a)$, $(i = 1, \ldots, m)$, and

(5.6) $\Gamma_F(t, \hat{r}, r) = F(t, r) - F(t, \hat{r}) -$

$$\Sigma_{\nu=0}^{\ell}[r_\nu - \hat{r}_\nu] F_{r_\nu}[t, \hat{r}],$$

By argument entirely analogous to that of [6; Sect. 4], it follows that upon considering $y \in \underline{M}'_k: \gamma$ with $u(t) = -\gamma'(t) = \hat{u}(t) = -\hat{\gamma}'(t)$, then for $(y_\sigma(a))$ near $(\hat{y}_\sigma(a))$ the last term of the right-hand member of (5.5) is the dominant one, and consequently we have also $\zeta_\sigma(a) = 0$, $(\sigma = 1, \ldots, m+1)$. Also, again by argument similar to that of [6; Sect. 4], we have that $\hat{u}(t) = k$ a.e. on the set where $\zeta_{m+1}(t) < 0$ and $\hat{u}(t) = 0$ a.e. on the set where $\zeta_{m+1}(t) > 0$. The latter conditions embody the extremum "bang-bang" principle for the solution of the minimum problem considered, and the conclusion that $\zeta_\sigma(a) = 0$, $(\sigma = 1, \ldots, m+1)$, is the transversality, or natural boundary, condition resulting from the fact that there is no restriction of the end-values of the admissible functions y. Under stronger hypotheses of continuity on F and F_{r_ν} the results of the above theorem are deducible from general results of control theory, (see, for example, [4; Ch. 7], or [5; Chs. 2.4]). From these results it follows immediately that $(y; y_{m+1})$ and (ζ_σ) form in the sense of Filippov a solution of the differential equations of (5.1) with boundary conditions (5.2).

Corresponding to Theorem 4 of [6], we also have the following result.

THEOREM 5.2. *Suppose that* F *satisfies hypotheses* H_1), H_2), H_3), *while* $(y; y_{m+1})$ *and* (ζ_σ) *is a solution of* (5.1), (5.2). *If* F *is also convex in* y *for* $t \in [a, b]$, *then* y *is a minimizing function for* J[y] *in the class* M_k'. *Moreover, if* F *is strictly convex in* y, *then the minimizing function for* J[y] *in the class* M_k' *is unique; therefore, under these hypotheses there is a unique solution of the differential system* (5.1) *satisfying the boundary conditions* (5.2).

6. *A class of special functions* F. In particular, suppose that $\ell = 0$, so that F is a function of t, r_o, and that for $j = 0, 1, \ldots, m$ the coefficient function $a_j(t)$ of L[y] is of class $\underline{C}^{[j]}$. In this case, $L^*[\hat{\zeta}] = \Sigma_{j=o}^{m} (-1)^j \{a_j(t)\hat{\zeta}\}^{[j]}$ may be differentiated term by term to obtain an expression of the form $\Sigma_{j=0}^{m} b_j(t)\hat{\zeta}^{[j]}$. Moreover, it follows readily that $(y; y_{m+1})$, (ζ_σ) is a solution of (5.1), (5.2) if and only if $y, \hat{\zeta} = \zeta_m/a_m, z = \zeta_{m+1}$ is a solution of the differential system

$$\Sigma_{j=o}^{m} \{a_j(t)y^{[j]}(t)\}' = - G_k(z(t)),$$

$$(6.1) \qquad \Sigma_{j=o}^{m} (-1)^j \{a_j(t)\hat{\zeta}(t)\}^{[j]} = -F_{r_o}(t, y(t)),$$

$$-z'(t) - \hat{\zeta}(t) = 0,$$

$$(6.2) \qquad z^{[j]}(a) = 0 = z^{[j]}(b), \quad (j = 0, 1, \ldots, m).$$

Indeed, if $L_1[y]$ is the formal differential operator $L_1[y] = \Sigma_{j=o}^{m} \{a_j(t)y^{[j]}\}'$, then the formal adjoint of $L_1[y]$ is $L_1^*[z] = \Sigma_{j=o}^{m} (-1)^j \{a_j(t)[-z']\}^{[j]}$, and the differential system (6.1), (6.2) is reduced to (6.1′), (6.2) with

$$(6.1') \qquad L_1[y] = -G_k(z), \quad L_1^*[z] = -F_{r_o}(t, y).$$

If $h : [a, b] \to R$ is a bounded measurable function then $F = \frac{1}{2}[r_o - h(t)]^2$ satisfies hypotheses $\underline{H}_1$), $\underline{H}_2$), $\underline{H}_3$), and is strongly convex in r_o for fixed t. In this case system (6.1′) becomes

$$L_1[y] = -G_k(z), \quad L_1^*[z] + y = h.$$

Also, if $h \in \underline{C}^{[m+1]}$ then this system is equivalent to

$$L_1 L_1^*[z] - G_k[z] = L_1[h], \quad y = h - L_1^*[z].$$

Finally, if $L[y] = y^{[m]}$ and $h \in \underline{C}^{[m+1]}$, then $L_1[y] = y^{[m+1]}$, $L_1^*[z] = (-1)^{m+1} z^{[m+1]}$, and the boundary problem determining the solution to the problem of minimizing $J[y]$ on $\underline{M}'_k$ is $(-1)^{m+1} z^{[2m+2]} - G_k[z] = h^{[m+1]}$, with the boundary conditions (6.2).

REFERENCES

[1]. Barlow, R. E., Bartholomew, D. J., Bremner, J. M., Brunk, H. D., Statistical Inference under Order Restrictions, John Wiley and Sons, New York 1972.

[2]. Filippov, A. F., Differential equations with discontinuous righthand side, American Mathematics Society Translations, Series 2, Vol. 42, pp. 199-231.

[3] Graves, L. M., The Theory of Functions of Real Variables, Second Edition, McGraw-Hill, New York, 1956.

[4] Hestenes, M. R., Calculus of Variations and Optimal Control Theory, John Wiley and Sons, New York, 1966.

[5] Pontryagin, L. S., Boltyanskii, V. G., Gamkrelidze, R. V., and Mischenko, E. F., The Mathematical Theory of Optimal Processes, Wiley-Interscience, New York, 1962.

[6] Reid, W. T., A simple optimal control problem involving approximation by monotone functions, Journal of Optimization Theory and Applications, Vol. 2(1968), pp. 365-377.

[7] Reid, W. T., Ordinary Differential Equations, John Wiley and Sons, New York, 1971.

University of Oklahoma, Norman, Oklahoma

A SINGULAR POINT PROBLEM ARISING IN CONTROL THEORY*

David L. Russell**

1. L^2-Optimal Control. We shall consider a linear, constant coefficient control system

(1.1) $\dot{x} = Ax + Bu, \quad x \in E^n, \; u \in E^m,$

A and B being $n \times n$ and $n \times m$ matrices, respectively, which satisfy the controllability condition

(1.2) $\operatorname{rank}\,[B, AB, \ldots, A^{n-1}B] = n.$

The basic control problem is the following: given a positive real number T and points $x_0, x_1 \in E^n$, we wish to find $u \in L^2[0, T; E^m]$ such that the solution of (1.1) which satisfies

$$x(0) = x_0$$

will at the end of the time interval [0, T] satisfy

$$x(T) = x_1.$$

In general it can be seen [1] that there are infinitely many

* Supported in part by the Office of Naval Research under contract NR-041-404.

** Department of Mathematics, University of Wisconsin, Madison, Wisconsin 53706. Research Consultant, Honeywell, Inc., Minneapolis, Minn.

such control functions u, forming an affine subspace $U(x_0, x_1) \subseteq L^2[0, T; E^m]$. It is easy to see that $U(x_0, x_1)$ is closed. If $U(x_0, x_1)$ is non-empty, which we will shortly demonstrate, it contains an element $\hat{u}$ of least $L^2[0, T; E^m]$ norm. The following proposition provides a complete characterization of this control function $\hat{u}$. A proof appears in [2] but we shall repeat it here to make our presentation self-contained.

Proposition 1. *The controllability condition (1.2) is equivalent to the condition that*

$$Z(T) = \int_0^T e^{A(T-t)} BB^* e^{A^*(T-t)} dt$$

should be positive definite for every $T > 0$. Assuming these equivalent conditions to be met, the closed subspace $U(x_0, x_1)$ is non-empty and the element $\hat{u}$ of least norm in $U(x_0, x_1)$ has the form

$$u(t) = B^* e^{A^*(T-t)} \xi, \tag{1.3}$$

where $\xi \in E^n$ is given by

$$\xi = Z(T)^{-1} (x_1 - e^{AT} x_0). \tag{1.4}$$

Proof. Clearly $Z(T)$ is hermitian and non-negative. Now for $x \in E^n$

$$x^* Z(T) x = \int_0^T \| B^* e^{A^*(T-t)} x \|^2 dt$$

and hence $= 0$ if and only if

(1.5) $$B^*e^{A^*(T-t)}x \equiv 0, \quad t \in [0, T].$$

Repeated differentiation of (1.5) shows that it can be valid for some non-zero $x \in E^n$ if and only if

$$\text{rank} \begin{bmatrix} B^* \\ B^*A^* \\ \vdots \\ B^*(A^*)^{n-1} \end{bmatrix} < n.$$

Hence we conclude that

$$x^*Z(T)x > 0, \quad x \neq 0$$

if and only if (1.2) is valid.

A control u steers x_0 to x_1 during the time interval $[0, T]$ if and only if

(1.6) $$x_1 = e^{AT}x_0 + \int_0^T e^{A(T-t)}Bu(t)\,dt.$$

Substituting (1.3) into (1.6) we have

$$x_1 - e^{AT}x_0 = Z(T)\xi$$

and if we determine ξ via (1.4) our equation is satisfied and $\hat{u}$, given by (1.3) and (1.4) is a solution of the control problem.

If u is an arbitrary element of $U(x_0, x_1)$ then

$$u(t) = \hat{u}(t) + \tilde{u}(t)$$

where $\tilde{u}$ steers 0 to 0 during the interval $[0, T]$. A necessary and sufficient condition for $\bar{u}$ to be of minimal

$L^2[0, T; E^m]$ norm is

$$(\hat{u}, \tilde{u})_{L^2[0, T; E^m]} = 0 \tag{1.7}$$

for all such $\tilde{u}$. If $\tilde{u}$ steers 0 to 0 during $[0, T]$ then

$$\int_0^T e^{A(T-t)} B \tilde{u}(t)\, dt = 0$$

whence

$$(\xi, \int_0^T e^{A(T-t)} B \tilde{u}(t)\, dt)_{E^n}$$

$$= \int_0^T (B^* e^{A^*(T-t)} \xi, \tilde{u}(t))_{E^m} dt = (\hat{u}, \tilde{u})_{L^2[0, T; E^m]} = 0$$

and thus $\hat{u}$, as given by (1.3), (1.4), satisfies (1.7) and is therefore of minimal norm in $L^2[0, T; E^m]$, which completes the proof.

2. Representation of $\hat{u}$ in Time-Varying Feedback Form

Having obtained an explicit solution of our L^2-optimal control problem, we proceed now to an alternate representation of the control function $\hat{u}$. It is the search for such a representation which will lead us to a certain singular point problem in linear ordinary differential equations. In this study it will be convenient-and sufficient- to consider the case wherein the final state $x_1 = 0$. That this is sufficient is demonstrated by the variable change

$$x_1 - e^{AT} x_0 = -e^{AT} y_0 \text{ or } y_0 = x_0 - e^{-AT} x_1$$

which shows that if u steers y_0 to zero, i.e.

$$0 = e^{AT} y_0 + \int_0^T e^{A(T-s)} Bu(s)ds$$

then it also steers x_0 to x_1, i.e.

$$x_1 = e^{AT} x_0 + \int_0^T e^{A(T-s)} Bu(s)ds.$$

We shall continue to use the variable x and assume $x_1 = 0$ in the rest of this section.

From (1.3), (1.4) the L^2-optimal control steering x_0 to 0 during $[0, T]$ is given by

$$\hat{u}(t) = -B^* e^{A^*(T-t)} Z(\Gamma)^{-1} e^{AT} x_0 \tag{2.1}$$

and the corresponding response $x(t)$ therefore satisfies

$$\dot{x}(t) = Ax(t) - BB^* e^{A^*(T-t)} Z(T)^{-1} e^{AT} x_0. \tag{2.2}$$

The variation of parameters formula gives

$$x(t) = e^{AT} x_0 - \int_0^t e^{A(t-s)} BB^* e^{A^*(T-s)} Z(T)^{-1} e^{AT} ds\, x_0$$

$$= e^{At}[I - (\int_0^T e^{-As} BB^* e^{-A^*s} ds) e^{A^*T} Z(T)^{-1} e^{AT}]\, x_0$$

$$= e^{At}[\int_t^T e^{-As} BB^* e^{-A^*s} ds] e^{A^*T} Z(T)^{-1} e^{AT} x_0$$

$$\equiv \Phi(t)\, x_o. \tag{2.3}$$

The foregoing calculations show that $\Phi(t)$ is invertible for $0 \leq t < T$ but reduces to the zero matrix for $t = T$, as we would anticipate since $x(T) = 0$ for every initial state x_0.

The control $\hat{u}$, given by (2.1) can now be written in time varying feedback form (recall we are taking $x_1 = 0$)

$$\hat{u}(t) = -B^* e^{A^*(T-t)} Z(T)^{-1} e^{AT} \Phi(t)^{-1} x(t) \tag{2.4}$$

and the equation (2.2) for $x(t)$ becomes

$$\dot{x}(t) = [A - BB^* e^{A^*(T-t)} Z(T)^{-1} e^{AT} \Phi(t)^{-1}] x(t). \tag{2.5}$$

Since $x(t) = \Phi(t) x_o$, $\Phi(t)$ is the fundamental matrix solution of (2.5), i.e.

$$\dot{\Phi}(t) = [A - BB^* e^{A^*(T-t)} Z(T)^{-1} e^{AT} \Phi(t)^{-1}] \Phi(t). \tag{2.6}$$

Now, from (2.3),

$$\Phi(t)^{-1} = e^{-AT} Z(T) e^{-A^*T} [\int_t^T e^{-As} BB^* e^{-A^*s}]^{-1} e^{-At}$$

so (2.6) can be rewritten as

$$\dot{\Phi}(t) = [A - BB^* e^{-A^*t} (\int_t^T e^{-As} BB^* e^{-A^*s})^{-1} e^{-At}] \Phi(t)$$

$$= [A - BB^* (\int_0^{T-t} e^{-As} BB^* e^{-A^*s} ds)^{-1}] \Phi(t). \tag{2.7}$$

We can also rewrite $\Phi(t)$ in the more convenient form (cf. (2.3))

$$\Phi(t) = (\int_0^{T-t} e^{-As} BB^* e^{-A^*s} ds) e^{A^*(T-t)} Z(T)^{-1} e^{AT}$$

$$= e^{-A(T-t)} (\int_0^{T-t} e^{A(T-t-s)} BB^* e^{A^*(T-t-s)} ds) Z(T)^{-1} e^{AT}$$

$$= e^{-A(T-t)} Z(T-t) Z(T)^{-1} e^{AT} \tag{2.8}$$

This form for $\Phi(t)$ enables us to describe rather easily the flow in R^n which corresponds to solutions $\Phi(t)x_0$ of (25). Let $R(t)$ denote the set

$$(2.9)\quad R(t) = \{x \in R^n \mid x = e^{-A(T-t)}Z(T-t)y,\ y \in R^n,\ \|y\| = 1\}.$$

Then

$$\Phi(t)R(0) =$$

$$\{x \in R^n \mid x = e^{-A(T-t)}Z(T-t)Z(T)^{-1}e^{AT}w,\ w \in R(0)\}$$

$$= \{x \in R^n \mid x = e^{-A(T-t)}Z(T-t)Z(T)^{-1}e^{AT}e^{-AT}Z(T)y,\ y \in R^n,\ \|y\| = 1\}$$

$$= \{x \in R^n \mid x = e^{-A(T-t)}Z(T-t)y,\ y \in R^n,\ \|y\| = 1\} = R(t),$$

which means that the flow associated with (2.5) carries $R(0)$ into $R(t)$ for each $t \in [0, T]$.

We shall undertake in this paper a study of the asymptotic behavior of $\Phi(t)$ as $t \to T$ and thus provide information on the manner in which the L^2-optimal controls operate. A further by-product of such a study will be explained after we state

Proposition 2. For each $t \in [0, T)$ the matrix

$$A_1(T-t) \equiv A - BB^*\left(\int_0^{T-t} e^{-As}BB^*e^{-A^*s}\,ds\right)^{-1}$$

is Hurwitzian, i.e. its eigenvalues all have negative real parts.

Proofs of this proposition were given, independently, by Lukes [2] and Kleinman [3]. It is shown that with

$$(2.10)\quad Q(T-t) \equiv \int_0^{T-t} e^{-As}BB^*e^{-A^*s}\,ds$$

(so that $A_1(T-t) = A-BB^* Q(T-t)^{-1}$) we have

$$A_1(T-t)Q(T-t)+Q(T-t)A_1(T-t)^*+BB^*+e^{-A(T-t)}BB^*e^{-A^*(T-t)}=0$$

and then an application of the invariance principle [4] shows that if

(2.11) $$\frac{dy}{ds} = A_1(T-t)y$$

we have

$$\lim_{s \Rightarrow \infty} y(s)^*Q(T-t)y(s) = 0.$$

Since $Q(T-t)$ is positive definite for $t \in [0, t)$ the result follows.

This proposition shows that the control system (1.1) is stabilized by each of the linear feedback control laws

(2.12) $$u(s) = - B^*Q(T-t)^{-1} x(s).$$

We have, then, as Kleinman says in the title of his paper, "an easy way to stabilize a linear constant system". It is certainly conceptually easier than the usual present day method [5] which involves the solution of a quadratic matrix equation. It does, however, have some serious shortcomings.

Since $A_1(T-t)$ is just one value of the coefficient matrix occurring in the time varying system (2.7) we may hope, in undertaking an asymptotic study of (2.7) to be able to obtain asymptotic expressions for the eigenvalues and eigenvectors of $A_1(T-t)$ as $t \Rightarrow T$. This should shed some light on what values of t should be used when (2.12) is employed in a given application.

3. Inversion of $Q(T-t)$ in the Case $m = 1$.

In order to be able to give a reasonably specific description of the coefficient matrix

$$A_1(T-t) = A - BB^*Q(T-t)^{-1}$$

of the system (2.11), where $Q(T-t)$ is given by (2.10), we need to be able to describe $Q(T-t)^{-1}$. We shall do that here in the case $m = 1$ so that

$$B = b, \; b \in R^n.$$

Since (1.2) gives

$$\text{rank } [b, Ab, \ldots, A^{n-1}b] = n$$

there is a nonsingular $n \times n$ matric P (see [1], e.g.) such that

$$(3.1) \quad P^{-1}b = e_n = \begin{pmatrix} 0 \\ \vdots \\ 0 \\ 1 \end{pmatrix},$$

$$(3.2) \qquad P^{-1}AP = \begin{pmatrix} a_{n-1} & 1 & \cdots & 0 \\ \vdots & \vdots & \ddots & \vdots \\ a_1 & 0 & \cdots & 1 \\ a_0 & 0 & \cdots & 0 \end{pmatrix}.$$

Let us assume this transformation to be carried out beforehand so that A has the form appearing on the right hand side of (3.2) and b has the form (3.1). Then

$$e^{-As}b = e_n - e_{n-1}s + e_{n-2}\frac{s^2}{2} + \cdots + (-1)^{n-1}e_1\frac{s^{n-1}}{(n-1)!} + \mathcal{O}(s^n)$$

and thus

$$e^{-As}bb^*e^{-A^*s} = \begin{pmatrix} (-1)^{n-1}\dfrac{S^{n-1}}{(n-1)!} + \mathcal{O}(s^n) \\ (-1)^{n-2}\dfrac{S^{n-2}}{(n-2)!} + \mathcal{O}(s^n) \\ \vdots \\ -s \quad + \mathcal{O}(s^n) \\ 1 \quad + \mathcal{O}(s^n) \end{pmatrix} (\text{''})^* \tag{3.3}$$

where $(\text{''})^* = b^*e^{-A^*s}$ is the row vector which is the transpose of the column vector shown in (3.3). Thus

$$e^{-As}bb^*e^{-A^*s} = (C^i_j(s)) = \frac{(-s)^{2n-(i+j)}}{(n-i)!(n-j)!} + \mathcal{O}(s^{2n-\max\{i,j\}}),$$

and, integrating, we have

$$Q(T-t) = q^i_j(T-t) = \left(\frac{(-1)^{2n-(i+j)}(T-t)^{2n+1-(i+j)}}{(2n+1-(i+j))(n-i)!(n-j)!}\right)$$

$$+ \mathcal{O}((T-t)^{2n+1-\max\{i,j\}}) \quad . \tag{3.4}$$

We observe then that $Q(T-t)$ can be factored in the form

$$Q(T-t) = (T-t)^{2n-1} D(T-t) \; H(T-t) \; D(T-t),$$

where

$$D(T-t) = \operatorname{diag}\left(\frac{(-1)^{n-1}}{(n-1!}\; \frac{(-1)^{n-2}(T-t)^{-1}}{(n-2)!} \;\ldots,\; (T-t)^{-n+1}\right).$$

and

$H(T-t) = H_0 + \mathcal{O}(T-t),$

H_0 being the Hilbert matrix

$$H_0 = (h_0)^i_j = \frac{1}{2n+1-(i+j)}$$

$$\begin{pmatrix} \frac{1}{2n-1} & \cdots & \frac{1}{n+2} & \frac{1}{n+1} & \frac{1}{n} \\ \vdots & & \vdots & \vdots & \vdots \\ \frac{1}{n+2} & \cdots & \frac{1}{5} & \frac{1}{4} & \frac{1}{3} \\ \frac{1}{n+1} & \cdots & \frac{1}{4} & \frac{1}{3} & \frac{1}{2} \\ \frac{1}{n} & \cdots & \frac{1}{3} & \frac{1}{2} & 1 \end{pmatrix}$$

Now an explicit inverse for the Hilbert matrix H_0 is known [6] and it is

$$G = (g^i_j) = \left(\frac{(-1)^{(2n-(i+j))}(2n-i)!(2n-j)!}{(2n-(i+j)+1)\,[(n-i)!]^2\,(i-1)!(j-1)!} \right)$$

Then

$$H(T-t) = G + \mathcal{O}(T-t)$$

and

$$Q(T-t)^{-1} = (T-t)^{-2n+1} D(T-t)^{-1} [G + \mathcal{O}(T-t)]\, D(T-t)^{-1}$$

and

$$(3.5)\qquad Q(T-t)^{-1} = (r^i_j) = (-1)^{2n-(i+j)} (T-t)^{-(2n+1)-(i+j))} (n-i)!(n-j)!\, g^i_j + \mathcal{O}((T-t)^{-(2n-(i+j))}).$$

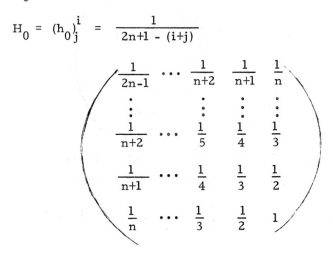

4. Behavior of $A_1(T-t)$ for $t \to T$ in the Case $m = 1$.

If we assume that A and b have the form indicated in (3.1) and (3.2) then the only non-zero entry of bb^* is a "1" in the last row and last column. Thus the only non-zero entries of $bb^*Q(T-t)^{-1}$ lie in the last row and that row consists of the number (cf. (3.5)

$$(T-t)^{-(n+1-j)}(-1)^{n-j}(n-j)!\, g_j^n + \mathcal{O}((T-t)^{n-j})$$

$$= \frac{(T-t)^{-(n+1-j)} n!(2n-j)!}{(n+1-j)\,(n-j)!(j-1)!(n-1)!} + \mathcal{O}((T-t)^{-(n-j)})$$

(4.1)

$$= (T-t)^{-(n+1-j)}\, n \frac{(2n-j)!}{(n-j+1)!(j-1)!} + \mathcal{O}((T-t)^{-(n-j)}) \quad .$$

The matrix

$$A_1(T-t) = A - bb^* Q(T-t)^{-1}, \tag{4.2}$$

still assuming that A has the form (3.2), now looks like this:

$A_1(T-t) =$

$$
\begin{bmatrix}
a_{n-1} & 1 & \cdots & 0 & 0 \\
\vdots & \vdots & & \vdots & \vdots \\
a_1 & 0 & \cdots & 0 & 1 \\
\begin{array}{c} -(T-t)^{-n} n \dfrac{(2n-1)!}{n!} \\ + \mathcal{O}(T-t)^{-(n-1)} \end{array} & \begin{array}{c} -(T-t)^{-(n-1)} n \dfrac{(2n-2)!}{(n-1)!\,1!} \\ + \mathcal{O}(T-t)^{-(n-2)} \end{array} & \cdots & \begin{array}{c} -(T-t)^{-2} n \dfrac{(n+1)!}{2!(n-2)!} \\ + \mathcal{O}(T-t)^{-1} \end{array} & \begin{array}{c} -(T-t)^{-1} n^2 \\ + \mathcal{O}(1) \end{array}
\end{bmatrix}
\tag{4.3}
$$

+2o

The equation satisfied by x(t), viz.

$$\dot{x}(t) = A_1(T-t)\, x(t) \qquad \text{(cf. (4.2), (3.5))}$$

is easily seen to have a regular singular point at $t = T$ if we employ the transformation

$$X(t) = \operatorname{diag}(1, (T-t)^{-1}, \ldots, (T-t)^{-n+1})\, y(t).$$

The same transformation, with t fixed, enables one to see that the characteristic equation of the coefficient matrix in the system (2.11) is

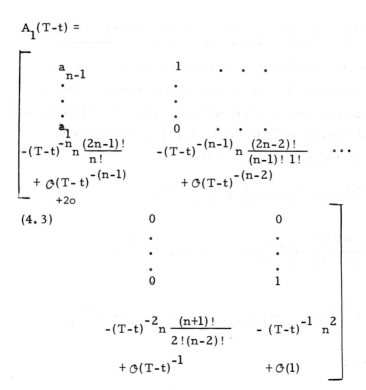

$$\lambda^n + [(T-t)^{-1} n^2 + \mathcal{O}(1)]\lambda^{n-1} + [(T-t)^{-2} n \frac{(n+1)!}{(n-2)!2!}$$

$$+ \mathcal{O}(T-t)^{-1}]\lambda^{n-2} + \cdots + [(T-t)^{-(n-1)} n \frac{(2n-2)!}{(n-1)!1!}$$

$$+ \mathcal{O}(T-t)^{-(n-2)}]\lambda + [(T-t)^{-n} \frac{(2n-1)!}{n!} + \mathcal{O}(T-t)^{-(n-1)}] = 0$$

The roots of this equation have the form

$$\lambda_k(T-t) = (T-t)^{-1}\lambda_k + \mathcal{O}(1), \quad t \Rightarrow T, \ k = 1, 2, \ldots, n$$

where $\lambda_1, \lambda_2, \ldots, \lambda_n$ are the roots of

$$\lambda^n + n^2\lambda^{n-1} + n\frac{(n+1)!}{(n-2)!2!}\lambda^{n-2} + \cdots + n\frac{(2n-2)!}{(n-1)!1!}\lambda$$

(4.3)

$$+ n\frac{(2n-1)!}{(n!)} = \lambda^n + \sum_{k=1}^{n} n\frac{(n+k-1)!}{(n-k)!\,k!}\lambda^{n-k} = 0$$

The following is a short table of values of λ_k, $k = 1, 2, \ldots, n$, for small values of n

Table 1

	λ_k, $k = 1, \ldots, n$
1	-1
2	$-2 \pm 2i$
3	$-2.681 \pm 3.050i$, -3.638
4	$-3.213 \pm 4.773i$, $-4.787 \pm 1.567i$
5	$-3.656 \pm 6.544i$, $-5.701 \pm 3.210i$, -6.287
6	$-4.039 \pm 8.346i$, $-6.471 \pm 4.900i$, $-7.491 \pm 1.622i$
7	$-4.379 \pm 10.17i$ $-7.141 \pm 6.623i$, $-8.512 \pm 3.281i$, -8.937
8	$-4.685 \pm 12.01i$, $-7.739 \pm 8.371i$, -9.406 ± 4.969, $-10.17 \pm 1.649i$

9 $-4.966 \pm 13.86i$, $-8.280 \pm 10.14i$, $-10.21 \pm 6.680i$, $-11.25 \pm 3.321i$, -11.59

10 $-5.225 \pm 15.73i$, $-8.776 \pm 11.92i$, $-10.93 \pm 8.409i$, $-12.23 \pm 5.013i$, $-12.84 \pm 1.666i$

It would be interesting, and perhaps challenging, to divine what pattern is emerging in this table and to give asymptotic formulae for, say, the N roots of smallest absolute real part, as $n \to \infty$

5. Behavior of Solutions of the Diffential Equations (2.6), (2.7)) for $t \to T$.

In this section we study the behavior near $t = T$ of solutions $x(t)$ of (2.6) which, in view of (2.7), (3.1), (3.2), we may rewrite as

$$(5.1)\qquad \begin{aligned} \dot{x}(t) &= (A - bb^* Q(T-t)^{-1})x(t) \\ &\equiv A_1(T-t)\, x(t) \quad \text{(cf. (2.10), (4.2))} \end{aligned}$$

Using the expression (4.3) for $A_1(T-t)$ this equation is seen to have a regular singular point at $t = T$. With the variable transformations

$$(5.2)\qquad \tau = T-t$$

$$(5.3)\qquad x(T-\tau) = \operatorname{diag}(\tau^{n-1}, \tau^{n-2}, \ldots, \tau, 1)y(\tau)$$

we can transform (5.1) into an appropriate canonical form for study of the behavior of solutions at the regular singular point $\tau = T-t = 0$, namely:

$$\tau \frac{dy}{d\tau} = \left[\begin{array}{ccccc} -(n-1) & -1 & \cdots & 0 & 0 \\ 0 & -(n-2) & \cdots & 0 & 0 \\ \vdots & \vdots & & \vdots & \vdots \\ 0 & 0 & \cdots & -1 & -1 \\ n\frac{(2n-1)!}{n!} & n\frac{(2n-2)!}{(n-1)!} & \cdots & n\frac{(n+1)!}{(n-2)!2!} & n^2 \end{array} \right] y$$

$$+ \mathcal{O}(\tau))y. \tag{5.4}$$

The indicial equation (cf. [7], [8]) is thus

$$\det \begin{pmatrix} \lambda+n-1 & 1 & & 0 & 0 \\ 0 & \lambda+n-2 & & 0 & 0 \\ \vdots & & & & \\ 0 & & & \lambda+1 & 1 \\ -n\frac{(2n-1)!}{n!} & -n\frac{(2n-2)!}{(n-1)!} & \cdots & -n\frac{(n+1)!}{(n-2)!2!} & \lambda-n^2 \end{pmatrix}$$

$$= \sum_{k=0}^{n} (-1)^{n-k} \, n\frac{(n+k-1)!}{(n-k)!k!} \prod_{\ell=1}^{n-k} (\lambda+n-\ell) = 0. \tag{5.5}$$

Inspection of several cases conviences one that the roots of this indicial equation are exactly the integers 1 through n. This conjecture is corroborated by observing that the trace of the matrix in (5.4) is

$$n^2 - \sum_{k=1}^{n-1} k = n^2 - \frac{(n-1)n}{2} = \frac{n(n+1)}{2} = \sum_{k=1}^{n} k$$

and by the observation that $\lambda = n$ is always a solution of (5.5)

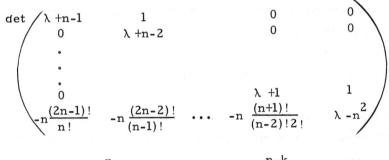

Indeed, with $\lambda = n$ (5.5) becomes

$$\sum_{k=0}^{n} (-1)^{n-k} \; n \; \frac{(n+k-1)!}{(n-k)!k!} \; \frac{(2n-1)!}{(n+k-1)!}$$

$$= \frac{n}{n!} (2n-1)! \sum_{k=0}^{n} (-1)^{n-k} \binom{n}{k} = \frac{n}{n!} (2n-1)! \, (1-1)^n = 0.$$

It appears to be a rather tedious combinatorial problem, however, to show directly that (5.5) also has the roots n-1, n-2, . . . , 2, 1.

In this particular case we are not wholly dependent upon the indicial equation in order to study the solutions. For (2.3), (2.8) give

$$\Phi(t) = e^{-A(T-t)} Z(T-t) \, Z(T)^{-1} e^{AT}$$

as a fundamental solution of (2.6), (2.7) and therefore, multiplying on the right by $Z(T)e^{-AT}$ and using (5.2), we see that

$$\psi(\tau) = e^{-A\tau} Z(\tau)$$

is a fundamental solution of

$$\frac{dx}{d\tau} = -\left(A - bb^* Q(\tau)^{-1}\right) x,$$

obtained from (5.1) via (5.2). Then

$$(5.6) \quad \theta(\tau) = \mathrm{diag}(\tau^{-(n-1)}, \tau^{-(n-2)}, \ldots, \tau^{-1}, 1) e^{-A\tau} Z(\tau)$$

is a fundamental solution matrix for (5.4). In general (5.6) is also rather difficult to analyze. Fortunately we do not need to consider the general case. Reflection upon the work already carried out shows that the coefficients $a_0, a_1, \ldots, a_{n-1}$, appearing in the assumed form (3.2) of the matrix A, have no influence on the lead matrix displayed in (5.4). These coefficients only affect the terms subsumed under the expression $\mathcal{O}(\tau)$ there.

Let us, then, consider the case wherein $a_0 = a_1 = \cdots = a_{n-1} = 0$ so that (cf. (3.2)) A has the form

$$(5.7) \qquad \begin{pmatrix} 0 & 1 & 0 & \ldots & 0 & 0 \\ 0 & 0 & 1 & \ldots & 0 & 0 \\ \vdots & \vdots & \vdots & & \vdots & \vdots \\ 0 & 0 & 0 & \ldots & 0 & 1 \\ 0 & 0 & 0 & \ldots & 0 & 0 \end{pmatrix}$$

Then, with $Z(\tau)$ defined as in Proposition 1, Section 1,

$$(5.8) \qquad \begin{aligned} e^{-A\tau} Z(\tau) &= e^{-A\tau} \int_0^\tau e^{A(\tau-t)} bb^* e^{A^*(\tau-t)} dt \\ &= \left(\int_0^\tau e^{-At} bb^* e^{-A^*t} dt\right) e^{A^*\tau} = Q(\tau) e^{A^*\tau} . \end{aligned}$$

When A has the form (5.7), $Q(\tau)$ is given by (3.4) with the terms $\mathcal{O}(\tau^{2n+1-\max\{i,j\}})$ deleted and

$$e^{A^*\tau} = \begin{pmatrix} 1 & 0 & 0 & \cdots & 0 & 0 \\ \tau & 1 & 0 & \cdots & 0 & 0 \\ \frac{\tau^2}{2} & \tau & 1 & \cdots & 0 & 0 \\ \vdots & \vdots & \vdots & & \vdots & \vdots \\ \frac{\tau^{n-2}}{(n-2)!} & \frac{\tau^{n-3}}{(n-3)!} & \frac{\tau^{n-4}}{(n-4)!} & \cdots & 1 & 0 \\ \frac{\tau^{n-1}}{(n-1)!} & \frac{\tau^{n-2}}{(n-2)!} & \frac{\tau^{n-3}}{(n-3)!} & \cdots & \tau & 1 \end{pmatrix}.$$

Then the matrix $\theta(\tau)$ in (5.6) can be written in the form

$$\theta(\tau) = \text{diag}\,(\tau^{-(n-1)}, \tau^{-(n-2)}, \ldots, \tau^{-1}, 1)\, Q(\tau) e^{A^*\tau}$$

$$= \Gamma_0 \Gamma_1 \quad \text{diag}\,(\tau^n, \tau^{n-1}, \ldots, \tau, 1)$$

where

$$(5.9) \quad (\Gamma_0)^i_j \quad \frac{(-1)^{2n-(i+j)}}{(2n+1-(i+j))\,(n-i)!(n-j)!}$$

and

$$(5.10)\quad \Gamma_1 = \begin{pmatrix} 1 & 0 & 0 & \cdots & 0 & 0 \\ 1 & 1 & 0 & \cdots & 0 & 0 \\ \frac{1}{2} & 1 & 1 & \cdots & 0 & 0 \\ \vdots & \vdots & \vdots & & \vdots & \vdots \\ \frac{1}{(n-2)!} & \frac{1}{(n-3)!} & \frac{1}{(n-4)!} & \cdots & 1 & 0 \\ \frac{1}{(n-1)!} & \frac{1}{(n-2)!} & \frac{1}{(n-3)!} & \cdots & 1 & 1 \end{pmatrix} .$$

Thus $\Gamma_0 \Gamma_1$ is nonsingular and the columns

$$(5.11)\quad \tau^n \theta_1, \tau^{n-1} \theta_2, \ldots, \tau^2 \theta_{n-1}, \tau \theta_n$$

of $\theta(\tau)$ are linearly independent solutions of (5.4), in the case where A has the form (5.7). Substituting each of the solutions (5.11) into (5.4) we see that the vectors $\theta_1, \theta_2, \ldots, \theta_{n-1}, \theta_n$ are all eigenvectors of the lead matrix in (5.4) with corresponding eigenvalues $n, n-1, \ldots, 2, 1$. Thus we have proved

Theorem 3. The lead matrix in (5.4) has eigenvectors which are the columns of $\Gamma_0 \Gamma_1$, where Γ_0 and Γ_1 are defined by (5.9) and (5.10). The corresponding eigenvalues, which are the roots of the polynomial equation (5.5), are $n, n-1, \ldots, 2, 1$.

Solutions of (5.1), still assuming that A has the form (5.7), are obtained with the help of (5.3) and are seen to have the form

$$x_k(t) = \text{diag}((T-t)^{n-1}, (T-t)^{n-2}, \ldots, (T-t), 1)(T-t)^{n-k+1} \theta_k$$

$$= \text{diag}((T-t)^{2n-k}, (T-t)^{2n-k-1}, \ldots, (T-t)^{n-k+1}) \theta_k.$$

A simple perturbation analysis of the type familiar in the study of regular singular points then yields the following theorem.

Theorem 4. *For a general matrix A in the form (3.2) and with $B = b$ in the form (3.1) the differential equation (5.1) (or, equivalently, (25)) has linearly independent solutions*

$$x_k(t) = \text{diag}((T-t)^{2n-k}, (T-t)^{2n-k-1}, \ldots, (T-t)^{n-k+1})$$

$$(\theta_k + \mathcal{O}(T-t)), \quad k = 1, 2, \ldots, n,$$

where θ_k, $k = 1, 2, \ldots, n$ *are the columns of* $\Gamma_0 \Gamma_1$ (cf. (5.9), (5.10)). *These vectors are also the columns of the matrix*

$$\psi(T-t) = e^{-A(T-t)} Z(T-t).$$

6. Concluding Remarks. In formula (2.9) the ellipsoids R(t), which are constant-time surfaces for the flow associated with the differential equation (2.5) (or (5.1)) are identified. Using the last statement in Theorem 4 together with (2.9) we see that

$$R(t) = \{x \in R^n \mid x = \sum_{k=1}^{n} y_k x_k(t)$$

$$= \sum_{k=1}^{n} y_k (\text{diag}((T-t)^{2n-k}, (T-t)^{2n-k-1}, \ldots, (T-t)^{n-k+1})$$

$$(\theta_k + \mathcal{O}(T-t))), \quad \sum_{k=1}^{n} (y_k)^2 = 1\}.$$

If we write the equation which appears here as

$$x = K(T-t)y \qquad \text{or} \quad y = K(T-t)^{-1} x$$

it is clear that the equation of the ellipsoidal surface $R(t)$ is

$$1 = \|y\|^2 = x^*(K(T-t)^{-1})^* K(T-t)^{-1} x.$$

One can calculate readily in the case $n=2$, $a_0 = a_1 = \cdots = a_{n-1} = 0$ that

$$(K(T-t)^{-1})^* K(T-t)^{-1} =$$

$$(6.1) = 144 \begin{pmatrix} (T-t)^{-6} + \frac{1}{4}(T-t)^{-4} & \frac{1}{2}(T-t)^{-5} + \frac{1}{12}(T-t)^{-3} \\ \frac{1}{2}(T-t)^{-5} + \frac{1}{12}(T-t)^{-3} & \frac{1}{4}(T-t)^{-4} + \frac{1}{36}(T-t)^{-2} \end{pmatrix}.$$

A necessary and sufficient condition for the $R(t)$ to be nested, i.e.

$$(6.2) \qquad R(t_2) \subseteq \text{Int } R(t_1), \quad t_2 > t_1$$

is that

$$\frac{d}{dt}[(K(T-t)^{-1})^* K(T-t)^{-1}] > 0, \tag{6.3}$$

i.e. is positive definite. It is easy to see that (6.3) fails for $(K(T-t)^{-1})^* K(T-t)^{-1}$ given by (6.1). We conclude then that (6.2) does not hold in general for $n > 1$. This means that the $L^2[0, T]$-optimal control (1.3) is not, in general, synthesizable, i.e. $T-t$ is not a univalent function of x.

A number of extensions of the above theory can be envisioned. For example, what is the relevant theory when m, the dimension of the control u, is greater than 1 and, letting $B = [b_1, b_2, \ldots, b_m]$, the subspaces E_j spanned by b_j, Ab_j, $\ldots$, $A^{n-1}b_j$ are not complementary ? Also it would be interesting to consider the time varying case

$$\dot{x} = A(t)\,x + B(t)\,u$$

under the assumption that the system is controllable on each interval $[t_1, T]$, $0 \leq t_1 < T$, but at the final instant we have

$$\text{rank}\,[B(T), A(T)B(T), \ldots, A(T)^{n-1}B(T)] < n.$$

REFERENCES

[1] E. B. Lee and L. Markus: "Foundations of Optimal Control Theory", John Wiley and Sons, Inc., New York, 1967.

[2] D. L. Lukes: "Stabilizability and optimal control.", Funk. Ekvac., Vol. 11(1968), pp. 39-50.

[3] D. Kleinman: "An easy way to stabilize a linear constant system", IEEE Trans. Auto. Control, Vol. AC-15(1970), p. 692.

[4] J. P. LaSalle and S. Lefschetz: "Stability by Liapounov's Direct Method; with Applications", Academic Press, Inc., New York, 1961.

[5] B. D. O. Anderson and J. B. Moore= "Linear Optimal Control", Prentice Hall, Inc., Englewood Cliffs, New Jersey, 1971.

[6] J. R. Westlake: "A Handbook of Numerical Matrix Inversion and Solution of Linear Equations", John Wiley and Sons, Inc., New York, 1968.

[7] W. R. Wasow: "Asymptotic Expansions for Ordinary Differential Equations", Interscience Pub. Co., New York, 1965.

[8] E. L. Ince: "Ordinary Differential Equations", Dover Publications, Inc., New York, 1956.

University of Wisconsin-Madison, Madison, WI 53706

A SPECTRAL THEORY FOR LINEAR ALMOST PERIODIC DIFFERENTIAL EQUATIONS

Robert J. Sacker*

and

George R. Sell**

I. Introduction. In this paper we would like to outline a theory describing the asymptotic behavior (as $t \to \pm \infty$) of solutions of linear differential equations with almost periodic coefficients. By using a theory on the existence of dichotomies and invariant splittings [1], a certain subset σ of the real line R turns out to be crucial in our theory. We call this subset the "spectrum." The main reason for our use of this term is the rather suggestive analogy between our methods, in the study of differential equations, with some of the classical methods of the perturbation theory of linear operators.

This paper is a preliminary report of our investigations. Complete details including proofs will be published elsewhere.

This research was supported in part by

(*) U.S. A.R.M.Y Contract DAHC-04-74-G-0013,

(**) N.S.F- Grant GP- 38955.

II. Linear Skew Product Flows. Let X denote a finite dimensional (real or complex) linear space with norm $\|\cdot\|$, let Y denote a topological space and R denote the real line $-\infty < t < \infty$. A mapping $\pi: X \times Y \times R \to X \times Y$ is said to be a linear skew product flow on $X \times Y$ if the following hold

(i) π is continuous and has the form

(1) $$\pi(x, y, t) = (\varphi(x, y, t), \sigma(y, t))$$

where $\sigma : Y \times R \to Y$ is a flow on Y.

(ii) $\pi(x, y, 0) = (x, y)$ for all $(x, y) \in X \times Y$.

(iii) $\pi(\pi(x, y, t), s) = \pi(x, y, t + s)$ for all $(x, y) \in X \times Y$ and $t, s \in R$.

(iv) $\varphi(x, y, t)$ is linear in x.

The prototype for a linear skew product flow arises in the study of linear ordinary differential equations [1]. Specifically consider the family of linear differential equations

(2) $$x' = A(t)x$$

where $A \in \mathcal{A}$, a translation-invariant space of locally-integrable matrix-valued functions A(t) defined for $t \in R$. The translation-invariance means that $A_\tau \in \mathcal{A}$ whenever $A \in \mathcal{A}$ and $\tau \in R$, where

$$A_\tau(t) = A(\tau + t)$$

In addition we shall assume that $\mathcal{A}$ is a topological space and that the topology on $\mathcal{A}$ is such that the solution

$\varphi(x_0, A_0, t)$, of the initial value problem

$$x' = A_0(t)x, \quad x(0) = x_0,$$

is continuous in $(x_0, A_0, t) \in X \times \mathcal{A} \times R$. Under these conditions, the mapping

$$\pi(x, A, t) = (\varphi(x, A, t), A_t) \tag{3}$$

defines a linear skew product flow on $X \times \mathcal{A}$.

Let π be a linear skew product flow on $X \times Y$ given by (1). Then for any $\lambda \in R$ define

$$\pi_\lambda(x, y, t) = (e^{-\lambda t}\varphi(x, y, t), \sigma(y, t)).$$

It is easy to verify that for every $\lambda \in R$, π_λ is a linear skew product flow on $X \times Y$. In the case of a linear skew product flow given by the differential equation (2), the flow π_λ is essentially the same flow one gets by making the change of variables $w = e^{-\lambda t}x$ in (2).

For each $\lambda \in R$ define the bounded set $\mathcal{B}_\lambda$ the stable set $\mathcal{S}_\lambda$ and the unstable set $\mathcal{U}_\lambda$ by :

$$\mathcal{B}_\lambda = \{(x, y): \|e^{-\lambda t}\varphi(x, y, t)\| \text{ is uniformly bounded in } t\}$$

$$\mathcal{S}_\lambda = \{(x, y): \|e^{-\lambda t}\varphi(x, y, t)\| \to 0 \text{ as } t \to +\infty\}$$

$$\mathcal{U}_\lambda = \{(x, y): \|e^{-\lambda t}\varphi(x, y, t)\| \to 0 \text{ as } t \to -\infty\}$$

Also for each $y \in Y$ define the fibers

$$\mathcal{B}_\lambda(y) = \{x \in X: (x, y) \in \mathcal{B}_\lambda\}$$

$$\mathcal{S}_\lambda(y) = \{x \in X: (x, y) \in \mathcal{S}_\lambda\}$$

$$\mathcal{U}_\lambda(y) = \{x \in X: (x, y) \in \mathcal{U}\}$$

It is not hard to see that for every λ, the sets $\mathcal{B}_\lambda$, $\mathcal{S}_\lambda$ and $\mathcal{U}_\lambda$ are invariant sets for π, and the fibers $\mathcal{B}_\lambda(y)$, $\mathcal{S}_\lambda(y)$ are linear subspaces of X.

As an application of [1] we have the following result.

Theorem 1. Let $\pi = (\varphi, \sigma)$ be a linear skew product flow on $X \times Y$ and assume that Y is a compact minimal set in the flow σ. Then for any $\lambda \in R$ the following two statements are equivalent:

I. $\mathcal{B}_\lambda$ is trivial, i.e. $\mathcal{B}_\lambda = \{0\} \times Y$.

II. $\mathcal{S}_\lambda$ and $\mathcal{U}_\lambda$ are closed subbundles of $X \times Y$ and $X \times Y = \mathcal{S}_\lambda + \mathcal{U}_\lambda$, as a Whitney sum.

The second statement above means that

(i) $\mathcal{S}_\lambda$ and $\mathcal{U}_\lambda$ are closed subsets of $X \times Y$.

(ii) There are integers k and ℓ such that

$\dim \mathcal{S}_\lambda(y) = k$ and $\dim \mathcal{U}_\lambda(y) = \ell$ for all $y \in Y$.

(iii) $\mathcal{S}_\lambda(y)$ and $\mathcal{U}_\lambda(y)$ vary continuously in y in the sense that there exist mappings $P, Q : X \times Y \to X$

$$P(x, y) = P(y)\cdot x, \text{ and } Q(x, y) = Q(y)x$$

where $P(y)$ and $Q(y)$ are projections on X with range of $P(y) = \mathcal{S}_\lambda(y)$ and range of $Q(y) = \mathcal{U}_\lambda(y)$, and P and Q are jointly continuous in $(x, y) \in X \times Y$, cf. [4].

Conditions (i), (ii) and (iii) above say that $\mathcal{S}_\lambda$ and

$\mathcal{U}_\lambda$ are closed subbundles of $X \times Y$. The final assertion that $X \times Y = \mathcal{S}_\lambda + \mathcal{U}_\lambda$, as a Whitney sum, means that $k + \ell = n$ where $n = \dim X$ and $X = \mathcal{S}_\lambda(y) + \mathcal{U}_\lambda(y)$ for all $y \in Y$. The condition that $X \times Y = \mathcal{S}_\lambda + \mathcal{U}_\lambda$, as a Whitney sum, is equivalent to the existence of an appropriate exponential dichotomy, cf. [1].

III. The Spectrum. Throughout this section we shall assume that $\pi \doteq (\varphi, \sigma)$ is a linear skew product flow on $X \times Y$ where $\dim X = n$ $(n \geq 1)$ and the base Y is a compact minimal set in the flow σ.

The resolvent set is defined by

$$\rho = \rho(y) = \{\lambda \in R : \mathcal{B}_\lambda \text{ is trivial}\}$$

and the spectrum is the complement

$$\sigma = \sigma(y) = \{\lambda \in R : \mathcal{B}_\lambda \neq \{0\} \times Y\}.$$

The following statement is our basic result.

Theorem 2. Under the above conditions, on Y the spectrum $\sigma(Y)$ is a nonempty compact subset of R consisting of the disjoint union of k intervals $[a_i, b_i]$ $(1 \leq i \leq k)$ where $k \leq n$. Moreover associated with each spectral interval $[a_i, b_i]$ there is a nontrivial closed subbundle $\mathcal{S}_i$ of $X \times Y$ which is invariant under π and

$$X \times Y = \mathcal{S}_1 + \cdots + \mathcal{S}_k$$

as a Whitney sum.

The proof of this result will appear elsewhere. However, we can define the subbundles $\mathcal{S}_i$ $i = 1, \ldots, k$. Assume that the intervals $[a_i, b_i]$ are ordered so that

$$a_1 \leq b_1 < a_2 \leq b_2 < \cdots < a_k \leq b_k.$$

Now choose points $\lambda_0, \ldots, \lambda_k$ in the resolvent $\rho(Y)$ so that

$$\lambda_0 < a_1 \leq b_1 < \lambda_1 < a_2 \cdots b_{k-1} < \lambda_{k-1} < a_k \leq b_k < \lambda_k.$$

Then $\mathcal{S}_i$ is given by

$$\mathcal{S}_i = \mathcal{U}_{\lambda_{i-1}} \cap \mathcal{S}_{\lambda_i} \qquad i = 1, 2, \ldots, k.$$

We should note that the fibers

$$\mathcal{S}_i(y) = \{x \in X: (x, y) \in \mathcal{S}_i\}$$

are disjoint, i.e. $\mathcal{S}_i(y) \cap \mathcal{S}_j(y) = \{0\}$ if $i \neq j$, and

$$n = \dim \mathcal{S}_1(y) + \cdots + \dim \mathcal{S}_k(y)$$

for all $y \in Y$.

The theory of the spectrum, as formulated in Theorem 2, generalizes the theory of the Lyapunov type numbers, or Lyapunov characteristic exponents. Recall that if $(x, y) \in X \times Y$ with $x \neq 0$, then the four Lyapunov type numbers are defined by:

$$\lambda_s^+(x, y) = \limsup_{t \to +\infty} \frac{1}{t} \log \|\varphi(x, y, t)\|$$

$$\lambda_i^+(x, y) = \liminf_{t \to +\infty} \frac{1}{t} \log \|\varphi(x, y, t)\|$$

$$\lambda_s^-(x, y) = \limsup_{t \to -\infty} \frac{1}{t} \log \|\varphi(x, y, t)\|$$

$$\lambda_i^-(x, y) = \liminf_{t \to +\infty} \frac{1}{t} \log \|\varphi(x, y, t)\| .$$

One has the following result.

Theorem 3. Let

$$\sigma(Y) = [a_1, b_1] \cup \cdots \cup [a_k, b_k]$$

be the spectral decomposition described by Theorem 2, *and let* $\mathcal{S}_i$ *denote the invariant subbundle associated with the spectral interval* $[a_i, b_i]$, $i = 1, \ldots, k$. *If* $(x, y) \in \mathcal{S}_i$ *with* $x \neq 0$ *then the four Lyapunov type numbers* $\lambda_s^+(x, y)$, $\lambda_i^+(x, y)$, $\lambda_s^-(x, y)$ *and* $\lambda_i^-(x, y)$ *lie in the corresponding spectral interval* $[a_i, b_i]$. *In particular, if* $a_i = b_i$ *then for all* $(x, y) \in \mathcal{S}_i$ *with* $x \neq 0$, *the two limits*

$$\lim_{t \to +\infty} \frac{1}{t} \log \|\varphi(x, y, t)\|, \quad \lim_{t \to -\infty} \frac{1}{t} \log \|\varphi(x, y, t)\|$$

exist and are equal (to $a_i = b_i$).

The final statement in Theorem 3 is of particular interest. In this case the interval $[a_i, b_i]$ is dengenerate, i.e., it consists of a single point. It seems natural to refer to $\lambda = a_i = b_i$ as part of the *point spectrum.* Also if $[a_i, b_i]$ is nondegenerate then we refer to $[a_i, b_i]$ as being in the *continuous spectrum.* We shall say that $\sigma(Y)$ is pure point spectrum if all the spectral intervals are degenerate.

The connection between the structure of $\sigma(Y)$ and possible extensions of the Floquet theory to aperiodic systems has been presented elsewhere, cf. [5, 6]. Suffice it to mention here that in the periodic case ($Y = S^1$) the spectrum $\sigma(Y)$ is pure point spectrum and the values $\lambda \in \sigma(\lambda)$ are simply the real parts of the Floquet **exponents.**

The central problem in the spectral theory of linear almost periodic differential equations is to determine whether the spectrum $\sigma(\lambda)$, where Y is an almost periodic minimal set, is always pure point spectrum. We shall describe some preliminary results on this problem in the next section, but before doing that we would like to describe an example due to R. McGehee.

Let $f: R \to R$ be a bounded uniformly continuous function with the property that the hull

$$H(f) = \text{cl}\,\{f_\tau : \tau \in R\}$$

is a compact minimal set, cf. [7] and

$$\limsup_{t \to +\infty} \frac{1}{T} \int_0^T f(s)ds = b > a = \liminf_{T \to +\infty} \frac{1}{T} \int_0^T f(s)ds.$$

The existence of such a function is a direct consequence of the theory of non-uniquely ergodic minimal sets, cf. [8] for example. Now let π be the linear skew product flow on $R \times H(f)$ described in (3) for the linear equations $x' = f^*(t)\,x$, $f^* \in H(f)$. Since

$$\varphi(x, f, t) = e^{\int_0^t f(s)ds}\, x$$

it follows from the characterization of f and Theorem 1 that the interval $[a, b]$ is contained in the spectrum $\sigma(H(f))$.

IV. Point Spectrum vs. Continuous Spectrum. The phenomenon of continuous spectrum, which was illustrated above, cannot occur with scalar-valved linear differential equations

$$x' = a(t)x \quad (x \in R^1 \text{ or } C^1)$$

with an almost periodic coefficient $a(t)$. In this case the spectrum σ consists of one point $m(a)$, the mean value of a, which is given by

$$m(a) = \lim_{t-s \to +\infty} \frac{1}{t-s} \int_s^t a(s)ds$$

For higher dimensional equations the situation is not at all settled. However some results can be described. Let X be a fixed real linear space of dimension n $(n \geq 1)$ and let $\mathcal{AP}$ denote the Banach space consisting of all continuous $n \times n$ matrices $A = A(t)$ with real-valued almost periodic coefficients, where $\mathcal{AP}$ has the sup-norm. For each $A \in \mathcal{AP}$ let

$$H(A) = \text{cl}\,\{A_\tau : \tau \in R\}$$

denote the hull of A. The spectrum $\sigma(H(A))$ is defined as in Section III and we shall write $\sigma(A) = \sigma(H(A))$ for any $A \in \mathcal{AP}$.

Now define

$\mathcal{AP}_p = \{A \in \mathcal{AP}: \sigma(A)$ consists only of point spectrum$\}$

$\mathcal{AP}_c = \{A \in \mathcal{AP}: \sigma(A)$ contains continuous spectrum$\}$.

The central problem described above is to determine whether or not $\mathcal{AP}_c$ is empty or not. Recall that every constant matrix A and every periodic matrix A(t) lies in $\mathcal{AP}_p$.

For $i = 1, \ldots, n$ define

$\mathcal{P}_i = \{A \in \mathcal{AP}: \sigma(A)$ contains exactly i intervals$\}$.

We can prove the following result concerning $\mathcal{P}_n$.

<u>Theorem 5.</u> <u>Under the above condition,</u> $\mathcal{P}_n$ <u>is an open set in</u> $\mathcal{AP}$ <u>and</u> $\mathcal{P}_n \subseteq \mathcal{AP}_p$.

The proof that $\mathcal{P}_n$ is open is based on a perturbation argument which uses the fact that exponential dichotomics are preserved under "small" perturbations. This argument would apply equally well for differential equations with complex coefficients. Our proof that $\mathcal{P}_n \subseteq \mathcal{AP}_p$ depends on the assumption that the coefficients be real-valued. We do not know whether it is valid for differential equations with complex coefficients.

One can show that if $\lambda \notin \sigma(A)$ for some $A \in \mathcal{AP}$, then there is a neighborhood $\mathcal{V}$ of A in $\mathcal{AP}$ for which $\lambda \notin \sigma(A^*)$ for all $A^* \in \mathcal{V}$. This means that $\sigma(A)$ is upper semi-continuous in A, i.e. if $A_n \to A$ in $\mathcal{AP}$ then

$$\limsup_{n \to \infty} \sigma(A_n) = \bigcap_{N=1}^{\infty} \bigcup_{n=N}^{\infty} \sigma(A_n) \subseteq \sigma(A).$$

It is not hard to show that $\sigma(A)$ is continuous on the set $\mathcal{P}_n$, which is a reformulation of a result of Lillo [9].

REFERENCES

[1] R. J. Sacker and G. R. Sell. Existence of Dichotomies and Invariant Splittings for Linear Differential Systems I. J. Diff. Eqns. (15) (1974), 429-458.

[2] R. J. Sacker and G. R. Sell. Existence of Dichotomies and Invariant Splittings for Linear Differential Systems II. (to appear)

[3] J. F. Selgrade. Isolated Invariant Sets for Flows on Vector Bundles. (to appear)

[4] G. R. Sell. Linear Differential Systems. Lecture Notes University of Minnesota, 1974.

[5] G. R. Sell. The Floquet Problem for Almost Periodic Linear Differential Equations. Proceedings of International Conference on Differential Equations, Dundee, Scotland, March, 1974. (to appear)

[6] R. K. Miller and G. R. Sell. Topological Dynamics and its Relation to Integral Equations and Nonautonomous Systems. Proceedings of International Conference on Dynamical Systems, Providence, R.I., August, 1974. (to appear)

[7] G. R. Sell. Nonautonomous Differential Equations and Topological Dynamics, I and II. Trans. Amer. Math. Soc. 127(1967), 241-283.

[8] J. C. Oxtoby. Ergodic Sets. Bull. Amer. Math. Soc. 58(1952), 116-136.

[9] J. C. Lillo. Continuous Matrices and the Stability Theory of Differential Systems. Math. Z. 73(1960), 45-58.

University of Southern California, Los Angeles, California
University of Minnesota, Minneapolis, Minnesota

SUBDOMINANT SOLUTIONS ADMITTING A PRESCRIBED STOKES PHENOMENON

Yasutaka Sibuya

1. Introduction. Consider a differential equation

$$(1.1) \qquad y'' - [x^m + \sum_{k=1}^{m} a_k x^{m-k}]\, y = 0,$$

where $a_1, \ldots, a_m$ are complex numbers. Let us denote by $\mathcal{S}_k$ a sector

$$(1.2) \qquad \left| \arg x - \frac{2k}{m+2}\pi \right| < \frac{\pi}{m+2} ,$$

where k is an integer.

Definition 1.1.: A nontrivial solutions $y(x)$ of (1.1) is said to be subdominant in $\mathcal{S}_k$, if $y(x)$ tends to zero as x tends to infinity along any direction in $\mathcal{S}_k$.

If we denote by $\bar{\mathcal{S}}_k$ the closure of $\mathcal{S}_k$, then $\bar{\mathcal{S}}_0, \bar{\mathcal{S}}_1, \ldots, \bar{\mathcal{S}}_{m+1}$ cover the x-plane completely.

The main results of this paper were presented at the 708-th meeting of American Mathematical Society, University of Minnesota, Minneapolis, Minn., November 3, 1973.
This research was partially supported by the United States Army under Contract No. DA-31-124-ARO-D-462 and NSF GP 27275 and GP-38955.

Definition 1.2: A set of m+2 nontrivial solutions $y_0, \ldots, y_{m+1}$ of (1.1) is called a complete set of subdomiant solutions of (1.1), if $y_0, \ldots, y_{m+1}$ are subdominant in $\mathscr{S}_0, \ldots, \mathscr{S}_{m+1}$ respectively.

Set

$$\omega = \exp\left\{\frac{2\pi i}{m+2}\right\} \tag{1.3}$$

and

$$C(\gamma, \rho) = \begin{bmatrix} \gamma & 1 \\ -\omega\rho & 0 \end{bmatrix}, \tag{1.4}$$

where γ and ρ are complex numbers. In this paper we shall prove the following two theorems.

Theorem A: Assume that m is odd, and that $\gamma_0, \ldots, \gamma_{m+1}$ are given complex numbers satisfying the condition

$$C(\gamma_{m+1}, 1) \cdots C(\gamma_1, 1) C(\gamma_0, 1) = I, \tag{1.5}$$

where I is the two-by-two unit-matrix. Then there exist m complex numbers $a_1, \ldots, a_m$ such that differential equation (1.1) has a complete set of subdominant solutions $y_0, \ldots, y_{m+1}$ which admit the connection formulas

$$y_k(x) = \gamma_k y_{k+1}(x) - \omega y_{k+2}(x) \quad (k = 0, \ldots, m+1), \tag{1.6}$$

where

$$y_{m+2} = y_0, \quad y_{m+3} = y_1. \tag{1.7}$$

Theorem B: Assume that m is even, and that $\gamma_0, \ldots, \gamma_{m+1}, \rho$ are given complex numbers satisfying the conditions:

(1.8) $$\gamma_j \neq 0 \text{ for some } j,$$

(1.9) $$\rho \neq 0,$$

(1.10) $$C(\gamma_{m+1}, \rho^{-1}) C(\gamma_m, \rho) \cdots C(\gamma_1, \rho^{-1}) C(\gamma_0, \rho) = I.$$

Then there exist m complex numbers $a_1, \ldots, a_m$ such that differential equation (1.1) has a complete set of subdominant solutions $y_0, \ldots, y_{m+1}$ which admit the connection formulas

(1.11) $$y_k(x) = \gamma_k \, y_{k+1}(x) - \omega \rho^{(-1)^k} \, y_{k+2}(x)$$

$$(k = 0, \ldots, m+1).$$

2. Lemmas. Differential equation (1.1) has a solution

(2.1) $$y = \mathcal{Y}_m(x, a_1, \ldots, a_m)$$

such that

(i) $\mathcal{Y}_m(x, a)$ is an entire function of $(x, a_1, \ldots, a_m)$,

and

(ii) $\mathcal{Y}_m(x, a)$ admits an asymptotic representation

(2.2) $$\mathcal{Y}_m(x, a) = x^{r_m}[1 + O(x^{-\frac{1}{2}})] \exp\{-E_m(x, a)\}$$

uniformly on each compact set in the a-space as x tends to infinity in any closed subsector of $\mathcal{S}_{-1} \cup \bar{\mathcal{S}}_0 \cup \mathcal{S}_1$, where

$$(2.3)\qquad E_m(x,a) = \frac{2}{m+2}\, x^{\frac{1}{2}(m+2)} + \sum_{1 \leq h < \frac{m}{2}+1} \frac{2}{m+2-2h}\, b_h(a) x^{\frac{1}{2}(m+2-2h)} ,$$

$$(2.4)\qquad r_m = \begin{cases} -\frac{1}{4}\, m & (m\text{: odd}), \\ -\frac{1}{4}\, m - b_{\frac{1}{2}m+1}(a) & (m\text{: even}), \end{cases}$$

and

$$(2.5)\qquad \left\{ 1 + \sum_{k=1}^{m} a_k\, x^{-k} \right\}^{\frac{1}{2}} = 1 + \sum_{h=1}^{\infty} b_n(a)\, x^{-h} .$$

Lemma 2.1: Let x, s, $a_1, \ldots, a_m$ be complex variables. Then the function $\mathcal{Y}_m(x, a)$ satisfies the identity

$$(2.6)\qquad \mathcal{Y}_m(x+s,\ a_1, \ldots, a_m) = K_m(x, a_1, \ldots, a_m)\, \mathcal{Y}_m(s, u_1, \ldots, u_m),$$

where K_m $u_1, \ldots, u_m$ are independent of s, and they are defined respectively by

$$(2.7)\qquad K_m(x,a) = \begin{cases} 1 & (m\text{: odd}), \\ \exp\{-E_m(x,a)\} & (m\text{: even}), \end{cases}$$

and

$$(2.8) \qquad (x+s)^m + a_1(x+s)^{m-1} + \cdots + a_{m-1}(x+s) + a_m$$

$$= s^m + u_1 s^{m-1} + \cdots + u_{m-1} s + u_m \quad .$$

(Cf. Y. Sibuya [5; p. 56].)

Set

$$(2.9) \qquad G^k(a) = (\omega^{-k} a_1, \omega^{-2k} a_2, \ldots, \omega^{-mk} a_m),$$

and

$$(2.10) \quad \mathcal{Y}_{m,k}(x, a) = \mathcal{Y}_m(\omega^{-k} x, G^k(a)) \quad (k = 0, \ldots, m+1).$$

Lemma 2.2: The $m+2$ functions $\mathcal{Y}_{m,k}(x, a)$ $(k = 0, \ldots, m+1)$ form a complete set of subdominant solutions of (1.1). (Cf. P.E. Hsieh and Y. Sibuya [3].) These subdominant solutions admit the connection formulas:

$$(2.11) \quad \mathcal{Y}_{m,k}(x, a) = C_k(a) \mathcal{Y}_{m,k+1}(x, a)$$

$$+ \tilde{C}_k(a) \mathcal{Y}_{m,k+2}(x, a) \quad (k=0, \ldots, m+1).$$

Set

$$(2.12) \qquad C_0(a) = C(a) \quad \tilde{C}_0(a) = \tilde{C}(a).$$

Lemma 2.3: The quantities $C_k(a)$ and $\tilde{C}_k(a)$ have the following properties:

(i) $$C_k(a) = C(G^k(a)), \quad \tilde{C}_k(a) = \tilde{C}(G^k(a)) \;;$$

(ii) *the functions* $C(a)$ *and* $\tilde{C}(a)$ *are entire in* a, *and*

$$(2.13) \qquad \tilde{C}(a) = \begin{cases} -\omega & (m: \text{ odd}), \\ -\omega^{1-2\nu(a)} & (m: \text{ even}), \end{cases}$$

where

$$(2.14) \qquad \nu(a) = b_{\frac{1}{2}m+1}(a) \;;$$

(iii) *if we define* $u_1, \ldots, u_m$ *by* (2.8), *then*

$$(2.15) \qquad C(u) = \begin{cases} C(a) & (m: \text{ odd}), \\ \exp\{2E_m(x, a)\}\, C(a) & (m: \text{ even}); \end{cases}$$

(iv) *if we set*

$$(2.16) \qquad S_k(a) = \begin{bmatrix} C_k(a) & 1 \\ \tilde{C}_k(a) & 0 \end{bmatrix} \qquad (k = 0, \ldots, m+1),$$

then

$$(2.17) \qquad S_{m+1}(a) S_m(a) \cdots S_1(a)\, S_0(a) = I \;;$$

(v) *for each fixed* a, *there exists an integer* k *such that*

(2.18) $$C_k(a) \neq 0.$$

Proof: Property (i) follows from (2.10) and (2.11). To prove (ii), express $C(a)$ and $\tilde{C}(a)$ as ratios of Wronskians, and compute Wronskians by using the asymptotic property of $\mathcal{Y}_m(x, a)$. (Cf. Y. Sibuya [6; III, § 6].) Property (iii) follows from Lemma 2.1 if we use the following properties of $b_h(a)$, $E_m(x, a)$ and u_j:

(2.19) $$b_h(G^k(a)) = \omega^{-hk} b_h(a),$$

(2.20) $$E_m(\omega^{-k}x, G^k(a)) = (-1)^k E_m(x, a),$$

and

(2.21) $$u_j(\omega^{-k}x, G^k(a)) = \omega^{-jk} u_j(x, a),$$

where $u_j(x, a)$ denotes u_j as a function of x and a. To prove (iv), note that $\mathcal{Y}_m(x, a)$ is single-valued in the (x, a)-space. Hence

(2.22) $$\mathcal{Y}_{m, m+2}(x, a) = \mathcal{Y}_{m, 0}(x, a)$$

$$\mathcal{Y}_{m,m+3}(x, a) = \mathcal{Y}_{m, 1}(x, a).$$

Then (iv) follows from (2.11) immediately. (Cf. Y. Sibuya [6; III, § 7].)

To prove (v), suppose for a contradicition that

(2.23) $$C_k(a) = 0 \qquad (k = 0, \ldots, m+1)$$

for some fixed a. Then it follows from (2.11) that $\mathcal{Y}_{m, k}(x, a)$

and $\mathcal{Y}_{m,k+2}(x, a)$ are linearly dependent. Hence m must be even, and two entire functions $\mathcal{Y}_{m,0}(x, a)\exp\{E_m(x, a)\}$ and $\mathcal{Y}_{m,1}(x, a)\exp\{-E_m(x, a)\}$ become polynomials in x. However, this is impossible, since

$$(2.24) \qquad \mathcal{Y}_{m,0}(x, a)\mathcal{Y}_{m,1}(x, a) = x^{-\frac{1}{2}m}[\omega^{-r_m} + 0(x^{-\frac{1}{2}})]$$

as x tends to infinity. In deriving (2.24), we used identity

$$(2.25) \qquad b_{\frac{1}{2}m+1}(G(a)) = -b_{\frac{1}{2}m+1}(a) \quad .$$

This completes the proof of Lemma 2.3.

The next lemma is based on a work done by R. Nevanlinna [4].

Lemma 2.4: Let $c_0, \ldots, c_{m+1}$ be given points on the Riemann sphere. Then, there exist $a_1, \ldots, a_m$ such that differential equation (1.1) has two linearly independent solutions $y_1(x)$ and $y_2(x)$ satisfying the conditions

$$(2.26) \qquad \frac{y_1(x)}{y_2(x)} \to c_k \quad \text{as } x \text{ tends to infinity in } \mathcal{S}_k$$

$$(k = 0, 1, \ldots, m+1),$$

if and only if

(I) $\quad c_k \neq c_{k+1} \quad (k = 0, \ldots, m+1;\ c_{m+2} = c_0)$

and

(II) the set $\{c_0, \ldots, c_{m+1}\}$ contains at least three distinct points on the Riemann sphere.

We shall need another lemma concerning matrices.

Lemma 2.5: Let $a, b, c, \alpha, \beta, \gamma, \xi, \eta$, and ζ be complex numbers. Assume that none of them is zero. Then, if

$$(2.27)\qquad \begin{bmatrix} a & 1 \\ \xi & 0 \end{bmatrix}\begin{bmatrix} b & 1 \\ \eta & 0 \end{bmatrix}\begin{bmatrix} c & 1 \\ \zeta & 0 \end{bmatrix} = \begin{bmatrix} \alpha & 1 \\ \xi & 0 \end{bmatrix}\begin{bmatrix} \beta & 1 \\ \eta & 0 \end{bmatrix}\begin{bmatrix} \gamma & 1 \\ \zeta & 0 \end{bmatrix},$$

it follows that

$$(2.28)\qquad a = \alpha, \quad b = \beta, \quad c = \gamma .$$

(Cf. Y. Sibuya [6; III, § 9, Lemma 9.2].)

3. Main problems. To prove Theorem A, we shall consider the following problem.

Problem A: Assume that m is odd and that $\gamma_0, \ldots, \gamma_{m+1}$ are given complex numbers. Find m complex numbers $a_1, \ldots, a_m$ so that we have

$$(3.1)\qquad C_k(a_1, \ldots, a_m) = \gamma_k \qquad (k = 0, 1, \ldots, m+1).$$

Note that, if m is odd, we have

$$(3.2)\qquad \tilde{C}_k(a) = -\omega \qquad (k = 0, \ldots, m+1),$$

and that the matrices $S_k(a)$ satisfy condition (2.17). Hence the complex numbers $\gamma_0, \ldots, \gamma_{m+1}$ must satisfy condition (1.5), in order that Problem A has a solution. In case when $m = 1$, condition (1.5) becomes

$$\begin{bmatrix} \gamma_2 & 1 \\ -\omega & 0 \end{bmatrix} \begin{bmatrix} \gamma_1 & 1 \\ -\omega & 0 \end{bmatrix} \begin{bmatrix} \gamma_0 & 1 \\ -\omega & 0 \end{bmatrix} = \begin{bmatrix} 1 & 0 \\ 0 & 1 \end{bmatrix} , \tag{3.3}$$

where

$$\omega = \exp\left\{\frac{2}{3}\pi i\right\} . \tag{3.4}$$

If we regard (3.3) as an equation for γ_0, γ_1 and γ_2, there is a solution given by

$$\gamma_0 = \gamma_1 = \gamma_2 = -\omega^2 . \tag{3.5}$$

Lemma 2.5 guarantees that (3.4) is the only solution for (3.3). In this way, we can prove Theorem A for $m = 1$. For $m \geqq 3$, we shall prove the following theorem.

Theorem A′ : Assume that

(i) $m \geqq 3$ and m is odd;

(ii) $\gamma_0, \ldots, \gamma_{m+1}$ are given complex numbers;

(iii) $\gamma_0, \ldots, \gamma_{m+1}$ satisfy condition (1.5).

Then there exist m complex numbers $a_1, \ldots, a_m$ such that we have

$$C_k(a) = \gamma_k \qquad (k = 0, 1, \ldots, m+1). \tag{3.6}$$

If $a_1, \ldots, a_m$ satisfy (3.6), the subdominant solutions $\mathcal{Y}_{m,k}(x, a)$ admit the connection formulas given by (1.6). Therefore, Theorem A follows.

To prove Theorem B, we shall consider the following problem.

Problem B: Assume that m is even and that $\gamma_0, \ldots, \gamma_{m+1}$ and ρ are given complex numbers. Find m complex numbers $a_1, \ldots, a_m$ so that we have

$$\omega^{-2\nu(a)} = \rho \tag{3.7}$$

and

$$C_k(a) = \gamma_k \qquad (k = 0, 1, \ldots, m+1). \tag{3.8}$$

Set

$$\beta(a) = \omega^{-2\nu(a)} \quad . \tag{3.9}$$

Then it follows from (2.13), (2.14), (i) of Lemma 2.3 and (2.19) that

$$\tilde{C}_k(a) = \begin{cases} -\omega\beta(a) & \text{if } k \text{ is even,} \\ -\omega\beta(a)^{-1} & \text{if } k \text{ is odd.} \end{cases} \tag{3.10}$$

Hence

$$S_k(a) = \begin{cases} C(C_k(a), \beta(a)) & \text{if } k \text{ is even,} \\ C(C_k(a), \beta(a)^{-1}) & \text{if } k \text{ is odd,} \end{cases} \tag{3.11}$$

where $C(\gamma, \rho)$ is the matrix defined by (1.4). Therefore, it follows from (2.17) that, in order that Problem B has a solution, it is necessary that $\gamma_0, \ldots, \gamma_{m+1}$ and ρ satisfy conditions (1.9) and (1.10). Condition (1.8) must be aldo satisfied, since the functions $C_k(a)$ have property (v) of Lemma 2.3. However, it can be shown that we need condition (1.8) only when

$$\rho^{\frac{1}{2}m+1} = (-1)^{\frac{1}{2}m} \quad . \tag{3.12}$$

Otherwise, (1.8) follows from (1.10).

In case m = 2, condition (1.10) becomes

$$\begin{bmatrix} \gamma_3 & 1 \\ -i\rho^{-1} & 0 \end{bmatrix} \begin{bmatrix} \gamma_2 & 1 \\ -i\rho & 0 \end{bmatrix} \begin{bmatrix} \gamma_1 & 1 \\ -i\rho^{-1} & 0 \end{bmatrix} \begin{bmatrix} \gamma_0 & 1 \\ -i\rho & 0 \end{bmatrix} = \begin{bmatrix} 1 & 0 \\ 0 & 1 \end{bmatrix} .$$

This is equivalent to

$$\begin{cases} (\gamma_3\gamma_2 - i\rho)\gamma_1 - i\rho^{-1}\gamma_3 = 0 , \\ \gamma_3\gamma_2 - i\rho = i\rho^{-1} , \\ -i\rho^{-1}\gamma_1\gamma_2 - \rho^{-2} = 1 , \\ -i\rho^{-1}\gamma_2 = -i\rho^{-1}\gamma_0 . \end{cases} \tag{3.13}$$

Therefore, we have

$$\gamma_0 = \gamma_2 \quad , \quad \gamma_1 = \gamma_3 \quad , \tag{3.14}$$

and

$$\gamma_1\gamma_2 = i(\rho + \rho^{-1}). \tag{3.15}$$

Conversely, (3.13) follows from (3.14) and (3.15). If one of the γ_j is zero, then

$$\rho^2 = -1.$$

On the other hand, it is known that, if m=2, the functions $C_k(a)$ and $\beta(a)$ are given by

$$\begin{cases} C_0(a_1, a_2) = C_2(a_1, a_2) = 2^b \, e^{\frac{1}{4}a_1^2} \, e^{-i\pi(\frac{1}{2}b - \frac{1}{4})} \dfrac{\sqrt{2\pi}}{\Gamma(\frac{1}{2}+b)}, \\ C_1(a_1, a_2) = C_3(a_1, a_2) = 2^{-b} \, e^{-\frac{1}{4}a_1^2} \, e^{i\pi(\frac{1}{2}b + \frac{1}{4})} \dfrac{\sqrt{2\pi}}{\Gamma(\frac{1}{2}-b)}, \\ \beta(a_1, a_2) = e^{-ib\pi}, \end{cases}$$

where

$$b = \frac{1}{2} a_2 - \frac{1}{8} a_1^2 .$$

(Cf. Y. Sibuya [6; III, § 6].)

To solve Problem B for $m = 2$, let us assume that the quantities $\gamma_0, \gamma_1, \gamma_2, \gamma_3$ and ρ satisfy the conditions

(i) $$\rho \neq 0,$$

(ii) $$\gamma_0 \text{ or } \gamma_1 \neq 0 ,$$

and (3.14) and (3.15). In case $\gamma_0 \neq 0$, choose b so that

$$e^{-ib\pi} = \rho \text{ and } \Gamma(\tfrac{1}{2}+b) \neq \infty .$$

Then, choose a_1 so that

$$C_0(a_1, a_2) = C_2(a_1, a_2) = \gamma_0 \; (= \gamma_2).$$

It is easily verified that, if we choose b and a_1 in this way, we have also

$$C_1(a_1, a_2) = C_3(a_1, a_2) = \gamma_1 \ (= \gamma_3).$$

We can treat the case $\gamma_1 \neq 0$ in the same way. Therefore, if $m = 2$, we can prove Theorem B by solving Problem B.

For $m \geqq 4$, we have not found any method of solving Problem B. Instead, we shall prove the following theorem.

Theorem B′: Assume that

(i) $m \geqq 4$ and m is even ;

(ii) $\gamma_0, \gamma_1, \ldots, \gamma_{m+1}$ and ρ are given complex numbers;

(iii) $\rho \neq 0$ and $\gamma_0 \neq 0$;

(iv) $\gamma_0, \gamma_1, \ldots, \gamma_{m+1}$ and ρ satisfy condition (1.10).

Then there exists m complex numbers $a_1, \ldots, a_m$ and an integer p such that

$$\beta(a) = \rho\omega^{-2p}$$

(3.16)
$$C_k(a) = \begin{cases} \omega^{2pk}\gamma_k & \text{if } k \text{ is even,} \\ \omega^{-2pk}\gamma_k & \text{if } k \text{ is odd,} \end{cases}$$

$$(k = 0, 1, \ldots, m+1).$$

If $p = 0$, then Theorem B′ would give a solution of Problem B. However, we still do not know how to prove $p = 0$.

If $a_1, \ldots, a_m$ and p satisfy condition (3.16), the subdominant solutions of (1.1) defined by

$$(3.17)\qquad y_k(x) = \begin{cases} \omega^{-pk}\mathcal{Y}_{m,k}(x,a) & \text{if } k \text{ is even,} \\ \omega^{p(k-1)}\mathcal{Y}_{m,k}(x,a) & \text{if } k \text{ is odd,} \end{cases}$$

admit connection formulas (1.11). Therefore, Theorem B follows from Theorem B′ if $\gamma_0 \neq 0$. The case when $\gamma_0 = 0$ but $\gamma_j \neq 0$ for some j can be reduced to the case $\gamma_0 \neq 0$ easily. This means that, in order to prove Theorem B, it is sufficient to prove Theorem B′.

4. Proof of Theorem A′: Let us define matrices $\Gamma_0, \ldots, \Gamma_{m+1}$ by

$$(4.1)\qquad \Gamma_k = C(\gamma_k, 1) = \begin{pmatrix} \gamma_k & 1 \\ -\omega & 0 \end{pmatrix},$$

where $C(\gamma, \rho)$ is the matrix defined by (1.4). Matrices (4.1) satisfy condition (1.5), i.e.

$$(4.2)\qquad \Gamma_{m+1}\Gamma_m \cdots \Gamma_1\Gamma_0 = I.$$

Set

$$(4.3)\qquad H_k = \Gamma_k\Gamma_{k-1} \cdots \Gamma_0$$

$$= \begin{bmatrix} \tau_k & \sigma_k \\ \lambda_k & \mu_k \end{bmatrix} \qquad (k = 0, \ldots, m-1).$$

Since H_k are non-singular, the quantities σ_k/τ_k ($k = 0, 1, \ldots, m-1$) are well-defined points on the Riemann sphere. Set

$$(4.4) \qquad \begin{cases} c_0 = \infty \quad , \\ c_1 = 0 \quad , \\ c_k = \dfrac{\sigma_{k-2}}{\tau_{k-2}} \qquad (k = 2, 3, \ldots, m+1). \end{cases}$$

We shall show that

$$(4.5) \qquad \begin{cases} c_2 \neq 0 , \\ c_{k+1} \neq c_k \qquad (k = 2, 3, \ldots, m), \\ c_{m+1} \neq \infty \quad . \end{cases}$$

Since $H_0 = \Gamma_0$, we have $\tau_0 = \gamma_0$ and $\sigma_0 = 1$. This shows that $c_2 \neq 0$. For $k \geqq 0$, we have

$$H_{k+1} = \Gamma_{k+1} H_k$$

and hence

$$\begin{cases} \tau_{k+1} = \gamma_{k+1} \tau_k + \lambda_k , \\ \sigma_{k+1} = \gamma_{k+1} \sigma_k + \mu_k . \end{cases}$$

Therefore

$$\begin{vmatrix} \tau_{k+1} & \tau_k \\ \sigma_{k+1} & \sigma_k \end{vmatrix} = \begin{vmatrix} \lambda_k & \tau_k \\ \mu_k & \sigma_k \end{vmatrix} = -\det H_k \neq 0.$$

This shows that $c_{k+1} \neq c_k$ for $k = 2, \ldots, m$.
Condition (4.2) can be written in the form

$$\Gamma_{m+1} \Gamma_m H_{m-1} = I.$$

Hence

$$H_{m-1} = [\Gamma_{m+1} \Gamma_m]^{-1} = \omega^{-2} \begin{bmatrix} -\omega & -\gamma_{m+1} \\ \omega\gamma_m & \gamma_{m+1}\gamma_m - \omega \end{bmatrix} .$$

This shows that $\tau_{m-1} = -\omega^{-1} \neq 0$. Therefore $c_{m+1} \neq \infty$.
Thus we have proved (4.5).

Since m is odd, (4.5) implies that the set $\{c_0, \ldots, c_{m+1}\}$ contains at least three distinct points on the Riemann sphere. Hence by virtue of Lemma 2.4

(Section 2), there exist m complex numbers $a_1, \ldots, a_m$ and two linearly indpendent solutions $y_1(x)$, and $y_2(x)$ of

(1.1) $$y'' - [x^m + a_1 x^{m-1} + \cdots + a_{m-1} x + a_m] y = 0$$

such that

(4.6) $$F(x) = \frac{y_1(x)}{y_2(x)} \to c_k \text{ as } x \to \infty \text{ in } \mathscr{S}_k$$

$$(k = 0, 1, \ldots, m+1).$$

Since $c_0 = \infty$ and $c_1 = 0$, we must have

(4.7) $$F(x) = \lambda \frac{\mathscr{Y}_{m,1}(x, a)}{\mathscr{Y}_{m,0}(x, a)},$$

where λ is a non-zero complex number. Let us define the matrices $S_k(a)$ by (2.16). Then connection formulas (2.11) are written as

$$[\mathscr{Y}_{m,k}(x, a), \mathscr{Y}_{m,k+1}(x, a)] = [\mathscr{Y}_{m,k+1}(x, a), \mathscr{Y}_{m,k+2}(x, a)] \, S_k(a)$$

$$(k = 0, 1, \ldots, m+1).$$

Let us set

(4.8) $$L_k(a) = S_k(a) S_{k-1}(a) \cdots S_0(a) = \begin{bmatrix} t_k(a) & s_k(a) \\ \ell_k(a) & m_k(a) \end{bmatrix}$$

$$(k = 0, 1, \ldots, m-1).$$

Then the connection formulas

$$[\mathcal{V}_{m,0}(x,a), \mathcal{V}_{m,1}(x,a)] = [\mathcal{V}_{m,k+1}(x,a), \mathcal{V}_{m,k+2}(x,a)]\ L_k(a)$$

$$(k = 0, 1, \ldots, m-1)$$

yield

(4.9) $$F(x) \to \lambda \frac{s_{k-2}(a)}{t_{k-2}(a)} \quad \text{as } x \to \infty \quad \text{in} \quad \mathcal{S}_k$$

$$(k = 2, 3, \ldots, m+1).$$

Hence from (4.4), (4.6) and (4.9) it follows that

(4.10) $$\lambda \frac{s_k(a)}{t_k(a)} = \frac{\sigma_k}{\tau_k} \quad (k = 0, 1, \ldots, m-1).$$

We shall now prove that

(4.11) $$C_k(a) = \begin{cases} \lambda \gamma_k & \text{if } k \text{ is even,} \\ \lambda^{-1} \gamma_k & \text{if } k \text{ is odd,} \end{cases}$$

where $k = 0, 1, \ldots, m-1$. For $k = 0$, we have

$$\begin{cases} t_0(a) = C_0(a), & s_0(a) = 1, \\ \tau_0 = \gamma_0, & \sigma_0 = 1. \end{cases}$$

Hence (4.10) yields

$$C_0(a) = \lambda \gamma_0$$

In general, (4.11) is equivalent to

$$S_k(a) = \begin{cases} \Lambda \Gamma_k \Lambda & \text{if } k \text{ is even,} \\ \Lambda^{-1} \Gamma_k \Lambda^{-1} & \text{if } k \text{ is odd,} \end{cases} \tag{4.12}$$

where

$$\Lambda = \begin{bmatrix} \lambda^{\frac{1}{2}} & 0 \\ 0 & \lambda^{-\frac{1}{2}} \end{bmatrix} . \tag{4.13}$$

Hence if (4.11) is true for $k = 0, \ldots, h$, then

$$L_h(a) = \begin{cases} \Lambda \; H_h \; \Lambda & \text{if } h \text{ is even,} \\ \Lambda^{-1} H_h \; \Lambda & \text{if } h \text{ is odd} \end{cases} . \tag{4.14}$$

Assume now that h is even. Then (4.14) becomes

$$L_h(a) = \begin{bmatrix} \lambda \tau_h & \sigma_h \\ \lambda_h & \lambda^{-1} \mu_h \end{bmatrix} . \tag{4.15}$$

Since

$$L_{h+1}(a) = S_{h+1}(a) \, L_h(a) \quad \text{and} \quad H_{h+1} = \Gamma_{h+1} H_h,$$

we have

$$\begin{cases} t_{h+1}(a) = C_{h+1}(a) \, [\lambda \tau_h] + \lambda_h , \\ s_{h+1}(a) = C_{h+1}(a) \sigma_h + [\lambda^{-1} \mu_h] \end{cases}$$

and

$$\begin{cases} \tau_{h+1} = \gamma_{n+1} \tau_h + \lambda_h , \\ \sigma_{h+1} = \gamma_{h+1} \sigma_h + \mu_h . \end{cases}$$

Hence (4.10) yields

$$\frac{\sigma_h[\lambda C_{h+1}(a)] + \mu_h}{\tau_h[\lambda C_{h+1}(a)] + \lambda_h} = \frac{\sigma_h \gamma_{h+1} + \mu_h}{\tau_h \gamma_{h+1} + \lambda_h} .$$

Since det $H_h \neq 0$, we obtain

$$\lambda C_{h+1}(a) = \gamma_{h+1} .$$

Similarly, we get

$$\lambda^{-1} C_{h+1}(a) = \gamma_{h+1}$$

if h is odd. Thus we have proved (4.11) (or (4.12)) for $k = 0, 1, \dots, m-1$.

The matrices $S_k(a)$ satisfy condition (2.17), Since (4.12) is true for $k = 0, 1, \dots, m-1$ and since $m-1$ is even, formula (2.17) becomes

$$S_{m+1}(a) S_m(a) \Lambda H_{m-1} \Lambda = I.$$

On the other hand, (4.2) can be written as

$$\Gamma_{m+1} \Gamma_m H_{m-1} = I.$$

Hence

$$\Lambda S_{m+1}(a) S_m(a) \Lambda = \Gamma_{m+1} \Gamma_m .$$

This means that

$$\text{(4.16)} \qquad \lambda = 1, \quad C_m(a) = \gamma_m, \quad C_{m+1}(a) = \gamma_{m+1} .$$

Now (4.11) and (4.16) yield

$$S_k(a) = \Gamma_k \qquad (k = 0, \ldots, m+1).$$

This completes the proof of Theorem A'.

5. Proof of Theorem B'. Let us define matrices $\tilde{\Gamma}_0, \ldots, \tilde{\Gamma}_{m+1}$ by

$$\text{(5.1)} \quad \begin{cases} \tilde{\Gamma}_k = C(\tilde{\gamma}_k, 1) & (k = 0, \ldots, m-1), \\ \tilde{\Gamma}_m = C(\tilde{\gamma}_m, \rho^{\frac{1}{2}m+1}), & \\ \tilde{\Gamma}_{m+1} = C(\tilde{\gamma}_{m+1}, \rho^{-\frac{1}{2}m-1}), & \end{cases}$$

respectively, where $C(\gamma, \rho)$ is the matrix defined by (1.4), and

$$\text{(5.2)} \quad \tilde{\gamma}_k = \begin{cases} \rho^k \gamma_0^{-1} \gamma_k & \text{if } k \text{ is even}, \\ \rho^{-k} \gamma_0 \gamma_k & \text{if } k \text{ is odd}, \end{cases} \qquad (k = 0, \ldots, m)$$

and

$$(5.3) \qquad \gamma_{m+1} = \rho^{-\frac{1}{2}m} \gamma_0 \gamma_{m+1} .$$

The matrices $\tilde{\Gamma}_k$ satisfy the condition

$$(5.4) \qquad \tilde{\Gamma}_{m+1} \tilde{\Gamma}_m \cdots \tilde{\Gamma}_1 \tilde{\Gamma}_0 = I.$$

In fact, if we set

$$(5.5) \qquad \Gamma = \begin{bmatrix} \gamma_0^{\frac{1}{2}} & 0 \\ 0 & \gamma_0^{-\frac{1}{2}} \end{bmatrix}, \qquad R = \begin{bmatrix} 1 & 0 \\ 0 & \rho \end{bmatrix},$$

and

$$(5.6) \qquad \Gamma_k = C(\gamma_k, \rho^{(-1)^k}) \qquad (k = 0, \ldots, m+1),$$

then

$$(5.7) \qquad \tilde{\Gamma}_k = \begin{cases} \rho^k R^{-k-1} \Gamma^{-1} \Gamma_k \Gamma^{-1} R^{-k} & \text{if } k \text{ is even}, \\ \rho^{-k} R^{k+1} \Gamma \Gamma_k \Gamma R^k & \text{if } k \text{ is odd}, \end{cases}$$

$$(k = 0, \ldots, m-1),$$

and

$$(5.8) \qquad \begin{cases} \tilde{\Gamma}_m = \rho^m R^{-\frac{1}{2}m} \Gamma^{-1} \Gamma_m \Gamma^{-1} R^{-m}, \\ \tilde{\Gamma}_{m+1} = \rho^{-\frac{1}{2}m} \Gamma \Gamma_{m+1} \Gamma R^{\frac{1}{2}m} . \end{cases}$$

Identity (5.4) follows from (1.10), (5.7) and (5.8). Note that $\tilde{\gamma}_0 = 1$.

Hence

$$(5.9) \qquad \tilde{\Gamma}_0 = \begin{bmatrix} 1 & 1 \\ -\omega & 0 \end{bmatrix} .$$

Set

$$(5.10) \qquad H_k = \tilde{\Gamma}_k \tilde{\Gamma}_{k-1} \cdots \tilde{\Gamma}_0 = \begin{bmatrix} \tau_k & \sigma_k \\ \lambda_k & \mu_k \end{bmatrix}.$$

$$(k = 0, \ldots, m-1).$$

and

$$(5.11) \qquad c_0 = \infty , \quad c_1 = 0, \quad c_k = \frac{\sigma_{k-2}}{\tau_{k-2}} \quad (k = 2, \ldots, m+1).$$

Then $c_0, \ldots, c_{m+1}$ are well defined points on the Riemann sphere, and they satisfy the condition

$$c_{k+1} \neq c_k \ (k = 0, \ldots, m+1; \ c_{m+2} = c_0).$$

Furthermore, since $c_2 = 1$, the set $\{c_0, \ldots, c_{m+1}\}$ contains three distinct points ∞, 0 and 1. Hence by virtue of Lemma 2.4 (Section 2), there exist m complex numbers $\tilde{a}_1, \ldots, \tilde{a}_m$ and two linearly independent solutions $y_1(x)$ and $y_2(x)$ of

$$(5.12) \qquad y'' - [x^m + \tilde{a}_1 x^{m-1} + \cdots + \tilde{a}_{m-1} x + \tilde{a}_m] y = 0$$

such that

$$(5.13)\qquad F(x) = \frac{y_1(x)}{y_2(x)} \rightarrow c_k \quad \text{as} \quad x \to \infty \quad \text{in} \quad \mathscr{S}_k$$

$$(k = 0, \ldots, m+1).$$

Since $c_0 = \infty$, $c_1 = 0$ and $c_2 = 1$, we must have

$$(5.14)\qquad C_0(\tilde{a}) \neq 0$$

and

$$(5.15)\qquad F(x) = C_0(\tilde{a}) \frac{\mathscr{Y}_{m,1}(x, \tilde{a})}{\mathscr{Y}_{m,0}(x, \tilde{a})} .$$

The subdominant solutions $\mathscr{Y}_{m,k}(x, \tilde{a})$ admit the connection formulas

$$[\mathscr{Y}_{m,k}(x, \tilde{a}), \mathscr{Y}_{m,k+1}(x, \tilde{a})] = [\mathscr{Y}_{m,k+1}(x, \tilde{a}), \mathscr{Y}_{m,k+2}(x, \tilde{a})]\, S_k(\tilde{a})$$

$$(k = 0, \ldots, m+1),$$

where the matrices S_k are defined by (2.16). Let us define another set of subdominant solutions by

$$(5.16)\qquad \tilde{y}_k(x) = \begin{cases} \beta(\tilde{a})^{\frac{1}{2}k}\, C_0(\tilde{a})^{-\frac{1}{2}}\, \mathscr{Y}_{m,k}(x, \tilde{a}) & \text{if } k \text{ is even,} \\ \beta(\tilde{a})^{-\frac{1}{2}(k-1)}\, C_0(\tilde{a})^{\frac{1}{2}} \mathscr{Y}_{m,k}(x, \tilde{a}) & \text{if } k \text{ is odd} \end{cases}$$

$$(k = 0, \ldots, m+1),$$

where the function β is defined by (3.9). Let

$$(5.17)\qquad [\tilde{y}_k(x), \tilde{y}_{k+1}(x)] = [\tilde{y}_{k+1}(x), \tilde{y}_{k+2}(x)]\, \tilde{S}_k(\tilde{a})$$

$$(k = 0, \ldots, m+1)$$

by the connection formulas for the subdominants solutions $\mathscr{y}_0(x), \ldots, \mathscr{y}_{m+1}(x)$, where $\mathscr{y}_{m+2}(x) = \mathscr{y}_0(x)$ and $\mathscr{y}_{m+3}(x) = \mathscr{y}_1(x)$. Then

$$
(5.18)\quad \begin{cases}
\tilde{S}_k(\tilde{a}) = \begin{bmatrix} Q_k(\tilde{a}) & 1 \\ -\omega & 0 \end{bmatrix} \quad (k = 0, \ldots, m-1), \\[2ex]
\tilde{S}_m(\tilde{a}) = \begin{bmatrix} Q_m(\tilde{a}) & 1 \\ -\omega\beta(\tilde{a})^{\frac{1}{2}m+1} & 0 \end{bmatrix}, \\[2ex]
\tilde{S}_{m+1}(\tilde{a}) = \begin{bmatrix} Q_{m+1}(\tilde{a}) & 1 \\ -\omega\beta(\tilde{a})^{-\frac{1}{2}m-1} & 0 \end{bmatrix},
\end{cases}
$$

where

$$
(5.19)\quad Q_k(\tilde{a}) = \begin{cases} \beta(\tilde{a})^k\, C_0(\tilde{a})^{-1}\, C_k(\tilde{a}) & \text{if } k \text{ is even}, \\ \beta(\tilde{a})^{-k} C_0(\tilde{a})\, C_k(\tilde{a}) & \text{if } k \text{ is odd}, \end{cases} \qquad (k = 0, \ldots, m)
$$

and

$$
(5.20)\quad Q_{m+1}(\tilde{a}) = \beta(\tilde{a})^{-\frac{1}{2}m} C_0(\tilde{a}) C_{m+1}(\tilde{a}).
$$

On the other hand, we have

$$
(5.21)\qquad F(x) = \frac{\mathscr{y}_1(x)}{\mathscr{y}_0(x)} .
$$

Let us put

$$(5.22)\quad L_k(\tilde{a}) = \tilde{S}_k(\tilde{a})\tilde{S}_{k-1}(\tilde{a}) \cdots \tilde{S}_0(\tilde{a}) = \begin{bmatrix} t_k(\tilde{a}) & s_k(\tilde{a}) \\ \ell_k(\tilde{a}) & m_k(\tilde{a}) \end{bmatrix}$$

$$(k = 0,\ 1, \ldots,\ m-1).$$

Then we get

$$(5.23)\qquad F(x) \to \frac{s_{k-2}(\tilde{a})}{t_{k-2}(\tilde{a})} \quad \text{as} \quad x \to \infty \quad \text{in} \quad \mathscr{S}_k$$

$$(k = 2, \ldots,\ m+1).$$

Hence

$$(5.24)\qquad \frac{s_k(\tilde{a})}{t_k(\tilde{a})} = \frac{\sigma_k}{\tau_k} \qquad (k = 0, 1, \ldots,\ m-1).$$

Note that we have

$$(5.25)\qquad \tilde{S}_0(\tilde{a}) = \tilde{\Gamma}_0 \ .$$

From (5.24) and (5.25) we can derive

$$(5.26)\qquad \tilde{S}_k(\tilde{a}) = \tilde{\Gamma}_k \qquad (k = 0,\ 1, \ldots,\ m-1)$$

in the same way as in Section 4. Since the matrices $\tilde{\Gamma}_k$ satisfy condition (5.4) and the matrices $\tilde{S}_k(\tilde{a})$ satisfy the condition

$$\tilde{S}_{m+1}(\tilde{a})\ \tilde{S}_m(\tilde{a})\tilde{S}_{m-1}(\tilde{a}) \ldots \tilde{S}_0(\tilde{a}) = I,$$

we have

$$\tilde{S}_{m+1}(\tilde{a})\tilde{S}_m(\tilde{a}) = \tilde{\Gamma}_{m+1}\tilde{\Gamma}_m \ .$$

It follows from Lemma 2.5 (Section 2) that

$$\tilde{S}_m(\tilde{a}) = \tilde{\Gamma}_m \quad \text{and} \quad \tilde{S}_{m+1}(\tilde{a}) = \tilde{\Gamma}_{m+1} .$$

Hence

$$(5.27) \quad \begin{cases} \beta(\tilde{a})^{\frac{1}{2}m+1} = \rho^{\frac{1}{2}m+1} , \\ Q_k(\tilde{a}) = \tilde{\gamma}_k \qquad (k = 0, \ldots, m+1). \end{cases}$$

Let s be an unknown quantity and define $a_1, \ldots, a_m$ by

$$(5.28) \quad x^m + a_1 x^{m-1} + \cdots + a_{m-1}x + a_m$$

$$= (x+s)^m + \tilde{a}_1(x+s)^{m-1} + \cdots + \tilde{a}_{m-1}(x+s) + \tilde{a}_m .$$

The quantities $a_1, \ldots, a_m$ are polynomials in s. In general

$$b_{\frac{1}{2}m+1}(a) = \frac{1}{2\pi i} \int_{\mathcal{C}} [x^m + \sum_{k=1}^{m} a_k x^{m-k}]^{\frac{1}{2}} dx$$

is $\mathcal{C}$ is a simply closed curve such that all roots of the polynomial are contained in its interior. Hence

$$(5.29) \quad \beta(a) = \beta(\tilde{a}) .$$

On the other hand, it follows from (iii) of Lemma 2.3 (Section 2) that

$$C_k(a) = \exp\{(-1)^k 2E_m(s, \tilde{a})\} C_k(\tilde{a}) \qquad (k=0, \ldots, m+1).$$

Therefore

$$(5.30) \qquad Q_k(\tilde{a}) = \begin{cases} \beta(a)^k C_0(a)^{-1} C_k(a) & \text{if } k \text{ is even,} \\ \beta(a)^{-k} C_0(a) C_k(a) & \text{if } k \text{ is odd,} \end{cases}$$

$$(k = 0, \ldots, m)$$

and

$$(5.31) \qquad Q_{m+1}(\tilde{a}) = \beta(a)^{\frac{1}{2}m} C_0(a) C_{m+1}(a).$$

Determine s so that

$$(5.32) \qquad C_0(a) = \gamma_0 .$$

Then Theorem B′ follows from (5.2), (5.3), (5.27), (5.30), (5.31) and (5.32).

6. <u>A remark.</u> In 1909, G. D. Birkhoff [1] proposed a general problem of constructing a system of linear differential equations which admits prescribed asymptotic behaviors of solutions and a prescribed Stokes phenomenon around an irregular singular point. Subsequently in 1913, he published a general method of solving such a problem. (Cf. G. D. Birkhoff [2; §§ 13 and 14, pp 545-551].) Our problem is differenti from that of Birkhoff, since, in our case, the prescribed asymptotic behavior of solutions depend on the parameters $a_1, \ldots, a_m$ which must be determined by the special form of the differential equation and the prescribed Stotes phenomenon. Note that our proofs of Theorems A and B are based on a work done by

R. Nevanlinna [4]. (Cf. Lemma 2.4 of Section 2.)

REFERENCES

[1] G. D. Birkhoff: Singular points of ordinary linear differential equations, Trans. Amer. Math. Soc., 10(1909) 436-470.

[2] G. D. Birkhoff: The generalized Riemann problem for linear differential equations and the allied problems for linear difference and q-difference equations, Proc. Amer. Acad. Arts. and Sci., 49(1913) 521-568.

[3] P. F. Hsieh and Y. Sibuya: On the asymptotic integration of second order linear ordinary differential equations with polynomial coefficients, J. Math. Ana. Appl., 16(1966) 84-103;

[4] R. Nevanlinna: Über Riemannsche Flachen mit Endlich Vielen Windungspunkten, Acta Math., 58(1932) 295-373.

[5] Y. Sibuya: Subdominant solutions of linear differential equations with polynomial coefficients, Mich. Math. Jour., 14(1967) 53-63.

[6] Y. Sibuya: Uniform simplification in a full neighborhood of a transition point, Amer. Math. Soc. Memoir 149, 1974.

University of Minnesota, Minneapolis, Minnesota 55455

NECESSARY CONDITIONS WITHOUT DIFFERENTIABILITY IN UNILATERAL CONTROL PROBLEMS

J. Warga

1. Introduction.

In this paper we continue our investigation of necessary conditions without differentiability assumptions, and state results generalizing those of [6] to unilateral control problems defined by ordinary differential equations. Specifically, we consider the relaxed optimal control problem defined, in its original (unrelaxed) version, by the cost functional $h^0(y(t_1))$ and the relations

(1.1) $\dot{y}(t) = f(t, y(t), u(t))$ a.e. in $T = [t_0, t_1]$,

(1.2) $y(t_0) \in A_0, \quad h^1(y(t_1)) \in A_1$,

(1.3) $u(t) \in U^{\#}(t)$ a.e. in T,

(1.4) $a^i(t, y(t)) \leq 0 \quad (t \in T,\ i=1, 2, \ldots, m_2)$,

where the functions $f(t, \cdot, u)$, h^0, h^1, and $a^i(t, \cdot)$ are assumed Lipschitz-continuous over bounded sets but not necessarily differentiable or with any particular convexity properties. This problem differs from the one investigated in [6] by addition of the control restriction (1.3) and the

Partially supported by Grant GP-37507X of the National Science Foundation.

unilateral restriction (1.4); and it differs from the problems previously investigated by the author [2, 3, 4, 5] and by Rockafellar [1] by the absence of differentiability or convexity assumptions.

Our present results are obtained in a manner similar to that of [6] and are based on the use of convolutions with mollifiers to approximate the functions $h^0, h^1, a^i(t, \cdot)$ and $f(t, \cdot, u)$ by C^1 functions. This yields a sequence of "approximating" unilateral problems of a type for which necessary conditions were previously derived [4], [5]. Our final results are obtained by investigating the behavior of these necessary conditions for ever finer approximations. The only phase of this research using techniques other than those encountered in [6] is the study of the convergence of the "dual" functions for the approximating unilateral problems.

Our basic results are presented in Theorems 2.2 and 2.3. The proofs are available from the author and will be published elsewhere.

2. Existence and necessary conditions.

2.1 Definitions and assumptions. We shall use the terms "a.e.", "a.a." (almost all) and "measurable" in the sense of the Borel-Lebesgue measure on T which we denote by μ. Our optimal control problem is relaxed by replacing relations (1.1) and (1.3) by

(2.1.1) $\quad \dot{y}(t) \in \text{co } f(t, y(t), \overline{U^\#}(t))$ a.e. in T,

where $\overline{U^\#}(t)$ is the closure of $U^\#(t)$ and co denotes the convex hull. We refer to an absolutely continuous function y that satisfies relations (1.2), (1.4) and (2.1.1) as an <u>admissible relaxed solution.</u> If $\bar{y}$ minimizes $h^0(y(t_1))$ among admissible relaxed solutions then it is a <u>minimizing relaxed solution.</u> For $x = (x^1, \ldots, x^\ell) \in \mathbb{R}^\ell$, we define the norm of x by $|x| = \operatorname{Max}_i |x^i|$. We make the following assumptions:

(a) V is an open subset of $\mathbb{R}^n$;

(b) A_0 and A_1 are closed convex subsets of $\mathbb{R}^n$ and $\mathbb{R}^m$, respectively;

(c) U is a compact metric space, $U^\#(t) \in U$ $(t \in T)$, the function $t \to U^\#(t)$ is measurable (that is, for every open subset G of U, the set $\{t \in T \mid U^\#(t) \cap G \neq \emptyset\}$ is measurable), and either $U^\#(t)$ is closed for all $t \in T$ or, for all $t \in T$, $U^\#(t)$ is contained in the closure of its interior and for every $\varepsilon > 0$ there exists $T_\varepsilon \subset T$ such that $\mu(T - T_\varepsilon) \leq \varepsilon$ and the set $\{(t,u) \in T_\varepsilon \times U \mid u \in \text{interior } U^\#(t)\}$ is open relative to $T_\varepsilon \times U$;

(d) There exist a number $c > 0$ and a compact set $D \subset V$ such that the functions $h^0 : V \to \mathbb{R}$, $h^1 : V \to \mathbb{R}^m$, $a = (a^1, \ldots, a^{m_2}) : T \times V \to \mathbb{R}^{m_2}$, and $f : T \times V \times U \to \mathbb{R}^n$ satisfy the following conditions for all $(t, v, u) \in T \times V \times U$:

(d1) The functions $f(t, \cdot, u)$, h^0, h^1 and $a(t, \cdot)$ admit c as a common bound (for the norms of their values) and a common Lipschitz-constant;

(d2) $f(\cdot, v, u)$ is measurable and both $f(t, \cdot, \cdot)$ and $a(\cdot, \cdot)$ are continuous;

(d3) $y([t_0,\tau]) \subset D$ for every $\tau \in T$ and every absolutely continuous $y: [t_0,\tau] \to V$ satisfying (2.1.1) a.e. in $[t_0,\tau]$ and with $y(t_0) \in A_0$.

Remark. As it is easy to see, we may weaken these assumptions as far as f is concerned by replacing c with an integrable function $\psi: T \to (0,\infty)$, and then choosing as the new independent variable $\theta = \int_{t_0}^{t} \psi(\tau)d\tau$. This transformation will cause the function $\tilde{f}(\theta,\cdot,u)$ of the transformed problem to admit 1 as a bound and a Lipschitz-constant and will not affect the other assumptions.

Now let e_k denote the k-th column of the unit matrix I of appropriate dimension, $S^F(x,r)$ the closed ball with center x and radius r, $d(x,y)$ ($d[x,A]$) the distance between two points (a point and a set), $S^F(A,r)$ the set $\{b \mid d[b,A] \leq r\}$, and $L(\mathbb{R}^a, \mathbb{R}^b)$ the space of real $b \times a$ matrices. If $\phi = (\phi^1,\ldots,\phi^b): G \to \mathbb{R}^b$ has a Lipschitz-constant $\tilde{c}$, G is an open subset of $\mathbb{R}^a$, $\varepsilon > 0$, $x \in G$, $i \in \{1,2,\ldots,b\}$, and $k \in \{1,2,\ldots,a\}$, we set $A_{i,k}(x) = [-\tilde{c},\tilde{c}]$ if $S^F(x,2\varepsilon) \not\subset G$ and

$$A_{i,k}(x) = \{\tfrac{1}{2\alpha}[\phi^i(\xi+\alpha e_k) - \phi^i(\xi-\alpha e_k)] \mid |\xi - x| \leq \varepsilon,\ 0 < \alpha \leq \varepsilon\}$$

if $S^F(x,2\varepsilon) \subset G$, and then define $\Delta^\varepsilon\phi(x)$ as the set of all $b \times a$ matrices $(M_{i,k})$ such that $M_{i,k} \in [\inf A_{i,k}(x), \sup A_{i,k}(x)]$ for all i and k. We write Δ_V^ε for the "partial" Δ^ε operator with respect to the argument in V; thus $\Delta_V^\varepsilon f(t,v,u)$, $\Delta_V^\varepsilon a^i(t,v)$ etc. denote $\Delta^\varepsilon\phi(v)$, where ϕ represents $f(t,\cdot,u)$, $a^i(t,\cdot)$ etc.

Finally, we denote transposition by T, treat each

element of $\mathbb{R}^a$ as a column vector, and write $M\mathcal{A}$ for $\{MA \mid A \in \mathcal{A}\}$ if $\mathcal{A}$ is a collection of matrices.

We can now state our first theorem combining existence with necessary conditions.

2.2 Theorem Assume that the set of admissible relaxed solutions is nonempty. Then there exist a minimizing relaxed solution $\bar{y}$ and a sequence $((y_j, u_j))_{j=1}^{\infty}$ such that each $y_j: T \to V$ is Lipschitz-continuous, each u_j is a measurable selection of $U^{\#}$, each (y_j, u_j) satisfies the differential equation (1.1), and $\lim_j y_j = \bar{y}$ uniformly. Furthermore, there exist

$\ell_0 \geq 0$, $\ell_1 \in \mathbb{R}^m$, $\tilde{h}^0 \in L(\mathbb{R}^n, \mathbb{R})$, $\tilde{h}^1 \in L(\mathbb{R}^n, \mathbb{R}^m)$, Borel measurable $\tilde{a}^i: T \to L(\mathbb{R}^n, \mathbb{R})$ and positive Radon measures ω^i on T $(i=1, 2, \ldots, m_2)$, Lipschitz-continuous $Z: T \to L(\mathbb{R}^n, \mathbb{R}^n)$, and $k: T \to \mathbb{R}^n$ such that

(1) $$\ell_0 + |\ell_1| + \sum_{i=1}^{m_2} \omega^i(T) > 0, \quad Z(t_1) = I;$$

(2) $\omega^i (i=1, 2, \ldots, m_2)$ is supported on the set

$\{t \in T \mid a^i(t, \bar{y}(t)) = 0\}$;

(3) $$\tilde{h}^\ell \in \bigcap_{\varepsilon > 0} \Delta^\varepsilon h^\ell(\bar{y}(t_1)) \quad (\ell=0, 1);$$

(4) $$\tilde{a}^i(t) \in \bigcap_{\varepsilon>0} \overline{co} \bigcup_{|\tau - t| < \varepsilon} \Delta_V^\varepsilon a^i(\tau, \bar{y}(t)) \quad \omega^i\text{-a. e.} \quad (i=1, 2, \ldots m_2);$$

(5) $$k(t)^T = [\ell_0 \tilde{h}^0 + \ell_1^T \tilde{h}^1 |^{m_2}| \int_{[t,t_1]} \tilde{a}^i(\tau) Z(\tau)^{-1} \omega^i(d\tau)] Z(t) \quad (t \in T);$$

(the maximum principle)

(6) $$\frac{d}{dt} (\bar{y}(t), Z(t)) \in \text{co} \bigcap_{\varepsilon>0} K_\varepsilon(t) \quad \text{a. e. in } T,$$

where

$$K_{\varepsilon}(t) = \text{closure}\,\{(f(t,\bar{y}(t),\tilde{u}), -W) \mid W \in Z(t)\Delta_V^{\varepsilon} f(t,\bar{y}(t),\tilde{u}),\ \tilde{u} \in U,$$

$$k(t) \cdot f(t,\bar{y}(t),\tilde{u}) = \underset{U^{\#}(t)}{\text{Min}}\ k(t) \circ f(t,\bar{y}(t),u)\ \} ;$$

and

$$(7) \qquad k(t_0) \cdot \bar{y}(t_0) = \underset{a_0 \in A_0}{\text{Min}}\ k(t_0) \cdot a_0,\ \ \ell_1 \cdot h^1(\bar{y}(t_1)) = \underset{a_1 \in A_1}{\text{Max}}\ \ell_1 \cdot a_1 .$$

Our second theorem, generalizing Theorem 2.2, provides a different set of necessary conditions for each representation of $f(t,\cdot,u)$, h^0, h^1, and $a^i(t,\circ)$ as compositions. 2.5] for a discussion of this matter.) In order to state this theorem, we must use the definition of derivate containers introduced in [6].

Let $\phi: V \to \mathbb{R}^b$ be Lipschitz-continuous. We say that the sets $\Lambda^{\varepsilon}(v)$ $(\varepsilon > 0,\ v \in V)$ determine a derivate container for ϕ if there exist positive integers ℓ, $k_0, \ldots, k_\ell$, open sets $V_i \subset \mathbb{R}^{k_i}$ $(i=0, 1, \ldots, \ell)$, $\gamma > 0$, and Lipschitz-continuous $\phi_i: V_i \to V_{i-1}$ $(i=1, \ldots, \ell)$ such that $V_\ell = V$, $V_0 = \mathbb{R}^b$, $\phi = \phi_1 \circ \phi_2 \circ \cdots \circ \phi_\ell$, $S^F(\phi_i(V_i),\gamma) \subset V_{i-1}$, and

$$\Lambda^{\varepsilon}(v) = \{M_1 M_2 \cdots M_\ell \mid w_\ell = v,\ w_i = \phi_{i+1} \circ \cdots \circ \phi_\ell(v),\ M_i \in \Delta^{\varepsilon}\phi_i(w_i)\}.$$

We say that $\Lambda_f^{\varepsilon}(t,v,u)$ $(\varepsilon > 0,\ t \in T,\ v \in V,\ u \in U)$ determine a derivate container for f if there exist positive integers ℓ, $k_0, \ldots, k_\ell$, open $V_i \subset \mathbb{R}^{k_i}$, $\gamma > 0$, and $\hat{f}_i: T \times V_i \times U \to V_{i-1}$ $(i=0, \ldots, \ell)$ such that $V_\ell = V$, $V_0 = \mathbb{R}^n$, $S^F(f_i(t,V_i,u),\gamma) \subset V_{i-1}$,

$$f(t,\cdot,u) = \hat{f}_1(t,\cdot,u) \circ \cdots \circ \hat{f}_\ell(t,\cdot,u) \quad (t \in T,\ u \in T),$$

$\Lambda_f^\varepsilon(t, v, u)$ is defined as $\Lambda^\varepsilon(v)$ with $\phi_i = \hat{f}_i(t, \cdot, u)$ and, furthermore, each $\hat{f}_i(\cdot, v, u)$ is measurable, each $\hat{f}_i(t, \cdot, \cdot)$ continuous, each $\hat{f}_i(t, \cdot, u)$ Lipschitz-continuous with a Lipschitz-constant independent of i, t and u, and each $\hat{f}_i$ bounded. We similarly define a derivate container $\Lambda_{a,i}^\varepsilon(t, v)$ for a^i which must be based on a composition

$$a^i(t, \cdot) = \hat{a}_1^i(t, \cdot) \circ \hat{a}_2^i(t, \cdot) \circ \ldots \circ \hat{a}_\ell^i(t, \cdot)$$

with each function $\hat{a}_j^i(\cdot, \cdot)$ continuous and each $\hat{a}_j^i(t, \cdot)$ admitting the same Lipschitz-constant (independent of i, j and t). (We might add, for the sake of clarity, that ℓ, k_i, V_i etc. may be different for each of the functions f, h^0, h^1, a^i).

2.3 Theorem Let the conditions of Theorem 2.2 be satisfies and $\Lambda_f^\varepsilon(t, v, u)$, $\Lambda_0^\varepsilon(v)$, $\Lambda_1^\varepsilon(v)$, $\Lambda_{a,i}^\varepsilon(t, v)$ $(i=1, \ldots, m_2)$ determine derivate containers for f, h^0, h^1, a^i, respectively. Then the conclusions of Theorem 2.2 remain valid with $\Delta_v^\varepsilon f, \Delta^\varepsilon h^0, \Delta^\varepsilon h^1, \Delta_\sigma^\varepsilon a^i$ replaced by $\Lambda_v^\varepsilon, \Lambda_0^\varepsilon, \Lambda_1^\varepsilon, \Lambda_{a,i}^\varepsilon$, respectively.

REFERENCES

[1] R. T. Rockafellar, State constraints in convex problems of Bolza, SIAM J. Control 10 (1972), 691-715.

[2] J. Warga, Minimizing variational curves restricted to a preassigned set, Trans. Amer. Math. Soc. 112 (1964), 432-455.

[3] ______________, Unilateral variational problems with several inequalities, Michigan Math. J. 12 (1965), 449-480.

[4] ______________, Unilateral and minimax problems defined by integral equations, SIAM J. Control 8 (1970), 372-382.

[5] ______________, "Optimal Control of Differential and Functional Equations", Academic Press, New York, 1972.

[6] ______________, Necessary conditions without differentiability assumptions in optimal control, J. Diff. Eqs., to appear.

Northeastern University, Boston, Massachusetts

SOME RECENT RESULTS IN THE THEORY OF ADIABATIC INVARIANTS

Wolfgang Wasow

1. Introduction.

The main purpose of this mostly expository article is to direct the attention of mathematicians with an interest in the asymptotic properties of differential equations to the intriguing and potentially rewarding mathematical questions connected with the physical concept of adiabatic invariance.

To a physicist an adiabatic invariant is a quantity that depends on the state of a physical system in such a way that it remains unchanged even under substantial modifications of that system, provided these are performed "infinitesimally slowly".

Since 1911 ([12], p. 450) a large body of literature on this subject has accumulated, most of it written by mathematically inclined physicists. These physically inspired, often very ingenious contributions have a purely mathematical content that is imperfectly brought out in those papers. I believe that both Physics and Mathematics will benefit, if more mathematicians take a look at such problems.

If I am going to concentrate here on linear problems it is only because I know little about the difficult nonlinear situations, and I believe that not much is known by anybody, if criteria of mathematical precision are imposed. Much of the literature does deal with nonlinear differential equations, but it is clear from studying these papers that a purely mathematical formulation of the arguments would require a careful global analysis beyond the power of the techniques employed by these writers. The recent remarkable advances in the global theory of differential equations ought to be brought into play.

In mathematical - but sketchy - language the essence of what the physicists mean by an adiabatic invariant has to do with n-dimensional autonomous systems of ordinary differential equations of the form

$$\frac{dy}{d\tau} = f(y, a) , \tag{1.1}$$

where a is an m-dimensional parameter. The "infinitesimally slow" change of the system can be mathematically expressed by introducing the "compressed", or "slow", time

$$t = \varepsilon\tau , \tag{1.2}$$

where ε is a positive parameter that tends to zero, and by replacing the constant parameter a with a function of t. Let $y(t, \varepsilon)$ be a solution of the modified differential equation

$$\frac{dy}{d\tau} = f(y, a(\varepsilon\tau)) , \tag{1.3}$$

or, equivalently,

$$\varepsilon \frac{dy}{dt} = f(y, a(t)) . \tag{1.4}$$

If, then, $I(y, a, t)$ is a scalar function with the property that

$$\lim_{\varepsilon \to 0+} I(y(t,\varepsilon),\ a(t),\ t) \tag{1.5}$$

exists and is independent of t for an appropriate class of solutions of (1.4), then $I(y, a, t)$ is called an adiabatic invariant for the differential equation (1.1).

Here is a more precise version of the description just given:

DEFINITION 1.1. Let $\mathcal{Y}$ and $\mathcal{A}$ be given subsets of $\mathbb{C}^n$ and $\mathbb{C}^m$, respectively, and denote by f a given continuous function from $\mathcal{Y} \times \mathcal{A}$ into $\mathbb{C}^n$. Let $\mathcal{F}$ be a family of continuous function from an interval $\mathcal{J} = \{t \mid t_1 \le t \le t_2\}$ into $\mathcal{A}$, and let $\mathcal{S}$ be a family of solutions of the differential equation (1.4) with $a(\cdot) \in \mathcal{F}$, and $0 < \varepsilon \le \varepsilon_0$. A function I from $\mathcal{S} \times \mathcal{F} \times \mathcal{J}$ into $\mathbb{C}$ is called an adiabatic invariant of the differential equation (1.1) with respect to $\mathcal{F}, \mathcal{J}$ and $\mathcal{S}$, if the limit (1.5) exists and is independent of t.

There is no generally accepted precise definition of adiabatic invariance in the literature. The one given above is my personal choice and differs in important respects from others that have been proposed (See, e.g. [5], [6], [9], [10], [11], [16], [17], [28].). It clearly can be extended to non-autonomous differential equations and to more general functional equations of evolution type. In the examples to be discussed below $I(y, a, t)$ does not depend explicitly on t. We will then use the simpler notation $I(y, a)$.

Invariably the ideas of the many papers on this subject are guided by one or the other of the two instances of adiabatic invariants which I am going to describe in the language of the preceding definition.

(A) <u>The Pendulum of Slowly Varying Length.</u> Here, $n = 2$, $m = 1$,

$$f(y, a) = \begin{bmatrix} 0 & 1 \\ -a^2 & 0 \end{bmatrix} y. \tag{1.6}$$

One may take $\mathcal{Y} = \mathbf{C}^2$, and $\mathcal{A}$ is best defined as the set of all positive numbers, if one wishes to avoid serious complications. Let $\mathcal{T} = \{t \mid -\infty \leq t \leq \infty\}$, and define $\mathcal{F}$ as the set of all positive-valued functions in $C^\infty[-\infty, \infty]$ all of whose derivatives are in $L^1(-\infty, \infty)$. Finally, we define $\mathcal{S}$ as the set of all solutions of (1.4) whose values are real and independent of ε at $t = 0$. Then, as has been more or less known since 1911 ([12], p. 450),

$$I(y, a) = ay_1^2 + a^{-1}y_2^2 \tag{1.7}$$

is an adiabatic invariant. Here y_j, $j = 1, 2$ are the components of $\mathbf{y}$. The first complete proof of this fact under the particular hypotheses above can be found in [29] and [31].

In more familiar notation one can write $y_1 = u$, $y_2 = \varepsilon\, du/dt = du/d\tau$. The differential equation becomes then $d^2u/d\tau^2 + a^2u = 0$, which is, indeed the differential equation for the linearized oscillations of a pendulum, and

$$\frac{1}{2}I(y, a) = \frac{1}{2a}[a^2u^2 + (du/d\tau)^2]$$

is --- for constant a --- the quotient of energy over frequency. The severe regularity requirements in the definition $\mathcal{F}$ above can be very much relaxed.

(B) The Time-Dependent Schrödinger Equation. In this example n is arbitrary, $\mathcal{Y} = \mathbf{C}^n$, $m = n^2$, the parameter a is the n-by-n matrix $H = \{h_{jk}\}$, and $\mathcal{A}$ is the set of all anti-Hermitian n-by-n matrices. In a notation better suited to this example, the differential equation (1.4) becomes here

$$\varepsilon \frac{dy_{jk}}{dt} = \sum_{\mu=1}^{n} h_{j\mu}(t) y_{\mu k}\ ; \quad j, k = 1, 2, \cdots, n. \tag{1.8}$$

Let $\mathcal{I}$ be again the whole t-axis. $\mathcal{J}$ is to be the set of anti-Hermitian matrix functions whose entries have the same regularity properties as in Example (A). Let $\Phi(H)$ be a unitary matrix whose columns are eigenvectors of H. For each $H \in \mathcal{J}$ the family $\mathcal{S}$ consists here of the one matrix solution $Y(t,\varepsilon)$ of (1.8) which satisfies the initial condition

$$Y(0,\varepsilon) = \Phi(H(0)). \tag{1.9}$$

If $\varphi_{jk}(H)$ denotes the entries of $\Phi(H)$ it can then be proved that each of the functions

$$I_{jk}(y, H) = |y_{jk}| - |\varphi_{jk}(H)|, \quad j, k = 1, 2, \cdots, n, \tag{1.10}$$

is an adiabatic invariant. (See [2], [3], [4], [20], [46].)

2. Connections with the Standard Asymptotic Theory.

Both examples inthe preceding section involve the analysis of matrix differential equations of the form

$$\varepsilon \frac{dy}{dt} = A(t)y \tag{2.1}$$

for small positive ε. There exists an extensive body of results on this and related equations. One of the simpler one is the following. (See, e.g., [41], p. 146.)

THEOREM 2.1. *Let* $A(t)$ *be holomorphic at* $t = 0$ *and assume that the eigenvalues* $\lambda_j(t)$, $j = 1, 2, \ldots, n$, *of* $A(t)$ *are distinct at* $t = 0$. *Then* (2.1) *possesses a particular fundamental matrix solution* $Y = U(t, \varepsilon)$ *of the form*

$$U(t,\varepsilon) = \hat{U}(t,\varepsilon) \exp\left\{\frac{1}{\varepsilon}\int_0^t \Lambda(s)ds\right\}, \tag{2.2}$$

where

$$\Lambda(t) = \operatorname{diag}[\lambda_1(t), \lambda_2(t), \ldots, \lambda_n(t)], \tag{2.3}$$

and $\hat{U}(t,\varepsilon)$ *possesses in a neighborhood* $|t| \le t_0$ *a uniformly asymptotic expansion*

$$\hat{U}(t,\varepsilon) \sim \sum_{r=0}^{\infty} U_r(t)\varepsilon^r, \quad \varepsilon \to 0+, \quad |t| \le t_0. \tag{2.4}$$

The functions $U_r(t)$ *are holomorphic in* $|t| \le t_0$, *and the columns of* $U_0(t)$ *are linearly independent eigenvectors of* $A(t)$.

The point $t = 0$ in the statement above can, of course, be replaced by any other point where A is holomorphic and its eigenvalues are distinct. The weakness of the theorem is its local character: t_0 may be very small, and it is not known explicitly. Nevertheless, we can easily deduce from Theorem 2.1 fairly general theorems about adiabatic invariants, at least for small intervals $\mathcal{J}$.

THEOREM 2.2. *Assume, in the notation of Definition* 1.1, *that* $t_1 = 0$ *and that for all* $a \in \mathcal{J}$ *and all* $t \in \mathcal{J}$,

(a) $f(y, a) = M(a)y$, *and* $M(a(t))$ *is holomorphic;*

(b) *The eigenvalues* $\lambda_j(t) = \mu_j(a(t))$ *of* $M(a(t))$ *are distinct;*

(c) t_2 is so small that Theorem 2.1 is valid for

(2.5) $$\varepsilon \frac{dy}{dt} = M(a(t))y \;;$$

(d) Re $\mu_j(a(t)) \leq 0$; $j = 1, 2, \ldots, n$;

(e) There exist constants c_j such that $\sum_{j=1}^{n} c_j \mu_j(a) \equiv 0$.

Then, if $P(a)$ is a matrix whose j^{th} column is an eigenvector of $M(a)$ corresponding to $\mu_j(a)$, $j=1, 2, \ldots, n$,

(2.6) $$I(y, a) = \prod_{j=1}^{n} [(P^{-1}(a)y)_j]^{c_j}$$

is an adiabatic invariant of

(2.7) $$\frac{dy}{d\tau} = M(a)y \;,$$

with respect to $\mathcal{F}, \mathcal{J}$ and the family $\mathcal{S}$ of solutions of (2.5) with values independent of ε at $t = 0$.

The notation "$(\ )_j$" above indicates the jth component of the vector inside the parentheses.

Proof: By Theorem 2.1 we may set $U_0(t) = P(a(t))$. Now,

$$y(t, \varepsilon) = \hat{U}(t, \varepsilon) \exp\left\{\frac{1}{\varepsilon}\int_0^t \Lambda(s)ds\right\}\hat{U}^{-1}(0, \varepsilon)\mathring{y}$$

is the solution of (2.5) with $y(0, \varepsilon) = \mathring{y}$. From Theorem 2.1 and Assumption (d) it follows then that

$$P^{-1}(a(t))y(t, \varepsilon) = (I+0(\varepsilon)) \exp\left\{\frac{1}{\varepsilon}\int_0^t \Lambda(s)ds\right\}\hat{U}^{-1}(0, \varepsilon)\mathring{y}$$

$$= \exp\left\{\frac{1}{\varepsilon}\int_0^t \Lambda(s)ds\right\}U_0^{-1}(0)\mathring{y} + 0(\varepsilon).$$

The proof is immediately completed by using assumptions (e) and (d).

This theorem can be applied to Example (A) of section 1,

provided $\mathcal{J}$ and $\mathcal{F}$ are appropriately changed, since the eigenvalues are $\pm ia$ in that example. One has, therefore, $c_1 = c_2 = 1$ in Assumption (e). The corresponding adiabatic invariant (2.6) becomes then, indeed, the same as (1.7). The restriction of $\mathcal{J}$ to small intervals of unknown size makes this, of course, a much weaker result than the one quoted in section 1. In many cases the validity of Theorem 2.1 can, however, be extended to large intervals $\mathcal{J}$ and larger families $\mathcal{F}$, and this is, in fact, essentially the procedure used in [44] and in [27].

In the article [27] by A. Leung and K. Meyer the matrix M is assumed to be of even order and Hamiltonian for all functions of the family $\mathcal{F}$. This implies that the negative of each eigenvalue is also an eigenvalue. Thus, there are n/2 linearly independent linear relations among the eigenvalues. Leung and Meyer add the further condition that all eigenvalues are purely imaginary. Theorem 2.2 then permits the construction of n/2 adiabatic invariants. The article [27] deals with the same invariants, but the results obtained are much stronger, since they assert global adiabatic invariance on the whole t-axis. The essential part of the proof amounts to an extension of Theorem 2.1 to the whole real line under suitable regularity hypotheses.

It would be interesting to know if there exist more adiabatic invariants of the form (2.6) that have some physical relevance.

If the restrictive conditions on $\mathcal{J}$ and $\mathcal{F}$ in Theorem 2.2 are accepted it is also quite easy to deduce from

Theorem 2.1 a local generalization of Example (B) in section 1.

THEOREM 2.3. In the hypotheses of Theorem 2.2 replace (d) and (e) by the following condition:

(d^*) The eigenvalues of $M(a)$ are pure imaginary. For each given $a \in \mathcal{I}$ let $\mathcal{S}$ consist of the one matrix solution of (2.5) which agrees with $P(a)$ at $t = 0$. Then

(2.8) $$I_{jk}(y, a) = |y_{jk}| - |(P(a))_{jk}|, \quad j, k = 1, 2, \ldots, n$$

are adiabatic invariants with respect to $\mathcal{I}, \mathcal{F}$ and $\mathcal{S}$.

Proof: One has

$$Y(t,\varepsilon) = U(t,\varepsilon)U^{-1}(0,\varepsilon)P(a(0)).$$

As one may take $U_0(t) = P(a(t))$, and Λ is pure imaginary, Theorem 2.1 implies that

$$Y(t,\varepsilon) = P(a(t))\exp\{\frac{1}{\varepsilon}\int_0^t \Lambda(s)ds\}+0(\varepsilon),$$

and the statement of the theorem is immediate.

In Example (B) of section 1 the matrix $M(a)$ is anti-Hermitian and its eigenvalues are therefore purely imaginary. In [46] a thorough analysis of this adiabatic invariant is based on an extension of Theorem 2.1 to larger intervals and less restricted functions.

3. The Size of the Adiabatic Variation.

After having established the adiabatic invariance of a function $I(y, a, t)$ it is natural to ask for the approximate change of this quantity for small ε. In other words, one wishes to calculate

(3.1) $$I(y(t,\varepsilon), a(t), t) - I(y(t_1,\varepsilon), a(t_1), t_1)$$

approximately for small ε. It turns out that often this quantity depends in a very sensitive manner on the smoothness of the data. If one is motivated primarily by applications to Physics, the question of the best mathematical model then becomes crucial. It seems to me that in most cases functions that are piecewise C^∞ with jumps in some derivative of low order are best suited to carry over to a mathematical treatment the pertinent features of physical systems. The methods outlined in the preceding sections can be supplemented so as to yield explicitly the leading term, for small ε, of the expression (3.1) in the linear problems under discussion. In [46] I have carried out the details of such a calculation for our example (B) in the case that $H(t)$ has exactly one jump discontinuity in its first derivative. The result is an expression of order $O(\varepsilon)$, whose size depends — predictably — on the size of the jump.

For a physicist it is natural to ask primarily for the total change of the adiabatic invariant during the process and to assume that outside the time interval $[t_1, t_2]$ the parameter a remains constant. For the two examples (A) and (B) under discussion here A. Lenard [26] found that the total change of the adiabatic invariant is $O(\varepsilon^N)$, for all N. (See also [15].) This is a very surprising phenomenon, because for values of t in $t_1 < t < t_2$ the change of the invariant can be shown to be $O(\varepsilon)$, and no better. An inspection of Lenard's reasoning shows, however, that he uses decisively the assumption that $a(t)$, when continued beyond t_1 and t_2 as a constant function, is C^∞ in

$-\infty \leq t \leq \infty$. In other words, the process must be switched on and off "infinitely gently". Such a function $a(t)$ satisfies, in particular, the more general hypotheses on the family $\mathcal{F}$ in our examples. Under the latter conditions Lenard's result was proved by Littlewood [29], [31], for Example (A). A proof for Example (B) is contained in [46].

As I mentioned before, I doubt that the condition of C^{∞} - smoothness for $-\infty < t < \infty$ is a satisfactory model for a physical process. The fact that such a system "remembers" at $t = +\infty$ where it came from at $t = -\infty$ and settles down to a value of the adiabatic invariant especially close to that starting value is, very likely, no more than a mathematical curiosity. Nevertheless, the question of the "true" value of the total change in the adiabatic invariant (1.7) has intrigued numerous physicists and mathematicians, including myself. To obtain such results the function $a(t)$ must be characterized still more specifically as belonging to some more restricted family $\mathcal{F}$. We refer to [1], [2], [13], [19], [23], [33], [44], [45] for more information. In all those papers $a(t)$ is either one specific analytic function or required to belong to some particular family of such functions. The expressions obtained for the total change of the adiabatic invariant $I(y, a)$ turns out to be of the order $O(e^{-c/\varepsilon})$ with some positive constant c.

In [39], G. Stengle has pursued this question still further. Among the C^{∞}-functions the analytic ones are distinguished by the property that the sequence of their successive derivatives grows at a sufficiently slow rate with the order of differentiation to produce a convergent Taylor series. Stengle has succeeded in finding expressions for the order of magnitude of the total change of $I(y, a)$ in (1.7) in terms of the rate of growth of the sequence $\{d^k a/dt^k\}$ with k, when $a(t)$ is not analytic. His result is too complicated to state here more precisely.

4. Coalescing Eigenvalues.

As soon as multiple eigenvalues of the coefficient matrix are allowed, the asymptotic theory of linear differential equations becomes very much more intricate. The two examples (A) and (B) of adiabatic invariants on which we have concentrated most of out attention in this paper have been analyzed for certain cases of such degeneracy, but the theory is still very incomplete. More such investigations would be very desirable. For our Example (B), Born and Fock [4] have shown that a simple crossing of two eigenvalues at some value of t does not destroy the property of adiabatic invariance, and T. Kato [20], [21] has obtained more general results in this direction, even for the case that H is an operator in a function space.

Friedrichs, inthe unpublished reports [14], has performed detailed calculations of the adiabatic invariants $I_{jk}(y, H)$ of (1.10) in the case that $n = 2$. He assumes that the family $\mathcal{J}$ of anti-Hermitian matrices $H(t)$ is holomorphic in the interval $\mathcal{J}$ and that at one interior point, say at $t = 0$, the difference of the two eigenvalues, $\lambda_1(t) - \lambda_2(t)$, has a simple zero. The initial values determining the family $\mathcal{S}$ of solutions are chosen so that

$I_{jk}(y(t_1, \varepsilon), H(t_1)) = 0$. Friedrichs then obtains explicit formulas for the asymptotically leading term of the adiabatic invariant in $\mathcal{J}$. It turns out that in $[t_1, 0)$ this expression is of order $0(\varepsilon)$. Upon crossing the point $t = 0$ one finds, however a larger value of order $0(\sqrt{\varepsilon})$ for the adiabatic invariant.

These results can also be obtained from the asymptotic formulas in [43]. (See also [40].) The problem belongs into the theory of turning points. Although the difference of the eigenvalues vanishes of the first order, the point $t = 0$ resembles a turning point of order two for a single second order equation in that the Stokes curves divide the complex neighborhood of $t = 0$ into four, not three, sectors.

The hypothesis of analyticity simplifies the investigations decisively, because one benefits then of Rellich's theorem [34] which asserts that the eigenvalues are holomorphic at $t = 0$, instead of having a branch point.

The restriction to systems of order two is rather drastic. In this connection a theorem of Sibuya [36] is of value, which makes possible the decomposition of higher order systems into a set of uncoupled second and first order equations, provided no more than two eigenvalues coalesce at the same point. In [42] it is shown that for anti-Hermitian coefficient matrices this decomposition can be performed without destroying that anti-Hermitian property. D. Russell [35] has extended the last mentioned result to operators in Hilbert space. In the light of this situation the study of the case $n = 2$ takes on a more general importance.

REFERENCES

[1] Backus, G., A. Lenard and R. Kulsrud, Bemerkungen über die adiabatische Invarianz des Bahnmomentes geladener Teilchen. Z. Naturforschung, 15a (1960), 1007-1009.

[2] Born, M., Vorlesungen über Atomdynamik. Berlin, 1925, p. 113.

[3] Born, M,. Das Adiabatenprinzip der Quantenmechanik. Zeitschr. für Phys. 40 (1926), 167-192.

[4] Born, M. and V. Fock, Beweis des Adiabatensatzes. Zeitschr. für Phys. 51 (1928), 165-180.

[5] Burgers, J.M., Adiabatic invariants of mechanical systems. Phil. Mag. 33 (1917), 514-520.

[6] Burgers, J.M., Die adiabatischen Invarianten bedingt periodischer Systeme. Ganzzahligkeit der Quantenzahlen. Ann. der Phys. 52 (1917), 195-202.

[7] Chandrasekhar, S., Adiabatic invariants in the motion of charged particles. In "Plasma in a Magnetic Field". K.M. Landshoff, ed., pp. 3-22, Stanford Univ. Press, California, 1958.

[8] Eckart, C., The penetration of a potential barrier by electrons. Phys. Rev. 35 (1930), 1303-1309.

[9] Ehrenfest, P. E., Adiabatische Invarianten und Quantentheorie. Ann. der Phys. 51 (1916), 327-352.

[10] Ehrenfest, P. E., Adiabatic invariants and the theory of quanta. Phil, Mag. 33 (1917), 500-513.

[11] Ehrenfest, P. E., Remarks on quantisation. British

Assoc. for Advancement of Science, Liverpool, 1923, p. 508, Report on the 91st Meeting.

[12] Einstein, A., Inst. intern. phys. Solvay, Conseil phys., Rapports I (1911).

[13] Epstein, P. S., Reflection of waves in an inhomogeneous absorbing medium. Proc. Nat. Acad. Sci. 16 (1930), 627-636.

[14] Friedrichs, K. O., On the adiabatic theorem in Quantum Theory. NYU Reports IMM-NYU # 218 (1955), # 230 (1956), New York Univ., New York.

[15] Gardner, C., Adiabatic invariants of periodic classical systems. Phys. Rev. 115 (1959), 791-794.

[16] Geppert, H., Theorie der adiabatischen Invarianten allgemeiner Differentialsysteme. Math. Ann. 102 (1930), 193-243.

[17] Geppert, H., Sugli invarianti adiabatici di un generico systema differenziale. Atti della Reale Accademia Dei Lincei, serie 6ta, Rendiconti classe di scienze fisiche mat. e nat., 8 (1928), 30-34, 191-198, 294-299.

[18] Hellwig, G., Über die Bewegung geladener Teilchen in schwach veränderlichen Magnetfeldern. Z. Naturforschung 10a (1955), 508-516.

[19] Hertweck, F. and A. Schlüter, Die "adiabatische Invarianz" des magnetischen Bahnmomentes geladener Teilchen. Z. Naturforschung 12a (1957), 844-849.

[20] Kato, T., On the adiabatic theorem of Quantum Mechanics. J. of the Phys. Soc. of Japan 5 (1950), 435-439.

[21] Kato, T., Integration of the equation of evolution in a Banach space, J. Math. Soc. of Japan 5 (1953), 208-234.

[22] Kneser, H., Die adiabatische Invarianz des Phasenintegrals bei einem Freiheitsgrad. Math. Annalen 91 (1927), 156-160.

[23] Knorr, G. and D. Pfirsch, The variation of the adiabatic invariant of the harmonic oscillator. Z. Naturforschung 21a (1966), 688-693.

[24] Krutko, G. and V. Fock, Über das Rayleighsche Pendel. Zeitschr. für Phys. 13 (1923), 195-202.

[25] Kulsrud, R., Adiabatic invariant of the harmonic oscillator. Phys. Rev. 106 (1957), 205-207.

[26] Lenard, A., Adiabatic invariance to all orders. Ann. of Phys. 6 (1959), 261-276.

[27] Leung, A. and Kenneth Meyer, Adiabatic invariants for linear Hamiltonian systems. To be published.

[28] Levi-Civita, T., Drei Vorlesungen über adiabatische Invarianten. Abh. des Math. Sem. Hamburg 6 (1928), 323-366.

[29] Littlewood, J. E., Lorentz's pendulum problem. Ann. of Phys. 21 (1963), 233-242.

[30] Littlewood, J. E., Adiabatic invariance II: Elliptic motion about a slowly varying center of force. Ann. of Phys. 26 (1964), 131-156.

[31] Littlewood, J. E., Adiabatic invariance IV: Note on a new method for Lorentz's pendulum problem. Ann. of Phys. 29 (1964), 13-18.

[32] Littlewood, J. E., Adiabatic invariance V: Multiple periods. Ann. of Phys. 29(1964), 138-153.

[33] Meyer, R. E., Adiabatic variation, Part I: Exponential property for the simple oscillator. ZAMP 24(1973), 293-303; Part II: Action change for the simple oscillator. ZAMP 24(1973), 517-524.

[34] Rellich, F., Störungstheorie der Spektralzerlegung, I. Mitteilung. Math. Annalen 113(1936), 600-619.

[35] Russell, D., An extended block-diagonalization theorem. Proceedings, U.S.-Japan sem. on Diff. and Functional Equations, Edited by W. A. Harris, Jr. and Y. Sibuya, W. A. Benjamin, Inc. New York, 1967, pp. 567-575.

[36] Sibuya, Y., Sur réduction analytique d'un système d'équations différentielles ordinaires linéaires contenant un paramètre. J. Fac. Sci. Univ. Tokyo, Sec. I, 7(1954), 299-241.

[37] Snyder, H. S,. Remarks concerning the adiabatic theorem and the S-matrix. Phys. Rev. 83(1951), 1154-1159.

[38] Sommerfeld, A., Atombau und Spektrallinien. 3. Aufl., 1922, p. 375, ff., p. 718.

[39] Stengle, G., Asymptotic estimate for the adiabatic invariance of a simple oscillator. To be published.

[40] Wasow, W., A turning point problem for a system of two linear differential equations. J. Math. Phys. 38(1959/60), 257-278.

[41] Wasow, W., Asymptotic Expansions for Ordinary Differential Equations, John Wiley-Interscience, New York, 1965.

[42] Wasow, W., Simplification of selfadjoint differential equations with a parameter. J. of Diff. Eqn. 2(1966), 378-390.

[43] Wasow, W., On turning point problems for systems with almost diagonal coefficient matrix. Funkcial. Ekvac. 8(1966), 143-171.

[44] Wasow, W., Adiabatic invariance of a simple oscillator. SIAM J. Math. Anal. 4(1973), 153-170.

[45] Wasow, W., Calculation of an adiabatic invariant by turning point theory. SIAM J. Math. Anal. 5(1974).

[46] Wasow, W., On the magnitude of an adiabatic invariant in Quantum Mechanics. MRC Technical Summary Report #1345 (1973); Math. Research Ctr., Univ. of Wisconsin - Madison.

University of Wisconsin - Madison, Madison, Wisconsin

THE RÔLE OF POSITIVE DEFINTE FUNCTIONS IN THE STUDY OF VOLTERRA INTEGRAL EQUATIONS

James S. W. Wong

§1. Introduction.

The purpose of this paper is to describe the rôle of positive definite functions in the study of Volterra integral equations. The study of positive definite functions originates from Bochner's theorem [1] in connection with harmonic analysis on the line. This very important theorem has been extended in a number of ways and has useful implications in many mathematical disciplines. Here, we like to investigate the qualitative behavior of solutions of Volterra integral equations when the convolution kernels are positive definite functions.

Let $a(t) \in C(0,\infty) \cap L^1_{loc}[0,\infty)$. We say that $a(t)$ is <u>positive definite</u> on $(0,\infty)$ if for any function $\varphi(t) \in C(0,\infty)$,

$$(1) \qquad Q_a[\varphi](T) = \int_0^T \varphi(t)\int_0^t a(t-\tau)\varphi(\tau)\,d\tau\,dt \geq 0$$

for all $T \geq 0$. Note that this definition is more general than that first introduced by Halanay [9] where it is assumed that $a(t) \in C[0,\infty)$ thus excluding functions such as $t^{-\alpha}$, $0 < \alpha < 1$. It is easy to see that this definition agrees with Bôchner's

---

Research supported in part by Army Research Office, Durham, through Contract No. DA-ARO-D31-24-72-G95.

original definition on the half line, for a change in order of integration yields

$$\int_0^T \varphi(t) \int_0^T a(t-\tau)\varphi(\tau)\,d\tau\,dt = 2Q_a[\varphi](T).$$

The importance of this quadratic form $Q_a[\varphi](T)$ in the study of Volterra integral equations was first demonstrated, through not explicitly, by Popov [35], [36] in connection with absolute stability of nonlinear control systems, see also [4]. It is well known from circuit theory (see, for example, König and Meixner [14]) that if the Laplace transform $\hat{a}(s)$ of $a(t)$ defined by

$$\hat{a}(s) = \lim_{T\to\infty} \int_0^T e^{-st} a(t)dt$$

exist for $\operatorname{Re} s > 0$ and satisfies

$$\text{(2)} \qquad \operatorname{Re} \hat{a}(s) > 0 \quad \text{for} \quad \operatorname{Re} s > 0,$$

then $a(t)$ is positive definite in the sense of (1). If $\hat{a}(s)$ exists for $\operatorname{Re} s = 0$ and is continuous in $\operatorname{Re} s \geq 0$ then it follows from (2) that

$$\text{(3)} \qquad \operatorname{Re} \hat{a}(s) \geq 0 \quad \text{for} \quad \operatorname{Re} s = 0.$$

Note that the Laplace transform $\hat{a}(s)$ on the imaginary axis, i.e., $\operatorname{Re} s = 0$, coincides with the Fourier cosine transform of $a(t)$. Thus, if $a(t)$ is an even function on $\mathbb{R}$, then its Fourier transform $\mathcal{J}(a)$ satisfies

$$\mathcal{J}(a)(\omega) = \int_{-\infty}^{\infty} e^{-i\omega t} a(t)dt = \tfrac{1}{2}\,\hat{a}(i\omega).$$

Invoking the Parseval theorem for Fourier transforms, Popov [35] showed that if $a(t) \in L^1 \cap L^2(0,\infty)$, then (3) alone implies that $a(t)$ is positive definite. This seems to be the key in Popov's work. The importance of Popov's results, particularly their wide usage in the analysis of servomechanisms can hardly be overstated. (See Lefschetz [15] and Halanay [8] for a more comprehensive exposition.) In the course of investigating the asymptotic stability of a nonlinear Volterra integral arising in dynamics of nuclear reactors, Levin and Nohel [19], [20] considered kernels $a(t)$ which are completely monotonic, i.e.,

$$(4) \qquad (-1)^k a^{(k)}(t) \geq 0, \quad k = 0, 1, 2, 3, \cdots,$$

where $a^{(k)}(t)$ denotes the kth derivative of $a(t)$. The basis of their analysis is the so-called Bernstein representation theorem for completely monotonic functions, which among other things yields the fact that completely monotone functions are positive definite. This is of course well known. In fact, if $a(t)$ is convex in the sense that

$$(5) \qquad (-1)^k a^{(k)}(t) \geq 0, \qquad k = 0, 1, 2,$$

(see Loeve [26; p. 217]) then $a(t)$ is positive definite. It is perhaps the works of Levin [16], Levin and Nohel [20] in their construction of very sophisticated Lyapunov functions that led to Halanay's investigation of the quadratic form $Q_a[\varphi](T)$ in [9]. This is certainly the motivation of the recent joint work with MacCamy [30]. In [30], we completed Halanay's idea concerning the use of $Q_a[\varphi](T)$ in the investigation of a nonlinear Volterra integral equation originally studied by Levin [16]. It turns out that the positivity of $Q_a[\varphi](T)$ leads to the boundedness of solutions whilst the boundedness of $Q_a[\varphi](T)$ yields the asymptotic stability of solutions.

Recent results of MacCamy [27], [28], [29], who extended the concepts introduced in [30] and obtained many far-reaching results concerning solutions of integro-partial differential equations, and of London [23], [24], [25], who improved another stability result of Levin [17] in various directions, further fortify the belief that positive definite functions are destined to play an increasingly important role in the study of the asymptotic theory of Volterra integral equations. We refer the interested reader to [37], [38] for further extension of these concepts to Hilbert spaces and their application in the study of integro-differential equations in connection with nonlinear control systems and nonlinear viscoelasticity.

In the next section, we study positive definite functions and a related class of functions, which we call D-positive definite functions; in particular, we discuss the various sufficient conditions such as (2), (3), (4) and (5) for positive definiteness and analogous results for D-positive definiteness. In section 3, we introduce the concept of strongly positive definite functions and strongly D-positive definite functions. The usefulness of positive, D-positive and strongly positive, D-positive definite functions in the analysis of Volterra integral equations was demonstrated in the last section. The author wishes to thank Professors R. C. MacCamy and S. O. London for the opportunity to read preprints of their results prior to publication.

§2. Positive and D-positive definite functions.

Let $a(t) \in C(0,\infty) \cap L^1_{loc}[0,\infty)$. We consider the linear

Volterra integral operator having $a(t)$ as its kernel defined by

$$(6) \qquad V_a[\varphi](t) = \int_0^t a(t-\tau)\varphi(\tau)\,d\tau .$$

If we assume in addition that $a(t) \in C^1[0,\infty)$ then we can differentiate (6) and obtain another integral operator $W_a[\varphi](t)$ defined by

$$(7) \qquad W_a[\varphi](t) = \frac{d}{dt}\int_0^t a(t-\tau)\varphi(\tau)\,d\tau .$$

Note that when $a(0_+)$ exists and is finite, then we can also express $W_a[\varphi]$ as

$$W_a[\varphi](t) = a(0)\varphi(t) + \int_0^t a'(t-\tau)\varphi(\tau)\,d\tau .$$

Otherwise, (7) does not define a function in the usual sense. However, we can still consider the following quadratic form,

$$R_a[\varphi](T) = \int_0^T \varphi(t)\frac{d}{dt}\int_0^t a(t-\tau)\varphi(\tau)\,d\tau \; dt ,$$

when $a(0_+)$ does not exist or is not finite by restricting the function $\varphi(t)$ to $C^1[0,\infty)$. Clearly, $V_a: C[0,\infty) \to C^1[0,\infty)$ and if $a(0_+)$ exists and is finite then $W_a: C[0,\infty) \to C[0,\infty)$. We say a function $a(t)$ is D-<u>positive definite</u> if

$$(8) \qquad R_a[\varphi](T) \geq 0 \qquad\qquad \text{for all} \quad T \geq 0$$

for all $\varphi \in C[0,\infty)$. When $a(0+)$ does not exist, we define $R_a[\varphi](T)$ for $\varphi \in C^1[0,\infty)$ to be the uniform limit of $R_a[\varphi_n](T)$ where $\varphi_n \in C[0,\infty)$ $\lim_{n\to\infty} \varphi_n(t) = \varphi(t)$ uniformly on $[0, T]$.

In terms of the operator V_a, the positive definiteness of $a(t)$ can be rephrased as the fact V_a defines a semi-definite operator on the Hilbert space $L^2(0, T)$ for every $T \geq 0$. When $a(0_+)$ exists and is finite, then D-positive definiteness means that W_a defines a semi-definite operator on $L^2(0, T)$ for every $T > 0$. On the other hand, when $a(0_+)$ does not exist or becomes infinite then $W_a[\varphi]$ induces a positive bilinear form on $\mathscr{S} \times \mathscr{S}'$ where $\mathscr{S}, \mathscr{S}'$ are Schwartz space and its associated space of tempered distributions.

Let $\hat{a}(s)$ denote the Laplace transform of $a(t)$ and we assume for simplicity that $a(t)$ in addition belongs to $L^\infty(0, \infty)$, hence $\hat{a}(s)$ exists for $\operatorname{Re} s > 0$. Suppose that $\hat{a}(s)$ also exists almost everywhere on the imaginary axis $\operatorname{Re} s = 0$, in particular if $\hat{a}(i\omega)$ exists except when $\omega = \omega_k$, $k = 1, 2, \cdots, N$. Our first result on the positive and D-positive definiteness of $a(t)$ is given by the following:

THEOREM 1. Suppose that $\hat{a}(i\omega)$ exists almost everywhere for $\omega \in \mathbb{R}$. Then for almost all $\omega \in \mathbb{R}$,

(a) $\operatorname{Re} \hat{a}(i\omega) \geq 0 \Rightarrow a(t)$ is positive definite, and

(b) $-\operatorname{Im} \omega\hat{a}(i\omega) \geq 0 \Rightarrow a(t)$ is D-positive definite.

PROOF. Fix $T > 0$, we define for $\varphi(t) \in C(0, \infty)$ the truncated function $\varphi_T(t)$ as follows:

$$\varphi_T(t) = \begin{cases} \varphi(t) & 0 \leq t \leq T \\ 0 & \text{otherwise.} \end{cases}$$

Since $\varphi_T(t)$ has compact support in $\mathbb{R}$, it belongs to the Schwartz space $\mathscr{S}$, the space of rapidly decreasing functions. Since $a(t) \in L^\infty(0, \infty) \subseteq \mathscr{S}'$, the dual space of $\mathscr{S}$, (the space of tempered distributions), we can invoke the Parseval's relation in distribution form (see, e.g., Hörmander [12; p. 19] to conclude that

$$(9) \qquad \mathcal{F}(V_a[\varphi_T])(\omega) = \mathcal{F}(a)(\omega)\mathcal{F}(\varphi_T)(\omega),$$

for all $\omega \in \mathbb{R}$, where $\mathcal{F}(a)(\omega) \in \mathcal{S}'$ exists in the distribution sense. Observe that

$$(10) \qquad Q_a[\varphi](T) = \int_{-\infty}^{\infty} \varphi_T(t) V_a[\varphi_T](t)dt.$$

Since φ_T and $V_a[\varphi_T]$ have compact support in $\mathbb{R}$, hence belong to $L^1 \cap L^2(-\infty, \infty)$, we can apply the Parseval's theorem in the classical form to (10) and obtain

$$(11) \qquad Q_a[\varphi](T) = \frac{1}{2\pi} \int_{-\infty}^{\infty} \mathcal{F}(\varphi_T)(\omega) \overline{\mathcal{F}(V_a[\varphi_T])(\omega)} d\omega,$$

where "bar" denotes complex conjugation. Note that $\mathcal{F}(a)(\omega)$ coincides with $\hat{a}(i\omega)$ whenever the latter exists in the classical sense. We can now substitute (9) into (11) and obtain

$$(12) \qquad Q_a[\varphi](T) = \frac{1}{2\pi} \int_{-\infty}^{\infty} \hat{a}(i\omega) |\mathcal{F}(\varphi_T)(\omega)|^2 d\omega.$$

Since the left hand side is always real, this together with the given hypothesis, yields the positive definiteness of $a(t)$, proving part (a).

To prove part (b), we proceed in a similar manner and obtain instead of (9) the following relation for $\mathcal{F}(W_a[\varphi_T])$, i.e.,

$$(13) \qquad \mathcal{F}(W_a[\varphi_T]) = -i\omega \mathcal{F}(a)(\omega)\mathcal{F}(\varphi_T)(\omega).$$

Using (13), we obtain the formula for $R_a[\varphi](T)$ (Note that the assumption of D-positive definiteness guarantees that $R_a[\varphi](T)$ is defined.) analogous to (12) as follows:

$$(14) \qquad R_a[\varphi](T) = \frac{1}{2\pi}\int_{-\infty}^{\infty} -i\omega\hat{a}(i\omega)\left|\mathcal{F}(\varphi_T)(\omega)\right|^2 d\omega .$$

Again noting the fact that the left hand side of (14) is always real and the assumption that $\omega \operatorname{Im} \hat{a}(i\omega) \leq 0$ almost everywhere, we obtain the desired conclusion.

As an immediate corollary, we have the following known facts:

PROPOSITION. (a) *Let* $a(t) \in C^1(0,\infty) \cap L^1_{loc}[0,\infty)$, $a(t)$ *non-negative,* $a'(t)$ *non-positive and non-decreasing, then* $a(t)$ *is positive definite;*

(b) *Let* $a(t) \in C^1(0,\infty) \cap L^1_{loc}[0,\infty)$, $a(t)$ *non-negative and non-decreasing, then* $a(t)$ *is* D-*positive definite.*

PROOF. (a) The present hypothesis on $a(t)$ implies that $\hat{a}(s)$ exists for $\operatorname{Re} s \geq 0$ except $s = 0$. The proof that $\hat{a}(s)$ satisfies (3), $s \neq 0$ is given in [30; p. 17], cf. also [10]. Thus, the positive definiteness of $a(t)$ follows.

(b) To see that $\omega \operatorname{Im} \hat{a}(i\omega) \leq 0$ for $\omega \neq 0$, we note that

$$-\operatorname{Im} \hat{a}(i\omega) = \int_0^\infty a(t)\sin \omega t \, dt = \sum_{k=0}^{\infty}\left(\int_{\frac{2k\pi}{\omega}}^{\frac{2k+1}{\omega}\pi} + \int_{\frac{2k+1}{\omega}\pi}^{\frac{2k+2}{\omega}\pi}\right) a(t)\sin \omega t \, dt$$

$$= \sum_{k=0}^{\infty} \int_0^{\frac{\pi}{\omega}} \left(a\left(s + \frac{2k\pi}{\omega}\right) - a\left(s + \frac{(2k+1)\pi}{\omega}\right)\right) \sin \omega s \, ds .$$

Since $a(t)$ is non-increasing, each term under the summation sign above is non-negative if $\omega > 0$. There is some problem in justifying this argument when $a(0_+)$ does not exist. This is overcome by the assumption that $a(t) \in L^1_{loc}[0,\infty)$, which implies that $\lim_{\varepsilon \to 0+} \varepsilon\alpha(\varepsilon) = 0$, see [30; p. 17]. The desired assertion now follows.

The criteria for positive and D-positive definiteness of a function given in Theorem 1 has several advantages over earlier versions. We know of a variety of important kernels of Volterra integral equations which satisfy these criteria. For example, any trigonometric polynomials in cosines with positive coefficients are positive definite, i.e.,

$$a(t) = \sum_{k=0}^{N} c_k \cos \lambda_k t, \quad c_k > 0, \quad k = 0, 1, \cdots, N. \tag{15}$$

Here $\hat{a}(i\omega)$ exists for all ω except $\omega = \lambda_k$, $k = 0, 1, \cdots, N$. However, since these are only a finite number of points, the theorem applies. Similarly, any trigonometric polynomials in sines with positive coefficients are D-positive definite. For a discussion of such kernels and their interpretation in linear viscoelasticity theory, we refer to Gurtin and Herrera [7].

§3. Strongly positive and D-positive definitive functions.

As a typical example of positive definite functions which appear often as kernel of Volterra integral equations is the negative exponential function, i.e., $a(t) = \epsilon e^{-\alpha t}$, $t \geq 0$ for some positive constants $\epsilon, \alpha > 0$. Such kernels arise often in applied problems, such as heat and radiation transfer problems [31], nuclear reactor dynamics [19], wave propagation in viscoelastic materials [2], [6], relaxing water waves [13], and also in biological problems [3]. This special class of positive definite functions possesses certain properties which are desirable in the study of asymptotic satbility of solutions. Following [9], we introduce the concept of a strongly positive definite function. A function $a(t)$ is called strongly positive definite if there exist positive constants $\epsilon, \alpha > 0$ such that $a(t) - \epsilon e^{-\alpha t}$ is positive definite. Clearly, strongly positive definite functions are positive definite.

Another example of important convolution kernel is given by $a(t) = \frac{1}{\sqrt{t}}$, which arises from the study of a nonlinear boundary value problem in the theory of superfluidity of Helium, see Lin [22], Levinson [21]. Extensive analysis of kernels which in particular include negative powers of t may be found in the works of Miller [32], Miller and Feldstein [33]. Although such kernels are also strongly positive as we soon see, they seem to fit more naturally in a class of kernels introduced by Levin [17], (London [23]). Denote by L the class of non-negative non-increasing functions which are not integrable in L^1 sense, i.e.,

$$L = \{b(t): b(t) \geq 0, \ b(t) \text{ non-increasing}, \\ b(0) < \infty, \ b(t) \notin L^1(0, \infty)\}.$$

Prototypes of functions in L are the function $(t+1)^{-\nu}$, $0 \leq \nu \leq 1$, $1+e^{-\alpha t}$, $\alpha \geq 0$ and $(\log(t+2))^{-\mu}$, $0 \leq \mu \leq \infty$. We now introduce a new class of strongly positive definite functions. A function $a(t)$ is called strongly D-positive definite if there exists a function $b(t) \in L$ such that $a(t) - b(t)$ is D-positive definite. Clearly, a strongly D-positive definite function is D-positive definite.

According to Theorem 1, a sufficient condition for strongly positive definiteness of $a(t)$ is the existence of positive constants $\epsilon, \alpha > 0$ so that

$$\text{Re}\; \hat{a}(i\omega) \geq \frac{\epsilon\alpha}{\alpha^2+\omega^2}, \quad \text{for almost all } \omega \in \mathbb{R}. \tag{16}$$

On the other hand, a sufficient condition for strongly D-positive definiteness of $a(t)$ is the existence of $b(t) \in L$ so that

$$-\text{Im}\; \omega\hat{a}(i\omega) \geq -\text{Im}\; \omega\hat{b}(i\omega), \quad \text{for almost all } \omega \in \mathbb{R}. \tag{17}$$

In terms of function $a(t)$ itself, it turns out that if $a(t)$ is convex, or more generally satisfies the hypothesis of Proposition 1(a), and if in addition

$$a(t) \not\equiv a(0), \tag{18}$$

then $a(t)$ is strongly positive definite. This is a rather difficult result; we refer the reader to [30; p. 16] for a proof. On the other hand, every function $b(t) \in L$ is, by definition, strongly D-positive definite. However, it is not essential that $a(t)$ be monotonic in any way to possess the property of strongly positive or D-positive definiteness. Thus, the function $e^{-\alpha t} \cos \lambda t$ is strongly positive definite so does any finite combination of such functions with positive coefficients. Moreover, the functions $t^{-\nu}$, $0 < \nu < 1$, are strongly positive definite. This is particularly useful because the special case $\nu = \frac{1}{2}$ as discussed in the last paragraph arises from important physical problems. To see this, we note that

$$\text{Re}\left(\int_0^\infty e^{-i\omega t}\, t^{-\nu}\, dt\right) = \int_0^\infty t^{-\nu} \cos \omega t\, dt$$

$$= \frac{\pi}{2}[\Gamma(\nu)]^{-1} \sec\frac{\nu\pi}{2}\, |\omega|^{\nu-1},\ \omega \neq 0,$$

from which it is easy to see that (16) holds for some appropriate positive constants $\epsilon, \alpha > 0$.

The class of strongly D-positive definite functions are less manageable. Clearly all positive constant functions belong to L, hence are strongly D-positive definite, but they are not strongly positive definite. Another example of strongly D-positive functions is the function $(t+1)^{-1}$ which clearly belongs to L. It turns out that the function $t^{-\frac{1}{2}}$, although not belonging to L, is strongly D-positive definite. In fact, the entire class of functions, $t^{-\nu}$, $0 < \nu < 1$, are strongly D-positive definite. To see this, we compare $t^{-\nu}$ with $(t+1)^{-1} \epsilon L$. Observe that

$$(19) \qquad -\operatorname{Im}\left(\int_0^\infty e^{-i\omega t}\, t^{-\nu} dt\right) = \Gamma(1-\nu)\cos\left(\frac{\nu\pi}{2}\right)\omega^{\nu-1}, \quad \omega > 0.$$

and

$$(20) \qquad -\operatorname{Im}\left(\int_0^\infty e^{-i\omega t}(t+1)^{-1}\, dt\right) = -\operatorname{Im}\Psi(1,1,i\omega),$$

where Ψ denotes the confluent hypergeometric function. For $s = \sigma + i\omega$ small, we know that $\Psi(1,1,s)$ has the valid expansion

$$(21) \qquad \Psi(1,1,s) = -\log s - \gamma + O(s \log s), \quad s \to 0,$$

where γ denotes the Euler constant (see [5; p. 262]). On the other hand, for large values of s, we have the expansion

$$(22) \qquad \Psi(1,1,s) = \frac{1}{s} + O(s^{-2}), \quad s \to \infty,$$

see [5; p. 278]. From (21) and (22), it is easy to see that there exists a positive constant $\epsilon > 0$ such that for $0 < \nu < 1$,

$$(23) \qquad -\mathrm{Im}\ \Psi(1,1,i\omega) \leq \epsilon^{-1}\omega^{\nu-1}, \qquad \omega > 0.$$

Using this together with (19) and (20), we find that the function $t^{-\nu}$ satisfies (17) with $b(t) = \epsilon(t+1)^{-1}$, where $\epsilon > 0$ is chosen so that (23) hold. This establishes that $t^{-\nu}$ is strongly D-positive definite. Note that the function $e^{-\alpha t}\sin \lambda t$, though clearly D-positive definite, is not strongly D-positive definite. This is to be expected since the definition of strongly positive definiteness is given in terms of some negative exponential function $\epsilon e^{-\alpha t}$, $\epsilon, \alpha > 0$, while such a function itself fails to belong to L. Finally, we remark that the sum of a strongly D-positive definite function and a D-positive definite function is strongly D-positive. Thus the function $t^{-\nu}+\sin \lambda t$, $0 < \nu < 1$, $\lambda \neq 0$, is strongly D-positive definite whilst it remains oscillatory on $[0, \infty)$.

§4. Nonlinear volterra integral equations.

Consider the following nonlinear Volterra integral equation

$$(24) \qquad u'(t) = -\int_0^t a(t-\tau)g(u(\tau))d\tau, \qquad t \geq 0,$$

where the nonlinear function $g(u)$ satisfies:

$$(25) \qquad \begin{cases} g(u) \in C(-\infty, \infty) \quad ug(u) > 0 \quad u \neq 0; \\ \lim_{|u|\to\infty} \left(G(u) = \int_0^u g(\xi)d\xi \right) = \infty. \end{cases}$$

Levin [16] proved that if $a(t)$ satisfies the monotonicity condition;

$$(26) \qquad (-1)^k a^{(k)}(t) \geq 0 \qquad k = 0, 1, 2, 3.$$

Then every solution of (24) is bounded. More surprisingly, if $a(t)$ satisfies in addition (18), then every solution $u(t)$ satisfies

$$(27) \qquad \lim_{t\to\infty} u(t) = 0 .$$

This result was later improved by Hannsgen by relaxing (26) to hold only for $k = 0, 1, 2$. Halanay [9] showed that if $a(t)$ is positive definite then every solution is bounded. Furthermore, if $a(t)$ is strongly positive then every solution satisfies (27). Halanay's original proof concerning the stability part seems incomplete and also contains a number of technical assumptions which with the exception that $a' \in L_1(0,\infty)$ can all be removed, see [30].

To illustrate the simplicity of Halanay's approach, we sketch the proof of his result in the stronger form as given in [30]. Multiplying (24) by $g(u(t))$ and integrating from 0 to T, we obtain

$$(28) \qquad G(u(T)) = G(u(0)) - Q_a[g(u)](T).$$

Since a is positive definite, so the boundedness of solutions follows from (28) and the hypothesis on (25). Moreover, we obtain from (28) the boundedness of $Q_a[g(u)]$:

$$(29) \qquad 0 \le Q_a[g(u)](T) \le G(u(0)), \qquad T \ge 0.$$

Now if we assume that $a(t)$ is in addition strongly positive definite, then the desired conclusion (27) follows from (29) and the following lemma due to Halanay [9]:

LEMMA. Let $a(t)$ be strongly positive definite and $|Q_a[\varphi](t)| \le B$ for all $t \ge 0$. If φ is uniformly continuous, then $\lim_{t\to\infty} \varphi(t) = 0$.

To apply Halanay's lemma, we need to establish that $g(u(t))$ is uniformly continuous. Since $g(u) \in C(-\infty,\infty)$ and $u(t)$ is bounded, it suffices to show that $u(t)$ is uniformly continuous. Differentiating (24), we obtain

$$u''(t) = -a(0)g(u(t)) - \int_0^t a'(t-\tau)g(u(\tau))d\tau,$$

which will imply boundedness of $u''(t)$ if we assume in addition that $a'(t) \in L^1(0,\infty)$. Now the boundedness of $u(t)$ and $u''(t)$ imply the boundedness of $u'(t)$, hence uniform continuity of $u(t)$. To summarize, we have

THEOREM 2. *Let* $a(t)$ *be strongly positive definite and* $a'(t) \in L^1(0,\infty)$. *Assume that* $g(u)$ *satisfies* (25), *then every solution of* (24) *satisfies* (27).

To illustrate the usefulness of Theorems 1 and 2, we present two examples which are not covered by any previous results. Consider

$$(30) \qquad u'(t) = -\int_0^t g(u(t-\tau))\sum_{k=0}^{N} c_k \cos \lambda_k \tau \, d\tau, \quad c_k > 0,$$

and

$$(31) \qquad u'(t) = -\int_0^t g(u(t-\tau))\sum_{k=1}^{N} c_k e^{-\alpha_k \tau} \cos \lambda_k \tau \, d\tau,$$

where $\alpha_k, c_k > 0$, $\lambda_k \neq 0$, and $g(u)$ satisfies (25). For the above discussion, we can conclude that all solutions of (30) are bounded and all solutions of (31) tend to zero as $t \to \infty$.

We now consider another Volterra integral equation originally studied by Levin [17]:

$$(32) \qquad u(t) = f(t) - \int_0^t a(t-\tau)g(u(\tau))\,d\tau ,$$

where $f'(t) \in L^1(0,\infty)$ and $g(u)$ satisfies (25). Equation (32) can be considered as an extension of the linear equation, when $g(u) \equiv u$, originally studied by Paley and Wiener [34] in connection with Mercer's theorems. Levin's result [17] on the asymptotic stability of solutions of (32) has recently been improved by London [23], [24], who shows among other things if $g(u)$ satisfies in addition to (25) the growth condition

$$(33) \qquad |g(u)| \le M(1 + G(u)), \quad \text{for all} \quad u \in \mathbb{R}$$

and $a(t) \in \mathcal{L}$, then every solution of (32) satisfies (27).

We can now improve London's result to read

THEOREM 3. *Let* $a(t)$ *be D-positive definite,* $f' \in L^1(0,\infty)$, *and* $g(u)$ *satisfies* (25) *and* (33). *Then every solution of* (32) *is bounded. If in addition that* $a(t)$ *is strongly D-positive definite, and* $a' \in L^1(0,\infty)$, *then every solution of* (32) *satisfies* (27).

PROOF. We differentiate (32) and obtain

$$(34) \qquad u'(t) = f'(t) - \frac{d}{dt}\int_0^t a(t-\tau)g(u(\tau))\,d\tau .$$

Form the inner product with $g(u(t))$ on $[0, T]$, we find as in the case of (28):

$$(35) \qquad G(u(T)) \le G(u(0)) + \int_0^T |f'(t)g(u(t))|\,dt + R_a[g(u)](T).$$

Since by hypothesis that $a(t)$ is D-positive definite, we may use (33) in (35) and obtain:

$$(36) \qquad G(u(T)) \le G(u(0)) + M\int_0^T |f'(t)|\,dt + M\int_0^T |f'(t)|G(u(t))dt.$$

Now a routine application of Gronwall's inequality to (36) yields the boundedness of $G(u(t))$, hence by (25) the boundedness of $u(t)$.

Since $u(t)$ is bounded and $g(u) \in C(-\infty, \infty)$, so $g(u(t))$ is also bounded. Using this fact in (35), we find that there exists a constant B such that

(37) $$0 \leq R_a[g(u)](T) \leq B, \qquad \forall T \geq 0.$$

By hypothesis, $a(t)$ is strongly D-positive definite, so there exists $b(t) \in L$ such that

(38) $$0 \leq R_b[g(u)](T) \leq R_a[g(u)](T) \leq B, \qquad T \geq 0.$$

We next show that $g(u(t))$ is uniformly continuous in t. Note that (34) is equivalent to

(39) $$u'(t) = f'(t) - a(0)g(u(t)) - \int_0^t a'(t-\tau)g(u(\tau))\,d\tau.$$

Since $g(u(t))$ is bounded and $f'(t)$, $a'(t) \in L^1(0,\infty)$ by hypothesis, (39) yields the fact that $u'(t)$ is bounded. Now, $u(t)$ is bounded and uniformly continuous, hence $g(u(t))$ is uniformly continuous. Now we may invoke the recent result of London [24] to complete the proof. However, for the sake of completeness, we sketch the ideas of London's very ingenious proof. In [24], London established the following identity:

$$2R_b[\varphi](T) = \int_0^T (b(T-\tau)+b(\tau))\varphi^2(\tau)\,d\tau - \int_0^T \int_0^t (\varphi(\tau)-\varphi(\tau-s))^2 db(s)\,d\tau.$$

Since $b(t)$ is non-negative and non-increasing, we obtain from (38)

(40) $$\int_0^t b(t-\tau)g^2(u(\tau))\,d\tau \leq B.$$

We now must quote a Lemma in [24; p. 110] which states that if (38) holds and $g(u(t))$ is uniformly continuous then for any sequence $\{t_n\}$ such that $\lim_{n\to\infty} t_n = \infty$, $\lim_{n\to\infty} g(u(t_n))$ exists and $T, \epsilon > 0$ arbitrary, there exists $N_0(T,\epsilon)$ such that

$$(41) \qquad g(u(t_n)) - \epsilon \le g(u(t)) \le g(u(t_n)) + \epsilon ,$$

for all $t_n - T \le t \le t_n$. Suppose that $\limsup_{t\to\infty} |g(u(t))| = \delta > 0$, then we can find a sequence $\{t_n\}$ such that $\lim_{n\to\infty} g(u(t_n)) = \delta$ as $t_n \to \infty$. For any $T > 0$, we can now invoke London's lemma to obtain N_0 so that (take $\epsilon = \frac{\delta}{2}$ in (41))

$$(42) \qquad g(u(t)) \ge \frac{\delta}{2}, \qquad t_n - T \le t \le t_n .$$

Consider (40) for large $t_n > T$, we note that

$$\int_0^{t_n} b(t_n-\tau)g^2(u(\tau))d\tau \ge \int_{t_n-T}^{t_n} b(t_n-\tau)g^2(u(\tau))d\tau \ge \frac{\delta^2}{4}\int_0^T b(\tau)d\tau ,$$

which is incompatible with $b(t) \notin L^1(0,\infty)$. This completes the proof.

The following equation has a kernel which cannot be treated by the result of London [24].

$$(43) \qquad u(t) = f(t) + \int_0^t g(u(t-\tau))(\tau^{-\nu} + \sin \lambda\tau)d\tau ,$$

where $f' \in L^1(0,\infty)$, $g(u)$ satisfies (25), (32), $0 < \nu < 1$ $\lambda \ne 0$. However, a simple application of Theorem 3 yields the result that every solution of (43) must tend to zero as $t \to \infty$.

We remark that the condition that $f' \in L^1(0,\infty)$ can be somewhat relaxed. London's more recent result [25] assumes only $f(t) \in C[0,\infty) \cap BV(0,\infty)$; see also Levin [18]. It is also interesting to note that the crucial argument in London's lemma is to consider the case when $b(t) \not\equiv b(0)$, (namely, condition (18)). This should be compared with the result in [30] which proves that $a(t)$ convex and condition (18) imply that $a(t)$ is strongly positive definite. Both of these proofs are rather tedious and difficult. We welcome any simpler proofs for either, preferably an argument which singles out the crucial rôle played by condition (18).

REFERENCES

[1] S. Bochner, "Monotone funktionen, Stieltjessche integrale und harmonische analyse", Math. Ann. 108 (1933), 378-410.

[2] B. T. Chu, "Stress waves in isotopic linear viscoelastic materials", J. Mécanique, 1 (1962), 439-462.

[3] K. L. Cooke and J. A. Yirke, "Some equations modelling growth process and Gonorrhea Epidemics", Math. Biosciences, 16 (1973), 75-101.

[4] C. Corduneanu, "Sur une équation intégrale de la théorie du reglage automatique", C. R. Acad. Sci. Paris, 256 (1963), 3564-3567.

[5] A. Erdélyi, W. Magnus, F. Oberhettinger, F. G. Tricomi, Higher Transcendental Functions, Vol. I, MacGraw-Hill, New York, 1953.

[6] M. E. Gurtin and E. Sternberg, "On the linear theory of viscoelasticity", Arch. Rational Mech. Anal. (11) 4 (1962), 291-356.

[7] M. E. Gurtin and I. Herrera, "On dissipation inequalities and linear viscoelasticity", Quart. J. Appl.

Math. 23 (1965), 235-245.

[8] A. Halanay, Differential Equations, Academic Press, New York, 1966.

[9] A. Halanay, "On the asymptotic behavior of the solutions of an integro-differential equation", J. Math. Anal. Appl. 10 (1965), 319-324.

[10] K. B. Hannsgen, "Indirect abelian theorems and a linear Volterra equation", Trans. Amer. Math. Soc. 142 (1969), 539-555.

[11] K. B. Hannsgen, "On a nonlinear Volterra equation", Michigan Math. J. 16 (1969), 365-376.

[12] L. Hörmander, Linear Partial Differential Operators, Springer, Berlin, 1963.

[13] D. Y. Hsieh, "A study of relaxing waves", J. Math. Phys. 13 (1972), 1769-1776.

[14] H. König and J. Meixner, "Lineare systeme und lineare transformationen", Math. Nachr. 19 (1958), 256-322.

[15] S. Lefschetz, Stability of Nonlinear Control Systems, Academic Press, New York, 1965.

[16] J. J. Levin, "The asymptotic behavior of the solution of a Volterra equation", Proc. Amer. Math. Soc. 14 (1963), 534-541.

[17] J. J. Levin, "The qualitative behavior of a nonlinear Volterra equation", Proc. Amer. Math. Soc. 16 (1965), 711-718.

[18] J. J. Levin, "On a nonlinear Volterra equation", J. Math. Anal. Appl. 39 (1972), 458-476.

[19] J. J. Levin and J. A. Nohel, "On a system of

integrodifferential equations occurring in reactor dynamics", J. Math. Mech. 9 (1960), 347-368.

[20] J. J. Levin and J. A. Nohel, "Note on a nonlinear Volterra equation", Proc. Amer. Math. Soc. 14 (1963), 924-929.

[21] N. Levinson, "Anonlinear Volterra equation arising in the theory of superfluidity", J. Math. Anal. Appl. 1 (1960), 1-11.

[22] C. C. Lin, "Hydrodynamics of liquid helium II", Phys. Rev. Letters, 2 (1959), 245-246.

[23] S. O. London, "On the solutions of a nonlinear Volterra equation", J. Math. Anal. Appl. 39 (1972), 564-573.

[24] S. O. London, "On a nonlinear Volterra integral equation", J. Differential Equations, 14 (1973), 106-120.

[25] S. O. London, "On the bounded solutions of a nonlinear Volterra equation", to appear.

[26] M. Loeve, Probability Theory. Foundations. Random Sequences, 2nd rev. ed., University Series in Higher Math., Van Nostrand, Princeton, N. J., 1960.

[27] R. C. MacCamy, "Approximations for a class of functional differential equations", SIAM J. Appl. Math., (1973), (to appear).

[28] R. C. MacCamy, "Asymptotic stability for a class of functional differential equations", SIAM J. Appl. Math., (to appear).

[29] R. C. MacCamy, "Nonlinear Volterra equations on a Hilbert space", J. Diff. Equ., (to appear).

[30] R. C. MacCamy and J. S. W. Wong, "Stability theorems for some functional equations", Trans. Amer. Math. Soc. 164 (1972), 1-37.

[31] W. R. Mann and F. Wolf, "Heat transfer between solids and gases under nonlinear boundary conditions", Quart. J. Appl. Math. 9 (1951), 163-184.

[32] R. K. Miller, Nonlinear Volterra Integral Equations, W. A. Benjamin, Menlo Park, California, 1971.

[33] R. K. Miller and A. Feldstein, "Smoothness of solutions of Volterra integral equations with weakly singular kernels", SIAM J. Math. Anal. 2 (1971), 242-258.

[34] R. E. A. C. Paley and N. Wiener, "Fourier transforms in the complex domain", American Mathematics Society Colloquium Publications, 1934.

[35] V. M. Popov, "On absolute stability of nonlinear systems of automatic control", Avtomat. Telemeh. 22 (1961), 961-979.

[36] V. M. Popov, "Solution of a new stability problem for control systems", Avtomat. Telemeh. 24 (1963), 1-23.

[37] J. S. W. Wong, "Positive definite functions and Volterra integral equations", Bull. Amer. Math. Soc., (to appear).

[38] J. S. W. Wong, "A nonlinear integro-partial differential equation arising from viscoelasticity", (forthcoming).

University of Iowa, Iowa City, Iowa

FAVARD'S CONDITION IN LINEAR ALMOST PERIODIC SYSTEMS

Taro Yoshizawa

Favard assumed some separation conditions which give the existence of an almost periodic solution of the linear almost periodic system

$$(1) \qquad x' = A(t)x + f(t), \qquad x, f \in R^n,$$

where $A(t)$ is an $n \times n$ almost periodic matrix and $f(t)$ is also almost periodic. Let $R = (-\infty, \infty)$, $R^+ = [0, \infty)$ and $R^- = (-\infty, 0]$, and denote by $H(p)$ the closed hull of an almost periodic function p.

One of Favard's results is the following theorem [2].

Favard's Theorem A. Suppose that for every $B \in H(A)$, every nontrivial bounded solution $x(t)$ on R of the system

$$(2) \qquad x' = B(t)x$$

satisfies the condition

$$(3) \qquad \inf_{t \in R} |x(t)| > 0.$$

If system (1) has a solution which is defined and bounded on R^+, there exists an almost periodic solution p of (1) and $m(p) \subset m(A, f)$, where m denotes the module of the almost periodic function.

More generally, for an almost periodic system

(4) $x' = f(t, x), \quad x, f \in R^n$,

where $f(t, x)$ is continuous on $R \times D$, D an open set in R^n, and is almost periodic in t uniformly for $x \in D$, we have the following result.

Theorem 1. Let K be a compact set in D. If the property P is an inherited property of solutions and for each $g \in H(f)$, the solution $x(t; g)$ in K for all t with the property P of the system $x' = g(t, x)$ exists and is unique, then these solutions are almost periodic and $m(x) \subset m(f)$.

In Favard's Theorem A, the minimal norm property satisfies the condition in Theorem 1.

In discussing relationships between stability properties and separation condition, Nakajima [4] has obtained the following result.

Theorem 2. If the linear almost periodic system (1) has a bounded solution on R^+ which is uniformly stable, then system (1) has an almost periodic solution p and $m(p) \subset m(A, f)$.

This theorem can be proved by showing that if system (1) has a bounded and uniformly stable solution on R^+, Favard's condition is satisfied.

Favard's Theorem A has been extended by Kato to the functional differential equation

(5) $\dot{x}(t) = A(t, x_t) + f(t)$,

where $A(t, \varphi)$ is continuous in (t, φ) on $R \times C$, $C = C([-h, 0], R^n)$, is linear in φ and is almost periodic in t

uniformly for $\varphi \in C$ and $f(t)$ is almost periodic.

Under condition (3), we can easily extend Favard's Theorem A to system (5). However, a linear autonomous equation $\dot{x}(t) = -x(t - \pi/2)$ has a solution $x(t) = \sin t$ which is bounded on R, but does not satisfy condition (3). On the other hand, every nontrivial bounded solution x on R satisfies

(6) $$\inf_{t \in R} \|x_t\| > 0,$$

where $\|\varphi\|$, $\varphi \in C$, is given by $\|\varphi\| = \sup_{-h \leqq \theta \leqq 0} |\varphi(\theta)|$ and here the vector norm is the Euclidean norm. Thus the replacement of condition (3) by a more desirable condition (6) is not obvious.

Theorem 3. Suppose that for every $B \in H(A)$, every nontrivial bounded solution x on R of system

(7) $$\dot{x}(t) = B(t, x_t)$$

satisfies condition (6). If system (5) has a solution bounded on R^+, system (5) has an almost periodic solution p and $m(p) \subset m(A, f)$.

To prove this theorem, we introduce a new norm $\|\cdot\|_*$ in C defined by

$$\|\varphi\|_* = \left(\int_{-h}^{0} |\varphi(s)|^2 ds\right)^{\frac{1}{2}}.$$

If system (5) has a bounded solution x on R, set

$$\lambda(x) = \sup\{\|x_t\|_* ;\ t \in R\}$$

and

$\Lambda(A+f) = \inf\{\lambda(x);$ x is a bounded solution on R of (5)$\}$.

Then, for every $B + g \in H(A+f)$, we can show that $\Lambda(B+g) = \Lambda(A+f)$. Moreover, if system (5) has a bounded solution on R, we can see the existence of a solution x of (5) such that $\lambda(x) = \Lambda(A+f)$. Thus Theorem 3 can be proved by the same idea as in ordinary differential equations. Namely, system (5) has a bounded solution x on R with the minimal norm $\Lambda(A+f)$, and this minimal property is inherited. Moreover, under the assumption in Theorem 3, we can show that for each $B + g \in H(A+f)$, the system

$$\dot{x}(t) = B(t, x_t) + g(t)$$

has a unique bounded solution on R with the minimal norm. Thus system (5) has an almost periodic solution p and $m(p) \subset m(A+f)$.

The other of Favard's conditions is that for each equation in the hull, only the trivial solution is bounded for all t. Favard gave the following theorem [2].

Favard's Theorem B. Consider the linear almost periodic system (1). Suppose that for each system (2), $B \in H(A)$, only the trivial solution is bounded for all $t \in R$. If system (1) has a bounded solution on R^+, then system (1) has an almost periodic solution.

If S(t) is an $n \times n$ continuous matrix on R and $\mathcal{D}$ is a given class of functions which contains the zero function, the homogeneous system

$$(8) \qquad x' = S(t)x$$

is said to be noncritical with respect to $\mathcal{D}$ if the only solution of (8) which belongs to $\mathcal{D}$ is the solution $x = 0$. Favard's condition is equivalent to saying that each system (2), $B \in H(A)$, is noncritical with respect to $\mathcal{B}$, where $\mathcal{B}$ is the class of continuous and bounded functions on R. Corresponding to system (8), consider

$$x' = S(t)x + h(t),$$

where h is a given function in some specified class. Let $\mathcal{P}_\omega$ denote the class of continuous and periodic functions of period ω and let $\mathcal{AP}$ denote the class of almost periodic functions. Then it is known that for S in $\mathcal{P}_\omega$, $(\mathcal{B}, \mathcal{B})$, $(\mathcal{AP}, \mathcal{AP})$, $(\mathcal{P}_\omega, \mathcal{P}_\omega)$ are admissible if and only if system (8) is noncritical with respect to $\mathcal{B}$, $\mathcal{AP}$, $\mathcal{P}_\omega$, respectively, (cf. [3]). It is also known that for S in $\mathcal{AP}$, $(\mathcal{AP}, \mathcal{AP})$ is admissible if and only if system (8) admits an exponential dichotomy.

Recently, Sacker and Sell [6] have shown through the linear skew-product flow that if for S in $\mathcal{AP}$ system (8) satisfies Favard's condition, then it admits an exponential dichotomy.

Sacker and Sell's Theorem. Assume that F(t) is an $n \times n$ matrix which is bounded and uniformly continuous on R. If for every system

$$(9) \qquad x' = F^*(t)x, \quad F^* \in H(F),$$

only the trivial solution is bounded for all $t \in R$, then there exists an $F_0 \in H(F)$ such that

$$(10) \qquad x' = F_0(t)x$$

admits an exponential dichotomy on R. If F(t) is almost periodic, system $x' = F(t)x$ admits an exponential dichotomy on R.

For ordinary differential equations, Nakajima has proved their result by using Perron's triangular form. For $u \in C(R, R^n)$, let H(u) be the closed hull of u in the compact-open topology, that is, $v \in H(u)$ if and only if there exists a sequence $\{\tau_k\}$ such that $u(t+\tau_k)$ converges to v(t) uniformly on any compact subset of R as $k \to \infty$.

Now consider a linear system

$$(11) \qquad x' = F(t)x,$$

where $x \in R^n$ and F(t) is an $n \times n$ continuous matrix on R.

Definition. The system (11) admits an exponential dichotomy on R^+ if there exists a projection Q and positive constants K and α such that

$$(12) \qquad \begin{aligned} |X(t)QX^{-1}(s)| &\leqq Ke^{-\alpha(t-s)}, \quad t \geqq s \geqq 0 \\ |X(t)(I-Q)X^{-1}(s)| &\leqq Ke^{-\alpha(s-t)}, \quad s \geqq t \geqq 0, \end{aligned}$$

where I is the identity matrix and X denotes the fundamental matrix solution of (11) such that $X(0) = I$. The system (11) admits an exponential dichotomy on R if (12) is valid for all $t, s \in R$.

To prove Sacker and Sell's theorem, we shall use the following lemmas.

Lemma 1. (Perron's triangular form [5]). For any bounded, uniformly continuous matrix F(t) on R, there exists a nonsingular matrix P(t) such that P(t), $P^{-1}(t)$,

$P'(t)$ are also bounded and uniformly continuous on R and that

$$[P(t)F(t) + P'(t)]P^{-1}(t) = A(t),$$

where $A(t)$ is triangular below and is bounded, uniformly continuous on R.

Consider a linear system

(13) $\quad x' = A(t)x.$

Lemma 2. Suppose that $A(t)$ is bounded and continuous on R. The system (13) admits an exponential dichotomy on R^+ if and only if for any bounded continuous function $f(t)$ on R^+

(14) $\quad x' = A(t)x + f(t)$

has a bounded solution on R^+. Furthermore, if system (13) admits an exponential dichotomy on R, system (14) with f bounded, continuous on R has a bounded solution on R.

Refer to [1].

In Lemmas 3, 4, 5 and 6, we assume that $A(t)$ in (13) is bounded and uniformly continuous on R.

Lemma 3. (a) If system (13) admits an exponential dichotomy on R^+, for some $B \in H(A)$ $x' = B(t)x$ admits an exponential dichotomy on R.

(b) If system (13) admits an exponential dichotomy on R, for all $A^* \in H(A)$ $x' = A^*(t)x$ admits an exponential dichotomy on R.

Lemma 4. Assume that for every system $x' = B(t)x$, $B \in H(A)$, only the trivial solution is bounded for all $t \in R$. Then, if the zero solution of system (13) is stable on $R^+(R^-)$, it is positively (negatively) uniformly asymptotically stable on R.

An essential idea for the proof is contained in the paper by Sacker and Sell. If we suppose that the zero solution is not uniformly stable, we have a nontrivial bounded solution on R for some system in the hull. Also the same argument proves uniformly asymptotic stability.

Lemma 5. Under the assumption in Lemma 4, let $n = 1$, that is, (13) be a scalar equation. Then there exists a $B \in H(A)$ such that $x' = B(t)x$ admits an exponential dichotomy on R.

We can see that there exists a $B \in H(A)$ such that $x' = B(t)x$ has a nontrivial bounded solution on R^+ or R^-. Since the equation is scalar, the existence of a nontrivial bounded solution on R^+ (R^-) implies that the zero solution is stable on R^+ (R^-). Thus, applying Lemma 4, we can prove this lemma.

For a matrix $A = (a_{ij})$, $1 \leqq i \leqq n$, $1 \leqq j \leqq n$, define a matrix A_r by $A_r = (a_{ij})$, $r \leqq i \leqq n$, $r \leqq j \leqq n$, and for a vector $x = (x_1, \cdots, x_n)$, let $\xi_r = (x_r, \cdots, x_n)$. Clearly $A_1 = A$.

Lemma 6. In system (13), we assume that $A(t)$ is triangular below, that is, $A(t) = (a_{ij}(t))$, $a_{ij}(t) \equiv 0$ if $i < j$. Moreover, assume that for every system $x' = B(t)x$, $B \in H(A)$, only the trivial solution is bounded for all t. If for some r, $2 \leqq r \leqq n$,

$$(15) \qquad \xi_r' = A_r(t)\xi_r$$

admits an exponential dichotomy on R, for some $B \in H(A)$

$\xi_{r-1}' = B_{r-1}(t)\xi_{r-1}$ admits an exponential dichotomy on R.

Proof. Denote by $a_{r-1}(t)$ the $(r-1, r-1)$ element of $A(t)$ and consider the scalar equation

(16) $\quad x_{r-1}' = a_{r-1}(t)x_{r-1}.$

Then equation (16) satisfies the assumption in Lemma 4. If not, there is a $c_{r-1} \in H(a_{r-1})$ such that $x_{r-1}' = c_{r-1}(t)x_{r-1}$ has a nontrivial bounded solution $x_{r-1}(t)$ on R. Let $C \in H(A)$ whose $(r-1, r-1)$ element is $c_{r-1}(t)$. Then, by Lemma 3, $\xi_r' = C_r(t)\xi_r$ admits an exponential dichotomy on R since system (15) admits an exponential dichotomy on R and $C_r \in H(A_r)$. By Lemma 2, $\xi_r' = C_r(t)\xi_r + f(t)$ has a bounded solution $\xi_r(t)$ on R if $f = (f_r, \cdots, f_n)$, $f_j(t) = c_{j,\,r-1}\, x_{r-1}(t)$, $j = r, \cdots, n$. Thus $x(t) = (0, \cdots, 0, x_{r-1}(t), \xi_r(t))$ is a nontrivial bounded solution on R of system $x' = C(t)x$, $C \in H(A)$, which contradicts the assumption. Therefore, equation (16) satisfies the assumption in Lemma 4, and hence, by Lemma 5, there exists a $d_{r-1} \in H(a_{r-1})$ such that

(17) $\quad x_{r-1}' = d_{r-1}(t)x_{r-1}$

admits an exponential dichotomy on R. Let $D \in H(A)$ whose $(r-1, r-1)$ element is $d_{r-1}(t)$. Since $D_r \in H(A_r)$ and system (15) admits an exponential dichotomy on R, by Lemma 3

(18) $\quad \xi_r' = D_r(t)\xi_r$

admits an exponential dichotomy on R.

Now we shall see that

$$(19) \quad \xi_{r-1}' = D_{r-1}(t)\xi_{r-1}$$

admits an exponential dichotomy on R. For any bounded $g(t) = (g_{r-1}, \cdots, g_n) \in C(R^+, R^{n-r+2})$, consider

$$(20) \quad \xi_{r-1}' = D_{r-1}(t)\xi_{r-1} + g(t).$$

Since equation (17) admits an exponential dichotomy on R, $x_{r-1}' = d_{r-1}(t)x_{r-1} + g_{r-1}(t)$ has a bounded solution $x_{r-1}(t)$ on R^+. Letting $h_j(t) = d_{j,r-1}(t)x_{r+1}(t) + g_j(t)$, $j = r, \cdots, n$,

$$\xi_r' = D_r(t)\xi_r + h, \quad h = (h_r, \cdots, h_n),$$

has a bounded solution $\xi_r(t)$ on R^+ since system (18) admits an exponential dichotomy on R and h is bounded. Therefore, $(x_{r-1}(t), \xi_r(t))$ is a bounded solution on R^+ of system (20). This shows that system (19) admits an exponential dichotomy on R^+, and thus, by Lemma 3, there exists a $B \in H(D)$ such that $\xi_{r-1}' = B_{r-1}(t)\xi_{r-1}$ admits an exponential dichotomy on R.

Now we are ready to prove Sacker and Sell's theorem. By Lemma 1, system $x' = F(t)x$ is transformed by $y = P(t)x$ into

$$(21) \quad y' = A(t)y, \; A(t) = [P(t)F(t) + P'(t)]\,P^{-1}(t),$$

and $A(t)$ is bounded, uniformly continuous on R and is triangular below. For any system

$$(22) \quad y' = A^*(t)y, \; A^* \in H(A),$$

only the trivial solution is bounded for all t. In fact, since P, P^{-1} and P' are bounded, uniformly continuous, there exist $P^* \in H(P)$ and $F^* \in H(F)$ such that

$$A^* = [P^*F^* + P^{*'}]P^{*-1},$$

and a nontrivial bounded solution $y^*(t)$ on R of system (22) gives a nontrivial bounded solution $P^{*-1}(t)y^*(t)$ on R of $x' = F^*(t)x$, which is a contradiction. Therefore system (21) satisfies the assumption in Lemma 4.

For the (n, n) element $a_{nn}(t)$ of $A(t)$,

$$(23) \quad y_n' = a_{nn}(t)y_n$$

satisfies the assumption in Lemma 4. In fact, if for some $b_{nn} \in H(a_{nn})$ $y_n' = b_{nn}(t)y_n$ has a nontrivial bounded solution $y_n(t)$ on R, then for $B \in H(A)$ whose (n, n) element is $b_{nn}(t)$) $y' = B(t)y$ has a nontrivial bounded solution $(0, 0, \cdots, 0, y_n(t))$ since B is triangular below, which is a contradiction. Therefore, by Lemma 5, for some $c_{nn} \in H(a_{nn})$, $y_n' = c_{nn}(t)y_n$ admits an exponential dichotomy on R. If we let $C \in H(A)$ whose (n, n) element is $c_{nn}(t)$, the above means that for $y' = Cy$, the assumption in Lemma 6 is satisfied for $r = n$. Thus, by proceeding successively, we can see that the conclusion in Lemma 6 is valid for $r = 2$, that is, there exists a $D \in H(A)$ such that

$$(24) \quad y' = D(t)y$$

admits an exponential dichotomy on R. Namely, there exists a projection Q and positive constants K, α such that

$$(25)\qquad \begin{aligned} |Y(t)QY^{-1}(s)| &\leqq Ke^{-\alpha(t-s)}, \quad t \geqq s \\ |Y(t)(I-Q)Y^{-1}(s)| &\leqq Ke^{-\alpha(s-t)}, \quad s \geqq t, \end{aligned}$$

where $Y(t)$ is the fundamental matrix solution of (24) such that $Y(0) = I$.

Then there exist $P_0 \in H(P)$ and $F_0 \in H(F)$ such that $D = (P_0F_0 + P_0{}')P_0^{-1}$. Therefore $X(t) = P_0^{-1}(t)Y(t)P_0(0)$ is the fundamental matrix solution of

$$(26)\qquad x' = F_0(t)x$$

and $X(0) = I$. Since Q is a projection, $R = P_0^{-1}(0)QP_0(0)$ is also a projection. It follows from (25) that

$$\begin{aligned} |X(t)RX^{-1}(s)| &\leqq K_1 Ke^{-\alpha(t-s)}, \quad t \geqq s \\ |X(t)(I-R)X^{-1}(s)| &\leqq K_1 Ke^{-\alpha(s-t)}, \quad s \geqq t, \end{aligned}$$

where K_1 is some constant, which shows that system (26) admits an exponential dichotomy on R.

If $F(t)$ is almost periodic, $F \in H(F_0)$ and hence the system admits an exponential dichotomy on R.

Remark. As was seen, if $F(t)$ is almost periodic and system (11) admits an exponential dichotomy on R, then for every $F^* \in H(F)$ system (9) admits an exponential dichotomy on R. It is also known that if system (9) admits an exponential dichotomy on R, it is noncritical with respect to B (cf. [1]). Therefore, $(\mathcal{AP}, \mathcal{AP})$ is admissible if and only if every system in the hull is noncritical with respect to B.

REFERENCES

[1] W. A. Coppel, Stability and Asymptotic Behavior of Differential Equations, Heath, Boston, 1965.

[2] J. Favard, Lecons sur les Fonctions Presque-périodiques, Gauthier-Villars, Paris, 1933.

[3] J. K. Hale, Ordinary Differential Equations, Wiley-Interscience, New York, 1969.

[4] F. Nakajima, Separation conditions and stability properties in almost periodic systems, Tohoku Math. J., 26 (1974), 305-314.

[5] O. Perron, Über eine Matrixtransformation, Math. Z., 32 (1930), 465-473.

[6] R. J. Sacker and G. R. Sell, Existence of dichotomies and invariant splittings for linear differential equations 1, J. Differential Eqs., 15 (1974), 429-458.

Tohoku University, Sendai, Japan

STABILITY OF NON-COMPACT SETS

P. N. Bajaj

Stability of compact sets has been studied extensively. In this paper stability and asymptotic stability of non-compact sets is discussed. Results are obtained on a rim-compact-rather a locally compact-Hausforff space.

Wichita State University, Wichita

A GENERALIZED FUCHSIAN THEORY AND SOME APPLICATIONS

K. Boehmer

In the neighborhood of ∞ the solutions of the linear differential equation (d. e.)

$$(D) \qquad w^{(m)} + \cdots + a_j w^{(j)} + \cdots + a_0 w = 0$$

$$a_j \text{ holomorphic in } \mathring{R} := \{z \in \mathbb{C} \mid R_0 < |z| < \infty\}$$

are linear combinations of solutions of the form

$$(*) \qquad w(z) = z^{\rho} \{d_j(z) + \cdots + d_{\ell}(z)(\log z)^{j-\ell} + \cdots + d_1(z)(\log z)^{j-1}\},$$

$$\rho \in \mathbb{C}, \; d_{\ell} \text{ holomorphic in } \dot{R}, \; \ell = 1(1)j_n \leq m.$$

With a fundamental system $\mathcal{J} = \{w_1, \ldots, w_m\}$ of solution (*) of (D) and

$$M(r, d_\ell) := \max_{|z|=r} \{|d_\ell(z)|\}, \quad M(r, w) := \max_{\ell=1}^{j} \{M(r, d_\ell)\},$$

$$M(r, \mathcal{T}) := \max_{n=1}^{m} \{M(r, w_n)\}, \quad M(r, a_j) := \max_{(z)=r} \{|a_j(z)|\},$$

$$M(r, A) := \max_{j=0}^{m-1} \{ r^{m-j} M(r, a_j) \}$$

and

$$\log^+_{\ell+1} \rho := \max\{0, \log_{\ell+1} \rho\}, \ \rho \in \mathbb{R}_+ \text{ big enough}, \ \ell \in \mathbb{N}_0$$

we define the following order of growth

$$\lambda^{(\ell)}(M(r, A)) := \lim_{r \to \infty} \frac{\log^+_{\ell+1} M(r, A)}{\log r}, \quad \lambda^{(\ell+1)}(M(r, W)) :$$

$$= \overline{\lim_{r \to \infty}} \frac{\log^+_{\ell+2} M(r, \mathcal{T})}{\log r} .$$

Then

$$\lambda^{(\ell)}(M(r, A)) = \sigma \Leftrightarrow \lambda^{(\ell+1)}(M(r, W))$$

$$= \sigma, \ \ell \in \mathbb{N}_0, \ 0 \leq \sigma \leq \infty .$$

For $\ell = \sigma = 0$ we have the classical FUCHSian theory. In a quite analogous way to RIEMANNs characterization of the hypergeometric d.e. it is now possible to characterize all d.e.s (D) (a_j singular only in finitely many points) of mathematical physics and for $m > 2$ treated until now in the literature.

University of Karlsruhe and
MRC, University of Wisconsin-Madison

DIFFERENTIAL EQUATIONS WITHOUT LINEAR TERMS

S. N. Busenberg
and
C. S. Coleman

The n-dimensional systems

(1) $\dot{x} = f(x)$ and (2) $\dot{z} = f(z) + g(z, t)$,

where f is a homogeneous form of integral degree $m > 1$ and $g = o(\|x\|^m)$, occur occasionally in applications [biological models, gas dynamics, kinetics] and are a natural extension of the linear and perturbed linear case when $m=1$. Current and recent work by Ladis and Reizin′ in the Soviet Union, Gerber at IBM and Jaderberg and ourselves has outlined the beginnings of an analytic, algebraic and topological structure theory for (1) and (2). There are both parallels and contrasts with the well-known theory for the case $f(x) = Ax$ $(m = 1)$. The results are harder to derive since solutions of (1) are no longer additive if $m > 1$, even though a kind of homogeneity remains [i.e., if $x(t, x_o)$ is a solution of (1), so is $cx(c^{m-1}t, x_o)$].

Type Numbers and Asymptotic Stability: For each $x_o \varepsilon R^n$ a type number $\lambda(x_o)$ is defined for the solution $x(t, x_o)$ of (1). The origin 0 is an asymptototically stable point of (1) and only if $\lambda(x_o) < 0$ for every x_o in a certain distinguished subset of the unit sphere. In this event 0 is also an asymptotically stable point of (2). If $x(t, x_o) \to 0$ as $t \to \infty$, then $\|x(t(t, x_o)\| t^{1/m-1} \to [(1-m)\lambda(x_o)]^{1/1-m}$. Solutions of (2) obey bounds of the form $\|z(t, z_o)\| \leqq at^{1/1-m}$ iff all type numbers are negative. These results do not extend to the infinite dimensional case [i.e., $n = \infty$], but do [in part] to the case where f depends periodically on t in addition to being homogeneous in x. Type numbers and estimates can also be obtained for (2).

Topological Structure: It has been conjectured that systems (1) and (2) [g independent of t] are locally topologically equivalent at 0 if 0 is a "canonical" hyperbolic point of (1). This has now been shown if f satisfies additional conditions; there are counterexamples if these conditions are not met.

Algebraic Structure: L. Markus [1959] attached a non-associative algebra to (1) in the quadratic case ($m = 2$). This has now been extended to give an algebraic characterization of the linearizability of a matrix Riccati system. Algebras for the general case $m \geqq 2$ have also been defined and the nilpotents and idempotents used to give a topological algebraic characterization of (1) in the two-dimensional case ($n = 2$). The problem remains open if $n > 2$.

Harvey Mudd College and Claremont, California

A NOTE ON THE EXISTENCE OF PERIODIC SOLUTIONS OF

G. J. Butler

Consider the second order scalar ordinary differential

$$x''(t) + f(t, x(t)) = 0, \tag{1}$$

where $f(t, x)$ is ω-periodic in t and $f(t, 0) = 0$ for all t. The usual existence results for periodic solutions, employing degree theory or other fixed point arguments, are generally unhelpful in this case, since the periodic solution that they predict may well be the trivial solution.

Jacobowitz has recently succeeded in applying the Poincare-Birkhoff "twist" theory to demonstrate that (1) has infinitely many periodic solutions when f satisfies a suitable "strong nonlinearity" condition with respect to x. Essential to his method of proof, however, is the condition $xf(t, x) > 0$ for all $t(x \neq 0)$. In this note we show how this hypothesis may be relaxed, by modifying a technique used by the author when considering the problem of the global existence of solutions of (1), which arises when the sign condition is removed.

In common with the paper of Jacobowitz, this not makes use of the rapidly oscillating nature of second order equations with a strongly nonlinear restoring force. The methods used throughout are quite elementary.

University of Southampton and University of Alberta

GEOMETRIC VIEWS IN EXISTENCE THEORY

Lamberto Cesari

For the sake of simplicity we refer only to closure theorems for Mayer problems. Let G be some subset of E^{ν} of finite measure, $\nu \geq 1$, and let $x(t) = (x^1, \ldots, x^n)$ and $u(t) = (u^1, \ldots, u^m)$ denote state and control functions, under usual constraints $x(t) \in A(t) \subset E^n$, $u(t) \in U(t, x(t)) \subset E^m$.

Let $f(t, x, u) = (f_1, \ldots, f_r)$ be given functions defined for all $t \in G$, $x \in A(t)$, $u \in U(t, x)$ satisfying usual Carathéodory continuity conditions. Let $Q(t, x)$ denote the sets $Q(t, x) = f(t, x, U(t, x))$, which we assume in general to be unbounded. We shall assume that (α): There is a bounded measurable function $p(t)$, $t \in G$, $p: G \to E^r$, say $|p(t)| \leq \sigma$, such that $p(t) \in Q(t, x)$ for all $t \in G$, $x \in A(t)$. This condition is certainly satisfied if, say $0 \in Q(t, x)$ for all t, x, as usually required in controllability theory. We shall also assume that (β): For every $t_o \in G$, $|f(t_o, x, u)| \to \infty$ as $|u| \to \infty$, $u \in U(t_o, x)$, uniformly on any compact subset of $A(t_o)$. This is also a natural condition. Then the following closure statement holds with no seminormality requirement:

Theorem. If $\xi_k(t)$, $\xi(t)$, $x_k(t)$, $x(t)$, $u_k(t)$, $t \in G, k = 1, 2, \ldots,$ are given measurable functions, $\xi_k, \xi \in (L_1(G))^r$, with $\xi_k(t) = f(t, x_k(t), u_k(t))$, $u_k(t) \in U(t, x_k(t))$ a.e. in G, if $\xi_k \to \xi$ weakly in $(L_1(G))^r$, $x_k \to x$ in measure in G as $k \to \infty$, if the sets $Q(t, x)$ are closed and convex, and (α), (β) hold, then there is some measurable function $u(t)$, $t \in G$, such that $\xi(t) = f(t, x(t), u(t))$, $u(t) \in U(t, x(t))$ a.e. in G.

The situation is similar to the one in Filippov's existence theorem. If only condition (α) hold but not (β), then the conclusion of our statement is still valid provided we know that for any $N \geq \sigma$ the equibounded convex closed sets $Q(t, x) \cap V(0, N)$ are upper semicontinuous by set inclusion with respect to x only (equivalently, they have property (K), or (Q) with respect to x only). Here $V(0, N) = [z \in E^r \mid |z| \leq N]$.

However, condition (α) is not needed if we know that $\xi_k \to \xi$ in measure. If neither condition holds, we need to know that the controls $u_k(t)$, $t \in G$, belong to $L_1(G)$ with $\|u_k\|_1 \leq L$. Finally, condition (α) can be replaced by (α'): There are $p_k(t)$, $p(t)$, $t \in G$, $p, p_k \in (L_1(G))^r$, $p_k(t) \in Q(t, x_k(t))$, with $p_k \to p$ strongly in $L_1(G)$. These and more general results concerning closure and lower closure theorems are found in L. Cesari, (a) Closure theorems for orientor fields and weak convergence, Archive Rat. Mech. Anal. 56, 1975; (b) Lower semicontinuity and lower closure theorems without seminormality conditions, Annali Mat. Pura Appl. 98, 1974, 381-397; L. Cesari and M. B. Suryanarayana, (a) Convexity and property (Q) in optimal control theory, SIAM J. Control 12, 1974; (b) Closure theorems without seminormality conditions, Journ. Optim. Theory Appl.; (c) Nemitsky's operators and lower closure theorems, Ibid.; M. B. Suryanarayana, Remarks on lower semicontinuity and lower closure. Ibid.

University of Michigan

POINTWISE DEGENERACY OF LINEAR TIME-INVARIANT DELAY-DIFFERENTIAL SYSTEMS

A. K. Choudhury

It is an important property of delay-differential system of the form

(1) $\dot{x}(t) = Zx(t) + Bx(t-h),\ h > 0,\ t > 0$

where (A and B are constant nxn matrices, $n > 2$) that

there exists A and B such that solutions of eq. (1) after a certain time do not span the whole space R^n for all choices of initial functions (continuous). Such property is called pointwise degeneracy. Negation of this property (i.e., if solutions of eq. (1) span the whole space R^n for some choice of initial functions) is called pointwise completeness. This property was first defined in connection with the study of controllability of delay-differential systems. Several authors, including the present author studied the above problem.

In this paper we shall study the necessary and sufficient conditions of pointwise degeneracy of delay-differential systems with distributed lag of the form

$$(2) \qquad \dot{x}(t) = Ax(t) + Bx(t-h) + C\dot{x}(t-h) + \int_0^h Kx(t-s)ds$$

The result will be obtained by constructing a new representation for solution of the system (2) in different intervals in terms of the operator $L(t;\cdot)$ associated with eq. (2) and its successive powers. By performing a large number of simple operations of differentiation and integration on the matrices A, B, C, and K, this method will determine whether the system (2) is degenerate or not.

Howard University

ASYMPTOTIC EQUIVALENCE OF AN ORDINARY AND A FUNCTIONAL DIFFERENTIAL EQUATION

K. L. Cooke

We prove an asymptotic equivalence theorem for systems

(1) $$v'(t) = A(t)u(t) + F(t, u_t)$$

(2) $$v'(t) = A(t)v(t)$$

where u, v are n-vectors, $A(t)$ is nxn, and $F(t, \phi)$ is a function on $R^+ \times C$ into R^n where C is the space of continuous functions on $[-\tau, 0]$. The perturbation F is assumed to satisfy an inequality

$$|F(t, \phi)| \leq (t, \|\phi\|) \text{ for } \phi \text{ in } C$$

or

$$|F(t, \phi)| \leq \psi(t, \|\phi\|_\infty) \text{ for } \phi \text{ in } W^1_\infty[-\tau, 0]$$

where $|\cdot|$ denotes a norm in R^n, $\|\cdot\|$ denotes the maximum form in C, and $\|\cdot\|_\infty$ denotes a norm in $W^1_\infty[-\tau, 0]$. The scalar nonnegative function $\psi(t, r)$ is nondecreasing in r for each t and satisfies

$$\int_0^\infty \psi(t, c)dt < \infty, \quad c > 0.$$

Under these and certain other conditions, asymptotic equivalence of bounded solutions of (1) and (2) is proved, using the Schauder-Tychonoff theorem. The theorem is applied to numerous examples, including equations with state dependent such as

$$u'(t) = Au(t) + \gamma(t)\{u(t)-u(t-r[t, u_t])\}.$$

Pomona College

STARTING IN MAXIMUM POLYNOMIAL DEGREE NORDSIECK-GEAR METHODS

R. Danchick

The intent of this paper is to show that the maximum k+1 polynomial degree Nordsieck-Gear methods first described in [1] admit of matched starting methods which are exact for all polynomials of degree $\leq k+1$. In general, it is shown that these starting methods yield starting errors of the required order $0(h^{k+2})$ for all initial value problems:

$$y^{(p)}(x) = f(x, y, y^{(1)}, y^{(2)}, \ldots, y^{(p-1)})$$

where f is k+1 times continuously differentiable in $x, y, y^{(1)}, \ldots, y^{(p-1)}$, in a neighborhood of the graph of the exact solution $(x, \overline{y}(x))$, $x \in [0, X]$.

Two theorems are proved. The first is the constructive existence of an algorithm which requires $(k-p+1)(k-p+2)/2$ evaluations of the function f to obtain approximations of the method's required initial higher order scaled derivatives:

$$h^{p+1}\overline{y}^{(p+1)}(0)/(p+1)!, \ldots, h^{k}\overline{y}^{(k)}(0)/k!$$

which are accurate to $0(h^{k+2})$. The second less general theorem shows that, when f is a polynomial in x, y and its higher order derivatives $y^{(1)}, y^{(2)}, \ldots, y^{(p-1)}$, an algorithm can be constructed for obtaining the higher order scaled derivatives exactly.

These results lay to rest once and for all any heuristic arguments against varying corrector minus predictor coefficients for preserving maximal order (polynomial degree) because starting values are in exact. Furthermore, and perhaps most importantly, the maximum polynomial degree Nordsieck-Gear methods are shown to have unique zero starting truncation error properties for an important class of differential equations.

REFERENCES

[1] Danchick, R., "On the Non-Equivalence of Maximum Polynomial Degree Nordsieck-Gear and Classical Methods," Proceedings of the Conference on the Numerical Solution of Ordinary Differential Equations, 19-20, October 1972, Lecture Notes in Mathematics 362, Springer-Verlag, New York, 1972.

Pan Science Systems

A CRITERION FOR THE UNIFORM ASYMPTOTIC STABILITY OF A CLASS OF DIFFERENTIAL-DIFFERENCE EQUATIONS

R. Datko

Consider the n-dimensional nonautonomous systems of linear differential-difference equations described by

1) $\dot{x}(t) = \sum_{i=o}^{m} A_i(t)x\,(t-h_i),\ o = h_o < h_1 < \cdots < h_m = h$, for $t \geq t_o$ and

2) $x(t) = \phi(t-t_o)$

if $t \in [t_o - h,\ t_o]$. Here ϕ is a continuous n-vector function defined on $[-h, 0]$ and the matrices $A_i(t)$, $0 \leq i \leq m$, are continuous and uniformly bounded on $[0, \infty]$. Let $x(t, t_o, \phi)$ denote any solution of (1) - (2) with initial values t_o and ϕ and $x_t(t_o, \phi) = \{x(s, t_o, \phi):\ t-h \leq s \leq t\}$.

The following two theorems give necessary and sufficient conditions for the uniform asymptotic stability of the system (1) - (2). Theorem 2 gives a condition in terms of bilinear functional and is obtained from Theorem 1.

<u>Theorem 1.</u> A necessary and sufficient condition for the system (1) - (2) to be uniformly asymptotically stable is the existence of a positive constant M such that for all ϕ

3) $$\int_{t_o}^{\infty} |x(t, t_o, \phi)|^2\, dt \leq M\ |\phi|^2,$$ where

4) $$|\phi| = \sup\left\{ \left(\sum_{i=1}^{n} \phi_i^2(t)\right)^{1/2} :\ t \in [-h, 0] \right\}.$$

<u>Theorem 2.</u> Let Y denote the Banach space of continuous n-vector functions defined on $[-h, o]$ whose norm is given

by equation (4). Then a necessary and sufficient condition for the system (1) - (2) to be uniformly asymptotically stable is the existence of a continuous bilinear mapping $H(\cdot)$ from $(0,\infty) \Rightarrow L(Y, Y)$ which satisfies the following two conditions.

(i) Theorem exists a positive constant M_o such that for all $t \in [0,\infty)$, $0 < H(t)(\phi,\phi) \leq M_o |\phi|^2$ if $|\phi| > 0$.

(ii) There exists a constant $a > 0$ such that for all $\phi \in Y$ and $t \geq t_0 \geq \infty$

5) $$\frac{d}{dt}[H(t)(x_t(t_0,\phi), x_t(t_0,\phi))] \leq -a\,|x(t, t_0, \phi)|^2.$$

Georgetown University

NONOSCILLATION AND INTEGRAL INEQUALITIES

S. Friedland

Consider the system

(1) $$\frac{dy}{dt} = A(t)\,y$$

where $A = (a_{jk})_1^n$ is an $n \times n$ real (complex) valued matrix and $y = (y_1, \ldots, y_n)$ is an n column real (complex) valued vector. The variable t varies on a domain D where $D \subset R^1$ if t is real and $D \subset R^2$ if t is complex. The system (1) is called nonoscillatory if the nonhomogeneous system

(2) $$\frac{dy}{dt} = A(t)\, y + f(t)$$

has a unique solution for the initial value problem

(3) $$y_j(t_j) = \eta_j$$

where $t_1, \ldots, t_n$ are n arbitrary points in D and $\eta_1, \ldots, \eta_n$ are n arbitrary given numbers. The purpose of this paper is to characterize the best possible constant s of nonoscillation $c_{1,2}$ such that the condition

(4) $$\int_{\partial D} \| A(t) \|_{1,2}\, dt < c_{1,2}$$

would imply that (1) is nonoscillatory. Here $\| A \|_{1,2}$ is the matrix norm $\sup \frac{\| Ax \|_2}{\| x \|_1}$. We show that the constants $c_{1,2}$ are infima of the appropriate integral functionals. This approach enables us to compute the constants $c_{1,2}$ for many norms $\| x \|_1$ and $\| x \|_2$. Finally the variational problem leads to the nonlinear Eule-Lagrange equations with interesting boundary values.

REFERENCE

[1] S. Friedland, Nonoscillation and Integral Inequalities, Bull. Amer. Math. Soc. (to appear).

Standford University

EQUIVALENCE CLASSES OF FUNCTIONS AND SOLVABILITY OF OPTIMAL CONTROL PROBLEMS

C. Galperin

The set of functionals connected with a optimal control problem can be divided into classes of equivalence concerning optimal solutions being generated. The classification of functionals obtained in this way gives the basis for investigation of optimal control problems with several criteria and under limitations on features of solutions and kinds of input information.

We consider in detail the classification of functionals for the linear optimal control problem with feedback controllers. This classification is the spectral one.

The single-single-valued reflection of equivalence classes of functionals into the set of spectral vectors of the closed optimal system is constructed.

Applications to solvability of optimal stabilization problems for systems with essential incompleteness of information and under limitations on the spectrum of the closed system are investigated.

Technion-Israel Institute of Technology

FROM "MULTIPLE BALAYAGE" TO FUZZY SETS

G. S. Goodman

Halkin's notion of "multiple balayage" (Amer. Math. Monthy. 73, 1966). when interpreted in terms of weak* convergence in $\mathcal{L}_\infty$, can be used to show that every measurable function which carries the unit interval into itself is the weak* limit of characteristic functions of ordinary sets. Thus, any semi-metric that induces the weak* topology on the unit ball in $\mathcal{L}_\infty$ can be pulled back, via the characteristic functions, to the measurable subsets of the unit interval, which thereby becomes a semi-metric space whose completion is just the class of measurable fuzzy sets.

Istituto Matematico "U. Dini", Firenze

CONTRAST ENHANCEMENT, SHORT TERM MEMORY AND ATTENTION IN REVERBERATING NEURAL NETWORKS

S. Grossberg

This talk will describe mechanisms for parallel processing of patterned information in the presence of noise, showing how significant information can be given greater weight and unimportant information can be suppressed. It shows how a channel can be defined by interactions within a field of cell populations in such a way that varying the number of active cells and intensity of activation does not disrupt decision making downstream in the network. This mechanism has been used to help synthesize networks in which drives, reinforcement and motivation interact to focus attention on cues important to the network at any given time. Also this mechanism can be used to generate various perceptual invariances, such as color and brightness constancy and contract. It is well-known to occur in physiological data, but the way in which it achieves the above psychological effects requires a rigorous analysis of network dynamics.

The results are global theorems about the limit and oscillations of some classes of nonlinear differential equations that arise in psychophysiological processes. One such system is

$$\dot{x}_i = -Ax_i + (B_i - x_i)\, f(x_i) - x_i \sum_{k \neq i} f(x_k) + I_i(t) \tag{1}$$

$i = 1, 2, \ldots, n$, where the constants A and B_i are positive, the functions $f(w)$ are nonnegative, continuous, and monotone increasing, the initial data is nonnegative, and the inputs $I_i(t)$ are nonnegative and continuous. Another such system is

$$\begin{aligned} \dot{x}_i &= -Ax_i + (B_i - x_i) f(x_i) - x_i \sum_{k \neq i} g(y_k) + I_i(t) \\ \dot{y}_i &= -Cy_i + (D - Ey_i) f(x_i) \end{aligned} \tag{2}$$

$i = 1, 2, \ldots, n$, with the constants, functions, initial data, and inputs similarly constrained. System (1) was studied jointly with Daniel Levine. System (2) was studied jointly with Samuel Ellias.

Both systems describe cell populations which excite themselves and inhibit other populations- a so-called on-center off-surround anatomy- with multiplicative mass action interactions, or shounts. The global behavior of these systems depends on the functions $F(w) = w^{-1} f(w)$ and

$G(w) = w^{-1}g(w)$. The results on system (1) generalize results in Grossberg (1973) where all B_i are equal. For example, let $B = \max_k B_k$ and let all inputs equal zero. Given $F(w)$ constant, the relative activities of all states such that $B_i = B$ is preserved, and all other activities converge to zero. The total activity $x(t) = \Sigma_k x_k(t)$ converges to a limit independent of initial data; hence in the absence of inputs, noise will be amplified if reverberation in short term memory (STM) is ever possible. If $F(w)$ is strictly increasing, then $x(t)$ can have any number of limit points if $f(w)$ is suitably chosen, and noise can be suppressed. Also a choice can be made among populations: only activities corresponding to one value of B_i are stored in STM, and among these activities, only the states with maximal initial activity are stored. If $F(w)$ is increasing for small w and constant at larger w values then a quenching threshold exists: all activities small then the threshold converge to zero, and the pattern of activities above threshold is contrast enchanced. A slow increase of B_i increases the probability that the corresponding activity will be stored in STM; applications to the development of neural networks and to focussing of attention on particular classes of feactures in such networks can be drawn from these examples.

System (2) unlumps the excitatory and inhibitory populations of system (1). New phenomena emerge. For example, limit cycles have been studied in various cases. Also, by varying the sign of (D- C), one can switch the networks between contrast enhancement and uniformizing tendencies in particular cases.

REFERENCE

[1] Grossberg, S. Studies in Applied Math., Vol. LII, No, 3, September, 1973, 213-257.

Massachusetts Institute of Technology,
Cambridge, Massachusetts

ASYMPTOTIC STABILITY IN SYSTEMS WITH SEVERAL PARAMETERS MULTIPLYING THE DERIVATIVES

P. Habets

Consider the system of equations

$$\dot{x} = f(t, x, y_1, \ldots, y_m),$$

$$\varepsilon_i \dot{y}_i = g(t, x, y_1, \ldots, y_m), \quad (i = 1, \ldots, m),$$

where $t \in]0, \infty[$, $x, y_1, \ldots, y_m$ are real vectors of dimension $n, k_1, \ldots, k_m$ respectively and the ε_i are positive real parameters. Conditions are given which ensure the uniform asymptotic stability of the origin for ε_1 and $\varepsilon_{i+1}/\varepsilon_i$ small enough. As an example, the stability of a plane is considered.

Universite Catholique de Louvain

APPLICATIONS OF AN ALTERNATIVE THEOREM FOR SINGULAR DIFFERENTIAL SYSTEMS

L. M. Hall

Let $\mathfrak{a}$ denote the space of n-vector functions $y(z)$ with components analytic for $|z| < 1$, and continuous for $|z| \leq 1$. Consider the operator L defined by

$$Ly(z) = z^{D} y'(z) + A(z)y(z)$$

where D and A are $n \times n$ matrices, $D = \operatorname{diag}(d_1, \ldots, d_n)$ with d_i integers ≥ 0, and each column of $A(z)$ belongs to $\mathfrak{a}$. The following theorem can be proved:

Theorem 1. The nonhomogeneous system $Ly = u$ has a solution in $\mathfrak{a}$ if and only if u belongs to $(\ker L^*)^{\perp}$ where L^* is the conjugate operator. The system $L^*f = g$ has a solution in $\mathfrak{a}^*$ if and only if g belongs to $(\ker L^{**})^{\perp}$.

By using by characterization of the space $\mathfrak{a}^*$ due to A. E. Taylor we obtain:

Theorem 2. The function $u(z)$ belongs to $(\ker L^*)^{\perp}$ if and only if $\lim_{r \to 1^-} B(u, F; r) = 0$ for every $F(z)$ analytic in $|z| < 1$ such that $\lim_{r \to 1^-} B(Ly, F; r) = 0$.

University of Nebraska-Lincoln

ON APPROXIMATE AND TRUE SOLUTIONS OF A NONLINEAR SINGULAR PERTURBATION PROBLEM

F. A. Howes

The existence and asymptotic behavior as $\varepsilon \to 0^+$ of solutions of the singularly perturbed boundary value problem $\varepsilon y'' = f(t, y, y', \varepsilon)$, $y(0, \varepsilon) = A(\varepsilon)$, $y(1, \varepsilon) = B(\varepsilon)$, are studied when there exists a known approximate solution $u = u(t, \varepsilon)$, in the sense that $\varepsilon u'' = f(t, u, u', \varepsilon) + O(\eta) + O(\eta\varepsilon^{-1} e^{-mkt\varepsilon^{-1}})$, $u(0, \varepsilon) = A(\varepsilon) + O(\eta)$, $u(1, \varepsilon) = B(\varepsilon) + O(\eta)$. Here $0 < \varepsilon << 1$, $m > 0$, $\eta = \eta(\varepsilon)$ is a gauge function, and k is a positive constant such that $f_{y'}(t, u(t, \varepsilon), u'(t, \varepsilon), \varepsilon) \leq -k$, for $0 \leq t \leq 1$. D. Willett, A. Erdelyi, K. W. Chang and P. Habets have considered this problem under the principal assumptions that $f_{y'y'} = O(\varepsilon)$ and $\eta = O(\varepsilon)$. In this report, differential inequalities for second order boundary value problems are employed to prove results similar to theirs under the weaker assumptions that $f_{y'y'} = O(1)$ and $\eta = O(1)$ as $\varepsilon \to 0^+$.

University of Southern California, Los Angeles, California
and
Courant Institute of Mathematical Sciences, New York Univ.

ON THE VARIATION OF THE SOLUTIONS OF A NONLINEAR INTEGRAL EQUATION

Sti-Olof Londen

Consider the nonlinear Volterra equation

$$(1) \qquad x'(t) + \int_0^t g(x(t-\tau))da(\tau) = f(t), \quad 0 \leqq t < \infty,$$

where g, a, f are prescribed real functions and suppose $g \in C(-\infty, \infty)$, $xg(x) > 0$, $x = 0$, $a \in NBV[0, \infty)$, $a(\infty) > 0$, $f(t) \to 0$, $t \to \infty$, and let $f, x \in L^{\infty}(0, \infty)$ where $x(t)$ is locally a.c. and satisfies (1) a.e. on $(0, \infty)$. Finally assume

$$(2) \qquad \mathrm{Re} \int_0^{\infty} e^{i\omega t} da(t) \geqq 0, \quad -\infty < \omega < \infty .$$

We show that under these hypotheses either $\lim_{t \to \infty} x(t) = 0$ holds or there exist $\varepsilon > 0$ and $t_n, \tau_n \to \infty$ such that

$$(3) \qquad \int_{t_n - \tau_n}^{t_n} |x(\tau)| d\tau \geqq \varepsilon \tau_n .$$

We also obtain that if (2) is strengthened to a strict inequality then (3) cannot hold.

Helsinki University of Technology

OSCILLATION CRITERIA FOR A FOURTH ORDER SELF-ADJOINT LINEAR DIFFERENTIAL EQUATION

D. L. Lovelady

An oscillation criterion is given for (*) $(ru'')'' + (pu')' + qu = 0$, where p is "small" in the sense that (**) $(rv')' + pv = 0$ is nonoscillatory. If we further restrict p to ensure that (**) has bounded nonoscillatory solutions, then a known oscillation criterion for $(ru'')'' + qu = 0$ ensures oscillation for (*).

Florida State University

COMPLEX FUNCTIONAL EQUATIONS WHOSE SOLUTIONS HAVE NATURAL BOUNDARIES

G. R. Morris

If $\beta(z) = z^2$, formal substitution of a power series into

$$(*) \qquad f'(z) = f[(z)), \qquad f(0) = 1$$

immediately gives the lacunary series

$$1 + \frac{z}{1} + \frac{z^3}{1.3} + \cdots + \frac{z^{2^n-1}}{1.3.7\cdots(2^n-1)} + \cdots$$

whose sum has the circle $|z| = 1$ as a natural boundary. This behavior has the intuitive explanation that for $|z| < 1$ we have $|\beta(z)| < |z|$ and hence (*) is a retarded equation whose solutions should be well-behaved, whilst for $|z| > 1$ we have $|\beta(z)| > |z|$ whence the equation is advanced and can be expected to have ill-behaved solutions.

If β is a finite Blaschke product it is shown that the solutions of (*) are C^∞ and have $|z| = 1$ as a natural boundary. Further, if A is analytic in a suitably large domain, the solutions of

$$f'(z) = A(z)f(\beta(z)), \quad f(0) = 1,$$

in general have $|z| = 1$ as a natural boundary, but there are exceptional possibilities.

University of New England, New South Wales, Australia

UNBOUNDED INVARIANT CURVES FOR FLOWS IN THREE-DIMENSIONAL SPACE

A. A. Newton

A vector field X, of class C^2 on some $R^n (n \geq 2)$ and with all first-order partials uniformly bounded on R^n, is given. Let $m \in \{1, \ldots, n-1\}$ be fixed but arbitrary and write $X = (Y, Z) = Y^1, \ldots, Y^m, Z^1, \ldots, Z^{n-m})$. If $x = (y, z)$, where $y \in R^m$, is a generic point of R^n, then clearly X defines a flow $\pi = \pi(x, t) = \pi(y, z, t)$ on R^n. It is assumed that there exists a constant $c_0 > 0$ for which the Euclidean inner product $\langle z, Z(y, z)\rangle < 0$ whenever $\|z\| = c_0$, so that the solid circular cylinder D defined by $\|z\| \leq c_0$ is positively invariant under π. Let F be any subset of the boundary of D of the form $R \times S$ where $\emptyset \neq S \subset \{z \in R^{n-m} : \|z\| = c_0\}$. Let $x_t = \pi(x, t)$. The

Jacobian matrix $[X^i_j]$ is required to satisfy

$$(1)\qquad \overline{\lim_{t\to\infty}}\ \sup_{x\in F} X^{m+1}_{m+1}(x_t) < 0,$$

$$(2)\qquad \sup_{t\in F} \{(X^i_i - X^j_j)(x_t)\} < 0 \text{ if } i < j,$$

$$(3)\qquad \sup_{\substack{x\in F\\ i\le m< j}} |X^i_j(x_t)| \le Ke^{-\lambda t}\ \forall\ t \ge 0 \text{ for some } \lambda > K.$$

Define

$$\beta = \overline{\lim_{t\to\infty}}\ t^{-1} \log \sup_{\substack{x\in F\\ i>j}} |X^i_j(x_t)|.$$

Then if $-\beta > 0$ and $\lambda - K > 0$ are sufficiently large, there exists, for each $z_0 \in S$, an unbounded invariant set $E_\infty = \text{Image } G \subset D$ where $G: R^m \to R^n$ is continuous, Image $(G^1, \ldots, G^m) = R^m$, and

$$\sup_{y\in R^m} \text{dist } (\pi(y, z_0, t), E_\infty) \to 0 \text{ as } t \to +\infty .$$

It is not difficult to give criteria for β and $\lambda - K$.

In the special case in which $n = 3$ and $m = 1$, if (3) is replaced by

$$(3')\qquad \text{sgn } X^i_j(x_t) = \pm 1 = \text{const. (indep. of } (i, j),\ x, t)$$
$$\text{whenever } x \in F,\ t \ge 0, \text{ and } i \ne j,$$

then the foregoing result holds as before ; moreover, if for some constant T and some $z_0 \in S$ we have

$$\inf_{y\in R} X^1_1(\pi(y, z_0, t)) \ge 0 \quad \text{for} \quad t \ge T$$

then there is a family of functions $g_t : R^1 \Rightarrow R^2$, with a common Lipschitz constant on R^1, such that the sets $E_t = \{\pi(y, z_0, t) : y \in R\}$ satisfy

$$E_t = \text{Graph } g_t, \quad 0 \leq t \leq \infty .$$

A subsequent result gives conditions on the set of rest points of π which imply that E_∞ is a C^1 curve which is the image of an embedding of R^1 in R^3.

University of Southern California

LIMIT SOLUTIONS OF (E) $\ddot{x} + g(x) = kf(x, \dot{x})\dot{x} + \epsilon P(t)$, k, ϵ SMALL

C. Obi

Equation (E) is supposed to satisfy sufficient conditions for the existence, uniqueness and continuity of solutions and $P(t)$ has the least period $2\pi/\omega$. We prove inter alia the following basic theorem for limit solutions (' central motions ' according to Birkhoff) and deduce from it a number of theorems for some cases of (E):

Let $\alpha_0^* > 0$, $\gamma > 0$ be two numbers satisfying $\omega = \gamma \rho_0(\alpha_0^*)$ where $2\pi/\rho_0(\alpha_0)$ is the least period of the solution $\phi_0(t) = \phi_0(\alpha_0, t)$ of $\ddot{x} + g(x) = 0$ with the stationary value α_0 at $t = 0$, and suppose that the first of the derivatives $d^r\rho_0(\alpha_0)/d\alpha_0^r$ which is not 0 when

$\alpha_0 = \alpha_0^*(\gamma)$ is $d^{2m+1}\rho_0(\alpha_0)/d\alpha_0^{2m+1}$ where $m \geq 0$ is an integer. Let $\Psi_0(t) = \Psi_0(t;\gamma) = \phi_0(\alpha_0^*(\gamma), t)$. Let $m[f(t)] = \lim \frac{1}{T}\int_0^T f(t)dt \ (T \Rightarrow \infty)$. Let $\phi(t)$ be the solution of the $t \Rightarrow t + \beta$-transform of (E) with the stationary value α at $t = 0$. If $\phi(t)$ is limiting with its stationary value α near α_0^* and if $m[f(\Psi_0, \dot{\Psi}_0)\dot{\Psi}_0^2] \neq 0$ then (C_1) there exists one or more pairs of functions $K(\beta, \epsilon; \gamma)$, $A(\beta, \epsilon;\gamma)$ such that

(1) $$k = \epsilon\ K(\beta,\epsilon;\gamma), \quad \alpha = \alpha_0^* + A(\beta,\epsilon;\gamma),$$

near $\alpha - \alpha_0^* = \epsilon = 0$; (C_2) each function of each pair $K(\beta,\epsilon;\gamma)$, $A(\beta,\epsilon;\gamma)$ is continuous in β, ϵ near $\epsilon = 0$ and has a period $2\pi\omega^{-1}$ in β; each $A(\beta,\epsilon;\gamma) \equiv 0$ when $\epsilon = 0$; and each $K(\beta,\epsilon;\gamma)$ satisfies

Lagos University

STABILITY OF UNDERDETERMINED SYSTEMS

V. M. Popov

One studies a category of systems described by relations of the form

$$y = L(x),\ x \in E,\ y \in F,$$
$$(x, y) \in N,$$

where E and F are some spaces of functions, L is a linear transformation and N is a (large) subset of $E \times F$, such that there are many "solutions" (x, y) satisfying the considered relations. (Relations of the above form arise in Control Theory and in many other fields, as a natural way of taking into account the uncertainty of our knowledge of the physical systems examined.) In the spirit of the classical qualitative theory of differential equations, one looks for conditions on E, F, L and N, under which all the solutions (x, y) exhibit a certain type of behaviour-corresponding to various properties of stability and conditional stability. In particular, one obtains conditions which extend the frequency-domain criteria of stability for nonlinear systems.

University of Florida

ON CONTROLLING SHOCKS IN SOLUTIONS OF NONLINEAR PARTIAL DIFFERENTIAL EQUATIONS

R. S. Printis
and
L. Weiss

By applying techniques of a generalized bounded-input bounded-output stability criteria, feedback regulartors for systems described by quasilinear partial differential equations of the conservation law class are found as bounded nonlinear operators mapping the solution, with Cauchy data of bounded variation in the sense of Tonelli and Cesari, into the Sobolev space, $W_p^1(R^n)$. The operator is of the form,

$H(x) = \int_E h(y) \mid x(t, z+y) \mid^k sy$, $k < 1$, and E, a compact subset of R^n, the spatial domain of the system. The closed-loop system $\partial x/\partial t + \Sigma \frac{\partial}{\partial z_k} f_k(x) + g(x) = H(x)$ is stable on arbitrary intervals $[0, T]$.

IBM T. J. Watson Research Laboratories
and
University of Maryland, College Park, MD.
Naval Research Laboratory, Washington, D. C.

ASYMPTOTIC SOLUTIONS AND STOKES PHENOMENON FOR A CERTAIN ORDINARY DIFFERENTIAL EQUATION

T. K. Puttaswamy

In order to introduce the investigations of this paper, let us take for consideration the linear homogeneous differential equation of third order,

$$(1) \quad a_0 z^3 \frac{d^3y}{dz^3} + z^2(b_0+b_1 z) \frac{d^2y}{dz^2} + z(c_0+c_1 z) \frac{dy}{dz} + (d_0+ d_1 z)\ y = 0 .$$

Let the variable z be regarded as complex as likewise the constants $a_0, b_i, c_i, d_i (i = 0,1)$ with $a_0 \neq 0$. Then, in the language of Fuch's theory, (1) will have a regular singular point at $z = 0$ and an irregular singular point at $z = \infty$. The indicial equation about $z = 0$ is found to be

(2) $$a_0 h(h-1)(h-2) + b_0 h(h-1) + c_0 h + d_0 = 0$$

we shall also assume that the roots of h_i, $i = 1, 2, 3$ of (2) are such that no two of them differ by an integer. The theory assures that (1) will have three linearly independent fundamental solutions of the form

$$\text{with} \quad \begin{array}{l} y_i(z) = z^{h_i} \sum_{n=0}^{\infty} g_i(n)\, z^n \\ g_i(0) = 1 \end{array} \qquad i = 1, 2, 3$$

in which $g_i(n)$, $n = 1, 2, \ldots$, are determinate functions of n and we know that each of these solutions will converge in the whole open plane of the extended complex z-plane. However, the behaviour of y_i, $i = 1, 2, 3$ in the neighbourhood of the point $z = \infty$ is not available, since the point $z = \infty$ is an irregular singular point. But, we know from the established theory of linear differential equations that in such case, there will exist three linearly independent solutions $\mathfrak{y}_i$, $i = 1, 2, 3$ that may be expressed asymptotically by means of "normal series". It is also know that there exists a set of connection coefficients, Stokes multipliers c_{ik} such that

$$y_i = \sum_{k=1}^{3} c_{ik}\, \mathfrak{y}_k \quad (i = 1, 2, 3)$$

Until some convergent method of computing these connection coefficients is devised, the solution of the problem in the large can not be considered as solved. It is to the solution of this problem in the large that the present paper has been addressed.

Ball State University

THEORY OF PARTIAL STABILITY AND PARTIAL BOUNDEDNESS OF FUNCTIONAL DIFFERENTIAL EQUATIONS

D. R. K. Rao

We consider systems of functional differential equations

(1a) $x'_1(t) = f_1(t, x_1(t), x_1(h(t))$

(1b) $x'_2(t) = f_2(t, x_2(t), x_1(h(t)), x_2(h(t)))$

and

(2a) $y'_1(t) = g_1(t, y_1(t), y_1(h(t)))$

(2b) $y'_2(t) = g_2(t, y_2(t), y_1(h(t)))$

where x_1, y_1, f_1, g_1 are n-vectors, x_2, y_2, f_2, g_2 are m-vectors, $h(t) \geq t$ for all $t \in [0, \infty)$. Systems (1) and (2) can also be written as

$$x'(t) = f(t, x(t), x(h(t)))$$

and

$$y'(t) = g(t, y(t,) y(h(t)))$$

where $x = \text{col}(x_1, x_2)$ and $y = \text{Col}(y_1, y_2)$. If $x(t; t_o, x_o)$ is a solution of (1), then $x(t; t_o, x_o) = \text{Col}(x_1(t; t_o, x_{10}), x_2(t; t_o, x_{20}))$, where $x_o = \text{Col}(x_{10}, x_{20})$. Let $V(t, x_1, x_2, y_1, y_2)$ be a scalar function defined on

$[0,\infty) \times R^n \times R^m \times R^m \times R^n \times R^m$ and taking values in R^+, Suppose the derivative of the scalar function along the partial solutions $x_1(t; t_o, x_{10})$ and $y_1(t; t_o, y_{10})$ of systems (1) and (2) respectively is

$$V^*_{(1)x_1,\ (2)y_1}(t, x_1, x_2, y_1, y_2)$$

$$(3) \quad = \overline{\lim_{p \to 0^+}} \frac{1}{p} [V(t+p,\ x_1 + pf_1,\ x_2,\ y_1 + pg_1,\ y_2) - V(t,\ x_1,\ x_2,\ y_1,\ y_2)]$$

Imposing suitable conditions upon the functions V and V* defined in (3), we obtain a series of results on relative partial stability boundedness of solutions of (1) with respect to (2). Lastly examples are constructed to illustrate the results.

Fundi Shapur Unibersitp

THE SINGULARLY PERTURBED INITIAL-VALUE PROBLEM WHEN THE REDUCED PATH ENCOUNTERS A POINT OF BIFURCATION

R. J. Schaar

The singularly perturbed initial-value problem $\dot{x} = f(x, y)$, $\varepsilon\, \dot{y} = g(x, y)$, $x(0) = \alpha$, $y(0) = \beta$ is considered in the neighborhood of the point $(x, y) = (0, 0)$ through which pass two solutions, $y = \phi(x)$, $x = \psi(y)$, of the reduced problem $g(x, y) = 0$. Under the assumptions that

$\Phi'(0) \neq 0$, $\psi'(0) = 0$, and $\psi'(0) > 0$, the path consisting of the curve $y = \phi(x)$ when $x \leq 0$ and $x = \psi(y)$ when $x \geq 0$ is the asymptotically stable family of solutions of the boundary layer equation under simple and natural assumptions concerning the function g. Under these assumptions, the solution of the full orbital problem approximates the orbit of the reduced problem to $(\varepsilon^{1/4} |\log \varepsilon|^{1/4})$ in an $(\varepsilon^{1/2} |\log \varepsilon|^{1/2})$ neighborhood of the origin. This can be shown to be best possible. If $\phi'(0) > 0$, the lower branch of $x = \psi(y)$ is approximated; if $\phi'(0) < 0$, the upper branch is approximated. $\phi'(0) \not< 0$ is shown to be essential.

University of Southern California

NATURAL CARATHÉODORY CONDITIONS FOR LINEAR RETARDED EQUATIONS

J. J. Schäffer

We consider the "initial-value problem" $\dot{u} + Mu = r$ (for $t \geqq a$) and u agrees with v for $t \leqq a$, where $a \in R$ is given, the solution u and the given function v are continuous functions from R to a Banach space E, r is a given locally integrable function from R to E, and the "memory" M maps continuous functions from R to E linearly into locally integrable ones in such a way that if $t \in R$ and u_1, u_2 agree on $]-\infty, t]$, then so do Mu_1, Mu_2. Under a mild and very local boundedness condition on M, strong existence and uniqueness results hold. A reasonable "global" condition on M ensures exponential bounds for the growth of solutions. If, in addition, E is finite dimensional and the memory is uniformly limited in its recall to a delay of 1, then the transition operators for time lapses not less than 1 are compact. The assumptions are effectively less restrictive than those usually imposed, even for time-independent memories and one-dimensional E.

(This abstract refers to work done jointly with C. V. Coffman.)

Carnegie-Mellon University

DIFFERENTIAL EQUATIONS OF ADVANCED TYPE

R. Suarez C.

Given a differential-difference equation with advanced argument

(1) $$\dot{x}(t) = f(t, x(t), x(t+T)), \quad T > 0$$

and a sequence $\{\bar{x}_k\}_{k=0}^{\infty}$, $\bar{x}_k \in R^n$, a function x is called a solution of (1) with initial condition $\{\bar{x}_k\}_{k=0}^{\infty}$, if x satisfies (1) and, $x(kT) = \bar{x}_k \;\forall\; k$. Theorems on existence and uniqueness for solutions of (1) have been obtained. Moreover, if $F: [0, T) \times S \to S$ (S is a space of sequences)

$$F(t, \{y_k\}_{k=0}^{\infty}) = \{f(t+kT,\ y_k\, y_{k+1})\}_{k=0}^{\infty}$$

it turns out that the equation (1) with the initial condition $\{\bar{x}_k\}_{k=0}^{\infty} \in S$ is equivalent to the ordinary equation

$$\text{(2)} \qquad \dot{y}(t) = F(t, y(t))$$

with the initial condition $y(0) = \{x_k\}_{k=0}^{\infty}$. Thus theorems on existence, uniqueness and continuous dependence on initial data for equation (1) can be obtained from corresponding theorems for equation (2).

The continuous solutions of (1) correspond to the solutions of (2) which satisfy the following boundary condition

$$\pi y(0) = y(T)$$

where π is the operator defined by

$$\pi\{y_k\}_{k=0}^{\infty} = \{y_{k+1}\}_{k=0}^{\infty}.$$

Using this, theorems on existence, uniqueness and continuous dependence on initial data are proved for continuous and bounded solutions of (1) under weaker weaker assumptions on f than those found in the existing literature.

Escuela Superior de Fisica y Mathematicas
del Instituto Politecnico Nacional

A 5
B 6
C 7
D 8
E 9
F 0
G 1
H 2
I 3
J 4